Proceedings of the

10th ASIAN LOGIC CONFERENCE

Proceedings of the

10th ASIAN LOGIC CONFERENCE

Kobe, Japan 1 – 6 September 2008

Editors

T Arai, J Brendle, H Kikyo
Kobe University, Japan

C T Chong
National University of Singapore, Singapore

R Downey
Victoria University of Wellington, New Zealand

Q Feng
Chinese Academy of Science, China

H Ono
Japan Advanced Institute of Science and Technology, Japan

World Scientific

NEW JERSEY • LONDON • SINGAPORE • BEIJING • SHANGHAI • HONG KONG • TAIPEI • CHENNAI

Published by

World Scientific Publishing Co. Pte. Ltd.
5 Toh Tuck Link, Singapore 596224
USA office: 27 Warren Street, Suite 401-402, Hackensack, NJ 07601
UK office: 57 Shelton Street, Covent Garden, London WC2H 9HE

British Library Cataloguing-in-Publication Data
A catalogue record for this book is available from the British Library.

PROCEEDINGS OF THE 10TH ASIAN LOGIC CONFERENCE

ISBN-13 978-981-4293-01-3
ISBN-10 981-4293-01-6

Printed in Singapore by B & JO Enterprise

Preface

The 10th Asian Logic Conference was held September 1–6, 2008 in Kobe University, Japan. Major funding was provided by the Association for Symbolic Logic, Japan Society for the Promotion of Science, Inoue Foundation for Science and Kobe University. The Program Committee consisted of Chong Chi Tat, Qi Feng, Rod Downey, Hiroakira Ono, Hirotaka Kikyo, Jörg Brendle and Toshiyasu Arai (Chair). The members of the Local Organizing Committee were Mutsunori Banbara, Hiroaki Minami, Ichiro Nagasaka, Akira Suzuki and Makoto Kikuchi (Chair).

The program consisted of three four-hour tutorial lectures given by Byunghan Kim, Greg Hjorth and Peter Cholak, four one-hour plenary talks given by Jeremy Avigad, Justin Moore, Kazushige Terui and Yang Yue, three parallel special sessions (one for model theory, non-classical logics, proof theory and constructive mathematics, one for set theory and one for recursion theory) of 23 invited talks (30 minutes each), and sessions of 23 contributed talks (20 minutes each).

There were 121 registered participants from Australia, Austria, China, Colombia, Czech Republic, India, Israel, The Netherlands, New Zealand, Singapore, South Korea, Spain, United Kingdom, USA and Japan.

Authors were invited to submit papers to the present volume. All of the submitted papers were fully refereed.

We wish to thank all those who help to review the submitted papers.

Editors
Toshiyasu Arai, Jörg Brendle, Chong Chi Tat,
Rod Downey, Feng Qi, Hirotaka Kikyo, Hiroakira Ono

Contents

Proceedings of the

10th ASIAN LOGIC CONFERENCE

AN APPLICATION OF NONSTANDARD ARGUMENTS TO THE STUDY OF GENERIC STRUCTURES

YUKI ANBO

Department of Mathematics, University of Tsukuba
Tsukuba, Ibaraki 305-8571, Japan
E-mail: anbo@math.tsukuba.ac.jp

1. Introduction

Fraïssé's construction is a method with which one can construct a universal homogeneous countable relational structure from a given class of finite structures. Around 1990, E. Hrushovski[9] invented a refined version of Fraïssé's construction, and succeeded to construct several counterexamples to open problems in stability theory. Later, many papers including Refs. 5, 7, and 10 applied the method to get model-theoretically interesting structures. A structure constructed in this way is called a generic structure. In this thesis, we propose a new approach for analyzing properties of generic structures.

As a generic structure is a limit of its finite substructures, epsilon-delta type proofs are often required to analyze certain properties of the structure. Motivated by this, we apply a nonstandard method in the study of generic structures. By using nonstandard arguments, we give simpler proofs to known facts and prove some new propositions on generic structures.

Stable generic structures are studied in many papers.[6,15,16] All known examples of stable generic structures are ω-stable or strictly stable. In 1993, Baldwin[4] asked whether there is a strictly superstable generic structure. In 1994, Wagner[16] presented a sufficient condition for saturated generic structures to be stable and a sufficient condition for them to be ω-stable. Ikeda and the author[3] obtained a partial answer to the problem by putting a new assumption.

In stable generic structures, we can define two notions of independence: forking independence and d-independence. Wagner[16] showed that for two closed sets A and B, if $A \cap B$ is algebraically closed, then A and B are forking independent over $A \cap B$ if and only if A and B are d-independent over $A \cap B$. He proved the result under the finite closure axiom. Verbovskiy and Yoneda[15] removed the finite closure axiom from Wagner's theorem and put an another axiom. Moreover, Ikeda and the author[3] showed a similar theorem. They removed the assumption that $A \cap B$ is algebraically closed from Wagner's theorem and put a new assumption instead.

In this paper, we apply nonstandard arguments to the study of generic structures. Nonstandard arguments were introduced as a new method of infinitesimal analysis besides the epsilon-delta argument by A. Robinson.[13] In Section 2, we review basic definitions and facts on generic structures and stability theory. In Section 3, we introduce a way to apply nonstandard arguments to the study of generic structures. In Section 4, we apply nonstandard arguments to the analysis of the strength of the stability of generic structures. We add some new propositions on Wagner's analysis.[16] In Section 5, we apply nonstandard arguments to classification of independence in generic structures. We give a nonstandard proof of the theorem which was proved by Wagner,[16] Verbovskiy and Yoneda.[15]

Conventions. We write AB for $A \cup B$ omitting the symbol of the union. $A \subset_{\mathrm{fin}} B$ means that A is a finite subset of B.

2. Preliminaries

2.1. *Generic structures*

We review basic definitions and facts on generic structures. Let L be a countable relational language. Assume that $\mathbb{K}$ is a nonempty class of finite L-structures closed under isomorphisms and substructures (we consider the empty set as an L-structure). Let $\overline{\mathbb{K}}$ be the class of L-structures M such that any finite substructure of M belongs to $\mathbb{K}$.

Definition 2.1. Let δ be a function from $\mathbb{K}$ to $\mathbb{R}_{\geq 0}$. We say that δ is a predimension function on $\mathbb{K}$ if for any members A and B of $\mathbb{K}$,

(1) (Submodularity) $\delta(AB) + \delta(A \cap B) \leq \delta(A) + \delta(B)$,
(2) $A \cong B \Longrightarrow \delta(A) = \delta(B)$, and
(3) $\delta(\emptyset) = 0$.

Example 2.2.

- (*ab initio* case) In a language $\{R\}$ of graphs, $\delta(A) = |A| - |R^A|$ for any finite graph A. Let $\mathbb{K} = \{A : \forall A' \subset A, \delta(A') \geq 0\}$.

- (the fusion of strongly minimal theories[8]) Let T_i be a strongly minimal L_i-theory for each $i = 1, 2$. Suppose that $L_1 \cap L_2 = \emptyset$. For any $A \subset_{\text{fin}} N \models T_1 \cup T_2$, put $\delta(A) = \dim_1(A) + \dim_2(A) - |A|$ where dim_i denotes the dimension in T_i defined by strong minimality for each $i = 1, 2$. Let $\mathbb{K} = \{A : \forall A' \subset A, \delta(A') \geq 0\}$.

We define $\delta(A/B)$ as follows: For finite A, B, $\delta(A/B) = \delta(AB) - \delta(B)$. For finite A and arbitrary B, $\delta(A/B) = \inf\{\delta(A - B/B_0) : B_0 \subset_{\text{fin}} B\}$. By submodularity, it is easy to check that these two definitions have the same value in the case A and B are both finite.

Definition 2.3. Let δ be a predimension function on $\mathbb{K}$. Let A and B be members of $\mathbb{K}$ with $A \subseteq B$ and ϵ a positive real number. We say that A is ϵ-closed in B if $\delta(A) \leq \delta(B') + \epsilon$ whenever $A \subseteq B' \subseteq B$. Simply we say that A is closed in B, if A is 0-closed in B. If A is closed in B, we write $A \leq B$. For any members A and B of $\overline{\mathbb{K}}$ with $A \subseteq B$, we say that $A \leq B$ if $A \cap B' \leq B'$ for any finite $B' \subset B$.

By submodularity, it is easy to check that these two definitions are the same in the case A and B are both finite.

Fact 2.4. *Let M be a member of $\overline{\mathbb{K}}$ and A a subset of M. Then there exists the smallest superset $\overline{A}$ of A in M such that $\overline{A} \leq M$. We call $\overline{A}$ the closure of A in M.*

Fact 2.5. *For any subset A of M, $\overline{A}$ is contained in the algebraic closure* $\mathrm{acl}(A)$ *of A.*

Definition 2.6. Let M be an L-structure. We say that M is a $\mathbb{K}$-generic structure if the following conditions are satisfied:

(1) M is countable;
(2) M is a member of $\overline{\mathbb{K}}$;
(3) for any members A and B of $\mathbb{K}$, if $A \leq B$ and $A \leq M$, then there is a copy B' of B over A with $B' \leq M$.

Example 2.7. A countable graph $G = (G, R)$ is called a random graph if it satisfies the following property: for each $m, n \in \omega$ and all $a_1, \ldots, a_m, b_1, \ldots, b_n \in G$, if $\{a_1, \ldots, a_m\}$ and $\{b_1, \ldots, b_n\}$ are disjoint, then there exists $c \in G$ such that $G \models \bigwedge_{1 \leq i \leq m} R(c, a_i) \wedge \bigwedge_{1 \leq j \leq n} \neg R(c, b_j)$. It is well known that a random graph uniquely exists. Let $\mathbb{K}$ be the class of all finite graphs and $\delta(A) = |A|$ for any finite graph A. Then the random graph is the $\mathbb{K}$-generic structure.

Let M be a member of $\overline{\mathbb{K}}$ and δ be a predimension function on $\mathbb{K}$. Put $d^M(A) = \inf\{\delta(A')|A \subseteq A' \subset_{\text{fin}} M\}$ for any finite subset A of M. We call d^M a dimension function for M. We often omit $*^M$ and write d. We define $d(A/B)$, the relative dimension of A over B as follows. For finite subsets A and B of M, we define $d(A/B) = d(AB) - d(B)$. For finite A and arbitrary B, put $d(A/B) = \inf\{d(A - B/B_0) : B_0 \subset_{\text{fin}} B\}$. Submodularity of d is proved in Section 3. Then we have that these two definitions have the same value in the case A and B are both finite.

Fact 2.8.

(1) Let B be a closed subset of M. Then for any finite subset A of B and any positive real number ϵ, there is a finite ϵ-closed set B_ϵ with $A \subseteq B_\epsilon \subseteq B$.
(2) Let A be a finite subset of M. Then for any $\epsilon > 0$ and any $A \subseteq B \subset_{\text{fin}} \overline{A}$, there exists $B \subseteq C \subset_{\text{fin}} \overline{A}$ such that $\delta(C) \leq d(A) + \epsilon$.
(3) Let A be a finite subset of M. Suppose that $(B_n)_{n<\omega}$ is an increasing sequence of subsets of $\overline{A}$ such that $\bigcup_{n<\omega} B_n = \overline{A}$ and B_n is $\frac{1}{n}$-closed for each $n < \omega$. Then $\lim_{n\to\omega} \delta(B_n) = d(A)$.
(4) Let $A \subseteq B \subset_{\text{fin}} \overline{A}$. Then $d(A) = d(B)$.

Definition 2.9. For any subsets A and B of M, we say that A and B are free over $A \cap B$ if the relations in AB is the union of the relations in A and the relations in B.

2.2. *Stability*

Now, we review stability theory very briefly. A more detailed explanation is given in many books of stability theory, e.g., Ref. 12.

Definition 2.10. Let T be a theory and κ be an infinite cardinal. Let $\mathcal{M}$ be a big saturated model of T.

(1) We say that T is κ-stable if for any $A \subset \mathcal{M}$ with $|A| = \kappa$, the cardinality of the set of complete types over A is equal to κ.
(2) We say that T is stable if T is κ-stable for some κ.
(3) Let M be an L-structure. We say that M is (κ-) stable if $\text{Th}(M)$ is (κ-) stable.

Fact 2.11. *Let T be a stable theory. Then one of the following holds:*

(1) T is κ-stable for all κ,
(2) T is κ-stable for all $\kappa \geq 2^{\aleph_0}$,

(3) T is κ-stable if and only if $\kappa^{\aleph_0} = \kappa$.

If (1) holds, we say that T is ω-stable. If (2) holds, we say that T is superstable. If T is superstable but not ω-stable, we say that T is strictly superstable. If (3) holds, we say that T is strictly (or merely) stable. If T is not stable, we say that T is unstable.

Definition 2.12. Let κ be an infinite cardinal and $(\bar{a}_i)_{i<\kappa}$ be a sequence of n-tuples in $\mathcal{M}$. We say that $(\bar{a}_i)_{i<\kappa}$ is an indiscernible sequence over A if for any $k < \omega$, $i_1 < \cdots < i_k$ and $j_1 < \cdots < j_k$, we have $\mathrm{tp}(a_{i_1}, \ldots, a_{i_k}/A) = \mathrm{tp}(a_{j_1}, \ldots, a_{j_k}/A)$.

Fact 2.13. *Suppose that T is stable. Let $(\bar{a}_i)_{i<\kappa}$ be an indiscernible sequence. For any $k < \omega$, if $i_1, \ldots, i_k$ are distinct and $j_1, \ldots, j_k$ are distinct, then we have $\mathrm{tp}(a_{i_1}, \ldots, a_{i_k}/A) = \mathrm{tp}(a_{j_1}, \ldots, a_{j_k}/A)$.*

Definition 2.14. Suppose that T is stable. Let $\mathcal{M}$ be a big saturated model of T.

(1) Let $\bar{a}$ be a finite tuple in $\mathcal{M}$ and $\varphi(x, \bar{a})$ be an $L(\bar{a})$-formula. We say that $\varphi(x, \bar{a})$ forks over A if there is an indiscernible sequence $(\bar{a}_i)_{i<\omega}$ over A with $\bar{a}_0 = \bar{a}$ such that $\{\varphi(x, \bar{a}_i) | i < \omega\}$ is inconsistent.
(2) Let $A \subseteq B \subset \mathcal{M}$ and $\Gamma(x)$ be a set of $L(B)$-formulas. We say that $\Gamma(x)$ forks over A if $\varphi(x)$ forks over A for some $\varphi(x) \in \Gamma(x)$.
(3) Let A, B, and C be subsets of $\mathcal{M}$. We say that A and B are forking independent over C and write $A \mathop{\smile\!\!\!\!\!\!|}_C B$ if $\mathrm{tp}(\bar{a}/BC)$ does not fork over C for any $\bar{a} \in A$.

Fact 2.15. *Suppose that T is stable. Then the following are equivalent:*

(1) T is superstable.
(2) For any $B \subset \mathcal{M}$ and type p over B, there is finite $A \subseteq B$ such that p does not fork over A.

3. Nonstandard Arguments

In this section, we introduce a nonstandard argument to the study of generic structures.

Suppose that $\mathbb{K}$ is a class of L-structures closed under isomorphisms and substructures and δ is a predimension function on $\mathbb{K}$. Let $M \in \overline{\mathbb{K}}$. We define the closure relation $\leq$ on $\overline{\mathbb{K}} \times \overline{\mathbb{K}}$ and the dimension function d for M as in Section 2. We consider M to be a 3-sorted structure:

$$(M \cup P \cup \mathbb{R}; F, \in, \delta, \leq, d, \cdots)$$

where P is the powerset of M, F is a unary relation on P such that for any subset A of M, $F(A)$ holds if and only if A is a finite set, $\in$ is the membership relation on $M \times P$, and "$\cdots$" contains L and $(+, \cdot, <)$ in $\mathbb{R}$.

We define the nonstandard model M^* of M by a sufficiently saturated elementary extension of this structure:

$$(M \cup P \cup \mathbb{R}; F, \in, \delta, \leq, d, \cdots) \prec (M^* \cup P^* \cup \mathbb{R}^*; F^*, \in^*, \delta^*, \leq^*, d^*, \cdots).$$

Notation 3.1.

- For any set variables X, Y, and Z, we define $X \subseteq^* Y$ (resp., $X \cap^* Y = Z$, $X \cup^* Y = Z$) as an abbreviation for $\forall x(x \in^* X \to x \in^* Y)$ (resp., $\forall x(x \in X \wedge x \in Y \leftrightarrow x \in Z)$, $\forall x(x \in X \vee x \in Y \leftrightarrow x \in Z)$).
- Note that a function from P to P which maps each member of P to its closure is defined by some formula $\varphi(X, Y)$. Let A^* be any member of P^*. We write $\overline{A^*}$ for the realization of $\varphi(A^*, Y)$. We say that A^* is closed if $\overline{A^*} = A^*$ holds.
- We denote F^* (resp., $\in^*$, $\subseteq^*$, $\cap^*$, $\cup^*$, δ^*, $\leq^*$, d^*) simply by F (resp., $\in$, $\subseteq$, $\cap$, $\cup$, δ, $\leq$, d) if there is no confusion.
- Ler r and s be any elements of $\mathbb{R}^*$. We write $r \approx s$ if $-a < r - s < a$ holds in M^* for all positive real numbers a.

Definition 3.2.

(1) Let A^* be any member of P^*. We say that A^* is a hyperfinite set if $F^*(A^*)$ holds in M^*.

(2) Let A be any subset of M and A^* be any hyperfinite set. We say that A^* is a hyperfinite extension of A and write $A^* \supset_{\text{hf}} A$ if

- $M^* \models a \in^* A^*$ for each member a of A, and
- $M^* \models A^* \subseteq^* A$.

Remark 3.3. For any subset A of M, there exists a hyperfinite extension of A.

Proof. It is enough to prove that the following set of formulas is satisfiable:

$$\Gamma(X) = \{a \in X | a \text{ is a member of } A\} \cup \{X \subseteq A\} \cup \{F(X)\}.$$

But for any finite subset A_0 of A, A_0 realizes the following set of formulas:

$$\Gamma_0(X) = \{a \in X | a \text{ is a member of } A_0\} \cup \{X \subseteq A\} \cup \{F(X)\}.$$

So, by compactness, $\Gamma(X)$ is satisfiable. □

Lemma 3.4. *Let r be a real number, $\bar{a}$ a finite tuple in M, and A a subset of M. Then the following are equivalent:*

(1) $\delta(\bar{a}/A) = r$.
(2) $\delta^*(\bar{a}/A^*) \approx r$ *for any hyperfinite extension* A^* *of* A.
(3) $\delta^*(\bar{a}/A^*) \approx r$, *for some hyperfinite extension* A^* *of* A.

Proof. We may assume that $\bar{a} \cap A = \emptyset$.

(1) $\Rightarrow$ (2). By submodularity of δ, for each $n < \omega$, there is a finite subset A_n of A such that

$$M \models \forall X(F(X) \wedge A_n \subseteq X \subseteq A \rightarrow r \leq \delta(\bar{a}/X) \leq r + 1/n).$$

The above formula holds also in M^* for each $n < \omega$. So if A^* is a hyperfinite extension of A, then we have

$$r \leq \delta^*(\bar{a}/A^*) \leq r + 1/n$$

for each $n < \omega$. So we have $\delta^*(\bar{a}/A^*) \approx r$.

(2) $\Rightarrow$ (3). Trivial.

(3) $\Rightarrow$ (1). We assume (3) and choose a witness A^*. Then $\delta^*(\bar{a}/A^*) \approx r$. Suppose (1) is not the case. Let $s = \delta(\bar{a}/A)$. Then $s \neq r$. By (1) $\Rightarrow$ (2), we have $\delta^*(\bar{a}/A^*) \approx s$. A contradiction. □

Definition 3.5. Let A^* be a hyperfinite set. We say that A^* is quasi-closed if $\delta(B^*/A^*) \gtrsim 0$ (that is, $\delta(B^*/A^*) \geq -\epsilon$ for all positive real numbers ϵ) for any hyperfinite set B^*.

Remark 3.6. For any hyperfinite set A^*, A^* is quasi-closed if and only if $d(A^*) \approx \delta(A^*)$.

Proof. Immediate. □

Lemma 3.7. *For any subset A of M, the following conditions are equivalent:*

(1) A is closed.
(2) There is a quasi-closed hyperfinite extension of A.

Proof. (1) $\Rightarrow$ (2). By Remark 2.8.1.

(2) $\Rightarrow$ (1). Suppose A is not closed. Then there exists finite subset B of M such that $\delta(B/A \cap B) < 0$. Take any hyperfinite extension A^* of A. Because $A \cap B \subseteq^* A^* \subseteq^* A$, we have $\delta(B/A^*) \leq \delta(B/B \cap A) < 0$. □

Lemma 3.8. *Let A be a finite subset of M and let B, C be any subsets of M. If $B \subseteq C$, then $d(A/B) \geq d(A/C)$.*

Proof. It is enough to show the statement in the case C is finite. Let D^* be a quasi-closed hyperfinite extension of $\overline{AC}$. Then we have the following conditions:

(1) $\delta(D^* \cap (\overline{AB} \cup \overline{C})) \leq \delta(D^* \cap \overline{AB}) + \delta(D^* \cap \overline{C}) - \delta(D^* \cap \overline{AB} \cap \overline{C})$
(2) $\delta(D^* \cap \overline{AB} \cap \overline{C}) \gtrsim \delta(D^* \cap \overline{B})$.
(3) $\delta(D^* \cap (\overline{AB} \cup \overline{C})) \gtrsim \delta(D^*)$.

The first condition holds by submodularity of δ. The second condition holds because $D^* \cap \overline{B}$ is quasi-closed. The last condition holds by Remark 2.8.4. Then again by 2.8.4, we have $d(A/C) \leq d(A/B)$. □

Lemma 3.8 is called submodularity of d. It shows that two definitions of $d(A/B)$ have the same value in the case A and B are finite.

Lemma 3.9. *Let r be a real number, $\bar{a}$ a finite tuple in M, and A a subset of M. Then the following are equivalent:*

(1) $d(\bar{a}/A) = r$.
(2) $d(\bar{a}/A^*) \approx r$ for any hyperfinite extension A^* of A.
(3) $d(\bar{a}/A^*) \approx r$, for some hyperfinite extension A^* of A.

Proof. By submodularity of d, we can prove this lemma in the same way as the proof of Lemma 3.4. □

Definition 3.10.[16]

(1) Let A and B be finite subsets of M and C be an arbitrary subset of M. Then we say that A and B are d-independent over C and write $A \mathop{\smile\!\!\!\!\!\!\,|}^{d}_{C} B$ if the following conditions are satisfied:
   - $d(A/BC) = d(A/C)$, and
   - $\overline{AC} \cap \overline{BC} = \overline{C}$.

(2) For arbitrary subsets A, B, and C of M, we say that A and B are d-independent over C if $A_0 \mathop{\smile\!\!\!\!\!\!\,|}^{d}_{C} B_0$ whenever $A_0 \subset_{\text{fin}} A$ and $B_0 \subset_{\text{fin}} B$.

Note that for closed sets A and B, A and B are d-independent over $A \cap B$ if and only if $d(A_0/B_0(A \cap B)) = d(A_0/A \cap B)$ for every finite subset A_0 of A and every finite subset B_0 of B.

Definition 3.11. Let A and B be closed subsets of M. Then we say that A and B are d^*-independent over $A \cap B$ if there exist a hyperfinite extension A^* of A and a hyperfinite extension B^* of B such that

- A^* and B^* are both quasi-closed and
- $d(A^*/B^*) = d(A^*/A^* \cap B^*)$.

Proposition 3.12. *Let A and B be closed subsets of M. Then the following are equivalent:*

(1) A and B are d-independent over $A \cap B$.
(2) A and B are d^*-independent over $A \cap B$.
(3) There exist a hyperfinite extension A^* of A and a hyperfinite extension B^* of B such that

- *A^* and B^* are both quasi-closed,*
- *$d(A^*/B^*) = d(A^*/A^* \cap B^*)$, and*
- *$(A^*B^*) \cap A \cap B = A^* \cap B^*$.*

Proof. (3) $\Rightarrow$ (2). Trivial.

(2) $\Rightarrow$ (1). Let A^*, B^* be witnesses of the d^*-independence. Take any finite subset A' of A and any finite subset B' of B. Then $d(A^*/B^*) \approx d(A^*/A^* \cap B^*)$. By transposition, $d(B^*/A^*) \approx d(B^*/A^* \cap B^*)$. By submodularity, $d(B^*/A'(A^* \cap B^*)) \approx d(B^*/A^* \cap B^*)$. Again by transposition, $d(A'/B^*) \approx d(A'/A^* \cap B^*)$. Again by submodularity, $d(A'/B'(A^* \cap B^*)) \approx d(A'/A^* \cap B^*)$. Finally, by Lemma 3.9, we have $d(A'/B'(A \cap B)) = d(A'/A \cap B)$.

(1) $\Rightarrow$ (3). Take $A^* \supset_{\mathrm{hf}} A$ and $B^* \supset_{\mathrm{hf}} B$ such that A^* and B^* are both closed and $(A^*B^*) \cap A \cap B = A^* \cap B^*$.

By compactness, it is enough to prove that for any finite subset A_0 of A, the following set of formulas are satisfiable:

(1) $F(X)$,
(2) $X \subseteq A$,
(3) $A_0 \subseteq X$,
(4) "X is quasi-closed",
(5) "$d(X/B^*) \approx d(X/X \cap B^*)$",
(6) $(XB^*) \cap A \cap B = X \cap B^*$.

Note that (4) and (5) are both expressed by an infinite set of formulas. We show that $A_0^* = \overline{A_0(A^* \cap B^*)} \cap A^*$ is a realization of the above set of formulas. (1), (2), (3), and (4) are clear.

(5). First,

$$\begin{aligned} d(A_0^*/B^*) &= d(A_0^*B^*) - d(B^*) \\ &= d(A_0B^*) - d(B^*) \\ &= d(A_0/B^*) \\ &\approx d(A_0/B). \end{aligned}$$

Second,

$$\begin{aligned} d(A_0^*/A_0^* \cap B^*) &= d(A_0^*) - d(A_0^* \cap B^*) \\ &= d(A_0(A^* \cap B^*)) - d(A_0^* \cap B^*) \\ &\leq d(A_0(A^* \cap B^*)) - d(A^* \cap B^*) \\ &= d(A_0/A^* \cap B^*) \\ &\approx d(A_0/A \cap B). \end{aligned}$$

Finally, by d-independence of A and B, $d(A_0/B) = d(A_0/A \cap B)$.

Hence, $d(A_0^*/A_0^* \cap B^*) \lesssim d(A_0^*/B^*)$. The other direction is clear by submodularity.

(6). Note that$(A_0^*B^*) \cap A \cap B = (A_0^* \cap A \cap B) \cup (B^* \cap A \cap B) \subseteq (A^* \cap B) \cup (A^* \cap B^*) \subseteq A^* \cap B^* \subseteq A_0^* \cap B^*$. □

4. The Strength of the Stability of Generic Structures

In 1993, Baldwin[4] asked whether there is a strictly superstable generic structure. In 1994, Wagner[16] presented a sufficient condition called DS for saturated generic structures to be stable, and a sufficient condition called DW for them to be ω-stable. Using nonstandard argument, we add some new propositions to Wagner's analysis. We simplify DS and clarify the relation between DS and DW. Our assumptions in this section follow Section 4 of Wagner's paper.[16] But a similar argument shows the same consequence of this section under the assumption in Section 3 of Ref. 16. For detail, see Ref. 2.

Let $\mathbb{K}$ be a class of finite L-structures closed under isomorphisms and substructures and δ be a predimension function on $\mathbb{K}$. Let M be a member of $\overline{\mathbb{K}}$. We define the closure relation $\leq$ on $\overline{\mathbb{K}} \times \overline{\mathbb{K}}$ and the dimension function d for M as in Section 2.

Definition 4.1. Let $A \subseteq B \in \mathbb{K}$. We say that B is a minimal extension of A if the following conditions are satisfied:

- $A \not\leq B$

- $A \leq B'$ for any $A \subseteq B' \subset B$.

Now, we add a new assumption: there is no infinite chain $(A_i)_{i<\omega}$ of elements of $\mathbb{K}$ such that A_{i+1} is a minimal extension of A_i for each $i < \omega$.

This assumption is called the finite closure condition. We say that a member N of $\overline{\mathbb{K}}$ has finite closures if for any finite $A \subseteq N$, the closure of A is also finite.

Fact 4.2.[6] *The following conditions are equivalent:*

(1) $\mathbb{K}$ satisfies the finite closure condition.
(2) Every member of $\overline{\mathbb{K}}$ has finite closures.
(3) Every ω-saturated member of $\overline{\mathbb{K}}$ has finite closures.
(4) Some ω-saturated member of $\overline{\mathbb{K}}$ has finite closures.

In this section, we assume that $\mathbb{K}$ satisfies the finite closure condition.

Wagner's definition[16] of DS, a sufficient condition for saturated generic structures M to be stable, is as follows:
For any closed subsets A and B of M, if for all $n \in \omega$, $A_0 \subset_{\text{fin}} A$, and $B_0 \subset_{\text{fin}} B$, there are finite closed sets A' and B' with $A_0 \subseteq A' \subseteq A$ and $B_0 \subseteq B' \subseteq B$ such that

$$d(A') + d(B') \leq d(A'B') + d(A' \cap B') + 1/n,$$

then A and B are free over $A \cap B$ and AB is closed.

Wagner's definition[16] of DW, a sufficient condition for saturated generic structures M to be ω-stable, is as follows:

(1) For any closed subsets A and B of M, if A and B are d-independent over $A \cap B$, then A and B are free over $A \cap B$ and AB is closed, and
(2) for any finite tuple $\bar{a}$ and set B in M, there is $B_0 \subset_{\text{fin}} B$ such that $d(\bar{a}/B_0) = d(\bar{a}/B)$.

Theorem 4.3. *For any closed subsets A and B of M, the following are equivalent:*

(1) For all $n \in \omega$, $A_0 \subset_{\text{fin}} A$, and $B_0 \subset_{\text{fin}} B$, there are finite closed sets A' and B' with $A_0 \subseteq A' \subseteq A$ and $B_0 \subseteq B' \subseteq B$ such that $d(A') + d(B') \leq d(A'B') + d(A' \cap B') + 1/n$.
(2) A and B are d^*-independent over $A \cap B$.
(3) A and B are d-independent over $A \cap B$.

Proof. (2) $\Leftrightarrow$ (3). By proposition 3.12.

(2) $\Rightarrow$ (1). Let A^*, B^* be witnesses of the d^*-independence. By the finite closure condition, we may assume that A^* and B^* are both closed. Take any $n \in \omega$, $A_0 \subset_{\text{fin}} A$, and $B_0 \subset_{\text{fin}} B$. Then A^* and B^* satisfy the following finite set of formulas:

- $A_0 \subseteq X \subseteq A \wedge B_0 \subseteq Y \subseteq B$,
- $F(X) \wedge F(Y)$,
- $X \leq A \wedge Y \leq B$,
- $d(A^*) + d(B^*) \leq d(A^*B^*) + d(A^* \cap B^*) + 1/n$.

Because M is an elementary substructure of M^*, we can take expected sets.

(1) $\Rightarrow$ (2). Assume (1). Then by saturatedness, there exist closed $A^* \supset_{\text{hf}} A$ and closed $B^* \supset_{\text{hf}} B$ such that for all $n \in \omega$,

$$d^*(A^*) + d^*(B^*) \leq d^*(A^*B^*) + d^*(A^* \cap B^*) + 1/n.$$

The other direction

$$d^*(A^*) + d^*(B^*) \geq d^*(A^*B^*) + d^*(A^* \cap B^*)$$

is clear by submodularity.

So we have

$$d^*(A^*) + d^*(B^*) \approx d^*(A^*B^*) + d^*(A^* \cap B^*),$$

equivalently,

$$d^*(A^*/B^*) \approx d^*(A^*/A^* \cap B^*).$$

□

By the theorem, we have the following corollaries:

Corollary 4.4. *DS is equivalent to condition (1) of DW. In particular, DW is a condition stronger than DS.*

Corollary 4.5. *Suppose that M is a saturated $\mathbb{K}$-generic structure satisfying DS. Suppose also that for any closed set A, B, $A \mathop{\smile\!\!\!\!\!\!|}^{d}_{A \cap B} B$ if and only if $A \mathop{\smile\!\!\!\!\!\!|}_{A \cap B} B$. Then $T = \mathrm{Th}(M)$ is ω-stable or strictly stable.*

Proof. By the theorem, it is enough to prove that if T is superstable, then condition (2) of DW holds. Take any finite tuple $\bar{a}$ and any set B in M. Because T is superstable, by Fact 2.15, there is $B_0 \subset_{\text{fin}} B$ with $\bar{a} \mathop{\smile\!\!\!\!\!\!|}_{B_0} B$. Because $\mathrm{cl}(*) \subseteq \mathrm{acl}(*)$, $\mathrm{cl}(\bar{a}B_0) \mathop{\smile\!\!\!\!\!\!|}_{B_0} \mathrm{cl}(B)$. Put $B_1 = \mathrm{cl}(\bar{a}B_0) \cap \mathrm{cl}(B)$. Then $\mathrm{cl}(\bar{a}B_0) \mathop{\smile\!\!\!\!\!\!|}_{B_1} \mathrm{cl}(B)$. By the assumption of the corollary, $\mathrm{cl}(\bar{a}B_0) \mathop{\smile\!\!\!\!\!\!|}^{d}_{B_1} \mathrm{cl}(B)$. Hence, we have $\bar{a} \mathop{\smile\!\!\!\!\!\!|}^{d}_{B_1} B$. □

This corollary is a partial solution to Baldwin's problem.

5. Characterization of Independence in Generic Structures

In stable generic structures, we can define two notions of independence: forking independence and d-independence. Wagner[16] showed that for closed sets A and B, if $A \cap B$ is algebraically closed, then A and B are forking independent over $A \cap B$ if and only if A and B are d-independent over $A \cap B$. He proved the result in the case that $\mathbb{K}$ satisfies finite closure condition. Verbovskiy and Yoneda[15] showed that the same result without assuming the finite closure condition. In this section, we do not assume the finite closure condition. To show the equivalence of forking independence and d-independence, we prove the following two statements:

(1) for closed sets A and B, A and B are d-independent over $A \cap B$ if and only if A and B are free over $A \cap B$ and AB is closed;
(2) for closed sets A and B, if $A \cap B$ is algebraically closed, then A and B are forking independent over $A \cap B$ if and only if A and B are free over $A \cap B$ and AB is closed.

Wagner proved (1) by using the epsilon-delta argument. In the first half of this section, we give a nonstandard proof of (1) by using Proposition 3.12.

In Refs. 16 and 15, (1) is used to prove (2). Tsuboi[14] proved (2) without using (1). In the second half of this section, we give a nonstandard proof of (2) by using an idea of Tsuboi.[14]

Let $L = \{R_i\}_{i<\omega}$ be a countable relational language and α_i be a positive real number for each $i < \omega$. For any finite L-structure A, we put $\delta(A) = |A| - \sum_{i<\omega} \alpha_i |R_i^A|$. Let $\mathbb{K} = \{A : \forall A' \subset A, \delta(A') \geq 0\}$. Then $\mathbb{K}$ is closed under isomorphisms and substructures and we can regard δ as a predimension on $\mathbb{K}$. Let $M \in \overline{\mathbb{K}}$ and $T = \mathrm{Th}(M)$. In this section, we assume that T is stable. Let $\mathcal{M}$ be a big model of T. Note that $\mathcal{M} \in \overline{\mathbb{K}}$. For if there is a finite substructure A of $\mathcal{M}$ with $\delta(A) < 0$, then the structure A' which is obtained by restricting A to some finite relations in L also satisfies $\delta(A') < 0$. But then, there is a substructure B of M satisfying $\delta(B) < 0$.

We define the closed relation $\leq$ and the dimension function d for $\mathcal{M}$ as in Section 2.

Note that for finite subsets A and B of $\mathcal{M}$, A and B are free over $A \cap B$ if and only if $\delta(A/B) = \delta(A/A \cap B)$.

Lemma 5.1. *For any subsets A and B of $\mathcal{M}$, the following conditions are equivalent:*

(1) A and B are free over $A \cap B$.

(2) For any $A^* \supset_{\mathrm{hf}} A$ *and any* $B^* \supset_{\mathrm{hf}} B$ *with* $(A^*B^*) \cap A \cap B = A^* \cap B^*$,

$$\delta(A^*/A^* \cap B^*) \approx \delta(A^*/B^*).$$

(3) There exist $A^* \supset_{\mathrm{hf}} A$ *and* $B^* \supset_{\mathrm{hf}} B$ *such that* $(A^*B^*) \cap A \cap B = A^* \cap B^*$ *and*

$$\delta(A^*/A^* \cap B^*) \approx \delta(A^*/B^*).$$

Proof. (1) $\Rightarrow$ (2). Suppose (2) is not the case. The following finite set of formulas are satisfiable:

- $F(X) \wedge F(Y)$,
- $X \subseteq A \wedge Y \subseteq B$,
- $(XY) \cap A \cap B = X \cap Y$,
- $\delta(X/X \cap Y) > \delta(X/Y)$.

Let A_0 and B_0 be subsets of $\mathcal{M}$ satisfying the above formulas. Then A_0 and B_0 witness that A and B are not free over $A \cap B$.

(2) $\Rightarrow$ (3). Trivial.

(3) $\Rightarrow$ (1). Suppose (1) is not the case. Then there exist finite $A' \subset A$ and finite $B' \subset B$ such that

$$(A'B') \cap A \cap B = A' \cap B' \text{ and } \delta(A'/A' \cap B') < \delta(A'/B').$$

On the other hand,

$$\begin{aligned} M \models \forall X, Y \in F[(&\exists X' \subset_{\mathrm{fin}} X, \exists Y' \subset_{\mathrm{fin}} Y \\ &(X'Y') \cap X \cap Y = X' \cap Y' \wedge \delta(X'/X' \cap Y') > \delta(X'/Y')) \\ &\rightarrow \delta(X/X \cap Y) > \delta(X/Y)]. \end{aligned}$$

So, for arbitrary $A^* \supset_{\mathrm{hf}} A$ and $B^* \supset_{\mathrm{hf}} B$ with $(A^*B^*) \cap A \cap B = A^* \cap B^*$, we have

$$\delta(A^*/A^* \cap B^*) > \delta(A^*/B^*).$$ □

Proposition 5.2. *Let* A *and* B *be closed subsets of* $\mathcal{M}$. *Then the following are equivalent:*

(1) AB *is closed and* A *and* B *are free over* $A \cap B$.
(2) A *and* B *are* d^**-independent over* $A \cap B$.
(3) A *and* B *are d-independent over* $A \cap B$.

Proof. (2) $\Leftrightarrow$ (3). By Proposition 3.12.

(1) $\Rightarrow$ (2). Take $D^* \supset_{\rm hf} AB$ such that D^* is quasi-closed and $\delta(D^*) = \delta(D^* \cap A) + \delta(D^* \cap B) - \delta(D^* \cap A \cap B)$. Then we have $d(D^*) = d(D^* \cap A) + d(D^* \cap B) - d(D^* \cap A \cap B)$.

(2) $\Rightarrow$ (1). Let A^*, B^* be witnesses of the d^*-independence. By Proposition 3.12, we may assume that $(A^*B^*) \cap A \cap B = A^* \cap B^*$. Then,

$$\begin{aligned}\delta(A^*B^*) &\leq \delta(A^*) + \delta(B^*) - \delta(A^* \cap B^*) \\ &\approx d(A^*) + d(B^*) - d(A^* \cap B^*) \\ &\approx d(A^*B^*).\end{aligned}$$

The other direction is trivial. So, we have $\delta(A^*B^*) \approx d(A^*B^*)$ and $\delta(A^*B^*) \approx \delta(A^*) + \delta(B^*) - \delta(A^* \cap B^*)$. By Lemma 3.7 and Lemma 5.1, AB is closed and A and B are free over $A \cap B$. □

The following fact is easy.

Fact 5.3.[16] *Let A and B be subsets of $\mathcal{M}$. Suppose A and B are free over $A \cap B$. Then for any $A' \leq A$ and $B' \leq B$ with $A' \cap B' = A \cap B$, $A'B'$ is closed in AB.*

Note that for any finite subset A of $\mathcal{M}$ and for any $n < \omega$, the relation $|A| = n$ is definable. So we may assume that the domain of the function $|*|$ is the set of all hyperfinite sets and its range is $\mathbb{R}^*$.

Proposition 5.4. *Let A and B be closed subsets of $\mathcal{M}$. Suppose that $A \cap B$ is algebraically closed. Suppose also that A and B are forking independent over $A \cap B$. Then*

(1) A and B are free over $A \cap B$ and
(2) AB is closed.

Proof. By the above fact, we can assume that A and B are both algebraically closed.

Claim A. *There are sequences $(A_i)_{i<\omega}$ and $(B_i)_{i<\omega}$ satisfying the following properties:*

- $\mathrm{tp}(A_iB_j/A \cap B) = \mathrm{tp}(AB/A \cap B)$ *for any $i, j \in \omega$;*
- $\{A_i : i < \omega\} \cup \{B_i : i < \omega\}$ *is an independent set.*

Proof of Claim A. Let $(B_i)_{i<\omega}$ be a Morley sequence over $A \cap B$ with $B_0 = B$. By $A \mathop{\smile\!\!\!\!\!\!\mid}_{A\cap B} B$, we can assume $\mathrm{tp}(AB_i/A\cap B) = \mathrm{tp}(AB/A\cap B)$ for any $i < \omega$. Let A_{n+1} be a realization of a nonforking extension of $\mathrm{tp}(A/\bigcup_{i<\omega} B_i)$ to $\bigcup_{j<n+1} A_j \cup \bigcup_{i<\omega} B_i$ for each n. □

(1) Freeness: Suppose not. For simplicity, we assume that there is a ternary relation $R \in L$ and elements $a \in A - (A\cap B)$, $b \in B - (A \cap B)$, and $c \in A \cap B$ such that $R(a,b,c)$ holds. By the above claim, there are $a_i \in A_i$ and $b_i \in B_i$ such that for any $i, j < \omega$, $\mathrm{tp}(a_i b_j c) = \mathrm{tp}(abc)$. In particular, $R(a_i, b_j, c)$ holds for any $i, j < \omega$. Then there are hyperfinite sets A^* with $|A^*| \geq n$ (for each $n < \omega$) and B^* with $|B^*| = |A^*|$ such that $\mathcal{M}^* \models \forall x \in A^* \forall y \in B^* R(x,y,c)$. Then we have

$$\begin{aligned}\delta^*(A^*B^*) &\leq |A^*| + |B^*| - \alpha(|A^*| \times |B^*|)\\ &= |A^*| + |A^*| - \alpha(|A^*| \times |A^*|)\\ &< 0.\end{aligned}$$

A contradiction.

(2) Closedness: Suppose not. For simplicity, we assume that there are elements $d \in \mathrm{acl}(AB) - AB$, $a \in A$, and $b \in B$ such that $\gamma := \delta(d/ab) < 0$. By the claim, for all $i, j \in \omega$, we can find $a_i \in A_i$, $b_i \in B_i$ and d_{ij} such that $\mathrm{tp}(abdAB) = \mathrm{tp}(a_i b_i d_{ij} A_i B_j)$.

Claim B. $(\bigcup_{(i,j)\in\omega^2} d_{ij}) \cap (\bigcup_{i\in\omega} A_i B_i) = \emptyset$

Proof of Claim B. Take any $e \in d_{i,j}$. By the choice of $d_{i,j}$, we have $e \notin A_i B_j$. By symmetry, it is enough to show that $e \notin A_m$ for any $m \neq i$. Note that $e \in \mathrm{acl}(A_i B_j)$ and $\mathrm{acl}(A_i B_j) \cap A_m = A \cap B$ by $A_i B_j \mathop{\smile\!\!\!\!\!\!\mid} A_m$. But $e \notin A \cap B$, so $e \notin A_m$. □

Claim C. *The d_{ij}'s are distinct.*

Proof of Claim C. By way of a contradiction, we assume $d_{i,j} = d_{i',j'}$ for some $(i,j) \neq (i',j')$. Then $d_{i,j} \in \mathrm{acl}(A_i B_j) \cap \mathrm{acl}(A_{i'} B_{j'})$. Note that $\mathrm{acl}(A_i B_j) \cap \mathrm{acl}(A_i B_{j'}) = A_i$, and if $i \neq i'$ and $j \neq j'$ then $\mathrm{acl}(A_i B_j) \cap \mathrm{acl}(A_{i'} B_{j'}) = A \cap B$. But $d_{i,j} \notin A_i$. A contradiction. □

Note that $\delta(\bigcup_{i<n} a_i b_i) = n\delta(a_0 b_0)$ and $\delta(\bigcup_{(i,j)\in n^2} d_{(i,j)} / \bigcup_{i<n} a_i b_i) \leq n^2\gamma$ for each $n < \omega$. So, there are hyperfinite sets A^* with $|A^*| \geq n$ (for each $n < \omega$), B^* with $|B^*| = |A^*|$, and D^* with $|D^*| = |A^*|^2$ such that $\delta^*(A^*B^*) = |A^*|\delta(a_0 b_0)$ and $\delta^*(D^*/A^*B^*) \leq |A^*|^2\gamma$.

Then we have

$$\begin{aligned}\delta(D^*A^*B^*) &\leq \delta(D^*/A^*B^*) + \delta(A^*B^*)\\ &\leq |A^*|^2\gamma + |A^*|\delta(a_0b_0)\\ &< 0.\end{aligned}$$

A contradiction. □

Definition 5.5. We say that T has amalgamation over closed sets if for any $N_0, N_1 \models T$, $A \in \overline{\mathbb{K}}$, and closed embeddings $f_i : A \to N_i$ $(i = 0, 1)$, there is $N \models T$ and elementary embeddings $g_i : N_i \to N$ $(i = 0, 1)$ such that $g_0 \circ f_0 = g_1 \circ f_1$ on A, equivalently for any $N \models T$ and $A_0, A_1 \subseteq N$, if $A_0 \cong A_1$ and A_0 and A_1 are both closed, then $\mathrm{tp}(A_0) = \mathrm{tp}(A_1)$.

Fact 5.6.[6,15] *If T has amalgamation over closed sets, then T is stable.*

Corollary 5.7. *Suppose T has amalgamation over closed sets, and A and B are closed sets such that $A \cap B$ is algebraically closed. Then the following are equivalent:*

(1) A and B are forking independent oner $A \cap B$.
(2) A and B are d-independent over $A \cap B$.
(3) A and B are free over $A \cap B$ and AB is closed.

Proof. We have already proved (1) ⇒ (3) and (2) ⇔ (3).

(3) ⇒ (1). Suppose (3). Take A' such that A' is independent from B over $A \cap B$ and satisfies $\mathrm{tp}(A/A \cap B)$. By (1) ⇒ (3), A' and B are free over $A \cap B$ and $A'B$ is closed. By amalgamation over closed sets, we have $\mathrm{tp}(AB/A \cap B) = \mathrm{tp}(A'B/A \cap B)$. Hence, A is also independent from B over $A \cap B$. □

Acknowledgments

The author is thankful to Professor Tsuboi for useful comments. This work is supported by JSPS Research Fellowships for Young Scientists.

References

1. Y. Anbo, On stability of saturated generic hypergraphs, RIMS Kokyuroku 1555, 9-17 (2007)
2. Y. Anbo, Nonstandard arguments and the stability of generic structures, RIMS Kokyuroku 1602, 10-16 (2008)
3. Y. Anbo and K. Ikeda, A Note on Stability Spectrum of Generic Structures, preprint

4. J. T. Baldwin, Problems on pathological structures, In Helmut Wolter Martin Weese, editor, Proceedings of 10th Easter Conference in Model Theory (1993) 1-9
5. J. T. Baldwin, An almost strongly minimal non-Desarguesian projective plane, Trans. Amer. Math. Soc. 342 (1994) 695-711
6. J. T. Baldwin and N. Shi, Stable generic structures, Annals of Pure and Applied Logic 79 (1996) 1-35
7. B. Herwig, Weight ω in stable theories with few types, J. Symbolic Logic 60 (1995) 353-373
8. K. Holland, An introduction to the fusion of strongly minimal sets: The geometry of fusions, Archive for Mathematical Logic, 6:395-413, 1995
9. E. Hrushovski, A new strongly minimal set, Annals of Pure and Applied Logic 46 (1990) 235-264
10. K. Ikeda, Minimal, non-strongly minimal structures with arbitrary finite dimensions, J. Symbolic Logic 66 (2001) 117-126
11. K. Ikeda, A remark on the stability of saturated generic graphs, Journal of the Mathematical Society of Japan, vol.57, no.4 (2005)
12. A. Pillay, An Introduction to Stability Theory, Oxford University Press, 1983
13. A. Robinson, Non-standard analysis, Revised edition, Princeton University Press, 1996
14. A. Tsuboi, Independence in generic structures, RIMS Kokyuroku 1602, 33-36 (2008)
15. V. Verbovskiy, I. Yoneda, CM-triviality and relational structures, Annals of Pure and Applied Logic, 122 (2003) 175-194
16. F. O. Wagner, Relational structures and dimensions, Kaye, Richard (ed.) et al., Automorphisms of first-order structures. Oxford: Clarendon Press. 153-180 (1994)

CONSTRUCTIVE GEOMETRY

MICHAEL BEESON

Department of Computer Science, San Jose State University,
San José, CA 95192, USA
E-mail: beeson@cs.sjsu.edu
www.MichaelBeeson.com/Research

Euclidean geometry, as presented by Euclid, consists of straightedge-and-compass constructions and rigorous reasoning about the results of those constructions. A consideration of the relation of the Euclidean "constructions" to "constructive mathematics" leads to the development of a first-order theory **ECG** of the "Euclidean Constructive Geometry", which can serve as an axiomatization of Euclid rather close in spirit to the *Elements* of Euclid. **ECG** is axiomatized in a quantifier-free, disjunction-free way. Unlike previous intuitionistic geometries, it does not have apartness. Unlike previous algebraic theories of geometric constructions, it does not have a test-for-equality construction. We show that **ECG** is a good geometric theory, in the sense that with classical logic it is equivalent to textbook theories, and its models are (intuitionistically) planes over Euclidean fields. We then apply the methods of modern metamathematics to this theory, showing that if **ECG** proves an existential theorem, then the object proved to exist can be constructed from parameters, using the basic constructions of **ECG** (which correspond to the Euclidean straightedge-and-compass constructions). In particular, objects proved to exist in **ECG** depend continuously on parameters. We also study the formal relationships between several versions of Euclid's parallel postulate, and show that each corresponds to a natural axiom system for Euclidean fields.*

1. Introduction

Euclid's *Elements*,[7] written down about 300 BCE, has been extraordinarily influential in the development of mathematics, and prior to the twentieth century was regarded as a paradigmatic example of pure reasoning. But during those 2300 years, most people thought that Euclid's theory was about something. What was it about? Some may have answered that it was about points, lines, and planes, and their relationships. Others may have

*We would like to thank Jeremy Avigad and Freek Wiedijk for productive and interesting discussions, and for careful reading and useful suggestions.

said that it was about methods for constructing points, lines, and planes with certain specified relationships to given points, lines, and planes, for example, constructing an equilateral triangle with a given side. In these two answers, we see the viewpoints of pure (classical) mathematics and of algorithmic mathematics represented. Hilbert's 1899 reworking of the theory[10] gave another answer, surprising at the time: Euclid's theories were *not about anything at all.* Instead of "points, lines, and planes", one should be able to read "tables, chairs, and beer mugs." All the reasoning should still be valid. The names of the "entities" were just place holders. That was the viewpoint of twentieth-century axiomatics.

In the late twentieth century, contemporaneously with the flowering of computer science, there was a new surge of vigor in algorithmic, or constructive, mathematics, beginning with Bishop's book.[4] In algorithmic mathematics, one tries to reduce every "existence theorem" to an assertion that a certain algorithm has a certain result. In the terminology of computer science, existence theorems should become correctness proofs of algorithms. The proof theory of arithmetic has provided many beautiful theorems to show that indeed, existence theorems in number theory (when constructively proved) contain algorithms that can be "extracted" from the proofs. In particular we mention the techniques of recursive realizability, the Dialectica interpretation of Gödel, and the extraction of algorithms from cut-free proofs as well-known examples of the phenomenon.

In this paper, we re-examine Euclidean geometry from the viewpoint of constructive mathematics. The phrase "constructive geometry" suggests, on the one hand, that "constructive" refers to geometrical constructions with straightedge and compass. On the other hand, the word "constructive" may suggest the use of intuitionistic logic. We investigate the connections between these two meanings of the word. Our method is to proceed by analogy with the extensive body of work that has been done on number theory and analysis, applying the relevant methodologies to the weaker theories of geometry. The basis for the work described here is the idea that in geometry, we can take "algorithm" in the restricted sense of "geometric construction." That is, we pursue the analogy

$$\frac{\text{formal number theory}}{\text{Turing computable functions}} = \frac{\text{intuitionistic geometry}}{\text{geometric constructions}}$$

To carry out this program, we need a suitable formal theory for intuitionistic geometry. It should be a theory with terms for the geometric constructions, so that there will be terms available to denote the means of constructing things that have been proved to exist. This leads us to look for a quantifier-free axiomatization. In formulating a suitable theory of geometry, another important consideration was the principle that "constructive proof implies continuity in parameters". This principle is familiar to those who have studied constructive mathematics; but it is easy to understand on an intuitive basis. If we wish to allow an interpretation of geometry in which points are given by approximations (as accurate as one may demand), for example, if they are given as pairs of real numbers (x, y), and we (constructively) prove $\forall x \exists y \ A(x, y)$, then we ought to be able to produce approximations to y when we are given approximations to x. This idea led us to look at the continuity of the Euclidean constructions. We looked for (and found) an axiom system **ECG** with intuitionistic logic, such that when an existential theorem is proved in **ECG**, the object asserted to exist can be constructed by straightedge and compass, continuously in parameters.

Decision functions, such as whether two points are equal or whether a given point lies on a given line, are discontinuous. Past work on the algebraic approach to constructions (see for example [13]) has always assumed decision functions, probably because they seem to be needed to define projection of a point on a line, which is needed in order to introduce coordinates in the Euclidean plane and connect geometry to field theory. But projection itself is continuous. In an early version of our theory **ECG**, we took it as a fundamental operation. In the version presented here, that operation is not necessary, because projection can be defined in terms of the operations of **ECG**, and its essential properties proved using the axioms of **ECG**.

Our interest in this subject began with a computer animation of Euclid's constructions that permits the user to drag the starting points, and see how the construction depends on the changed starting points. The results for Euclid's Book I, Proposition 2, were surprising and interesting. In Book I, Proposition 2, Euclid attempted to show that a rigid compass could be simulated by a collapsible compass. This famous construction shows how to use the rigid compass to construct a point $D = e(A, B, C)$ such that whenever $B \neq C$ and $A \neq B$ then $AD = BC$. The first problem with this construction is that it does not work when $A = C$. Of course, in that case we can just take $D = B$; but that case distinction requires classical logic. And the computer animation reveals that when we drag point C close to A, and then around A in a small circle, then the constructed point D moves around A in a circle of radius close to BC. Hence D does not depend continuously on C. This discontinuity, together with the need for a case distinction just mentioned, makes it clear that from the intuitionistic point of view, there are two different versions of Euclid I.2: the version in which (in addition to

$B \neq C$) we take $A \neq C$ as a hypothesis, or the "uniform" version in which we do not assume $A \neq C$, but assert that D can be constructed, whatever the values of A, B, and C, as long as $B \neq C$. Which version of Euclid I.2 is provable with intuitionistic logic turns out to depend on whether we take (a formalism corresponding to) a rigid compass, or Euclid's collapsible compass.

Past work on axiomatizations of constructive (intuitionistic) geometry, such [9] and [14], has replaced the law of the excluded middle by "apartness" (which is explained in the body of the paper). One can introduce an apartness "construction", but it is also not continuous.[a] Therefore we work in theories without apartness.

We are able to obtain the main metatheorem we want using cut-elimination; no new proof-theoretic techniques are developed here, so this paper should be accessible to geometers, not only to logicians.

Geometry is an ancient subject, but it is very much alive. There are a number of avenues of current research that are related to this work, but not dealt with in this paper; and rather than undertake to survey them, we prefer to stick to the specific aims outlined above. Similarly, there is a long history of axiomatizations of geometry. For a review of some of these, see [15]. Here we explain only how our axiomatization is different from others: (i) it uses constructive logic, rather than classical, which distinguishes it from all previous axiomatizations except Heyting's, and (ii) it does not use apartness, which enables it to have the property that points proved to exist can be constructed by Euclid's (continuous) constructions.

2. Euclid's Constructions as Algorithms

Euclid's five books present 48 two-dimensional constructions and about a dozen three-dimensional constructions. (For simplicity, we will not discuss the three-dimensional part of Euclid, contained mostly in Book 5). We consider the 48 two-dimensional constructions to be the world's first systematic collection of algorithms. (We do not say, "the world's first

[a]Even though apartness is intuitionistically acceptable, an apartness "constructor" must make use of the idea that points are not given "all at once" but by a sequence of approximations. The same point can be given by different sequences, and though an apartness constructor can be continuous in the approximating sequences, it cannot be continuous in the geometric topology on the points. Hence an apartness constructor needs algorithms beyond the Euclidean constructions. In other words, geometry with apartness goes beyond Euclid. We have extended our metamathematical work to theories with apartness, but we have not presented those extensions in this paper.

algorithms", because there certainly were a few number-theoretic algorithms known in China and India much earlier.)

If the constructions are considered as algorithms, then Euclid's *Elements* contained the first proofs of correctness of algorithms.

Euclid presents his readers with both "postulates" and "axioms". Modern mathematicians often treat these words as synonyms. For Euclid and his contemporaries, however, they had quite different meanings. Here is the difference, as explained by Pambuccian,[16] p. 12.

> For Proclus, who relates a view held by Geminus, a postulate prescribes that we construct or provide some simple or easily grasped object for the exhibition of a character, while an axiom asserts some inherent attribute that is known at once to one's auditors. And just as a problem differs from a theorem, so a postulate differs from an axiom, even though both of them are undemonstrated; the one is assumed because it is easy to construct, the other accepted because it is easy to know. That is, postulates ask for the production, the *poesis* of something not yet given ..., whereas axioms refer to the *gnosis* of a given, to insight into the validity of certain relationships that hold between given notions.

Euclid's famous "parallel postulate" states that if two lines L and M are traversed by another line T, forming adjacent interior angles on one side of T adding up to less than two right angles, then L and M will intersect on that side of T. Stated this way, the postulate can be viewed as a construction method for producing certain triangles. Nowadays, the parallel postulate is often stated as an axiom: Given a line L, and a point P not on L, there exists exactly one parallel to L through P. (Parallel lines are by definition lines that do not meet.) Written this way, the parallel postulate does not directly assert the existence of any specific point.[b]

3. The Algebraic Approach to Constructions

The geometrical theory that we shall eventually formulate is quantifier-free, with terms to denote the geometrical constructions. A model of such a theory can be regarded as a many-sorted algebra with partial functions representing the basic geometric constructions. Specifically, the sorts include at least *Point*, *Line*, and *Circle*, and they may possibly include *Arc*, *Ray*, *Triangle*, and *Square*. We have constants and variables of each sort.

[b]For example, it is not immediately clear whether this version implies the first version using only intuitionistic (constructive) logic, although of course it does in classical logic. This question is settled (in the negative) in [3].

This collection of data types is almost adequate to cover the return types and argument types of the 48 plane Euclidean constructions. The obvious exception is the construction of a regular pentagon or hexagon. More generally, some of the later constructions use the word "figure", which apparently means something like what a modern mathematician would call a "closed polygon". Some small fixed number of sides would suffice for the Euclidean constructions. In Euclid, no figure with more sides than an octagon is constructed, and no figure with more than four sides is an *input* to another construction, except for constructions that work on any "figure." The general concept of a closed polygon of any number of sides may be logically problematic as it drags the concept of integer into geometry.

Our algebras include function symbols for the basic constructors and accessors:

$\mathit{Line}\ (\mathit{Point}\ A, \mathit{Point}\ B)$	A and B lie on this line
$\mathit{pointOn1}(\mathit{Line}\ L)$	The points from which L was originally constructed
$\mathit{pointOn2}(\mathit{Line}\ L)$	
$\mathit{Circle}\ (\mathit{Point}\ A, \mathit{Point}\ B)$	A is the center, and the circle passes through B
$\mathit{center}(\mathit{Circle}\ C)$	
$\mathit{pointOnCircle}(\mathit{Circle}\ C)$	A point on circle C,

and for the "elementary constructions" (each of which has type *Point*):

$$\mathit{IntersectLines}(\mathit{Line}\ K, \mathit{Line}\ L)$$
$$\mathit{IntersectLineCircle1}(\mathit{Line}\ L, \mathit{Circle}\ C)$$
$$\mathit{IntersectLineCircle2}(\mathit{Line}\ L, \mathit{Circle}\ C)$$
$$\mathit{IntersectCircles1}(\mathit{Circle}\ C, \mathit{Circle}\ K)$$
$$\mathit{IntersectCircles2}(\mathit{Circle}\ C, \mathit{Circle}\ K)$$

Each of these has several "overloaded" variants, which can be defined from these using constructors and accessors. For example,

$$\begin{aligned}&\mathit{IntersectLines}(\mathit{Point}\ A, \mathit{Point}\ B, \mathit{Point}\ C, \mathit{Point}\ D)\\ &\quad = \mathit{IntersectLines}(\mathit{Line}\ (A, B), \mathit{Line}\ (C, D))\\ &\mathit{IntersectLineCircle1}(\mathit{Point}\ A, \mathit{Point}\ B, \mathit{Point}\ C, \mathit{Point}\ D)\\ &\quad = \mathit{IntersectLineCircle1}(\mathit{Line}\ (A, B), \mathit{Circle}\ (C, D))\\ &\mathit{IntersectLineCircle1}(\mathit{Point}\ A, \mathit{Point}\ B, \mathit{Circle}\ C)\\ &\quad = \mathit{IntersectLineCircle1}(\mathit{Line}\ (A, B), C)\end{aligned}$$

As these three examples illustrate, one can regard circles and lines as mere intermediaries; points are ultimately constructed from other points. In a lemma at the end of this section, we state and prove this principle precisely.

We can distinguish the study of such many-sorted algebras from the study of axiomatic first-order theories containing symbols for these algebraic operations. Some questions can be taken up without the consideration of axioms or logical inferences. We shall discuss a few of these.

There is a second constructor for circles, which we can describe for short as "circle from center and radius", as opposed to the first constructor above, "circle from center and point." Specifically *Circle3* (A, B, C) constructs a circle of radius BC and center A, provided $B \neq C$. These two constructors for circles correspond to a "collapsible compass" and a "rigid compass" respectively. The compass of Euclid was a collapsible compass: you cannot use it to "hold" the length BC while you move one point of the compass to A. You can only use it to hold the radius AB while one point of the compass is fixed at A, so in that sense it corresponds to *Circle* (A, B). The second constructor *Circle3* corresponds to a rigid compass.[c] In the next section we will have more to say about the relationship between these two constructors.

We introduce here a first example of a "construction" not considered by Euclid, the *test-for-equality* construction. This "construction" **D** takes four points, and tests its first two arguments for equality, producing the third or fourth argument depending on the outcome:

$$\mathbf{D}(a, b, c, d) = \begin{cases} c & \text{if } a = b \\ d & \text{if } a \neq b \end{cases}$$

The algebraic approach to constructions was pioneered by Kijne,[13] but all the systems he considered contained "decision functions" such as test-for-equality or test-for-incidence. In this paper we will not study systems containing decision functions.

We note in the following lemma that, as far as constructing points goes, the other types are mere conveniences. The elementary constructions can be expressed, as we have noted, in several ways using variables of different types. For example, we could have *IntersectLines* (K, L) where K and L have type *Line* , or *IntersectLines* (A, B, C, D), where A, B, C, and D have

[c] There is a word in Dutch, *passer*, for this type of compass, which was used in navigation in the seventeenth century. But there seems to be no single word in English that distinguishes either of the two types of compass from the other.

type *Point*, and

$$IntersectLines(Line(A, B), Line(C, D) = IntersectLines(A, B, C, D).$$

Lemma 3.1. *Let t be a term of type Point, whose variables are all of type Point. Then there is a term t^* with the same variables as t, also of type Point, such that in the standard plane t and t^* determine the same function, and t^* contains only function symbols of type Point having Point arguments.*

Proof. By induction on the complexity of t. Suppose that t has the form $IntersectLines(r, s)$ where r and s are terms of type *Line* . None of the elementary constructions has type *Line* , and r and s cannot be variables (since all variables in t are of type *Point*), so r must have the form *Line* (p, q) for some terms p and q, and s must have the form *Line* (u, v) for some terms u and v. Then t^* can be taken to be *Line* (p^*, q^*, u^*, v^*). The other elementary constructions are treated similarly. The basis case, when t is a variable or constant, is treated by taking $t^* = t$.

4. Models of the Elementary Constructions

The algebraic approach allows us to consider "models" without (yet) having formulated any axioms or logical theories. There are several interesting models, of which we now mention four. To define these models, we assume there are three constants α, β, and γ of type *Point.*

The *standard plane* is $\mathbf{R}^2$. Points, lines, and circles (as well as segments, arcs, triangles, squares, etc., in extended algebras in which such objects are considered) are interpreted as the objects that usually bear those names in the Euclidean plane. More formally, the interpretation of the type symbol "*Point* " is the set of points, the interpretation of "*Line* " is the set of lines, etc. In particular we must choose three specific non-collinear points to serve as the interpretations of α, β, and γ. Let us choose $\alpha = (0, 1)$, $\beta = (1, 0)$, and $\gamma = (0, 0)$. The constructor and accessor functions listed above also have standard and obvious interpretations. It is when we come to the five operations for intersecting lines and circles that we must be more specific. There are three issues to decide:

- when there are two intersection points, which one is denoted by which term?
- In degenerate situations, such as *Line* (P, P), what do we do?
- When the indicated lines and/or circles do not intersect, what do we do about the term(s) for their intersection point(s)?

We take up the last item first. When, for example, line L does not meet circle C, we say that the term *IntersectLineCircle1* (L, C) is *undefined.* In other words, the operations of these "algebras" do not have to be defined on all values of their arguments. The same issue, of course, arises in many other algebraic contexts, for example, division is not defined when the denominator is zero, and $\sqrt{x}$ is not defined (when doing real arithmetic) when x is negative.

Regarding the "which is which" issue, our guiding principle is continuity. We therefore make the following definitions: *IntersectCircles1* (C, K) is the intersection point P such that the angle from *center*(C) to *center*(K) to P makes a "left turn". This is defined as in computer graphics, using the sign of the cross product. Specifically, let A = *center*(C) and B = *center*(K). Then the sign of $(A - B) \times (P - B)$ determines whether angle ABP is a "left turn" or a "right turn". Thus $\alpha\beta\gamma$ is a left turn and $\gamma\beta\alpha$ is a right turn. In case the two intersection points are different, one of these cases must apply. This explanation has used a case distinction as well as the cross product; later we will show how to define "left turn" and "right turn" using only the axioms of Euclidean geometry and intuitionistic logic. For now we simply note that this notion is constructively appealing, because of continuity: there exists a unique continuous function of C and K that satisfies the stated handedness condition for *IntersectCircles1* when C and K have two distinct intersection points, and is defined whenever C and K intersect (at all).

The principle of continuity leads us to make *IntersectCircles1* (C, K) and *IntersectCircles2* (C, K) undefined in the "degenerate situation" when circles C and K coincide, i.e. have the same center and radius. Otherwise, as the center of C passes through the center of K, there is a discontinuity. It makes sense, anyway, to have them undefined when C and$K$ coincide, as the usual formulas for computing them get zero denominators, and there is no natural way to select two of the infinitely many intersection points.

We still need to settle the "which is which" issue for

$$\textit{IntersectLineCircle1}\,(\textit{Line}\ (A, B), C)$$

and

$$\textit{IntersectLineCircle2}\,(\textit{Line}\ (A, B), C).$$

Here the rule is that these two points must occur in the same order on *Line* (A, B) as A and B do. Again, we require continuity of the function

$$\textit{IntersectLineCircle1}\,(A, B, P, Q)$$

with four *Point* variables, i.e.

$$IntersectLineCircle1\,(Line\ (A,B),\ Circle\ (P,Q)).$$

There is a unique continuous extension from the set of (A, B, P, Q) where there are two intersection points to the set where there is at least one intersection point; that extension is the interpretation of *IntersectLineCircle1*.

Regarding the degenerate forms *Line* (P, P), we say that the term is undefined. With respect to degenerate circles, *Circle* (A, A), continuity and computability do not present the same obstacles as in the case of degenerate lines *Line* (A, A). Thus we have a choice to allow degenerate segments and circles, without destroying the continuity of the elementary constructions. We may choose to allow them or not. In the "standard model" R^2 we take *Circle* (A, A) to be defined, i.e. circles of zero radius are allowed. There is only one point on such a circle. Then *IntersectLineCircle1* $(L, Circle(A, A))$ is defined if A is on line L, and is equal to A. We also consider (briefly) the model R^{2-} in which degenerate circles are not allowed, so *Circle* (A, A) is undefined.

The *recursive plane* ${\mathbf{R}_{\mathbf{rec}}}^2$ consists of points in the plane whose coordinates are given by "recursive reals". We write $\{e\}(n)$ for the result, if any, of the e-th Turing machine at input n. Rational numbers are coded as certain integers, and modulo this coding we can speak of recursive functions from $\mathbf{N}$ to $\mathbf{Q}$. A "recursive real number" x is the index of a Turing machine e such that $|\{e\}(n) - x| \le 1/n$ for each $n \in \mathbf{N}$. The real number to which the approximations $\{x\}(n)$ converge is sometimes also called a "recursive real number", but we call it the "value of x". It is a routine exercise to show that the recursive points in the plane are closed under the Euclidean constructions. In particular, given approximations to two circles (or to a circle and a line), we can compute approximations to their "intersection points", even though it may turn out that when we compute better approximations to the circles, we see that they do not intersect at all.

In the recursive plane, there is no computable test-for-equality function, that is, no computable function D that operates on two Turing machine indices x and y, and produces 0 when x and y are recursive real numbers with the same value, and 1 when they are recursive real numbers with different values. Proof, if we had such a D, we could solve the halting problem by applying D to the point $(E(x), 0)$, where $\{E(x)\}(n) = 1/n$ if Turing machine x does not halt at input x in fewer than n steps, and $\{E(x)\}(n) = 1/k$ otherwise, where x halts in exactly k steps. Namely, $\{x\}(x)$ halts if and only if the value of $E(x)$ is not zero, if and only if

$D(Z, E(x)) \neq 0$, where Z is an index of the constant function whose value is the (number coding the) rational number zero.

Readers familiar with recursion theory may realize that there are several ways to define computable functions of real numbers. The model we have just described is essentially the plane version of the "effective operations". It is a well-known theorem of Tseitin, Kreisel, LaCombe, and Shoenfield, known traditionally as KLS (and easily adapted to the plane) that effective operations are continuous. Of course, in the case at hand we can check the continuity of the elementary constructions directly.

The *algebraic plane* consists of points in the plane whose coordinates are algebraic. Since intersection points of circles and lines are given by solutions of algebraic equations, the algebraic plane is also closed under these constructions. Since algebraic numbers can be computed, this is a submodel of the recursive plane.

In the algebraic plane, there *is* a computable test-for-equality function D. We assume algebraic numbers are given by means of a rational interval (a, b) and a square-free polynomial $f \in \mathbf{Q}[x]$ such that f has only one root in (a, b). To determine if (a, b) and f determine the same or a different real number than (p, q) and g, first check if the two rational intervals overlap. If not, the two reals are different. If so, let (r, s) be the intersection. Now we have to determine if f and g have a common zero on (r, s). There is a simple recursive algorithm to do that: Say g has degree greater than or equal to that of f. Then write $f = gh + r$ with r of lower degree than g. Then f and g have a common zero on (r, s) if and only if f and r have a common zero. Recurse until both polynomials are linear, when the decision is very easy to make.

The *Tarksi model* is $K \times K$, where $K = \mathbf{Q}(\sqrt{\ })$ is the least subfield of the reals containing the rationals and closed under taking the square root of positive elements. This is a submodel of the algebraic plane.

In such models, the elementary constructions are interpreted as functions from M to M, and of course M must be closed under those functions, but they may also have other interesting properties. For example, in the examples given above, the (interpretations of the) elementary constructions are all computable functions; indeed they are algebraic functions of low degree. In particular, they are all continuous on their domains. Hence, for example, there is no algebraic test-for-equality function in the algebraic plane, even though there is a computable one (since a test-for-equality function would enable us to define a function $f(x)$ such that $f(0) = 0$ but $f(x) = 1$ when $x \neq 0$; since f is not continuous, it cannot be algebraic).

5. Euclid's Book I, Proposition 2

Book I, Proposition 2 has been discussed in the introduction. The question it addresses concerns the constructor *Circle3*(A, B, C), which constructs a circle with center A and radius BC. As discussed in the introduction, Euclid gives a term that accomplishes this aim under the assumptions, not only that $B \neq C$, but also that $A \neq C \vee B \neq C$. We consider the stronger theorem "uniform Euclid I.2", which asserts that for every A and BC with $B \neq C$, there exists a D with $AD = BC$. The "uniformity" refers to the missing assumption $A \neq B$. To "realize" uniform Euclid I.2, we would need a term $e(A, B, C)$ that produces D uniformly, whether $A = B$ or not. If we had such an e, then of course we could define

$$\mathit{Circle3}(A, B, C) = \mathit{Circle}\ (A, e(A, B, C)).$$

Conversely, such an e can be defined from *Circle3* like this:

$$e(A, B, C) = \mathit{pointOnCircle}(\mathit{Circle3}(A, B, C))$$

But the mere fact that Euclid's own construction does not suffice to define *Circle3* does not show that some other construction won't do the job. However, we are able to prove that no other construction can define *Circle3*, as a corollary to the following theorem, which shows that without *Circle3*, no total function is definable, at least if we insist that *Circle* (x, x) is undefined.

Theorem 5.1 (with Freek Wiedijk). *No total unary point-valued function (other than the identity) is definable in the standard model R^{2-} (with degenerate circles undefined) from the elementary geometrical constructions excluding Circle3. More precisely, let t be a compound term containing exactly one variable A of type Point, and no other variables, but possibly containing some constants $\alpha, \beta, \ldots$ of type Point. Suppose that t does not contain Circle3. Let the constants be interpreted as certain (fixed) distinct points in the standard plane. Then for some value of A, t is not defined in the standard plane. In fact, we can make t undefined by assigning A the same value as any constant occurring in t, or if t has no constants, t is always undefined.*

Proof. We start by eliminating the "overloaded" versions of the elementary constructions from t. For instance, if t contains a subterm $IntersectLine(a, b, c, d)$, we replace it by a term using the "fundamental" form of the construction, *IntersectLines*(*Line* (a, b), *Line* (c, d)). The result of such replacements is a term with the same value as t under any assignment of a value to the variable A, and containing no variables of types other than *Point.*

We proceed by induction on the complexity of terms t as in the theorem, but also containing only the fundamental versions of the elementary constructions (no overloaded versions).

Since the theorem only applies to compound terms, the basis case occurs when t has only variables or constants for arguments. We note that *Circle* (A, A), *Line* (A, A), *Ray* (A, A), *Segment* (A, A), *Arc* (A, A, A) are undefined. But we also have to consider the possibility that the constants β or γ occupy one of the argument places. For example, *Circle* (A, β) is undefined when A takes the value $\bar{\beta}$ that is assigned to the constant β. In the rest of the proof we shorten this kind of statement to "is undefined when $A = \beta$." Similarly, *Circle* (β, A) is undefined when $A = \beta$; *Line* (A, β) and *Line* (β, A) are undefined when $A = \beta$, and the same for *Ray* , *Arc* , and *Segment* . Note that this argument does not work for *Circle3*, since *Circle3* (A, β, γ) is always defined, but by hypothesis, t does not contain *Circle3.*

Since t does not contain any overloaded constructions, the basis case is finished, as there is no fundamental construction that takes only arguments of type *Point.* Specifically, there are just five fundamental constructions for producing the intersections of lines and lines, lines and circles, or circles and circles, and they each need arguments of type *Line* or *Circle.*

Now consider the induction step. If the main symbol of t is a constructor, such as *Line* , then t has the form *Line* (a, b). One of a or b must contain a variable, and hence be somewhere undefined. Hence t is also somewhere undefined, and indeed the same assignment of a value to A that makes a or b undefined will work. Similarly for the other constructors (since *Circle3*, to which this argument does not apply, is not allowed).

Next consider the case when t is *IntersectLines* (p, q). Then p and q have type *Line.* The only terms of type *Line* are those of the form *Line* (a, b), or variables of type *Line.* But t is not allowed to contain variables of type *Line.* Therefore t must have the form,

$$\mathit{IntersectLines}(\mathit{Line}\ (a, b), \mathit{Line}\ (c, d))$$

Since t contains a variable, one of $s =$ *Line* (a, b) or $s =$ *Line* (c, d) must contain a variable, and by induction hypothesis the term $s\sigma$ is undefined when substitution σ assigns the variable of s to one of the constants in s (or any constant if s has no constants). Hence $t\sigma$ is also undefined.

Similarly for the other elementary constructions. That completes the proof.

Corollary 5.1. *Let $e(A, B, C)$ be any term built up from the elementary constructions, not containing Circle3, having type Point, and containing exactly three variables A, B, and C of type Point. Then it is not the case that in R^{2-}, whenever $B \neq C$ then $e(A, B, C)$ is defined and is a point D such that $AD = BC$.*

Proof. Let us invent two constants β and γ, and interpret them as two distinct points $\bar{\beta}$ and $\bar{\gamma}$ (fixed for the rest of the proof). Then let $f(A) = e(A, \beta, \gamma)$. Suppose, for proof by contradiction, that $e(A, B, C)$ is defined whenever $B \neq C$ and is a point D such that $AD = BC$. Then $f(A)$ is defined for all A and is always different from A, contradicting the previous theorem.

Corollary 5.2. *Circle3 is not definable in the model R^{2-} from the (other) elementary constructions.*

Proof. *Circle3* is a term e fulfilling the hypotheses of the previous corollary.

Circle3 is intimately connected with the "compass" of "straightedge and compass constructions". Euclid's compass is supposed to be "collapsible", so that you cannot use it to measure BC and then move it to draw a circle of that same radius centered at A. Therefore Euclid proved Euclid I.2, showing how you can accomplish this with a collapsible compass; but for that we need to assume $A \neq B$. In effect we need a test-for-equality construction. When Proclus criticized Euclid for omitting arguments by cases, perhaps this is what he had in mind. The theorem "for every A, B, and C with $B \neq C$ there exists a point D with $AD = BC$" is (apparently) not "realized" by an elementary construction $D = f(A, B, C)$.

The theorem above shows that without *Circle3*, we cannot even define a construction f such that for each point A, $f(A)$ is a point different from A. We also cannot define a construction $f(A, L)$ that takes a point A and line L into a point on L, different from A.

The fact that *Circle3* is not definable in R^{2-} means that, if we do not include *Circle3* as a primitive construction, we shall not be able to define it in any axiomatic theory that has R^{2-} for a model. It seems clear that Euclid's book does have R^{2-} as a model; so the uniform version of Book I, Proposition 2 is essentially non-constructive, using Euclid's non-rigid compass. We therefore add *Circle3* as a fundamental construction (rendering I.2 a triviality) and give a constructive theory that works for the rest of Euclid.

There is still an unresolved technical issue here: is *Circle3* undefinable in R^2? The above theorem does not extend to R^2, because, for example, *IntersectLineCircle2*(*Line* (A, B), *Circle* (B, x)) is a total function if *Circle* is total. We conjecture that *Circle3* is undefinable in R^2 as well as in R^{2-}.

6. Circles of Zero Radius and *Circle3*

In this section, we address the two issues of whether our basic theory ought to include *Circle3* and whether we ought to allow circles of zero radius. We conclude that both should be allowed. Our approach is to consider what is required to achieve a formalization of Euclid using intuitionistic logic, with no test-for-equality construction needed.

Here are some possible constructions we wish to consider.

Euclidean Extension. One of the Euclidean axioms says that we can extend a given segment AB by a segment CD. More precisely, we can construct a point $P = \mathit{Extend}(A, B, C, D)$ such that $BP = CD$ and B is between A and P. The assumptions here are that $A \neq B$ and $C \neq D$, but it is not assumed that $B \neq C$ or $B \neq D$.

Strong Extension. $\mathit{ExtendA}(A, B, C, D)$ is the unique continuous extension of $\mathit{Extend}(A, B, C, D)$ that is defined when $A \neq B$, i.e. without assuming $C \neq D$. When $C = D$, we have

$$\mathit{ExtendA}(A, B, C, D) = B.$$

We will show below how to define *ExtendA* in terms of *Circle3*.

Projection. The construction $\mathit{project}(P, L)$ takes a point P and line L and produces a point Q on L such that P lies on the perpendicular to L at Q. The well-known Euclidean construction for the projection applies only if P is known not to be on L. To define *project* using that construction, we would require a test-for-incidence that allows us to test whether point P is on line L or not. But no such test-for-incidence construction is computable over the computable plane, so the Euclidean projection construction does not lead in any obvious way to a definition of *project*. (That does not, however, constitute a proof that *project* is not definable in terms of the elementary constructions, which we give below.)

Projection is absolutely necessary in order to reduce geometry to algebra. We want to pick a line, call it X, and erect a perpendicular Y to X, and project each point P onto its coordinates $x = \mathit{project}(P, X)$ and $y = \mathit{project}(P, Y)$.

Lemma 6.1. *Strong extension and projection are definable from Circle3 (with circles of zero radius allowed). Also one can, using those primitives,* *construct the perpendicular to line L passing through point P, without conditions as to whether P is or is not on L.*

Proof. First we define

$$ExtendA(A, B, C, D) = IntersectLineCircle2(Line\,(A, B), Circle3(B, C, D))$$

Since circles of zero radius are allowed, we do not need to assume $C \neq D$.

To construct the projection of point P on line L, we just need some circle with center P that intersects L in two distinct points Q and R. Then the projection of P on L is the midpoint of segment QR. If L is *Line* (A, B), then a suitable radius would be the sum of the lengths of AB and PA. Thus the circle we need can be constructed as *Circle3* $(P, A, ExtendA(A, B, P, A))$. Note that we need strong extension, not just Euclidean extension, because we cannot rule out $P = A$. That completes the definition of projection.

Now, to construct the perpendicular to L at P, we simply erect the perpendicular to L at the projection of P on L, using the usual Euclidean construction. That completes the proof of the lemma.

The following lemma helps make the case for allowing circles of zero radius and for allowing *Circle3*.

Lemma 6.2. *In R^{2-} (where circles of zero radius do not exist), projection is not definable in terms of the elementary constructions without Circle3.*

Proof. Let α and β be constants, whose values will be two distinct (fixed) points. Let $t(A)$ be the term *project* $(A, Line\,(\alpha, \beta))$. Then $t(A)$ is defined for all values of A. By Theorem 5.1, $t(A)$ is not definable in terms of the elementary constructions without *Circle3*, in the model R^{2-}. That completes the proof.

There are also some "constructions" that go "beyond Euclid." We have already mentioned the test-for-equality function **D**. In addition, there is a notion of *apartness* introduced by Heyting.[9] The apartness axiom says that if $B \neq C$, then for any other point A, either $A \neq B$ or $A \neq C$. An "apartness construction" would be an operation $\#$ such that if $B \neq C$ then $\#(A, B, C)$ is defined[d], and is equal to B or to C, and satisfies $A \neq \#(A, B, C)$. In other words, when B and C are different, A must be different from one of them, and this constructor picks one from which A is different.

[d] Strict constructivists will want to assume B is apart from C in the definition of an apartness construction, but we haven't defined that yet and the discussion here makes sense with classical logic.

If we have an apartness construction, then $\#(\#(A, B, C), B, C)$ is the *other* point of B and C, so it would not be necessary to add a second apartness construction.

The following lemma just summarizes what the recursive model R^{2-} tells us about apartness and test-for-equality.

Lemma 6.3. *(i) No apartness construction is definable in terms of the elementary constructions and Circle3. (ii) Test-for-equality is not definable in terms of the elementary constructions and Circle3, even with the aid of an apartness construction.*

Proof. Ad (i): No apartness construction is definable in terms of the elementary constructions and *Circle3*, since all such terms define continuous functions, and # is not continuous. Ad (ii): We have shown in a previous section that no test-for-equality function exists in the recursive plane. However, the recursive plane does have *Circle3* and an apartness construction. Here is how to compute a point apart from one of two distinct recursive points B and C. Namely, let n be a positive integer such that $1/n$ is less than the length of segment BC, and compute rational approximations a, b, and c to A, B, and C within $1/(4n)$. Then a cannot be within $1/(4n)$ of both b and c, and the answer is C if $|a - b| < 1/(4n)$ and B otherwise. That completes the proof.

We investigated a system in which *Circle3* is included, but circles of zero radius are not allowed, and instead *project* is taken as primitive. This is a workable system, but it is a bit more complex, and there seems to be no intuitive justification for projection in terms of the compass that does not also suffice to justify allowing circles of zero radius. In fact it can be argued that there is no good reason *not* to allow the two points of the compass to coincide. We therefore chose to allow *Circle3*, and to allow circles of zero radius by requiring that *Circle3*(A, B, C) is always defined.

7. Continuous Coordinatization and Arithmetic

Nowadays we usually think of analytic geometry as coordinatizing a plane and translating geometrical relations between points and lines into algebraic equations and inequalities. But the converse is also possible: translating algebra into geometry, and this is important for lower estimates on the power of geometric constructions, for example for showing that the models of the geometry of constructions are planes over Euclidean fields.

In modern books (such as [5]) arithmetic is geometrized as operations on congruence classes of segments. We operate instead on points on some fixed line $X = \mathit{Line}\ (0,1)$, where 0 and 1 are two arbitrarily fixed points. As far as I can tell, past work on coordinatization has always assumed some discontinuous constructions, such as test-for-equality or at least apartness. Since coordinatization itself is patently computable and continuous, it is in some sense "overkill" to appeal to discontinuous and non-computable "constructions" to achieve coordinatization and arithmetization. Although coordinatization is standard, old, and not complicated, we need to check that it can in fact be done from the specified primitives, without using apartness or test-for-equality, by definitions that apply without (for example) case distinctions as to whether numbers being multiplied are equal to 0 or 1 or not. We note that it is crucial that circles of zero radius be allowed (or else we need to take projection as another primitive).

In this section, we give the definitions of constructions that serve to implement coordinatization and the arithmetic operations in a continuous way. We show that, in the models discussed above, these operations are defined and behave as desired. In a later section, we will give an axiomatic theory capable of formalizing these correctness proofs without reference to models.

Recall that the concepts of "right turn" and "left turn" have been introduced and discussed in Section 4. For example, in Fig. 1, AOZ is a "right turn", because the sign of the cross product $OA \times OB$ is positive. Intuitively, traveling from A to O to Z requires one to turn right at O. This definition of "right turn" is adequate for this section, since we are only concerned with models where cross product makes sense. (Later, we will expend considerable effort defining "right" and "left" in an axiomatic context.)

To perform addition geometrically we suppose given a line $L = \mathit{Line}\ (R,S)$ and an "origin", a point O on L with S between R and O. We need to define a construction $\mathit{Add}(A,B)$, which also depends, of course, on S, R, and O, such that $\mathit{Add}(A,B)$ is a point C on L representing the (signed) sum of A and B, with O considered as origin. In particular if A and B are on the same side of O then $\mathit{Add}(A,B) = \mathit{ExtendA}(O,A,O,B)$, but that does not suffice to define Add. It may be instructive to see another failed attempt to define $\mathit{Add}(O,A,B)$:

$$\mathit{IntersectLineCircle2}(\mathit{Line}\ (O,B), \mathit{Circle3}(A,0,B))$$

which is correct when $O \neq B$, but is undefined when $O = B$.

Lemma 7.1. *Given line L = Line(R,S), and a point O on L with S between R and O, we can construct a point $\mathit{Add}(A,B)$ on L representing the* *signed sum of A and B, with O considered as origin, using the elementary constructions and Circle3.*

Proof. We first define an auxiliary construction *rotate*, which requires as inputs three distinct, non-collinear points P, O, and Q (think of angle POQ), as well as a point A on *Line* (O, P). The desired result of *rotate* (P, O, Q, A) is a point Z on *Line* (O, Q) such that $OZ = OA$ and if $A \neq O$ then AOZ is a right turn. (See Fig. 1.) The point is that Z is defined even when $A = O$ (in which case it is just O, of course), and if A moves along *Line* (O, P) through O, then Z moves along *Line* (O, Q), passing through O as A does. To construct Z, we first bisect the angle POQ (by the usual Euclidean construction, which is not problematic since the three points are not collinear). Let the angle bisector be line L. Then let line K be the perpendicular to *Line* (O, P) at A, let B be the intersection point of K and L, and let Z be the projection of B on *Line* (O, Q) (which is defined no matter whether $O = A$ or not).

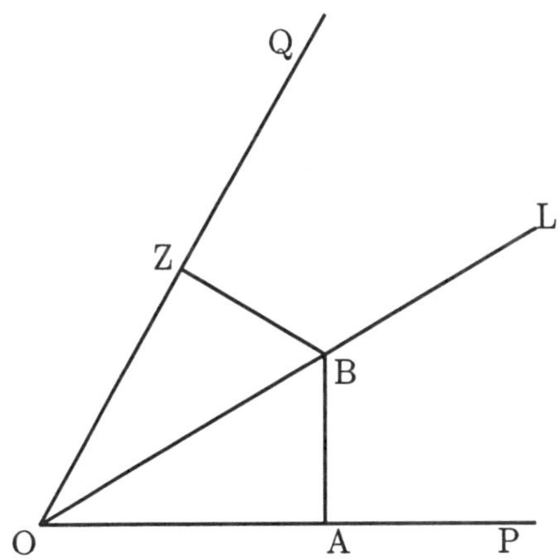

Fig. 1. $Z = rotate(P, O, Q, A)$

Note that there are, if $A \neq O$, two points Z on OQ such that $OZ = OA$. The one constructed by *rotate* (P, O, Q, A) is such that, if POQ is a right turn, then AOZ is a right turn when $A \neq O$, regardless of whether A is between O and P or not. Similarly, if POQ is a left turn, so is AOZ.

With *rotate* in hand, we can give a construction for *Add*(A, B) (depending also on R, S, and O). (The construction is illustrated in Fig. 2.) First, we replace R and S with new points on $L = $ *Line* (R, S), farther away from

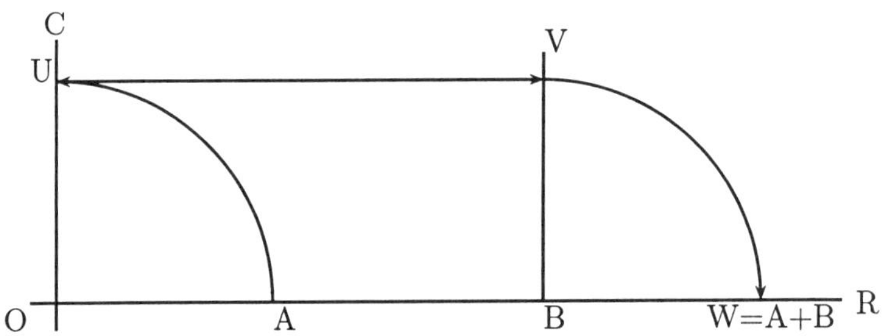

Fig. 2. Signed addition without test-for-equality

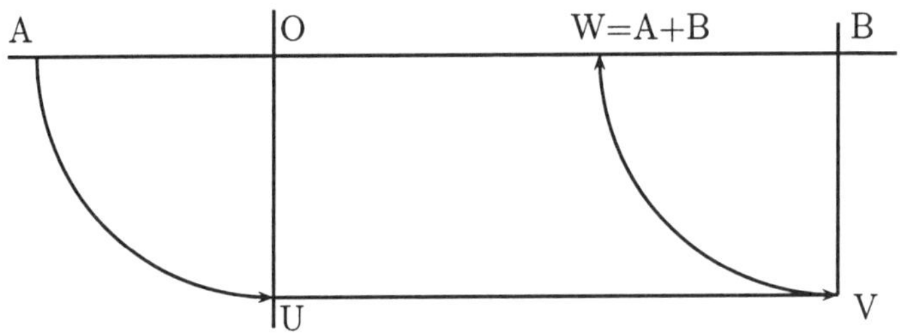

Fig. 3. Signed addition when A is negative

O, so that O, A, and B are all on the same side of R and S, and the new R and S are in the same order on line L as before. (This can be done using *Extend.*) Now erect the perpendicular K to L at O, and the perpendicular H to L at B. In the process of erecting these perpendiculars, we will have constructed points C on K and D on H such that ROC is a right turn. Then let

$$\begin{aligned} U &= \mathit{rotate}(R, O, C, A) \\ V &= \mathit{project}(U, H) \\ W &= \mathit{rotate}(D, B, R, V) \end{aligned}$$

We set $\mathit{Add}(A, B) = W$. Then $\mathit{Add}(A, B)$ is defined for all A, B. Suppose $A \neq O$. Then UV is perpendicular to both K and H. Then U and V are on the same side of L, since if UV meets L at a point X, then XU and XO are both perpendicular to K, which implies $U = O$, which implies $A = O$, contradicting $A \neq O$. It then follows from the property of *rotate* that B and W occur on line L in the same order that O and A occur. Refer to Fig. 3 for an illustration of the case when A is negative. This implies that

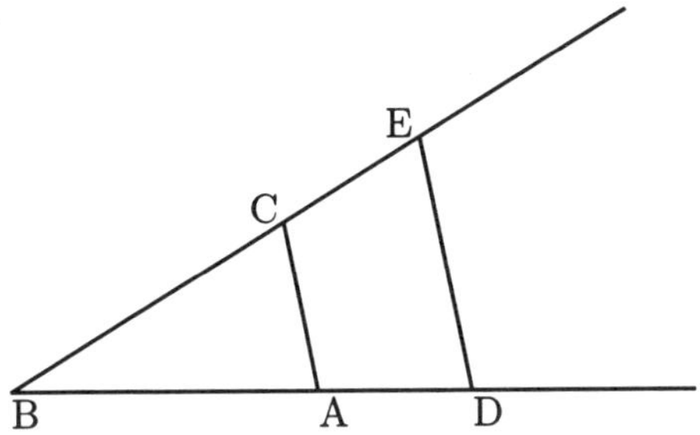

Fig. 4. *La Multiplication* according to Descartes

Add(A, B) represents the algebraic sum of A and B, since in magnitude $BW = OA$.

Having defined addition, we now turn to multiplication, division, and square root. The geometrical definitions of these operations go back to Descartes. On the second page of *La Geometrie*,[6] he gives constructions for multiplication and square roots. We reproduce the drawings found on page 2 of his book [6] in Figures 4 and 5. Here is Descartes' explanation of these figures:

> *(1) Let AB be taken as unity.*
> *(2) Let it be required to multiply BD by BC. I have only to join the points A and C, and draw DE parallel to CA; then BE is the product of BD and BC.*
> *(3) If it be required to divide BE by BD, I join E and D, and draw AC parallel to DE; then BC is the result of the division.*
> *(4) If the square root of GH is desired, I add, along the same straight line, FG equal to unity; then, bisecting FH at K, I describe the circle FIH about K as a center, and draw from G a perpendicular and extend it to I, and GI is the required root.*

From the point of view of constructive geometry, there is a problem with the construction. Namely, Descartes has only told us how to multiply two segments with non-zero lengths, and at least one of whose lengths is not 1 (the length of unity–he needs this when constructing AC parallel to DE), while we want to be able to multiply in general, without a test-for-equality construction. To solve this problem, we recall from Lemma 6.1 that we can define *perp*(P, L), the perpendicular to L passing through P, without regard to whether P is or is not on L. Then we can define a construction *para* such that, for any line L and any point P (which may or may not be on L), $para(P, L)$ passes through P, and if P is not on L then $para(P, L)$ is parallel to L, while if P is on L, then $para(P, L)$ has the same points as L. The definition of *para* is

$$para(P, L) = perp(P, perp(P, L)).$$

In words: First find the perpendicular to L passing through P. Then erect the perpendicular to that line at P.

Using the *para* construction where Descartes calls for "drawing DE parallel to CA", we no longer have a problem multiplying numbers near 1 or 0. We now give a construction for multiplication (which of course could be written as a single, much less readable, term). The construction assumes that 0 and 1 are two distinct points on line X, and D and Q are points on line X to be multiplied, and Y is another line through X, meeting X at 0 and distinct from X. We could, for example, take Y to be the perpendicular to X at 0, although that does not match the illustration from Descartes' book.

```
Multiply(Point D, Point Q)
{ I = IntersectLineCircle1(Y,Circle(0,1))
  C = rotate(1,0,I,Q)    // f is as in the lemma, so OC = OQ
  // now we have to multiply BD by BC  per Descartes
  L = Line(1,C)         // AC in Descartes' diagram.
                        // A there is 1 here; B is 0.
  K = para(1,C,D)       // parallel to AC through D  (or AC itself if D=1)
  E = IntersectLines(K,Y)  // defined because K is not parallel to Y
  N = rotate(I,0,1,E)   // rotate length OE back to line X from line Y
  return N
}
```

Note that since I is defined by *IntersectLineCircle1*, which side of 0 it lies on (on Y) is determined arbitrarily, by the unknown order on Y of the two points *pointOn1*(Y) and *pointOn2*(Y). That side of 0 on Y becomes the "positive" side. If Q is positive (lies on the same side of $Zero$ as One does) then C is positive, i.e. lies on the same side of 0 as I, and vice-versa, if Q is negative (lies on the opposite side of 0 as 1 does), then C lies on the opposite side of $Zero$ as U. This makes multiplication of signed numbers come out correctly without needing a test-for-equality constructor.

It remains, of course, to prove in some geometrical theory that multiplication and addition satisfy the field laws. We do not take that up at this point since we have not yet discussed theories and axioms.

Descartes' division method is handled similarly, using *para* where Descartes constructs a parallel. Turning to Descartes' square root construction, we show that it can be carried out uniformly, without needing to know that the point whose square root is required is different from *Zero*. We carry out Descartes's construction, but then we need to find a point on X whose distance from 0 is IG, without dividing into cases as to whether $G = 0$ or not.

Now we take up the geometrical construction of square roots. Fig. 5 shows Descartes' construction.

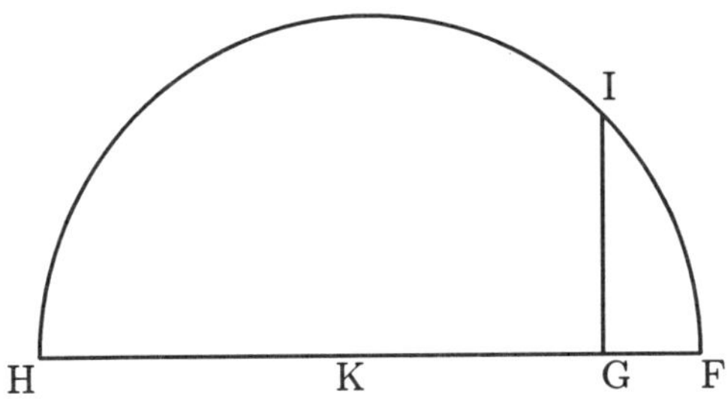

Fig. 5. Square roots according to Descartes

```
SquareRoot(Point G)
{ // H in Descartes' diagram is 0
  F = Add(0,G,0,1)  // so FG has unit length
  K = Midpoint(F,0)
  C = Circle(K,F)
  L = perp(G,Line(0,F))
  I = IntersectLineCircle1(L,C)
  U = IntersectLineCircle1(L,Circle(G,F))  // rotate unit length to line L
  R = rotate(U,G,F,I) // so now RG = IG but R is on X, on the same side of G as F
  // now we need N so that NO = RG
  MinusOne = IntersectLineCircle2(1,0,Circle(0,1))
  N = ExtendA(MinusOne,0,G,R)
  return N
}
```

The point here is that we do not need to assume $G \neq 0$ for this construction to work; we only need that 0 is not between G and 1, i.e. loosely speaking $G \geq 0$. This works because *perp* is total.

8. Euclidean Constructive Geometry ECG

In this section we develop a first order axiomatic theory of geometry as close as possible to Euclid. We call it **ECG**, for "Elementary Constructive

Geometry". Euclid worked with the following data types: *Point*, *Line*, *Circle*, *Segment*, and *Arc*.[e] For foundational purposes, it seems simplest to use only *Point*, *Line*, and *Circle*, and that is what we do in **ECG**. We therefore choose a multi-sorted theory, with "sorts" corresponding to those types. We use the words "sort" and "type" synonymously in this paper. It is, of course, not difficult to add sorts *Ray* , *Arc* and *Segment* , and axioms making the extended theories conservative over **ECG**, but we do not do so here. We take function symbols corresponding to constructors and accessors for those types, described in detail below. The relation symbols we use are standard in axiomatic geometry, **B** for (strict) betweenness and δ for equidistance. We emphasize that **B** is used for strict betweenness; as Hilbert put it, if C is between A and B, then A, B, and C are three distinct points.

We use $on\,(P, L)$ for the incidence of point P on line L, and $On\,(P, C)$ for the incidence of point P on circle C. There is a complete list of the axioms of **ECG** in the Appendix, for reference. In this section, we introduce these axioms one by one, with discussion and explanation.

ECG has five basic function symbols, shown here with arguments:

$$\begin{aligned}&\mathit{IntersectLines}\,(L, K)\\&\mathit{IntersectLineCircle1}\,(L, C)\\&\mathit{IntersectLineCircle2}\,(L, C)\\&\mathit{IntersectCircles1}\,(C, K)\\&\mathit{IntersectCircles2}\,(C, K)\end{aligned}$$

The intuitive meaning of these symbols has been discussed above. **ECG** does not have "overloaded" versions of these functions; in other words, we just write

$$\mathit{IntersectLines}\,(\mathit{Line}\ (A, B), \mathit{Line}\ (P, Q))$$

instead of having an overloaded version of *IntersectLines* that takes four points.

Our underlying logic is intuitionistic. We first give the specifically intuitionistic parts of our theory, which are very few in number. We do not adopt decidable equality, nor even the substitute concept of "apartness"

---

[e]Euclid also worked with triangles, squares, pentagons, hexagons, and "figures". By "figure" he seems to have meant, "closed polygon". One cannot work with arbitrary figures without introducing variables for integers, which in the modern view takes us beyond geometry. We therefore view those theorems of Euclid that mention "figure" as geometrical theorem *schemata*, which result in a theorem about polygons of N sides, for each particular integer N.

introduced by Heyting (and discussed above), primarily because we aim to develop a system in which definable terms (constructions) denote continuous functions, but also because we wish to keep our system closely related to Euclid's geometry, which contains nothing like apartness. Our first four axioms express our intuition that there is nothing asserting existence in the meaning of equality or incidence; hence assertions of equality or incidence can be constructively proved by contradiction.

$$\neg\neg x = y \supset x = y \qquad \text{(Axiom 1)}$$
$$\neg\neg\delta(A, B, C, D) \supset \delta(A, B, C, D) \qquad \text{(Axiom 2)}$$
$$\neg\neg on(P, L) \supset on(P, L) \qquad \text{(Axiom 3)}$$
$$\neg\neg On(P, C) \supset On(P, C) \qquad \text{(Axiom 4)}$$

We will take care to formulate our axioms without quantifiers and without disjunction, which will be key to our applications of proof theory. What we aim to do in this section is to formulate such a theory, which we feel is quite close in spirit to Euclid. In formulating this theory, we made use of the famous axioms of Hilbert,[10] which have been given a more modern and detailed formulation in the textbook of Greenberg.[8] Of course, we do not take the full continuity axioms of Hilbert, but only the line-circle and circle-circle continuity axioms. Where possible, we formulate our axioms as correctness statements about the constructions; in that form they are automatically quantifier-free. Some axioms, which are not about constructions, are inherently quantifier-free. The only question of serious interest is whether disjunction can be completely avoided. It can, as it turns out. The details of this axiomatization may seem somewhat tedious, but the system must of course be specified in complete detail in order to use it in metamathematical proofs. Moreover, some of the details, as far as I can determine, are actually new. In particular, we show how to define the relations "*ABC* is a left turn" and "*ABC* is a right turn" in our theory; the experts we consulted thought this was new. We need this in order to distinguish the two intersection points of two circles.

Since many of our function symbols denote "partial functions", i.e. functions that are not always defined, we will use the "logic of partial terms" **LPT**in our theories. This is a modification of first-order logic, in which the formation rules for formulas are extended by adding the following rule: If t is a term then $t\downarrow$ is a formula. Then in addition the quantifier rules are modified so instead of $\forall x(A(x) \supset A(t))$ we have $\forall x(t\downarrow \wedge A(x) \supset A(t))$, and instead of $A(t) \supset \exists x\, A(x)$ we have $A(t) \wedge t \downarrow \supset \exists x\, A(x)$. Details of LPT can be found in [2], p. 97.

We could try to deal with partial terms, such as $\sqrt{x}$, by simply using an ordinary function symbol for $\surd$, but not saying anything in the axioms about $\surd$ of negative numbers. Thus $\sqrt{-1}$ would some real number, but we would not know or care which one, and we would not be able to prove that its square is -1. This approach rapidly becomes awkward when complicated terms involving square roots of different quantities are used, and you must add extra hypotheses to every theorem asserting that what is under every square root is positive, and we choose to use **LPT** instead.

LPT includes axioms $c\downarrow$ for all constants c of any theory formulated in **LPT**; this is in accordance with the philosophy that terms denote things, and while terms may fail to denote (as in "the King of France"), there is no such thing as a non-existent thing. Thus $1/0$ can be undefined, i.e. fail to denote, but if a constant ∞ is used in **LPT**, it must denote something.

The meaning of $t = s$ is that t and s are both defined and they are equal. We write $t \cong s$ to express that if one of t or s is defined, then so is the other, and they are equal. **LPT** contains the axioms of "strictness", which are as follows (for each function symbol f and relation symbol R in the language):

$$f(t_1, \ldots, t_n) \downarrow \supset t_1 \downarrow \wedge \ldots \wedge t_n \downarrow$$
$$R(t_1, \ldots, t_n) \supset t_1 \downarrow \wedge \ldots \wedge t_n \downarrow$$

The following axioms express the meaning of the five main function symbols of **ECG**. They do not, however, make any assertions of geometrical content, nor do they distinguish one intersection point (of a line and circle, or of two circles) from the other.

$$P = \mathit{IntersectLines}(L, K) \supset on(P, L) \wedge on(P, K) \qquad \text{(Axiom 5)}$$
$$\mathit{IntersectLines}(L, K) \cong \mathit{IntersectLines}(K, L) \qquad \text{(Axiom 6)}$$
$$P = \mathit{IntersectLineCircle1}(L, C) \supset on(P, L) \wedge On(P, C) \qquad \text{(Axiom 7)}$$
$$P = \mathit{IntersectLineCircle2}(L, C) \supset on(P, L) \wedge On(P, C) \qquad \text{(Axiom 8)}$$
$$P = \mathit{IntersectCircles1}(C, K) \supset On(P, C) \wedge On(P, K) \qquad \text{(Axiom 9)}$$
$$P = \mathit{IntersectCircles2}(C, K) \supset On(P, C) \wedge On(P, K) \qquad \text{(Axiom 10)}$$
$$on(P, L) \wedge \neg on(P, K) \supset \mathit{IntersectLines}(L, K) \downarrow \qquad \text{(Axiom 11)}$$
$$\mathit{IntersectLines}(L, K) \downarrow \wedge on(P, L) \wedge on(P, K) \supset P = \mathit{IntersectLines}(L, K) \qquad \text{(Axiom 12)}$$
$$on(P, L) \wedge On(P, C) \supset \mathit{IntersectLineCircle1}(L, C) \downarrow \qquad \text{(Axiom 13)}$$
$$on(P, L) \wedge On(P, C) \supset \mathit{IntersectLineCircle2}(L, C) \downarrow \qquad \text{(Axiom 14)}$$
$$On(P, C) \wedge On(P, K) \supset \mathit{IntersectCircles1}(C, K) \downarrow \qquad \text{(Axiom 15)}$$
$$On(P, C) \wedge On(P, K) \supset \mathit{IntersectCircles2}(C, K) \downarrow \qquad \text{(Axiom 16)}$$

Axioms to distinguish between the two intersection points in each case will be given below, but that must await further developments.

In order to rule out "degenerate" lines, we need an axiom saying that they don't exist; but we *do* allow degenerate circles. The following axioms also provide for lines and circles to exist when they ought to.

$Line\ (A, B) \downarrow \leftrightarrow A \neq B$ (Axiom 17)

$Circle\ (A, B) \downarrow$ (Axiom 18)

There are functions symbols corresponding to the constructor and accessor functions for each of the sorts. The argument and value types of these symbols are obvious, and hence not specified here. Here are the axioms (20 through 27) relating the constructors and accessors.

$Line\ (pointOn1(L), pointOn2(L)) = L$ (Axiom 19)

$A \neq B \wedge pointOn1(Line\ (A, B)) = A$ (Axiom 20)

$A \neq B \wedge pointOn2(Line\ (A, B)) = B$ (Axiom 21)

$pointOn1(L) \neq pointOn2(L)$ (Axiom 22)

$Circle\ (center(C), pointOnCircle\ (C)) = C$ (Axiom 23)

$center(Circle\ (A, B)) = A$ (Axiom 24)

$pointOnCircle\ (Circle\ (A, B)) = B$ (Axiom 25)

$center(C) \neq pointOnCircle\ (C)$ (Axiom 26)

For readers unfamiliar with the logic of partial terms, we point out that Axiom 20 could have been written

$$pointOn1(Line\ (A, B)) \cong A$$

where the relation $t \cong s$ means that if either side is defined, both sides are defined, and they are equal. Also, written the way it is written, with equality instead of $\cong$, Axiom 20 implies that *Line* (A, B) is defined when $A \neq B$, making half of Axiom 17 superfluous.

There are two incidence relations, for $on\,(P, L)$ for points lying on lines, and $On\,(P, C)$ for points lying on and circles. There are three constants α, β, and γ with axioms saying that these three points are non-collinear. Specifically

$\neg on\,(\alpha, Line\ (\beta, \gamma))$ (Axiom 27)

$\neg on\,(\beta, Line\ (\alpha, \gamma))$ (Axiom 28)

$\neg on\,(\gamma, Line\ (\alpha, \beta))$ (Axiom 30)

We do not have to say explicitly that $\alpha \downarrow$, because it is part of the logic of partial terms that every constant is defined–it is nothing special to any particular theory.

The other axioms of incidence are

$on(A, Line\ (A, B))$ (Axiom 30)

$on(B, Line\ (A, B))$ (Axiom 31)

$on(P, L) \wedge on(Q, L) \wedge on(R, Line\ (P, Q)) \supset on(R, L)$ (Axiom 32)

We do not need $On(B, Circle(A, B))$ because that will follow axiom 44 below.

We use the equality symbol between points to mean "identically equal". Between lines, equality means "intensional equality". In the spirit of constructive mathematics, lines "come equipped" with two associated distinct points. Thus, $Line\ (A, B) = Line\ (P, Q)$ if and only if $A = B$ and $P = Q$. This does not need to be assumed, as it follows from the axioms given above for the accessor and constructor functions. It may, however, be confusing to those not accustomed to constructive mathematics. The notion of "extensional equality" refers to the defined relation between two lines, that the same points are on both lines. In practice, to avoid confusion, we rarely if ever mention equality between lines. It should be noted, however, that our theory does depend on this view of lines, since the order of the two intersection points of line $Line\ (A, B)$ with circle C is opposite to the order of the two intersection points of $Line\ (B, A)$ with C. In other words, when considering $IntersectLineCircle1\ (L, C)$, it is essential that L is given by two points.

The basic relations of our theories are equidistance and betweenness, which have been recognized as fundamental at least since Hilbert's famous 1899 book.[10] All the arguments of these two relations have sort *Point*. The (strict) betweenness relation is written $\mathbf{B}(a, b, c)$. We read this "b is between a and c". It implies that the three points are collinear. The first betweenness axiom is

$$\mathbf{B}(a, b, c) \supset \mathbf{B}(c, b, a) \qquad \text{(Axiom 33)}$$

Before giving a constructive version of the remaining betweenness axioms, we discuss a related principle. By "Markov's principle for betweenness" we mean

$$\neg\neg\mathbf{B}(A, B, C) \supset \mathbf{B}(A, B, C) \qquad \text{(Axiom 34)}$$

Markov's principle expresses the idea that by computing two points P and Q etc. to greater and greater accuracy, if they are not identical we will eventually find that out. We want it to be provable in **ECG** for several reasons: it is used in Euclid (e.g. I.6 and I.26, as is discussed below); it is needed for some fundamental theorems (see for example Lemma 8.2); and it makes for a smooth metatheory. While some may consider Markov's principle in number theory to be of questionable constructivity, we consider that geometry without Markov's principle is awkward. As we shall see in other sections, Euclid does use it, and including it does not harm our ability to construct things that are proved to exist. In terms of order, it expresses the principle that $\neg x \leq 0 \supset x > 0$.

Hilbert's second axiom for betweenness is, "given three distinct points, one and only one of the points is between the other two." That formulation is too strong, constructively. (For example its translation into the recursive plane is not provable in **HA** plus Markov's principle.) Instead, we consider the following version, which says that for three distinct points, if two of the alternatives fail then the third must hold, and no two can hold.

$$\begin{aligned}
&a \neq b \wedge a \neq c \wedge b \neq c \supset && \text{(Axiom 35*)}\\
&\quad (\neg \mathbf{B}(a,b,c) \wedge \neg \mathbf{B}(b,c,a) \supset \mathbf{B}(c,a,b)) \wedge \\
&\quad (\neg \mathbf{B}(b,c,a) \wedge \neg \mathbf{B}(c,a,b) \supset \mathbf{B}(a,b,c)) \wedge \\
&\quad (\neg \mathbf{B}(c,a,b) \wedge \neg \mathbf{B}(a,b,c) \supset \mathbf{B}(b,c,a)) \wedge \\
&\quad \neg(\mathbf{B}(a,b,c) \wedge \mathbf{B}(b,c,a)) \wedge \neg(\mathbf{B}(a,b,c) \wedge \mathbf{B}(b,a,c)) \wedge \\
&\quad \neg(\mathbf{B}(b,c,a) \wedge \mathbf{B}(b,a,c))
\end{aligned}$$

In the presence of Markov's principle for betweenness (Axiom 34), we can weaken Axiom 35* by replacing the unnegated betweenness formulas by their double negations. We obtain the following:

$$\begin{aligned}
&a \neq b \wedge a \neq c \wedge b \neq c \supset && \text{(Axiom 35)}\\
&\quad (\neg \mathbf{B}(a,b,c) \wedge \neg \mathbf{B}(b,c,a) \supset \neg\neg \mathbf{B}(c,a,b)) \wedge \\
&\quad (\neg \mathbf{B}(b,c,a) \wedge \neg \mathbf{B}(c,a,b) \supset \neg\neg \mathbf{B}(a,b,c)) \wedge \\
&\quad (\neg \mathbf{B}(c,a,b) \wedge \neg \mathbf{B}(a,b,c) \supset \neg\neg \mathbf{B}(b,c,a)) \wedge \\
&\quad \neg(\mathbf{B}(a,b,c) \wedge \mathbf{B}(b,c,a)) \wedge \neg(\mathbf{B}(a,b,c) \wedge \mathbf{B}(b,a,c)) \wedge \\
&\quad \neg(\mathbf{B}(b,c,a) \wedge \mathbf{B}(b,a,c))
\end{aligned}$$

In fact, Axioms 34 and 35 together are exactly equivalent to Axiom 35*, as the following lemma shows. We separate Axiom 35* into two axioms to facilitate the discussion of different verions of Euclid's parallel postulate, whose relations depend on Markov's principle.

Lemma 8.1. *Markov's principle for betweenness (Axiom 34) is provable from Axiom 35*.*

Proof. Suppose $\neg\neg\mathbf{B}(a,x,b)$. We want to prove $\mathbf{B}(a,x,b)$. From $\neg\neg\mathbf{B}(a,x,b)$ we immediately have $a \neq b$ and $a \neq x$ and $x \neq b$. By Axiom 35* it suffices to prove $\neg\mathbf{B}(x,a,b)$ and $\neg\mathbf{B}(a,b,x)$. To prove $\neg\mathbf{B}(x,a,b)$, suppose $\mathbf{B}(x,a,b)$. Then $\neg\mathbf{B}(a,x,b)$, by Axiom 35*. But that contradicts $\neg\neg\mathbf{B}(a,x,b)$. That proves $\neg\mathbf{B}(x,a,b)$. Similarly we have $\neg\mathbf{B}(a,b,x)$. That completes the proof of the lemma.

There is a betweenness axiom that says, in Hilbert's formulation, that *Line* (P, Q) contains a point between P and Q. We call this the "density" axiom. Since we want a quantifier-free axiomatization, we would like to specify the point asserted to exist. The natural candidate for a point between P and Q is the result of Euclid's segment-bisection construction. We therefore take the following axiom:

$$\mathbf{B}(P, \mathit{IntersectLines}(\mathit{Line}(P,Q), \mathit{Line}(\mathit{IntersectCircles1}(\mathit{Circle}(P,Q), \mathit{Circle}(Q,P)), \mathit{IntersectCircles2}(\mathit{Circle}(P,Q), \mathit{Circle}(Q,P)))), Q) \quad \text{(Axiom 36)}$$

Note that, in view of the strictness axioms of **LPT**, this axiom implies that the circles and intersection points involved are defined. Similarly, there is a betweenness axiom that asserts that *Line* (A, B) contains points outside the segment AB. Again, we want to specify such points so that our axiomatization is quantifier-free:

$$\mathbf{B}(\mathit{IntersectLineCircle1}(\mathit{Line}(P,Q), \mathit{Circle}(P,Q)), P, Q) \quad \text{(Axiom 37)}$$

$$\mathbf{B}(P, Q, \mathit{IntersectLineCircle2}(\mathit{Line}(P,Q), \mathit{Circle}(Q,P))) \quad \text{(Axiom 38)}$$

The remaining betweenness axiom is called the "plane separation axiom". To make its statement more readable, we introduce the usual definitions of two points P and Q being on opposite sides of, or on the same side of, line L:

$$\mathit{OppositeSide}(P, Q, L) := \mathbf{B}(P, Q, \mathit{IntersectLines}(\mathit{Line}\ (P,Q), L) \quad \text{(Definition 39)}$$

$$\mathit{SameSide}(P, Q, L) := \neg\mathbf{B}(P, Q, \mathit{IntersectLines}(\mathit{Line}\ (P,Q), L)) \quad \text{(Definition 40)}$$

When we use the symbol :=, we mean that the symbol on the left is regarded as an abbreviation at the meta-level, rather than a symbol of the formal language. When it is used in subsequent formulas, it stands for the formal equivalent given by the right hand side.

Note that if *Line* (P, Q) does not meet L, then the argument of B is undefined, so by the strictness axioms P and Q are on the same side of L. This formulation, however, does not require us to be able to decide whether L is or is not parallel to *Line* (P, Q). Using these definitions we can give the plane separation axiom(s):

$$\mathit{SameSide}(A, B, L) \wedge \mathit{SameSide}(B, C, L) \supset \mathit{SameSide}(A, C, L) \quad \text{(Axiom 41)}$$

$$\mathit{OppositeSide}(A, B, L) \wedge \mathit{OppositeSide}(B, C, L) \supset \mathit{SameSide}(A, C, L) \quad \text{(Axiom 42)}$$

Rays. Although we have not included rays in **ECG**, we do want to support our claim that a conservative extension including rays can easily be introduced; and also, we sometimes make informal arguments using rays with the implication that they can be formalized in **ECG**. We now show that the use of intuitionistic logic does not cause a problem about incidence on rays or segments. Using betweenness, we can define incidence for rays. However, there is a technicality: the origin O of *Ray* (O, B) is considered to lie on the ray, i.e. "rays are closed", while betweenness means "strictly between." It is thus easier to define the "opposite ray": P is on the opposite ray to *Ray* (O, B) if P is on *Line* (O, B) and O is between P and B. Then Q is on *Ray* (O, B) if it is on *Line* (O, B) but not on the opposite ray:

$$on(Q, Ray\ (O, B)) := on(Q, Line\ (O, B)) \wedge \neg\mathbf{B}(P, O, B) \quad \text{(Definition 43)}$$

This definition can be used to express informal arguments about rays in **ECG** without needing to introduce an explicit sort and axioms for rays. *Segments.* In a similar way we can define incidence for segments, so that "segments are closed".

$$On(P, Segment\ (Q, R)) \leftrightarrow On(P, Line\ (Q, R)) \wedge \neg\mathbf{B}(P, Q, R) \wedge \neg\mathbf{B}(Q, R, P)$$

Recall that *Segment* (R, R) and *Line* (R, R) are both undefined, so the fact that $\neg\mathbf{B}(P, P, P)$ does not make P lie on *Segment* (P, P). We do not number this definition, since it is not used in any further axioms or proofs.

The equidistance relation is written $\delta(A, B, C, D)$. We will often express this using the informal notation $AB = CD$. Axioms for equidistance are sometimes called "congruence axioms" since equidistance can be thought of as congruence of segments. Sometimes, following a tradition that goes back to Euclid, we write $AB = CD$ instead of $\delta(A, B, C, D)$.

Using equidistance, we define incidence for circles:

$$on(P, Circle\ (A, Q)) \leftrightarrow \delta(A, P, A, Q) \wedge A \neq Q \quad \text{(Axiom 44)}$$

The $A \neq Q$ part is needed to avoid conflict with our axiom that *Circle* (A, Q) is undefined if $A = Q$.

Remark. Again, we want to show that, if desired, arcs can be correctly handled in a natural extension of **ECG**:

$$\begin{aligned} &on(P, Arc\ (A, C, Q)) \\ &\quad := On(P, Circle\ (C, A)) \ \wedge\ \delta(A, C, C, Q) \\ &\quad \wedge\ on(IntersectLines(Line\ (A, Q), Line\ (C, P)), Segment\ (A, Q)) \end{aligned}$$

If any two of the three points A, C, and Q are equal, we have undefined terms both on the left and the right. Note that this definition makes arcs "closed", in that A and Q will be on *Arc* (A, C, Q), because segments are closed. No further use of arcs will be made in this paper.

Greenberg's first congruence axiom, paraphrased from [8], is closely related to the uniform version of Euclid's Book I, proposition 2:

$$A \neq B \wedge C \neq D \supset \quad \text{(not an axiom of } \mathbf{ECG}\text{)}$$
$$\exists R(on(R, Ray\ (A, B)) \wedge \delta(A, R, C, D)$$

This axiom permits us to "lay off" segment CD along *Ray* (A, B). Since we are seeking a quantifier-free axiomatization, we want to specify the point R. This we do by taking

$$R(A, B, C, D) = IntersectLineCircle1\,(A, B, Circle3\,(A, C, D)).$$

Our version of the axiom is thus

$$A \neq B \wedge C \neq D \supset (on\,(R(A, B, C, D), Ray\ (A, B)) \wedge \delta(A, R(A, B, C, D), C, D)$$

Note that the strictness axioms then will imply that $R(A, B, C, D)$ is defined when $A \neq B$ and $C \neq D$. The official version, with R replaced by is definition, is Axiom 45:

$$A \neq B \wedge C \neq D \supset$$
$$(on\,(IntersectLineCircle1\,(Line\ (A, B), Circle3\,(A, C, D)), Ray\ (A, B))$$
$$\wedge\ \delta(A, IntersectLineCircle1\,(Line\ (A, B), Circle3\,(A, C, D)), C, D)$$

The second congruence axiom is

$$\delta(A, B, C, D) \wedge \delta(A, B, E, F) \supset \delta(C, D, E, F) \quad \text{(Axiom 46)}$$

The third congruence axiom can be thought of as saying that addition is well-defined on congruence classes of segments:

$$\mathbf{B}(A, B, C) \wedge \mathbf{B}(P, Q, R) \wedge \delta(A, B, P, Q) \wedge \quad \text{(Axiom 47)}$$
$$\delta(B, C, Q, R) \supset \delta(A, C, P, R)$$

With the aid of the congruence axioms considered so far, we can formalize the notion "$AB < CD$". This is defined to mean that the point B' on ray CD such that $CB' = AB$ lies between C and D. More formally, we have to define B' by a term:

$$AB < CD := \quad \text{(Definition 48)}$$
$$\mathbf{B}(C, IntersectLineCircle2\,(Line\ (C, D), Circle3\,(C, A, B)), D)$$

In constructive mathematics, we cannot define $x \leq y$ as $x < y \vee x = y$. Instead, we define $x \leq y$ as $\neg y < x$. Writing out the definition of $y < x$ to make the definition of $x \leq y$ directly, we have

$$CD \leq AB := \quad \text{(Definition 49)}$$
$$\neg\mathbf{B}(C, IntersectLineCircle2\,(Line\ (C, D), Circle3\,(C, A, B)), D)$$

and as remarked earlier we also use the traditional abbreviation

$AB = CD := \delta(A, B, C, D)$ (Definition 50)

We have not included angles as a fundamental data type. Instead, statements about angles can be formalized as statements about two non-collinear rays with the same origin, or about three distinct points. Angles are thus always less than π–there is no such thing as a "straight angle". In this we follow Greenberg.[8] But Greenberg takes congruence of angles as a fundamental notion. Instead, we define it, essentially using the principle SAS to do so. Given three points (thought of as an angle) ABC, and three points PQR, by the first congruence axiom we can find P' on ray QP and R' on ray QR with $QP' = BA$ and $QR' = BC$. Then we define angle ABC to be congruent to angle PQR if and only if $AC = P'Q'$. This is a 6-ary relation between points. Note that it can be expressed in quantifier-free, disjunction-free form, since P' and R' are given by terms.

From the definition of congruence it follows that if A' distinct from B lies on *Ray* (B, A) and C' distinct from B lies on *Ray* (B, C), then angle $A'BC'$ is congruent to angle ABC. This is essentially the reflexivity of congruence viewed as relation between rays; that is the first half of Greenberg's Congruence Axiom 5, which becomes unnecessary. The same observation can be used to prove the symmetry of angle congruence. The second half of Congruence Axiom 5 is the transitivity of congruence of angles, which we take as our fifth congruence axiom, formulated with nine point variables instead of three angle variables. It really comes down to the transitivity of the congruence of triangles. We want to say that if angle ABC is congruent to triangle PQR and triangle PQR is congruent to triangle UVW then triangle ABC is congruent to triangle UVW. The congruence of triangle ABC and triangle PQR is expressed by $AB = PQ \wedge BC = QR \wedge AC = PR$. Hence the transitivity axiom we need is

$$AB = PQ \wedge BC = QR \wedge AC = PR \wedge PQ = UV \quad \text{(Axiom 51)}$$
$$\wedge QR = VW \wedge PR = UV \supset$$
$$AB = UV \wedge BC = VW \wedge AC = UW$$

We want to show that congruence of angles really only depends on the four rays involved, not on the six points. To that end we suppose that A' distinct from B lies on *Ray* (B, A) and C' distinct from B lies on *Ray* (B, C), angle ABC is congruent to angle PQR, and P' distinct from Q lies on *Ray* (Q, P), and R' distinct from R lies on *Ray* (Q, R). We wish to show that angle $A'BC'$ is congruent to angle $P'QR'$. Let P_1 on *Ray* (Q, P) have $BA = QP_1$, and R_1 on *Ray* (Q, R) have $BC = QR_1$. Then angle $A'BC'$ is congruent to angle ABC (as observed above), which is congruent to angle PQR by hypothesis, which is congruent to angle $P'QR'$. Hence by the transitivity of congruence, we are finished.

The sixth congruence axiom is the SAS criterion for triangle congruence. With our definition of congruence for angles, this axiom is provable. In Greenberg's system, it simply serves in place of a definition of angle congruence.

Greenberg's fourth congruence axiom states that for any angle BAC and any ray $A'B'$ there is a unique ray C' on a given side of $A'B'$ such that angle $B'A'C'$ is congruent to angle BAC. Of course rays are not necessary here: this is a statement that for any five points satisfying certain conditions, there exists another point C' satisfying a certain condition. To make precise the part about "on a given side of", we have to mention another point P not on $A'B'$ and demand that PC' should not meet *Line* (A', B'). To express this in a quantifier-free way, we need to construct the point C' in question. This we do as follows:

$C = \mathit{Circle}(A', A, B)$

$B'' = \mathit{IntersectLineCircle1}\,(\mathit{Line}\ (A', B'), C)$ so B'' lies on the ray $A'B'$

$K_1 = \mathit{Circle3}\,(B'', B, C)$

$K_2 = \mathit{Circle3}\,(A'', A, C)$

$C' = \mathit{IntersectCircles1}\,(K_1, K_2)$

Now the desired point C' is one of the intersection points of K_1 and K_2. There seems to be no reason, based on the axioms given so far, why these circles intersect. We take as our fourth congruence axiom, the assertion that both their intersection points are defined. Writing it out formally we have

$$\begin{aligned}
&A \neq B \land A \neq C \land B \neq C \land A' \neq B' \land B'' \qquad \text{(Axiom 52)}\\
&= \mathit{IntersectLineCircle1}\,(\mathit{Line}\ (A', B'), C) \land K_1\\
&= \mathit{Circle3}\,(B'', B, C)) \land K_2 = \mathit{Circle3}\,(A'', A, C)) \supset\\
&\mathit{IntersectCircles1}\,(K_1, K_2) \downarrow \land \mathit{IntersectCircles2}\,(K_1, K_2) \downarrow\\
&\land \mathit{OppositeSide}\,(\mathit{IntersectCircles1}\,(K_1, K_2),\\
&\qquad \mathit{IntersectCircles2}\,(K_1, K_2), \mathit{Line}\ (A', B'))
\end{aligned}$$

Of course one can use fewer variables and more complicated terms to express this axiom, eliminating the variables B'', K_1, and K_2, at the cost of human legibility. Now one of the two intersection points asserted to be defined will be the point needed to verify Greenberg's fourth congruence axiom, since the last line of Axiom 51 says that the two points are on opposite sides of $A'B'$.[f] Hence, by Betweenness Axiom 3, one of them is on the same side as P.

---

[f]We do not know if it is necessary to take this assertion as part of the axiom. Perhaps it can be proved.

Having defined $AB < CD$, we are in a position to formulate the axioms of line-circle continuity. The first two of these just tell us when a line and circles intersect–namely, when there is a point on the line closer (or equally close) to the center than the radius of the circle.

$$AP \leq AB \wedge on(P, L) \supset \qquad \text{(Axiom 53)}$$
$$\mathit{IntersectLineCircle1}\,(L, \mathit{Circle}\ (A, B)) \downarrow$$

$$AP \leq AB \wedge on(P, L) \supset \qquad \text{(Axiom 54)}$$
$$\mathit{IntersectLineCircle2}\,(L, \mathit{Circle}\ (A, B)) \downarrow$$

But the intersection points depend extensionally on the circle:

$$\begin{aligned} &A = center(C) \wedge A = center(K) \wedge \qquad \text{(Axiom 55)} \\ &On(P, C) \wedge On(Q, K) \wedge AP = AQ \supset \\ &\mathit{IntersectLineCircle1}\,(L, C) \cong \mathit{IntersectLineCircle1}\,(K, C) \wedge \\ &\mathit{IntersectLineCircle2}\,(L, C) \cong \mathit{IntersectLineCircle2}\,(K, C) \end{aligned}$$

We next give the basic axioms about intersections of two circles.

$$\begin{aligned} &On(P, C) \wedge AP \leq AB \supset \qquad \text{(Axiom 56)} \\ &\quad \mathit{IntersectCircles1}\,(C, \mathit{Circle}\ (A, B)) \downarrow \\ &\quad \wedge\ \mathit{IntersectCircles2}\,(C, \mathit{Circle}\ (A, B)) \downarrow \end{aligned}$$

Our next axiom specifies that the intersection points of two circles depend extensionally on the circles:

$$\begin{aligned} &A = center(C_1) \wedge A = center(C_1) \wedge \qquad \text{(Axiom 57)} \\ &On(P, C_1) \wedge On(Q, C_1) \wedge AP = AQ \supset \\ &\mathit{IntersectCircles1}\,(C_1, K) \cong \mathit{IntersectCircles1}\,(C_2, K) \wedge \\ &\mathit{IntersectCircles2}\,(C_1, K) \cong \mathit{IntersectCircles2}\,(C_2, K) \wedge \\ &\mathit{IntersectCircles1}\,(K, C_1) \cong \mathit{IntersectCircles1}\,(K, C_2) \wedge \\ &\mathit{IntersectCircles2}\,(K, C_1) \cong \mathit{IntersectCircles2}\,(K, C_2) \wedge \end{aligned}$$

We now come to the expression of Euclid's parallel postulate. We first define $\mathit{Parallel}\,(L, K)$ for lines L and K to mean that the lines do not meet:

$$\mathit{Parallel}\,(L, K) := \neg \mathit{IntersectLines}\,(K, L) \downarrow .$$

Of course, in view of the other axioms for *IntersectLines*, we have

$$\mathit{Parallel}\,(L, K) \leftrightarrow \forall x \neg (on(x, L) \wedge on(x, K)),$$

but the form we took as the definition has the advantage of being quantifier-free. Most modern treatments of geometry formulate the parallel axiom in this way: if two lines K and M are parallel to L through point p, then $K = M$. In symbols:

$$\begin{aligned} &\mathit{Parallel}\,(K, L) \wedge \mathit{Parallel}\,(M, L) \qquad \text{(Playfair's Postulate)} \\ &\quad \wedge\ on(p, K) \wedge on(p, M) \supset K = M \end{aligned}$$

We call this the "Playfair's postulate", or for short just "Playfair", after John Playfair, who published it in 1795, although (according to Greenberg,[8] p. 19) it was referred to by Proclus. Euclid's postulate 5 is

> *If a straight line falling on two straight lines make the interior angles on the same side less than two right angles, the two straight lines, if produced indefinitely, meet on that side on which are the angles less than the two right angles.*

We do not, however, take this form of Euclid's parallel postulate as an axiom. Instead we take the following axiom:

$$\neg \mathit{IntersectLines}(K, L) \downarrow \qquad \text{(Axiom 58, the parallel postulate)}$$
$$\wedge\ on(p, K) \wedge on(p, M) \wedge M \neq K \supset \mathit{IntersectLines}(L, M)$$

In other words, if K is parallel to L through P, then any other line M through P must meet L. This differs from Euclid's version in that we are not required to know in what *direction* M passes through P; but also the conclusion is weaker, in that it does not specify *where* M must meet L. The relationships between these different parallel axioms are discussed in section 13 below and in [3].

To illustrate our reasons for including Markov's principle in **ECG**, we exhibit the following lemma.

Lemma 8.2. *(In **ECG**) Suppose neither point A nor point B lies on line T. (i) If A is not on the same side of T as B, then A is on the opposite side of T from B. (ii) If A is not on the opposite side of line T from B, then A is on the same side of T as B.*

Proof. Ad (i). Suppose A is not on the same side of line T as B. Then segment AB not not meets T. By Markov's principle for line intersections (which is a theorem of **ECG**), line AB meets T in some point q. Since segment AB not not meets T, point q is not not in segment AB. Since A and B do not lie on T, q is not not between A and B. Then, by Markov's principle for betweenness, the intersection point of AB and T lies between A and B. That is, A is on the opposite side of T from B, proving (i).

Ad (ii). Suppose that A is not on the opposite side of line T as B. To show A is on the same side of T as B, we must show that segment AB does not meet T. Suppose that segment AB does meet T. Since A and B do not lie on T, the point of intersection Q is not equal to A or B. Hence, by Markov's principle for betweenness, Q lies between A and B. Hence A and B are on opposite sides of T, contradiction. That completes the proof of the lemma.

We now have given a quantifier-free, disjunction-free axiom system that enables us to verify Greenberg's axioms, after translating angles as triples of points. But this axiom system has a shortcoming: It does not distinguish which of the two intersection points of two circles are defined by *IntersectCircles1* and *IntersectCircles2*, and we have not completely specified which intersection point of a line and circle is which, either, although we have done so when the line is a diameter of the circle, by specifying that *IntersectLineCircle1* $(Line(A, B), Circle(A, B))$ is on the same side of A as B. Hence the axioms given up to now do not suffice to prove the continuity of *IntersectCircles1* and *IntersectCircles2*. To put this matter another way, the axioms given so far have models in which *IntersectLineCircle1* and *IntersectCircles1* are discontinuous; indeed arbitrarily discontinuous. Given one model, we can arbitrary switch some of the values of *IntersectLineCircle1* (P, Q) and *IntersectLineCircle2* (P, Q), and we still have a model.

We wish to add new axioms in such a way that the constructions defined by these terms are continuous. Our first approach to this was based on the observations that explicit moduli of continuity can be defined for the intersection points of circles and lines, and the intersection points of circles and circles. The latter involve $\delta = \sqrt{\epsilon}$, so they are constructible using straightedge and compass. However, the axioms required are neither short nor elegant, and we discovered an axiomatization that, while still not very short, does qualify as "elegant". The axiomatization is based on a concept that is fundamental to computer graphics, although it does not receive much attention in classical geometry. Namely, the concept that an angle PQR is a "left turn" or a "right turn". Analytically, the quantity in question is the sign of the vector cross product of the vectors QP and QR. What we will exhibit is a geometric axiomatization of this concept.

We could introduce a new predicate *Left* (A, B, C), but there is no need to do that, since we can just define, for A, B, and C not collinear,

$$Left\ (A, B, C) := \qquad \text{(Definition 59)}$$
$$C = IntersectCircles1\,(Circle\ (A, C), Circle\ (B, C))$$
$$Right\ (A, B, C) := \qquad \text{(Definition 60)}$$
$$C = IntersectCircles2\,(Circle\ (A, C), Circle\ (B, C))$$

If we did introduce new symbols for *Left* and *Right* then these would be axioms instead of definitions. We turn to the axioms about *Left* and *Right.* Recall that α, β, and γ are (constant symbols for) three arbitrary distinct points.

$$Left\ (\alpha, \beta, \gamma) \qquad \text{(Axiom 61)}$$
$$Right\ (\alpha, \gamma, \beta) \qquad \text{(Axiom 62)}$$

These axioms arbitrarily specify the orientation of the plane. The next axiom says that "handedness" is a property of the rays involved, not just the points:

$$P \neq P' \wedge R \neq R' \wedge on(P', Ray\ (Q, P)) \wedge \\ on\,(R', Ray\ (Q, R)) \wedge Left\ (P, Q, R) \supset Left\ (P', Q', R') \qquad \text{(Axiom 63)}$$

The next axiom says that if PQR is a left turn, and we move P (in any direction) without crossing line QR, it is still a left turn:

$$Left\ (P, Q, R) \wedge \\ \neg\mathbf{B}(P, IntersectLines\,(Line\ (Q, R), Line\ (P, P')), P') \\ \supset Left\ (P', Q, R) \qquad \text{(Axiom 64)}$$

and similarly if we move R without crossing PQ:

$$Left\ (P, Q, R) \wedge \\ \neg\mathbf{B}(R, IntersectLines\,(Line\ (Q, P), Line\ (R, R')), P') \\ \supset Left\ (P, Q, R') \qquad \text{(Axiom 65)}$$

Together these axioms permit us to rotate the sides of a left turn PQR as long as they do not coincide or become opposite, and it remains a left turn.

The next axiom permits us to perform a translation:

$$Left\ (A, B, C) \wedge AB = PQ \wedge BC = QR \wedge \\ AC = PR \wedge AP = BQ \wedge AP = CR \supset Left\ (P, Q, R) \qquad \text{(Axiom 66)}$$

Those are all the axioms for *Left* . Here are similar axioms for *Right* :

$$P \neq P' \wedge R \neq R' \wedge on(P', Ray\ (Q, P)) \wedge \\ on\,(R', Ray\ (Q, R)) \wedge Right\ (P, Q, R) \\ \supset Right\ (P', Q', R') \qquad \text{(Axiom 67)}$$

$$Right\ (P, Q, R) \wedge \\ \neg\mathbf{B}(P, IntersectLines\,(Line\ (Q, R), Line\ (P, P')), P') \\ \supset Right\ (P', Q, R) \qquad \text{(Axiom 68)}$$

$$Right\ (P, Q, R) \wedge \\ \neg\mathbf{B}(R, IntersectLines\,(Line\ (Q, P), Line\ (R, R')), P') \\ \supset Right\ (P, Q, R') \qquad \text{(Axiom 69)}$$

$$Right\ (A, B, C) \wedge AB = PQ \wedge BC = QR \wedge \\ AC = PR \wedge AP = BQ \wedge AP = CR \supset Right\ (P, Q, R) \qquad \text{(Axiom 70)}$$

Repeated applications of Axioms 64 and 65 permit us to perform an arbitrary rotation on a left turn PQR, preserving the fact that it is a left turn. The reader who wishes to understand the motivation for these axioms about *Right* and *Left* should see Lemma 8.3 and its proof, below.

With the aid of *Right* and *Left*, we define the concept "P and Q have the same order on line L as A and B", constructively and without needing case distinctions. Of course, we assume $P \neq Q$ and $A \neq B$. First, we construct point E such that ABE is a left turn. By Definition 59, $E = \mathit{IntersectCircles1}\,(\mathit{Circle}\,(A,B), \mathit{Circle}\,(B,A))$ is such a point. Then P and Q have the same order as A and B if and only if $\mathit{Left}\,(P,Q,E)$. Formally,

$$\begin{aligned}&\mathit{SameOrder}\,(A,B,P,Q) := \qquad\qquad \text{(Definition 71)}\\ &\quad A \neq B \wedge P \neq Q \wedge \mathit{on}(P, \mathit{Line}\,(A,B)) \wedge \mathit{on}(Q, \mathit{Line}\,(A,B)) \wedge \\ &\quad \mathit{Left}\,(P,Q,\mathit{IntersectCircles1}\,(\mathit{Circle}\,(A,B), \mathit{Circle}\,(B,A)))\end{aligned}$$

With *SameOrder* in hand, it is easy to distinguish the two intersection points of a circle. However, we must be careful to allow for the case when the two intersection points coincide.

$$\begin{aligned}&P = \mathit{IntersectLineCircle1}\,(\mathit{Line}\,(A,B),C) \wedge \qquad\qquad \text{(Axiom 72)}\\ &Q = \mathit{IntersectLineCircle2}\,(\mathit{Line}\,(A,B),C) \wedge P \neq Q\\ &\supset \mathit{SameOrder}\,(A,B,P,Q)\end{aligned}$$

Next we give the remaining axioms for *IntersectCircles1* and *IntersectCircles2*. We want to say essentially that if P and Q are the two intersection points of circles C and K with centers A and B respectively, then ABP is a left turn and ABQ is a right turn. But there is also the possibility that the two circles are tangent, and the two intersection points coincide. Then neither ABP nor ABQ is a left or right turn. Therefore, instead of saying ABP is a left turn and ABQ is a right turn, we say that ABP is not a right turn, and ABQ is not a left turn. Here are the axioms in question:

$$\begin{aligned}&R = \mathit{IntersectCircles1}\,(\mathit{Circle}\,(A,P), \mathit{Circle}\,(B,Q)) \supset \qquad \text{(Axiom 73)}\\ &\neg \mathit{Right}\,(A,B,R)\\ &R = \mathit{IntersectCircles2}\,(\mathit{Circle}\,(A,P), \mathit{Circle}\,(B,Q)) \supset \qquad \text{(Axiom 74)}\\ &\neg \mathit{Left}\,(A,B,R)\end{aligned}$$

This completes our list of axioms of **ECG**. Note that these axioms are all quantifier-free and disjunction-free. We will consider one more axiom, which does contain disjunction, as a possible addition to **ECG**, in the next section.

Lemma 8.3. *(In **ECG**) Let ABC be any triangle. Then we can determine the handedness of the turn ABC, in the following sense. Let α, β, and γ be the three fixed non-collinear points mentioned in the axioms of* **ECG**, *so that $\alpha\beta\gamma$ is a left turn by definition. Let $L = \mathit{Line}\,(\alpha,\beta)$. Then we can construct a point R such that ABC is a left turn if and only if R is on the* *same side of L as γ, and a right turn if and only if R is on the opposite side of L from γ.*

Proof. The idea of the proof is this: by a series of "moves" (applications of the axioms for *Left* and *Right* , which correspond to translations, dilations, and rotations), the triangle ABC is "reduced" to either triangle $\alpha\beta\gamma$ or triangle $\alpha\gamma\beta$, preserving the handedness of the turns; but Axioms 61 and 62 directly specify the handedness of the turns $\alpha\beta\gamma$ and $\alpha\gamma\beta$. We now give the details.

By a "move", applied to a triple of non-collinear points PQR, we mean a construction of a new triple UVW such that, according to the axioms for *Left* and *Right*, if PQR is a left turn then so is UVW, and if PQR is a right turn then so is UVW. This latter condition we describe for short by saying that the move "preserves handedness". The axioms describe several types of moves that preserve handedness, specifically, moving P along the ray QP, moving R along the ray QR, rotating PR or QR in such a way that the points P,Q, and R never become collinear, and translating the whole triple. We first show that the number of moves (applications of these axioms) required to perform a given rotation is bounded by a fixed constant. Consider the following procedure: First move P to decrease the angle to less than a right angle. Then we can rotate PQR by any angle up to a right angle, using two moves (one moves PQ and one moves QR). An arbitrary rotation can be performed by performing at most four rotations of less than ninety degrees. Thus in four or fewer rotations (requiring eight or fewer moves) we can bring one of its sides onto the desired ("target") ray. Then we can move P by the same amount as in the first step resulting in the desired rotation. Hence ten or fewer applications of the above axioms suffice to perform any rotation.

Now consider three non-collinear points P, Q, and R. We give a procedure for determining whether PQR is a left turn or a right turn. First translate PQR so that Q coincides with the point β (given by a constant of **ECG**). Then rotate it so that P lies on *Ray* (β, α). Then move P to α. By the axioms above, all these steps preserve the handedness of PQR. Now, if R is on the same side of *Line* (α, β) as γ, then by Axioms 65 and 69, PQR is a left turn, since $\alpha\beta\gamma$ is a left turn by definition. And if R (after the moves described) is on the opposite side of *Line* (α, β), then we claim that PQR is a right turn. To see this, let γ' be a point on the same side of *Line* (α, β) as γ, and on the same side of *Line* (Q, R) as α. (Such a point can be constructed by bisecting the angle formed by *Ray* (β, α) and the opposite ray to *Ray* (Q, R).) Then we can move P to γ' without changing the handedness of PQR, and then we can move R to α without changing the handedness of PQR. But now PQR coincides with $\gamma\beta\alpha$, which by definition is a right turn. That completes the proof of the lemma.

Remark. With classical logic, we could prove that ABC is either a right turn or a left turn. To reach that conclusion, we would need to know that if point R is not on $L = Line\ (\alpha, \beta)$, then either R is on the same side of L as γ or on the opposite side. The constructive status of this statement is discussed below.

Lemma 8.4. *The predicates Right (A, B, C) and Left (A, B, C) are definable in Greenberg's theory G, relative to an arbitrary choice of Left (α, β, γ) and Right (α, γ, β) for some triple of non-collinear points α, β, and γ. This can even be done with intuitionistic logic.*

Proof. It will suffice to define the relation $T(A, B, C, P, Q, R)$ with the meaning "ABC and PQR have the same handedness." First we note that it is possible to define the notion of one triangle being a translation of another; namely, ABC is a translation of PQR if the two triangles are congruent and $AP = BQ = CR$. It is also possible to define the notion of ABC being a rotation of PBQ (when the two angles share vertex B). This requires twenty variables to express, so it is too complex to write down intelligibly, but the definition in question says there exist twenty points representing ten "moves" according to the **ECG** axioms for rotations given above. Then ABC and PQR have the same handedness if there exist P' and Q' such that $P'BR'$ is a translation of PQR and there is a rotation $P''BR''$ of PQR with R'' on *Ray* (B, C) and P'' is on the same side of *Line* (B, C) as A.

Theorem 8.1. ***ECG** with classical logic is equivalent to Greenberg's system G with only line-circle and circle-circle continuity.*

Remark. Greenberg's system is not completely formal. But we understand here a weak second-order theory, with the first-order variables ranging over points, and circles, lines, segments, rays, and angles treated as sets of points.[g] Incidence means membership. There is a weak form of the comprehension axiom, only for formulas with no set quantifiers.

Proof. It suffices to deal with the point, line, and circle fragment of **ECG**, since the whole theory is conservative over this fragment (as follows either

[g]Technically, in Greenberg's book, lines are primitive objects and the incidence relation *on*$(Point, Line)$ is undefined, but rays, segments, and circles are sets of points and the incidence relations are set membership. See p. 144 for his explanation.

by model theory or cut-elimination from the fact that the axioms of both theories can expressed in this fragment). First we show that set variables are irrelevant in Greenberg's theory G. There are no set variables in the axioms, except those ranging over lines, circles, segments, rays, and angles. Hence, in the axioms, those variables can be replaced with variables of the corresponding sorts of **ECG**, and set membership by the appropriate incidence relations. By Gentzen's cut-elimination theorem, if a formula A without set variables has a proof in **ECG**, then there is a cut-free proof of a sequent $\Gamma \Rightarrow A$, where Γ is a conjunction of axioms. This entire proof contains no set variables, by the cut-elimination theorem; hence it is a proof in the language of **ECG**. We showed above that the axioms of G (interpreted in **ECG** in this way) are provable in **ECG**. That is one direction of the proof.

For the other direction, we define a translation A' of **ECG** into G as follows. In A', variables over lines and circles are replaced by set variables, restricted to appropriate predicates defining those concepts. Incidence relation symbols are replaced by 'ϵ'. For each term $t(x_1, \ldots x_n)$ of **ECG**, there is a formula $G_t(y, x_1, \ldots, x_n)$ that expresses $y = t(x_1, \ldots, x_n)$. The definition is given inductively on the complexity of t. The most difficult part of this is to interpret the function symbols for the two intersection points of two circles. However, the difficulty, which is defining the concepts *Left* (A, B, C) and *Right* (A, B, C) required to distinguish the two intersection points of two circles, has been taken care of in Lemma 8.4.

To interpret the function symbols for the intersection points of a line and a circle we need to define the concept "P and Q occur in the same order on *Line* (A, B) as A and B do." That can be done without case distinctions, simply by saying that for some R, RAB is a left turn and RPQ is a left turn. That completes the proof.

9. Euclid's Reasoning

Euclid's proofs have been analyzed in detail by Avigad *et. al.* in [1], and they conclude:

> Euclidean proofs do little more than introduce objects satisfying lists of atomic (or negation atomic) assertions, and then draw further atomic (or negation atomic) conclusions from these, in a simple linear fashion. There are two minor departures from this pattern. Sometimes a Euclidean proof involves a case split; for example, if *ab* and *cd* are unequal segments, then one is longer than the other,

> and one can argue that a desired conclusion follows in either case. The other exception is that Euclid sometimes uses a *reductio*; for example, if the supposition that ab and cd are unequal yields a contradiction then one can conclude that ab and cd are equal.

It is our purpose in this section to argue that Euclid's reasoning can be supported in **ECG**, including the two types of apparently non-constructive reasoning just discussed. The type of *reductio* argument mentioned corresponds to the principle $\neg\neg x = y \supset x = y$, which we have shown follows from the betweenness axioms of **ECG**. The first type (based on the idea that if ab and cd are unequal then one of them is longer) will be studied metamathematically. But first, we give two examples.

A typical example of such an argument in Euclid is Prop. I.6, whose proof begins

> Let ABC be a triangle having the angle ABC equal to the angle ACB. I say that the side AB is also equal to the side AC. For, if AB is unequal to AC, one of them is greater. Let AB be greater, ...

The same proof also uses an argument by contradiction in the form $\neg\neg x = y \supset x = y$. This principle, the "stability of equality", is included in **ECG**, and is universally regarded as constructively acceptable. The conclusion of I.6, however, is negative (has no $\exists$ or $\vee$), so we can simply put double negations in front of every step, and apply the stability of equality once at the end.

Prop. I.26 is another example of the use of the stability of equality: "...DE is not unequal to AB, and is therefore equal to it." While we have proved that classical arguments can be eliminated from proofs of Euclid's theorems, in fact it seems that the only classical arguments that *occur* in Euclid are applications of the principle "if ab and cd are unequal then one of them is longer."

In the examples above, this principle is not really needed to reach Euclid's desired conclusion. Since the conclusion concerns the equality of certain points, we can simply double-negate each step of the argument, and then add one application of the stability of equality at the end. The double negation of "if ab and cd are unequal then one is longer" is provable, since (intuitively speaking) $\neg(p < q)$ is $q \leq p$ and $p \leq q \wedge q \leq p$ implies $p = q$. In fact, this *had* to happen: we prove metatheorems below explaining why the principle in question, and indeed, any uses of classical logic whatsoever, are in principle eliminable from proofs of theorems of the form found in Euclid.

In order to arrive at such metatheorems, we first formulate the principle in question in the language of **ECG**, which does not contain $<$ as a primitive. Our formulation is as follows: If two unequal points B and C are both between A and D, then either B is between A and C or C is between A and B. Formally that is

$$B \neq C \wedge \mathbf{B}(A, B, D) \wedge \mathbf{B}(A, C, D) \qquad \text{(Axiom 75)}$$
$$\supset \mathbf{B}(A, B, C) \vee \mathbf{B}(A, C, B)$$

The point D is a matter of convenience; the axiom is really about A, B, and C and their positions on a ray emanating from A, but **ECG** does not have rays as primitive, so we need point D to express this in **ECG**.

We defined $AB < CD$ in Definition 48. We write $AB > CD$ to mean $CD < AB$.

Lemma 9.1. *(in **ECG**) Axiom 75 implies that if $AB \neq CD$, then either $AB > CD$ or $AB < CD$.*

Proof. Given AB and CD, let Q be the intersection point mentioned in the definition of $AB > CD$. Then according to Axiom 75, if $Q \neq B$ then either Q is between A and B or B is between Q and A. In the first case we have $AB > CD$. In the second case, B is inside the circle of radius CD about A. It follows that D is outside the circle of radius AB and center C. That completes the proof of the lemma.

The theory **ECG** has quantifier-free, disjunction-free axioms. It follows (as we will prove in Theorem 13.3) that no non-trivial disjunction can be proved in **ECG**. That is, if P is negative and **ECG** proves $\forall x\, P(x) \supset A(x) \vee B(x)$, then **ECG** proves $\forall x\, P(x) \supset A(x)$ or **ECG** proves $\forall x\, P(x) \supset B(x)$. Hence, Axiom 75 is not provable in **ECG**.

When we add Axiom 75, we will also introduce a new construction term, which we write $if(AB > CD, P, Q)$. This abbreviates the official version, which is $if(A, B, C, D, P, Q)$. Provided $AB \neq CD$, this term constructs a point which is equal either to P or to Q, depending on whether $AB > CD$ or $AB < CD$. The axiom expressing this is

$$(AB > CD \supset if(AB > CD, P, Q) = P) \qquad \text{(Axiom 76)}$$
$$\wedge\, (AB < CD \supset if(AB > CD, P, Q) = Q)$$

Note that Axiom 76 does not contain disjunction, but that by Axiom 75, we have

$$AB \neq CD \supset if(AB > CD, P, Q) \downarrow .$$

Definition 9.1. The theory **ECGD** is **ECG** plus the new function symbol "*if*", with Axioms 75 and 76.

Remark. The "D" in **ECGD** is for "disjunction".

The following lemmas give two appealing theorems of **ECGD** that cannot be proved in **ECG** (because they are non-trivial disjunctions).

Lemma 9.2. *(in* ***ECGD****) Let P be point not on line L. Then any point Q is either on the same side of L as P, or on the opposite side.*

Proof. Drop a perpendicular K from P to L, meeting L at point R. Projecting Q to point Q' on K. Extend segment PR past R by the amount RQ' twice, arriving at point D on K. Then both R and Q' are between P and D, so by Axiom 75, either R is between P and Q' or Q' is between P and R. In the first case, Q' and P are on the same side of L, and in the second case, they are on opposite sides. But Q and Q' are on the same side of L. Hence Q and P are on the same side, or on opposite sides, of L. That completes the proof of the lemma.

Lemma 9.3. *(in* ***ECGD****) In any triangle ABC, either ABC is a left turn or ABC is a right turn.*

Proof. In Lemma 8.3, we have already shown how to construct a point Q not on line $L = Line\ (\alpha, \beta)$, such that ABC is a left turn or a right turn according as Q lies on the same side of L as γ, or on the opposite side. In **ECGD** by Lemma 9.2, Q must lie on one side or the other of L. Hence ABC is either a left turn or a right turn. That completes the proof.

Terms of **ECGD** that involve the new symbol *if* represent geometrical constructions that can proceed by cases, with comparisons between constructed (unequal) lengths determining the next construction steps. Given Euclid's cavalier approach to case splits, the fact that such constructions are not explicitly mentioned in Euclid does not necessarily mean that they are not required to give a correct and complete version of Euclid. The question thus arises, whether Axiom 75 (or more generally, disjunctive axioms of any kind) are required to formalize Euclid. But because the *theorems* of Euclid do not mention disjunction in any essential way, we can simply take the double-negation interpretation, and eliminate Axioms 75 and 76, as will be shown below. Thus what happened in the example of Proposition I.6 happens necessarily in all examples of similar logical form.

10. Euclid's Parallel Postulate Proved in ECG

Let P be a point not on line L. We consider lines through P that do not meet L (i.e., are parallel to L). Playfair's version of the parallel postulate says that two parallels to L through P are equal. Our Axiom 58 says that if K is a parallel to L through P and M is another line through P, with $M \neq K$, then M meets L. Recall that Euclid's postulate 5 is

> *If a straight line falling on two straight lines make the interior angles on the same side less than two right angles, the two straight lines, if produced indefinitely, meet on that side on which are the angles less than the two right angles.*

We consider the formal expression of Euclid's parallel axiom. Suppose P is a point not on line L, and K is parallel to P through L, M is another line through P, and Q any point on L. Let A be any point on M not equal to P and not on L. Euclid's condition that M "make the interior angles less than two right angles" on the side of PQ where A lies can be conveniently expressed by saying that A is between Q and the intersection of QA with K. Thus a formal version of Euclid's parallel axiom is

$$\begin{array}{l} \mathit{Parallel}(L,K) \wedge \mathit{on}(P,K) \wedge \mathit{on}(P,M) \wedge \mathit{on}(A,M) \wedge \qquad \text{(Euclid's Postulate 5)} \\ \neg\, \mathit{on}(A,L) \wedge A \neq P \wedge \mathbf{B}(Q,A,\mathit{IntersectLines}(\mathit{Line}\ (Q,A),K)) \\ \quad \supset \mathbf{B}(P,A,\mathit{IntersectLines}(L,M)) \end{array}$$

Note that the logical axioms of LPT make it superfluous to state in the conclusion that $\mathit{IntersectLines}(L,M)$ is defined. That follows automatically.

Axiom 58 has a weaker conclusion than Euclid's Postulate 5, because it does not specify on which side of P the intersection point will lie. On the other hand, Axiom 58 also has a weaker hypothesis than Euclid's Postulate 5, so its exact relationship to Postulate 5 is not immediately clear. One direction is settled by the following theorem:

Theorem 10.1. ***ECG*** *proves Euclid's Postulate 5.*

Proof. Suppose Axiom 58, and let L be a line, P a point not on L, K parallel to L through P, M another line through P, Q be a point on L, A be a point on M not on PQ, and B the intersection of QA with K. Suppose that the interior angles made by L, M, and PQ make less than two right angles, which formally means that A is between Q and and B. Then by Axiom 58, since $M \neq K$, M does meet L at some point R. It remains to show that A is between P and R. By Markov's principle (Axiom 34) it suffices to prove that P is not between R and A, and R is not between P and A. Suppose first that P is between R and A. Then R is on the opposite side of line K from A. But R is not not on the same side of line K as Q, since if R is not on the same side of K as Q, then RQ (which is L) would meet K (which it does not). By Lemma 8.2, R and Q are on the same side of line

K. Since R is on the opposite side of K from A, it follows that A and Q are on opposite sides of line K. Hence point B, the intersection of AQ with K, must be between A and C. But that contradicts the fact that A is between Q and B. Hence the assumption that P is between R and A has led to a contradiction. Now suppose instead that R is between P and A. Then A and P are on opposite sides of L.

But B and P are on the same side of line L, since $BP = K$ does not meet L. Hence A and B are on opposite sides of L. But then the intersection point of AB and L, which is Q, lies between A and B, contradicting the fact that A lies between Q and B. Hence the assumption that R is between P and A has also led to a contradiction. Hence, as noted already, by Axiom 34, A is between P and R. That completes the proof of Euclid's Postulate 5 from Axiom 58.

Now we consider the converse problem, of deriving Axiom 58 from Euclid's Postulate 5. The obvious proof attempt works only if we assume Axiom 75.

Lemma 10.1. *Let T be the theory **ECG** without any parallel postulate. Then in T, Axiom 75 ("of two unequal segments one is longer") plus Euclid's Postulate 5 implies Axiom 58.*

Proof. Assume Euclid's postulate and let p be a point not on line L, and K parallel to L through p, and line M another line through p as in Axiom 58. Let q be the point on L at the foot of the perpendicular to L from p. Let A on line M and B on line L be on the same side of pq. Then K makes the interior angles Apq and pqB equal to two right angles. Hence M does not, or it would coincide with K. If angle Apq is less than a right angle, then by Euclid's Postulate 5, M meets L. Similarly, if angle Apq is more than a right angle, we can show by Euclid's Postulate that M meets L (on the other side of q from B). By Axiom 75, one of these alternatives must hold. More formally, let S be the intersection point of *Line* (A, B) with K. Then angle Apq is less than a right angle if A is between S and B, and more than a right angle if S is between A and B. By Axiom 75, one of these alternatives holds. Hence by Euclid's Proposition 5, M meets L. That completes the proof of the lemma.

11. Constructive Geometry and Euclidean Fields

We know (from Theorem 8.1) that **ECG** plus classical logic is not only "reasonable" but equivalent to textbook theories. But when we use only intuitionistic logic, is it still a "reasonable" theory? There are two possible ways to answer that question:

- Can Euclid be formalized in **ECG**?
- Are the models of **ECG** characterizable in some elegant way?

We have argued in a previous section that probably **ECG** suffices to formalize Euclid, and certainly **ECGD** does suffice. In this section we take up the second approach. Classically, the models of G are all planes over a Euclidean field, that is, an ordered field in which every positive element has a square root. Is that same thing true constructively for **ECG**? The main point of this section is to answer that question in the affirmative. The detailed answer is somewhat surprising, though. There turns out to be more than one natural set of axioms for constructive Euclidean fields, and these different versions correspond directly to **ECG**, to **ECGD**, and to **ECG**with our parallel postulate (Axiom 58) replaced by the (apparently) weaker version, Euclid's Postulate 5.

Before turning to the proofs of those correspondences, we must prove some elementary lemmas about intuitionistic geometry. We begin with a two-dimensional version of Markov's principle.

Lemma 11.1. *ECG proves that, if point P does not lie on line L, then some circle with nonzero radius and center P lies entirely on the same side of L as P.*

Proof. Let point P not lie on line L; we will construct a circle with center P lying on the same side of L as P. Let point K on L be the foot of the perpendicular from P to L. Then $K \neq P$. Hence the two circles $C_1 = \mathit{Circle}\ (K, P)$ and $C_2 = \mathit{Circle}\ (P, K)$ that are used to bisect PK have different centers and each contains a point inside the other. Hence the points $X = \mathit{IntersectCircles1}\,(C_1, C_2)$ and $Y = \mathit{IntersectCircles2}\,(C_1, C_2)$ are defined. If $X = Y$ then X is between P and K, contradicting $PX = PK$. Hence the midpoint of PK is given by $M = \mathit{IntersectLines}\,(\mathit{Line}\ (X, Y), \mathit{Line}\ (P, K))$. Hence $\mathit{Circle}\ (P, M)$ lies on the same side of L as P, because K is nearer to P than any other point of L. That completes the proof of the lemma.

Next one may wonder whether the axioms of betweenness of **ECG** are actually sufficient to establish a reasonable theory of order of points on a line. Instead of basing our definition of order on betweenness, we base it instead on *Right* and *Left*. Fix a line L and a point 0 on L. Let K be the perpendicular to L at K and let I be a point on K that is not on L. Define $x < y$ for points x and y on L to mean that xyI is a left turn, and $x < y$

to mean that xyI is a right turn. Then we can establish the fundamental facts about intuitionistic order in **ECG**.

Lemma 11.2. *With notation as above,* ***ECGD*** *proves* $x \neq 0 \supset x > 0 \vee x < 0$.

Proof. Since x, 0, and I are distinct and not collinear, xyI is either a left turn or a right turn, by Lemma 9.3. That completes the proof.

Lemma 11.3. *(in* ***ECG****) With notation as above,* ***ECG*** *proves* $0 \not< x \wedge x \not< 0 \supset x = 0$.

Proof. Suppose $0 \not< x$ and $x \not< 0$. Then $0xI$ is not a left turn and $x0I$ is not a left turn. Suppose $x \neq 0$. Then $x0I$ is a triangle and by Lemma 8.3, not not (one of $x0I$ and $0xI$ is a right turn and the other is a left turn). But neither is a left turn; hence $\neg\neg x = 0$. Since in **ECG** we have the axiom $\neg\neg x = y \supset x = y$, we can conclude $x = 0$. That completes the proof.

But note that **ECG** cannot prove $a < b \supset a < x \vee x < b$, since the decision as to which alternative holds cannot be made continuously in x. This theorem is essentially the axiom of apartness.

For points on L, we say that x is on the same side of y as z if y does not lie on *Segment* (x, z). Note that this is equivalent to saying that x and y are on the same side of the line perpendicular to L at y. Then we have

Lemma 11.4. *(in* ***ECG****) With notation as above, fix a point* $1 > 0$ *on* L*. Then* $x > 0$ *if and only if* x *is on the same side of 0 as 1.*

Proof. Since $0 < 1$, $01I$ is a left turn. Assume x is on the same side of 0 as 1. Then, by the axioms for *Left* , $0xI$ is a left turn. Hence $0 < x$. Conversely, assume $0 < x$. Then $0xI$ is a left turn. We must show x is on the same side of 0 as 1; that is, we must show that 0 does not lie on *Segment* $(x, 1)$. Suppose 0 does lie on *Segment* $(x, 1)$. Since $0xI$ is a left turn, $0 \neq x$, since $00I$ is not a left turn. Since $0 \neq 1$, we have $\mathbf{B}(x, 0, 1)$. But $01I$ is a left turn; hence both $0xI$ and $01I$ are left turns, contradicting Lemma 8.3. That completes the proof.

With the basic properties of order established, we are ready to turn to the characterization of models of **ECG** and **ECGD** as planes over Euclidean fields. We first discuss the axiomatization of Euclidean fields with intuitionistic logic. We use a language with symbols + for addition and $\cdot$ for multiplication, and a unary predicate $P(x)$ for "x is positive". We take the usual axioms for fields, except the axiom for multiplicative inverse, which says that positive elements have multiplicative inverses. If positive elements have inverses, it is an easy exercise to show that negative elements do too. We define a *Euclidean field* to be a commutative ring satisfying the following additional axioms:

$$x \neq 0 \supset \exists y\,(x \cdot y = 1) \qquad \text{EF1}$$

$$P(x) \wedge P(y) \supset P(x+y) \wedge P(x \cdot y) \qquad \text{EF2}$$

$$x + y = 0 \supset \neg(P(x) \wedge P(y)) \qquad \text{EF3}$$

$$x + y = 0 \wedge \neg P(x) \wedge \neg P(y) \supset x = 0 \qquad \text{EF4}$$

$$x + y = 0 \wedge \neg P(y) \supset \exists\, z(z \cdot z = x) \qquad \text{EF5}$$

$$\neg\neg P(x) \supset P(x) \qquad \text{EF6, or Markov's principle}$$

Axiom EF5 says that non-negative elements have square roots. This is a stronger axiom, intuitionistically, than simply specifying that positive elements have square roots. We could conservatively extend field theory by a unary function symbol for negation, $-x$, in which case the last axiom could be more readably written $\neg P(-x) \supset \exists z\,(z^2 = x)$.

As usual, we define $x < y$ to mean $\exists z(P(z) \wedge x + z = y$, or informally, $y - x$ is positive; and $x \leq y$ means $\neg(y < x)$. Then Markov's principle is equivalent to $\neg(x \leq 0) \supset 0 < x$.

We also consider *weakly Euclidean fields*, which instead of EF1, are required only to satisfy the weaker axiom

$$P(x) \supset \exists y\,(x \cdot y = 1) \qquad \text{EF0}$$

EF1 implies EF0, because $P(x) \supset x \neq 0$, but to derive EF1 from EF0, we would have to know that each nonzero element is either positive or negative. In the language without a symbol for additive inverse, this can expressed as

$$x + y = 0 \wedge x \neq 0 \supset P(x) \vee P(y) \qquad \text{EF7}$$

Fields that satisfy EF0 and EF2 through EF7 then automatically satisfy EF1 as well. We call these fields *strongly Euclidean.* To recap: In a Euclidean field, all nonzero elements have multiplicative inverses, while in a weakly Euclidean field, it is only required that elements known to be positive or negative have multiplicative inverses. In a strongly Euclidean field, nonzero elements are either positive or negative, so the distinction doesn't matter.

Remark. Since the double negation of EF7 follows from EF4, every Euclidean field is not not strongly Euclidean; in other words, one cannot give an example of a Euclidean field that is not strongly Euclidean. The standard plane is strongly Euclidean if and only if Markov's principle (in the formulation $\neg\neg x > 0 \supset x > 0$) holds in the reals.

Theorem 11.1. *The models of* **ECG** *are all isomorphic to planes* F^2, *where* F *is a Euclidean field. and the relations and function symbols of* **ECG** *are interpreted as usual in reducing geometry to field theory. The models of* **ECGD** *are all isomorphic to planes* F^2, *where* F *is a strongly Euclidean field. The models of* **ECG** *with the parallel Axiom 58 replaced by Euclid's Postulate 5 are all isomorphic to planes* F^2, *where* F *is a weakly Euclidean field. In fact, each of these geometrical theories interprets, and is interpretable in, the corresponding field theory.*

Remark. Theory A is interpretable in theory B if there is a map from the syntax of A (variables, terms, formulas) to that of B that preserves provability. Thus the theorem implies, for example, that Euclid's Postulate 5 implies Axiom 58 if and only if axiom EF0 implies EF1 with the help of EF2 through EF6.

Proof. We show how to interpret field theory in our geometrical theories. We have three fixed points α, β, and γ, pairwise unequal. Let 0 be another name for α and let 1 be another name for β. Let L be *Line* (α, β), and let K be the perpendicular to L at 0. Let I be a point on K but not on L such that $01I$ is a left turn, and define $P(x)$ to mean that $0xI$ is a left turn. The function symbols $+$ and $\cdot$ of field theory are interpreted by terms *Add* and *Multiply* of **ECG**, defined as shown earlier.

Technically, we ought to exhibit formal proofs in **ECG** of the (interpretations of) the ring axioms, corresponding to the informal proofs in the section on arithmetization, and it may well be possible to exhibit such proofs using a theorem-prover or proof-checker, but here we rely on the reader to be convinced that such proofs exist based on an examination of the proofs and the axioms of **ECG**.

Now we will check axiom EF3. Suppose $x + y = 0$ holds in this model. That means that *Add*$(x, y) = 0$. Suppose also that $\neg P(x)$ and $\neg P(y)$, i.e. neither $0xI$ nor $0yI$ is a left turn. Then x is not between 0 and 1, since $01I$ is a left turn, and if x is between 0 and 1, $0xI$ would be a left turn too. Similarly y is not between 0 and 1. Now assume $x < 0$. then *Add*$(x, y) = 0$ implies that x and y are on opposite sides of 0. Since $x0I$ is a left turn, by Lemma 8.3 $0xI$ is a right turn. Hence $0yI$ is a left turn, so $0 < y$, contradiction. Hence $\neg x < 0$. Since we have proved x is neither negative nor positive, we have $x = 0$ by Lemma 11.3. That completes the verification of axiom EF3.

Now we turn to axiom EF5. Suppose $x + y = 0$. Suppose also $\neg P(y)$; that is, it is not that case that $y \neq 0$ and $0yI$ is a left turn. Verifying this axiom amounts to checking that in **ECG**, Descartes' square root construction can be extended to a function defined for $x \geq 0$, without requiring a case distinction as to whether or not the argument is zero. We have given just such an extension of Descartes' construction in the section on arithmetization, and now it only remains to remark that the argument given there can be carried out in **ECG**.

Next we check axiom EF7 in **ECGD**. Let $x + y = 0$ and $x \neq 0$. Then x and y are on line L, and we can find a point D on the same side of 0 as 1 and a point A on the opposite side of 0 from 1, x between A and D. Then either 0 is between A and x, in which case x is positive (that is $0xI$ is a left turn), or x is between A and 0, in which case y is on the opposite side of 0 from x, i.e. the same side as 1, and hence y is positive.

Next we verify Markov's principle $\neg\neg P(x) \supset P(x)$. Suppose $\neg\neg P(x)$, i.e. $\neg\neg x > 0$. Let $y = \mathit{Add}(x, 1)$, so that $y > x$. Then $x > 0$ is equivalent to $\mathbf{B}(0, x, y)$. Hence we have $\neg\neg B(0, x, y)$. Then by Markov's principle in **ECG**, we have $B(0, x, y)$, i.e. $x > 0$, i.e. $P(x)$, as desired.

We now turn to the verifications of the parallel Axiom 58 and Euclid's Postulate 5. In order to define the reciprocal $1/x$ in geometry, we use Descartes's method. That is, we fix a line L to serve as the x-axis, and a point 0 on L and a point $1 \neq 0$ on L. Then let X be a point on L, and suppose $X \neq 0$. We wish to define a point $1/X$. Erect the perpendicular K to L at X, and find a point on K at distance 1 from X, for example $Q = \mathit{IntersectLineCircle1}\,(K, \mathit{Circle}\ (X, 0, 1))$. Then $Q \neq 0$, since Q is not on L, so we can form $M = \mathit{Line}\ (0, Q)$. Erect the perpendicular H to L at 1. Then K and H are parallel, since they are both perpendicular to L. Line M does not coincide with K, since 0 lies on M but not on K (since $X \neq 0$). Hence M is a line through Q that is not parallel to H. Then, by Axiom 58, M meets H in a point R. The segment $R1$ has the desired length. The desired point $1/X$ on line L is one of the intersection points of *Circle* $(0, 1, R)$ with L. Which one it is depends on the sign of X, which we do not know; but the selection is made automatically by the definition of *IntersectLineCircle2*:

$$1/X = \mathit{IntersectLineCircle2}\,(\mathit{Line}\ (0, X), \mathit{Circle}\ (0, 1, R))$$

since the two intersection points are numbered in the same order as 0 and X occur on L. Hence arithmetic on L satisfies the axioms of a Euclidean field. Similarly, if we only have Euclid's Postulate 5, we can still construct $1/X$ if we know that $X > 0$, as follows. Consider the interior angles made by M and H with the perpendicular dropped from Q to H. They will make less than two right angles if $X > 0$, and more than two right angles if $X < 0$. Hence by Euclid's Postulate 5, if $X > 0$, M meets H as shown in the figure, while if $X < 0$, M meets H at a point south of L. One can then verify that arithmetic on L satisfies the weak Euclidean field axioms. Finally, Axiom 75 erases the distinction in question, and enables us to verify axiom EF7 as well.

Conversely, assume F is a Euclidean field. We will show how to turn F^2 into a model of **ECG** (or, to describe the construction more formally, we will show how to interpret geometry in field theory). As usual in the corresponding classical theories, we take the points to be elements of F^2, and let lines, circles, arcs, and segments have their usual analytic definitions; in particular we define circles so that circles of zero radius are allowed. Hence *Circle3* can be interpreted. Markov's principle in F allows us to verify that axiom of **ECG**. The intersection points of circles and lines, and the intersection points of circles and circles, can be defined by the solution of quadratic equations; and which one is which (i.e. the concept "ABC is a left turn") can be defined as usual in computer graphics, by the cross product, which can be defined in Euclidean field theory (note that division is not required). Then one has to verify that the axioms of **ECG** about handedness are valid. The details follow the sketch given in an earlier section, and are omitted. The verification of the parallel Axiom 58 requires that the reciprocal $1/x$ be defined when $x \neq 0$; the verification of Euclid's Postulate 5 only requires that the reciprocal be defined when $x > 0$. If F is strongly Euclidean (satisfies EF7) then we can also verify Axiom 75, and in that case there is on obvious interpretation of the function symbol *if* of **ECGD** validating Axiom 76. That completes the proof of the theorem.

In [3] it is shown that the different constructive version of Euclidean field theory are not only apparently different, but really are not equivalent with constructive logic; and as a corollary, the different versions of the parallel postulate are also not equivalent.

12. Classical Logic Not Needed for Negative Theorems

Our plan in this section is to investigate the double-negation interpretation for geometric theories. Since the double-negation interpretation applies a double-negation to atomic formulas, we need to have $\neg\neg A \supset A$ for each atomic formula A. We first consider the case when A has the form $t\downarrow$.

The following schema seems initially to have the character of Markov's principle, since $t\downarrow$, in number theory or even in the recursive model $\mathbf{R}^2$, involves an existential quantifier.

$$\neg\neg t\downarrow \supset t\downarrow \qquad \text{for all terms } t$$

As it turns out, however, this schema is provable in **ECG**; however, we require Axiom 58 for the proof, and Euclid's Postulate 5 apparently does not suffice.

Lemma 12.1. *Let t be any term of **ECG**. Then **ECG** proves $\neg\neg t \downarrow \supset t \downarrow$. Moreover, Axiom 34 (Markov's principle for betweenness) is not needed in the proof.*

Proof. We proceed by induction on the complexity of the term t. If t is a variable or constant then $t \downarrow$ is an axiom of **LPT**, so we need consider only compound terms. Consider the induction step when t is a term $f(q, r)$, for terms q and r. By the induction hypothesis, **ECG** proves $\neg\neg q \downarrow \supset q \downarrow$. Argue in **ECG** as follows: Suppose $\neg\neg f(q, r) \downarrow$ We claim $\neg\neg q \downarrow$. For suppose $\neg q \downarrow$. Then by the strictness axioms of **LPT**, we have $\neg f(q, r) \downarrow$, contradicting $\neg\neg f(q, r) \downarrow$. That contradiction completes the proof that $\neg\neg q \downarrow$. Using the proof in **ECG** that $\neg\neg q \downarrow \supset q \downarrow$, we conclude $q \downarrow$. Similarly, we have $r \downarrow$. Hence, to complete the proof, it suffices to prove for each function symbol f of **ECG** that

$$s_1 \downarrow \wedge \ldots \wedge s_n \downarrow \wedge \neg\neg f(s_1, \ldots, s_n) \downarrow \supset f(s_1, \ldots, s_n) \downarrow$$

We will prove this by an exhaustive consideration of each function symbol of **ECG**. First consider the case when f is *IntersectLines*. Suppose that *IntersectLines*(K, L) is not undefined. By Axiom 12, if $K = L$ then *IntersectLines*(K, L) is undefined; hence $K \neq L$. Then by Axiom 58 (the parallel axiom), K meets L. Note that neither Playfair's version of the parallel axiom nor Euclid's would seem to suffice here: Playfair's does not guarantee the existence of the intersection point, and Euclid's guarantees it only when we know on which side of some transversal K makes interior angles less than two right angles.

Consider the case when f is *IntersectLineCircle1* or *IntersectLineCircle2*. Then we have a line L and a circle C that do not fail to meet, and we must show that they do indeed meet. Let P be the center of C, and let K be the line through P perpendicular to L (which we have shown how to construct without knowing whether P is on L or not). Let F be the foot of this perpendicular, i.e. the intersection point of L and K. Then L meets C if and only if PF is less than or equal to the radius r of C. So we have $\neg\neg PF \leq r$. Then (even without Markov's principle) we have $PF \leq r$, so C does meet L as desired.

Next consider the case when f is *IntersectCircles1* or *IntersectCircles2*. Then we have two circles C and K that do not fail to intersect, and we must show that they do intersect. The relevant geometrical fact is that two circles intersect if and only if the distance d between their centers is less than or equal to the sum of their radii $r_1 + r_2$. So if C and K do not fail to intersect implies $\neg\neg d \leq r_1 + r_2$. Even without Markov's principle we then have $d \leq r_1 + r_2$, so C and K do intersect.

Next we consider the constructors. *Circle* (P, Q) is always defined, since we allow zero-radius circles; similarly for *Circle3*(P, Q, R). *Line* (P, Q) is defined if and only if $P \neq Q$. Hence we need $\neg\neg P \neq Q \supset P \neq Q$. This follows from the general intuitionistic logical principal that triple negation is equivalent to single negation. The constructors for segments and arcs can be treated similarly.

Finally we consider the accessors, such as *center*, *pointOn1*, etc. These are all total, so there is nothing to prove. That completes the proof of the lemma.

Let A^- be the Gödel double-negation interpretation of A, obtained by replacing $\exists$ by $\neg\forall\neg$ and $A \vee B$ by $\neg(\neg A \wedge \neg B)$. We do not replace A by $\neg\neg A$ for atomic A since these are equivalent in intuitionistic **ECG**.

Theorem 12.1 (Double negation interpretation). *Suppose* ***ECG*** *with classical logic proves A. Then* ***ECG*** *with intuitionistic logic proves A^-.*

Proof. First we observe that $\neg\neg A$ is equivalent to A for atomic A. This is an axiom for all atomic formulas not of the form $t \downarrow$, and also for those of that form when t has the form *IntersectLines*(u, v). For other formulas of the form $t \downarrow$, we have proved it in Lemma 12.1. Since the axioms of **ECG** are quantifier-free and disjunction-free, it follows that so A^- is equivalent to A for axioms A of the **ECG**. Now the theorem follows as soon as we check the soundness of the double-negation interpretation in a multi-sorted logic with partial terms. But that is straightforward; sorts and partial terms offer no complications over the usual first-order case. That completes the proof.

Corollary 12.1. ***ECG*** *with classical logic is conservative over* ***ECG*** *with intuitionistic logic for negative formulae.*

Proof. For negative A, A^- is identical to A.

A typical theorem of Euclid has the form $H \supset A$, where A will be quantifier-free when formulated in **ECG**, and H is a collection of hypotheses that certain points are distinct, or certain incidence relations hold or do not hold. As Proclus pointed out, sometimes this implies a theorem formulated by cases. For example, Euclid I.2 has one proof if $A = C$ and another proof if $A \neq C$. Using the double-negation interpretation, we find a proof that $A = C \vee A \neq C$ implies the conclusion of Euclid 1.2, but without the law of the excluded middle we cannot conclude the "uniform version" of Euclid I.2.

Recall the example given above of Euclid's Prop. I.6, where Axiom 75 is used, but the same conclusion can be reached without it. Since Axiom 75 is classically tautological, the double-negation interpretation shows that it is *always* eliminable from proofs of negative theorems. But all the theorems in Euclid are either already negative, or assert the existence of some objects that can be constructed using the terms of **ECG**; when formulated more explicitly, they are negative in the sense that they say that the result of a certain construction has certain (quantifier-free) properties.

13. From Proofs to Geometric Algorithms

In this section we take up our plan of doing for **ECG** what cut-elimination and recursive realizability did for intuitionistic arithmetic and analysis, namely, to show that existence proofs lead to programs (or terms) producing the object whose existence is proved. In the case of **ECG** we want to produce geometrical constructions, not just recursive constructions (which could already be produced by known techniques, since **ECG** is interpretable in Heyting's arithmetic of finite types). Terms of **ECG** correspond in a natural way to straightedge and compass constructions.

Theorem 13.1 (Geometric constructions extracted from intuitionistic proofs). *(i) Suppose **ECG** proves $P(x) \supset \exists y\, \phi(x, y)$ where P is negative (does not contain $\exists$ or $\vee$). Then there is a term $t(x)$ of **ECG** (representing a geometric construction) such that $P(x) \supset (\phi(x, t(x))$ is also provable in **ECG**.*

*(ii) Same as (i) but with **ECGD** in place of **ECG**.*

*(iii) Let **ECG + DE** be **ECG**, augmented with a constant **D** and the axiom saying **D** is a test-for-equality function. Then the analogue of (i) holds for **ECG + DE**.*

Proof. We use cut-elimination.[h] Since our axiomatization is quantifier-free, if $\psi \rightarrow \exists y\, \phi$ is provable in constructive **ECG**, then there is a list Γ of

[h] In fact, we use cut-elimination for many-sorted logic with the logic of partial terms. The details of the cut-elimination theorem for such logics have not been published, but they are not significantly different from Gentzen's formulation for first-order logic.

quantifier-free axioms such that $\Gamma, \psi \Rightarrow \exists y\, \phi$ is provable by a cut-free (hence quantifier-free) proof. Since our axiomatization is disjunction-free, by [12] (or rather, by its adaptation to multi-sorted logic with the logic of partial terms) we can permute the inferences so that the existential quantifier is introduced at the last step. Then we obtain the desired proof just by omitting the last step of the proof. That completes the proof of part (i). All the work was in arranging the axiom system to be quantifier-free and disjunction-free. Part (iii) is proved in the same way, noting that the axioms for **D** are also disjunction-free.

Part (ii) requires a bit more work. Since Axiom 75 contains a disjunction, there is an issue about permuting the inferences in a cut-free proof. Suppose that we have a cut-proof of $\Gamma \Rightarrow \psi \rightarrow \exists y\, \phi$, where Γ is a list of axioms (or subformulas of axioms) of **ECGD**. Among Γ there may be occurrences of Axiom 75, which contains a disjunction. We prove by induction on the number of disjunctions in Γ that there exists a term t of **ECGD** and a list Δ of axioms of **ECGD** such that $\Gamma, \Delta \Rightarrow \phi(t)$ is provable. The basis case, when there are no disjunctions, is part (i) of the theorem. Now for the induction step. If the $\exists$ on the right is introduced at a lower level (nearer the end-sequent) than the lowest introduction of disjunction on the left, then we can complete the proof as above, since the line just before the $\exists$ is introduced will contain the desired term, and the $\exists$-introduction can just be postponed until the end. Otherwise there is a part of the proof that looks like this:

$$\frac{AB > AC, \Gamma_1 \Rightarrow \exists y\, \phi \qquad AB < AC, \Gamma_2 \Rightarrow \exists y\, \phi}{AB > AC \vee AB < AC, \Gamma_1, \Gamma_2 \Rightarrow \exists y\, \phi}$$

By induction hypothesis, there are terms t_1 and t_2 such that $AB > AC, \Gamma_1 \Rightarrow \phi(t_1)$ is provable and $AB < AC, \Gamma_2 \Rightarrow \phi(t_2)$ is provable. Let t be the term $\mathit{if}(AB > AC, t_1, t_2)$. Then **ECGD** proves $AB > AC \supset t = t_1$ (by Axiom 76), and **ECGD** proves $AB < AC \supset t = t_2$. Hence, for some list Δ of axioms of **ECGD**, there is a cut-free proof of $AB > AC, \Delta, \Gamma_1 \Rightarrow \phi(t)$, and a cut-free proof of $AB < AC, \Delta, \Gamma_2 \Rightarrow \phi(t)$. These two proofs can then be combined as follows:

$$\frac{AB > AC, \Delta, \Gamma_1 \Rightarrow \phi(t) \qquad AB < AC, \Delta, \Gamma_2 \Rightarrow \phi(t)}{AB > AC \vee AB < AC, \Delta, \Gamma_1, \Gamma_2 \Rightarrow \phi(t)}$$

That completes the induction step, and with it, the proof of the lemma.

Example 1. The "other intersection point". Many Euclidean constructions involve constructing one intersection point P of a line $L = \mathit{Line}\ (A, B)$

and a circle C, and then we say "Let Q be the other intersection point of L and C". Of course we can prove "if P lies on L and C and the two intersection points of L and C are not equal, then there exists an x such that $x \neq P \wedge on(x, L) \wedge On(x, C)$." Then by the theorem, there must be a term $t(A, B, C, P)$ such that, under the stated conditions, $t(A, B, C, P)$ is an intersection point of L and C that is not equal to P. It is not immediately obvious what this term t might be, and it would be interesting to extract it by computer from a proof. But we should be able to see directly how such a term could be constructed.

Here is a sketch of such a construction. First, project the center of C onto line L, obtaining point R on L. Since the two intersection points are distinct, $R \neq P$. Then we ask whether (A, B) is in the same order on L as (R, P) or not. Now Definition 71 shows how to construct a certain point E (given by a term involving A and B) such that (A, B) is in the same order as (R, P) if and only if ERP is a left turn. And the proof of Lemma 8.3 shows how to construct a (complicated) term $\ell(A, B, P, C)$ that will be equal to γ if and only if ERP is a left turn. If ERP is a right turn, then $\ell(A, B, P, C)$ can be arranged to be another specific point α' on the other side of $\alpha\beta$ from γ, the same distance from line $\alpha\beta$ along line $\beta\gamma$ as α, but on the other side. Combining this term ℓ with terms representing an appropriate dilation, translation, and rotations, we can construct a term $f(A, B, P, C)$ such that if (A, B) has the same order as (R, P) then $(f(A, B, P, C) = A$, and if (A, B) has the opposite order as (R, P), it is another point A' with $\mathbf{B}(A, B, A'$. Hence $(f(A, B, P, C), B)$ has the same or opposite order as (A, B), depending on whether (A, B) has the same or opposite order as (R, P). Note that if (A, B) has the same order as (R, P), then $P = \mathit{IntersectLineCircle2}(L, C)$, while if not, $P = \mathit{IntersectLineCircle1}(L, C)$. Hence we can take

$$t(A, B, C, P) = \mathit{IntersectLineCircle1}(\mathit{Line}(f(A, B, P, C), B), C)$$

as the "other intersection point" constructor.

In case one thinks, "this is not what Euclid had in mind!", that is of course true; but Euclid never tried to give uniform constructions of this type. Perhaps the complexity of this construction is one reason why not. This example shows that one could conservatively add a function symbol to **ECG** for "the other intersection point".

Theorem 13.2 (Geometric constructions extracted from classical proofs). *Suppose* ***ECG*** *with classical logic proves $P(x) \supset \exists y\, \phi(x, y)$ where P is quantifier-free and disjunction-free. Then there are terms*

*$t_1(x), \ldots, t_n(x)$ of **ECG** such that $P(x) \supset \phi(x, t_1(x)) \vee \ldots \vee \phi(x, t_n(x))$ is also provable in **ECG** with classical logic.*

Proof. This is a special case of Herbrand's theorem.

Example 2. Euclid's proof of Book I, Proposition 2 provides us with two such constructions, $t_1(A, B, C) = C$ and $t_2(A, B, C)$ the result of Euclid's construction of a point D with $AD = BC$, valid if $A \neq B$. Classically we have $\forall A, B, C\, \exists D(AD = BC)$, but we need two terms t_1 and t_2 to cover all cases.

Example 3. Let P and Q be distinct points and L a given line, and A, B, and C points on L, with A and B on the same side of C. Then there exists a point D which is equal to P if B is between A and C and equal to Q if A is between B and C. The two terms t_1 and t_2 for this example can be taken to be the variables P and Q. One term will not suffice, since D cannot depend continuously on A and B, but all constructed points do depend continuously on their parameters. This classical theorem is therefore not constructively provable.

We mentioned above that **ECG** cannot prove any non-trivial disjunctive theorem. That is a simple consequence of the fact that its axioms contain no disjunction. We now spell this out:

Theorem 13.3 (ECG cannot prove a nontrivial disjunctive theorem). *Suppose **ECG** proves $H(x) \supset P(x) \vee Q(x)$, where H is negative. Then either **ECG** proves $H(x) \supset P(x)$ or **ECG** proves $H(x) \supset Q(x)$. (This result depends only on the lack of disjunction in the axioms of **ECG**.)*

Proof. Consider a cut-free proof of $\Gamma, H(x) \supset P(x) \vee Q(x)$, where Γ is a list of some axioms of **ECG**. Tracing the disjunction upwards in the proof, if we reach a place where the disjunction was introduced on the right before reaching a leaf of the proof tree, then we can erase the other disjunct below that introduction, obtaining a proof of one disjunct as required. If we reach a leaf of the proof tree with $P(x) \vee Q(x)$ still present on the right, then it occurs on the left, where it appears positively. Its descendants will also be positive, so it cannot participate in in application of the rule for proof by cases (which introduces $\vee$ in the left side of a sequent); and it cannot reach left side of the bottom sequent, namely $\Gamma, H(x)$, as these formulas contain no disjunction. But a glance at the rules of cut-free proof, e.g. on p. 442 of [11], will show that these are the only possibilities. That completes the proof.

Theorem 13.4 (Disjunction Properties for ECG and ECGD). *Suppose* ***ECGD*** *proves*

$$H(x) \supset P(x) \vee Q(x)$$

where H *is negative. Then there is a term* $t(x)$ *of* ***ECGD*** *such that* ***ECG*** *proves*

$$H(x) \supset t(x) = \alpha \vee t(x) = \beta$$

and ***ECGD*** *proves*

$$t(x) = \alpha \supset P(x) \wedge t(x) = \beta \supset Q(x).$$

(Here α *and* β *are two constants of* ***ECGD****, with* $\alpha \neq \beta$ *an axiom.)*

Proof. Suppose **ECGD** proves $H(x) \supset P(x) \vee Q(x)$. Then also **ECGD** proves

$$H(x) \supset \exists y\, ((y = \alpha \supset P(x)) \wedge (y = \beta \supset Q(x))).$$

The formula on the right is disjunction-free, so by Theorem 13.1, there is a term t as required.

14. Conclusions

We have given a quantifier-free, disjunction-free axiomatization **ECG** of Euclidean Constructive Geometry, making use of the Logic of Partial Terms (**LPT**). We have verified that this theory is a reasonable version of intuitionistic geometry, by checking that its models are planes over Euclidean fields. Past versions of intuitionistic geometry have included either apartness or decidable equality. Both of these destroy the property of continuity that the terms of **ECG** possess. The terms of this theory correspond in a natural way to Euclidean straightedge-and compass constructions. Making use of more-or-less standard proof-theoretical tools, we have shown that proofs of existential theorems contain Euclidean constructions of the objects proved to exist, and that these constructions can be automatically extracted from such proofs. As a corollary, objects proved to exist in **ECG** depend continuously on parameters.

We set out to pursue the analogy

$$\frac{\text{formal number theory}}{\text{Turing computable functions}} = \frac{\text{intuitionistic geometry}}{\text{geometric constructions}}$$

and we think that we have found the correct theory **ECG** to place on the right side of this equation.

We have shown that Euclid is essentially constructive, in the process exposing these interesting facts:

- in constructive geometry, we need a rigid compass (for the uniform version of Prop. I.2).
- We need Markov's principle to prove the fundamental properties of same-side and opposite-side of a line. We therefore conclude that Markov's principle is fundamental to geometry; theories without it do not correspond to our geometrical intuition.
- Different versions of the parallel postulate correspond to different axiom systems for Euclidean fields: Euclid's postulate 5 amounts to assuming $1/x$ is defined when $x > 0 \vee x < 0$, while Axiom 58 amounts to assuming that $x \neq 0$ implies $1/x$ is defined. It may be simpler to determine the logical relations between these field theories than to work directly in geometrical theories.
- Axiom 58 implies Euclid's Postulate 5, and conversely with the aid of Axiom 75 (corresponding to $x \neq 0 \supset 0 < x \vee x < 0$). That the reverse implications are not constructively valid is shown in [3].
- Since **ECG** and **ECGD** correspond well to Euclid and to constructive field theories, we conclude that apartness is not necessary for constructive geometry.

15. Appendix: List of Axioms of ECG

In this section we list the axioms of **ECG** given above without comment, for reference. The underlying logic is three-sorted intuitionistic logic (sorts for points, lines, and circles) with the logic of partial terms **LPT**,[2] p. 97.

In the following list, the symbol := is used for a definition (macro). The symbols on the left side are to be replaced (with argument substitution) in their subsequent uses by the right hand side of the definition. α, β, and γ are the only constant symbols in **ECG**.

The underlying logic of LPT provides the following:

$$\forall x(t \downarrow \wedge A(x) \supset A(t))$$

$$A(t) \wedge t \downarrow \supset \exists x\, A(x)$$

$$f(t_1, \ldots, t_n) \downarrow \supset t_1 \downarrow \wedge \ldots \wedge t_n \downarrow$$

$$R(t_1, \ldots, t_n) \supset t_1 \downarrow \wedge \ldots \wedge t_n \downarrow$$

$$t \cong s := (t \downarrow \supset s \downarrow \wedge t = s) \wedge (s \downarrow \supset t \downarrow \wedge t = s)$$

$$x \downarrow \quad \text{for every variable } x$$

$$c \downarrow \quad \text{for every constant } c$$

In listing the axioms of **ECG**, it is not necessary to explicitly indicate the types (or sorts) of the variables, as this can be mechanically deduced from the signature of the relation and function symbols. (That is one reason for distinguishing *on* and *On*.) The signatures of the function symbols have also not been explicitly given here, as the names chosen for them convey that information. The following are the axioms of **ECG**, including the definitions used in stating the axioms.

$$\neg\neg x = y \supset x = y \tag{1}$$
$$\neg\neg\delta(A,B,C,D) \supset \delta(A,B,C,D) \tag{2}$$
$$\neg\neg on(P,L) \supset on(P,L) \tag{3}$$
$$\neg\neg On(P,C) \supset On(P,C) \tag{4}$$
$$P = IntersectLines(L,K) \supset on(P,L) \wedge on(P,K) \tag{5}$$
$$IntersectLines(L,K) \cong IntersectLines(K,L) \tag{6}$$
$$P = IntersectLineCircle1(L,C) \supset on(P,L) \wedge On(P,C) \tag{7}$$
$$P = IntersectLineCircle2(L,C) \supset on(P,L) \wedge On(P,C) \tag{8}$$
$$P = IntersectCircles1(C,K) \supset On(P,C) \wedge On(P,K) \tag{9}$$
$$P = IntersectCircles2(C,K) \supset On(P,C) \wedge On(P,K) \tag{10}$$
$$on(P,L) \wedge \neg on(P,K) \supset IntersectLines(L,K)\downarrow \tag{11}$$
$$IntersectLines(L,K)\downarrow \wedge on(P,L) \wedge on(P,K) \supset P = IntersectLines(L,K) \tag{12}$$
$$on(P,L) \wedge On(P,C) \supset IntersectLineCircle1(L,C)\downarrow \tag{13}$$
$$on(P,L) \wedge On(P,C) \supset IntersectLineCircle2(L,C)\downarrow \tag{14}$$
$$On(P,C) \wedge On(P,K) \supset IntersectCircles1(C,K)\downarrow \tag{15}$$
$$On(P,C) \wedge On(P,K) \supset IntersectCircles2(C,K)\downarrow \tag{16}$$
$$Line(A,B)\downarrow \leftrightarrow A \neq B \tag{17}$$
$$Circle(A,B)\downarrow \tag{18}$$
$$Line(pointOn1(L), pointOn2(L)) = L \tag{19}$$
$$pointOn1(Line(A,B)) = A \tag{20}$$
$$pointOn2(Line(A,B)) = B \tag{21}$$
$$pointOn1(L) \neq pointOn2(L) \tag{22}$$
$$Circle(center(C), pointOnCircle(C)) = C \tag{23}$$
$$center(Circle(A,B)) = A \tag{24}$$
$$pointOnCircle(Circle(A,B)) = B \tag{25}$$
$$center(C) \neq pointOnCircle(C) \tag{26}$$
$$\neg on(\alpha, Line(\beta,\gamma) \tag{27}$$
$$\neg on(\beta, Line(\alpha,\gamma) \tag{28}$$
$$\neg on(\gamma, Line(\alpha,\beta) \tag{29}$$
$$on(A, Line(A,B)) \tag{30}$$
$$on(B, Line(A,B)) \tag{31}$$

$$on(P,L) \wedge on(Q,L) \wedge on(R, Line\ (P,Q)) \supset on(R,L) \tag{32}$$

$$\mathbf{B}(a,b,c) \supset \mathbf{B}(c,b,a) \tag{33}$$

$$\neg\neg\mathbf{B}(a,b,c) \supset \mathbf{B}(a,b,c) \tag{34}$$

$$\begin{aligned} & a \neq b \wedge a \neq c \wedge b \neq c \supset (\neg\mathbf{B}(a,b,c) \wedge \neg\mathbf{B}(b,c,a) \supset \neg\neg\mathbf{B}(c,a,b)) \\ & \quad (\neg\mathbf{B}(b,c,a) \wedge \neg\mathbf{B}(c,a,b) \supset \neg\neg\mathbf{B}(a,b,c)) \\ & \quad (\neg\mathbf{B}(c,a,b) \wedge \neg\mathbf{B}(a,b,c) \supset \neg\neg\mathbf{B}(b,c,a)) \\ & \quad \neg\mathbf{B}(a,b,c) \wedge \mathbf{B}(b,c,a)) \wedge \neg(\mathbf{B}(a,b,c) \wedge \mathbf{B}(b,a,c)) \\ & \quad \neg(\mathbf{B}(b,c,a) \wedge \mathbf{B}(b,a,c)) \end{aligned} \tag{35}$$

$$\begin{aligned} & \mathbf{B}(P, IntersectLines\,(Line(P,Q), Line(\\ & \qquad IntersectCircles1\,(Circle(P,Q), Circle(Q,P)), \\ & \qquad IntersectCircles2\,(Circle(P,Q), Circle(Q,P))), Q) \end{aligned} \tag{36}$$

$$\mathbf{B}(IntersectLineCircle1\,(Line(P,Q), Circle(P,Q)), P, Q) \tag{37}$$

$$\mathbf{B}(P,Q, IntersectLineCircle2\,(Line(P,Q), Circle(Q,P))) \tag{38}$$

$$OppositeSide\,(P,Q,L) := \mathbf{B}(P,Q, IntersectLines\,(Line\ (P,Q), L) \tag{39}$$

$$SameSide\,(P,Q,L) := \neg\mathbf{B}(P,Q, IntersectLines\,(Line\ (P,Q), L)) \tag{40}$$

$$SameSide\,(A,B,L) \wedge SameSide\,(B,C,L) \supset SameSide\,(A,C,L) \tag{41}$$

$$OppositeSide\,(A,B,L) \wedge OppositeSide\,(B,C,L) \supset SameSide\,(A,C,L) \tag{42}$$

$$on(P, Ray\ (O,B)) := on(P, Line\ (O,B)) \wedge \neg\mathbf{B}(P,O,B) \tag{43}$$

$$On(P, Circle\ (A,Q)) \leftrightarrow \delta(A,P,A,Q) \wedge A \neq Q \tag{44}$$

$$\begin{aligned} & A \neq B \wedge C \neq D \supset (on(IntersectLineCircle1(\\ & \quad Line\ (A,B), Circle3(A,C,D)), Ray\ (A,B)) \\ & \quad \wedge\, \delta(A, IntersectLineCircle1\,(Line\ (A,B), Circle3(A,C,D)), C, D) \end{aligned} \tag{45}$$

$$\delta(A,B,C,D) \wedge \delta(A,B,E,F) \supset \delta(C,D,E,F) \tag{46}$$

$$\mathbf{B}(A,B,C) \wedge \mathbf{B}(P,Q,R) \wedge \delta(A,B,P,Q) \wedge \delta(B,C,Q,R) \supset \delta(A,C,P,R) \tag{47}$$

$$\begin{aligned} & AB < CD \\ & \quad := \mathbf{B}(C, IntersectLineCircle2\,(Line\ (C,D), Circle3(C,A,B)), D) \end{aligned} \tag{48}$$

$$\begin{aligned} & CD \leq AB \\ & \quad := \neg\mathbf{B}(C, IntersectLineCircle2\,(Line\ (C,D), Circle3(C,A,B)), D) \end{aligned} \tag{49}$$

$$AB = CD := \delta(A,B,C,D) \tag{50}$$

$$\begin{aligned} & AB = PQ \wedge BC = QR \wedge AC = PR \wedge PQ = UV \wedge QR = VW \wedge \\ & \quad PR = UV \supset AB = UV \wedge BC = VW \wedge AC = UW \end{aligned} \tag{51}$$

$$\begin{aligned} & A \neq B \wedge A \neq C \wedge B \neq C \wedge A' \neq B' \wedge \\ & \quad B'' = IntersectLineCircle1\,(Line\ (A',B'), C) \wedge \\ & \quad K_1 = Circle3(B'',B,C) \wedge K_2 = Circle3(A'',A,C) \supset \\ & \quad IntersectCircles1\,(K_1,K_2)\downarrow \wedge IntersectCircles2\,(K_1,K_2)\downarrow \wedge \\ & \quad OppositeSide\,(IntersectCircles1\,(K_1,K_2), \\ & \qquad\qquad IntersectCircles2\,(K_1,K_2), Line\ (A',B')) \end{aligned} \tag{52}$$

$$AP \leq AB \wedge on(P, L) \supset \mathit{IntersectLineCircle1}(L, \mathit{Circle}\ (A, B)) \downarrow \quad (53)$$

$$AP \leq AB \wedge on(P, L) \supset \mathit{IntersectLineCircle2}(L, \mathit{Circle}\ (A, B)) \downarrow \quad (54)$$

$$A = center(C) \wedge A = center(K) \wedge On(P, C) \wedge On(Q, K) \wedge AP = AQ \supset$$
$$\mathit{IntersectLineCircle1}(L, C) \cong \mathit{IntersectLineCircle1}(K, C)$$
$$\wedge\ \mathit{IntersectLineCircle2}(L, C) \cong \mathit{IntersectLineCircle2}(K, C) \quad (55)$$

$$On(P, C) \wedge AP \leq AB \supset \quad (56)$$
$$\mathit{IntersectCircles1}(C, \mathit{Circle}\ (A, B)) \downarrow \wedge\ \mathit{IntersectCircles2}(C, \mathit{Circle}\ (A, B)) \downarrow$$

$$A = center(C_1) \wedge A = center(C_1) \wedge On(P, C_1) \wedge On(Q, C_1) \wedge$$
$$AP = AQ \supset \mathit{IntersectCircles1}(C_1, K) \cong \mathit{IntersectCircles1}(C_2, K) \wedge$$
$$\mathit{IntersectCircles2}(C_1, K) \cong \mathit{IntersectCircles2}(C_2, K) \wedge$$
$$\mathit{IntersectCircles1}(K, C_1) \cong \mathit{IntersectCircles1}(K, C_2) \wedge$$
$$\mathit{IntersectCircles2}(K, C_1) \cong \mathit{IntersectCircles2}(K, C_2) \wedge \quad (57)$$

$$\neg \mathit{IntersectLines}(K, L) \downarrow \wedge on(p, K) \wedge on(p, M) \wedge M \neq K \quad (58)$$
$$\supset \mathit{IntersectLines}(L, M)$$

$$\mathit{Left}\ (A, B, C) := C = \mathit{IntersectCircles1}(\mathit{Circle}\ (A, C), \mathit{Circle}\ (B, C)) \quad (59)$$
$$\wedge\ \neg on(C, \mathit{Line}\ (A, B))$$

$$\mathit{Right}\ (A, B, C) := C = \mathit{IntersectCircles2}(\mathit{Circle}\ (A, C), \mathit{Circle}\ (B, C)) \quad (60)$$
$$\wedge\ \neg on(C, \mathit{Line}\ (A, B))$$

$$\mathit{Left}\ (\alpha, \beta, \gamma) \quad (61)$$

$$\mathit{Right}\ (\alpha, \gamma, \beta) \quad (62)$$

$$\mathit{Left}\ (P, Q, R) \wedge P \neq P' \wedge R \neq R' \wedge on(P', \mathit{Ray}\ (Q, P)) \wedge on(R', \mathit{Ray}\ (Q, R))$$
$$\supset \mathit{Left}\ (P', Q', R') \quad (63)$$

$$\mathit{Left}\ (P, Q, R) \wedge \neg \mathbf{B}(P, \mathit{IntersectLines}(\mathit{Line}\ (Q, R), \mathit{Line}\ (P, P')), P')$$
$$\supset \mathit{Left}\ (P', Q, R) \quad (64)$$

$$\mathit{Left}\ (P, Q, R) \wedge \neg \mathbf{B}(R, \mathit{IntersectLines}(\mathit{Line}\ (Q, P), \mathit{Line}\ (R, R')), P')$$
$$\supset \mathit{Left}\ (P, Q, R') \quad (65)$$

$$\mathit{Left}\ (A, B, C) \wedge AB = PQ \wedge BC = QR \wedge \quad (66)$$
$$AC = PR \wedge AP = BQ \wedge AP = CR \supset \mathit{Left}\ (P, Q, R)$$

$$P \neq P' \wedge R \neq R' \wedge on(P', \mathit{Ray}\ (Q, P)) \wedge on(R', \mathit{Ray}\ (Q, R)) \wedge$$
$$\mathit{Right}\ (P, Q, R) \supset \mathit{Right}\ (P', Q', R') \quad (67)$$

$$\mathit{Right}\ (P, Q, R) \wedge \neg \mathbf{B}(P, \mathit{IntersectLines}(\mathit{Line}\ (Q, R), \mathit{Line}\ (P, P')), P')$$
$$\supset \mathit{Right}\ (P', Q, R) \quad (68)$$

$$\mathit{Right}\ (P, Q, R) \wedge \neg \mathbf{B}(R, \mathit{IntersectLines}(\mathit{Line}\ (Q, P), \mathit{Line}\ (R, R')), P')$$
$$\supset \mathit{Right}\ (P, Q, R') \quad (69)$$

$$\mathit{Right}\ (A, B, C) \wedge AB = PQ \wedge BC = QR \wedge$$
$$AC = PR \wedge AP = BQ \wedge AP = CR \supset \mathit{Right}\ (P, Q, R) \quad (70)$$

$$\textit{SameOrder}\,(A,B,P,Q) := \tag{71}$$
$$A \neq B \wedge P \neq Q \wedge on(P, \textit{Line}\,(A,B)) \wedge on(Q, \textit{Line}\,(A,B)) \wedge$$
$$\textit{Left}\,(P,Q,\textit{IntersectCircles1}\,(\textit{Circle}\,(A,B), \textit{Circle}\,(B,A))$$

$$P = \textit{IntersectLineCircle1}\,(\textit{Line}\,(A,B),C) \wedge \tag{72}$$
$$Q = \textit{IntersectLineCircle2}\,(\textit{Line}\,(A,B),C) \wedge P \neq Q$$
$$\supset \textit{SameOrder}\,(A,B,P,Q)$$

$$R = \textit{IntersectCircles1}\,(\textit{Circle}\,(A,P), \textit{Circle}\,(B,Q)) \supset \neg \textit{Right}\,(A,B,R) \tag{73}$$

$$R = \textit{IntersectCircles2}\,(\textit{Circle}\,(A,P), \textit{Circle}\,(B,Q)) \supset \neg \textit{Left}\,(A,B,R) \tag{74}$$

The following two axioms are not part of **ECG**, but are used in **ECGD**:

$$B \neq C \wedge \mathbf{B}(A,B,D) \wedge \mathbf{B}(A,C,D) \supset \mathbf{B}(A,B,C) \vee \mathbf{B}(A,C,B) \tag{75}$$

$$(AB > CD \supset if(AB > CD, P, Q) = P) \wedge \tag{76}$$
$$(AB < CD \supset if(AB > CD, P, Q) = Q)$$

References

1. Avigad, J., Dean, Edward, and Mumma, John, A formal system for Euclid's *Elements*, *Review of Symbolic Logic* (to appear in 2009).
2. Beeson, M., *Foundations of Constructive Mathematics*, Springer-Verlag, Berlin/ Heidelberg/ New York (1985).
3. Beeson, M., Constructive geometry and the parallel postulate, to appear. Preprint available on the author's website.
4. Bishop, E., *Foundations of Constructive Analysis*, McGraw-Hill, New York (1967).
5. Borsuk, K. and Szmielew, W., *Foundations of Geometry*, Amsterdam, North Holland (1960).
6. Descartes, R., *La Geometrie*. Facsimile with English translation published as *The Geometry of Rene Descartes*, Dover, New York (1954).
7. Euclid, *The Thirteen Books of the Elements*, second edition of the Heath translation, Dover, New York (1956).
8. Greenberg, M. J., *Euclidean and Non-Euclidean Geometries*, fourth edition, Freeman, New York (2008).
9. Heyting, A., *Intuitionism, An Introduction*, North-Holland, Amsterdam (1956).
10. Hilbert, D., *Foundations of Geometry*, translated from the tenth German edition by Leo Unger, Open Court, La Salle, Illinois (1971).
11. Kleene, S. C., *Introduction to Metamathematics*, van Nostrand, Princeton (1952).
12. Kleene, S. C., Permutability of inferences in Gentzen's calculi LK and LJ, in: *Two Papers on the Predicate Calculus*, A.M.S. Memoirs **10** (1952), A. M. S., Providence, R. I.
13. Kijne, D., *Plane Construction Field Theory*, Ph. D. thesis, Rijksuniversiteit te Utrecht (1956).

14. von Plato, Jan, The axioms of constructive geometry, *Annals of Pure and Applied Logic* **76**, pp. 169–200 (1995) .
15. Pambuccian, V., Axiomatizing geometric constructions, *J. Applied Logic* **6**, pp. 24–46, 2008.
16. Pambuccian, V., Axiomatizations of hyperbolic and absolute geometries, in *Non-Euclidean geometries, János Bolyai memorial volume. Papers from the International Conference on Hyperbolic Geometry held in Budapest, July 6–12, 2002*, pp. 119–153. Edited by András Prékopa and Emil Molnár. Mathematics and Its Applications (New York), **581**. Springer, New York (2006).

A separation result for varieties of Brouwer's fan theorem

J. Berger

Mathematisches Institut, Universität München,
Theresienstraße 39, D-80333 München, Germany
E-mail: jberger@mathematik.uni-muenchen.de
www.mathematik.uni-muenchen.de

We recall the axioms FAN, c-FAN, and Π^0_1-FAN, which are versions of Brouwer's fan theorem. It is easy to see that Π^0_1-FAN implies c-FAN and that c-FAN implies FAN. We show that, in general, FAN does not imply c-FAN.

Keywords: Brouwer's fan theorem; constructive mathematics; subsystems of second order arithmetic.

We start with the description of a formal system FS. The language of FS contains variables for natural numbers $n, m \in \mathbb{N}$, for finite sequences of natural numbers $u, v, w \in \{0,1\}^*$ (we write $u = (u(0), \ldots, u(n-1))$ to denote the components), and for infinite binary sequences $\alpha, \beta \in \{0,1\}^{\mathbb{N}}$. Furthermore, we assume the existence of bijections $\theta : \mathbb{N} \times \mathbb{N} \to \mathbb{N}$ and $\eta : \mathbb{N} \to \{0,1\}^*$. With the help of θ, we can define functions from $\mathbb{N} \times \mathbb{N}$ into $\{0,1\}$, whereas η allows us to identify $\mathbb{N}$ with $\{0,1\}^*$. The system is based on intuitionistic predicate logic. Furthermore, it contains the defining axioms of the following functions:

- $u \mapsto |u|$, the length of $u \in \{0,1\}^*$;
- $(u, v) \mapsto u * v$, the concatenation of u and v ($u, v \in \{0,1\}^*$);
- $(u, k) \mapsto u(k)$, the projection onto the k-th component of $u \in \{0,1\}^*$, for $k < |u|$;
- $(\alpha, n) \mapsto \overline{\alpha}n$, the restriction of $\alpha \in \{0,1\}^{\mathbb{N}}$ to the first n components; that means that $\overline{\alpha}n = (\alpha(0), \ldots, \alpha(n-1))$.

Finally, we assume induction for quantifier-free predicates and the following function existence axiom: if $A(n)$ is a quantifier-free formula, then there exists a function $\alpha \in \{0,1\}^{\mathbb{N}}$ such that

$$\forall n\,(\alpha(n) = 0 \leftrightarrow A(n))\,.$$

The following axiom is Brouwer's fan theorem in its most common form.

FAN For every $f : \{0,1\}^* \to \{0,1\}$ we have

$$\forall \alpha \in \{0,1\}^{\mathbb{N}}\,\exists n\,(f(\overline{\alpha}n) = 0) \to$$

$$\exists N\,\forall \alpha \in \{0,1\}^{\mathbb{N}}\,\exists n\,(n \leq N \wedge f(\overline{\alpha}n) = 0)\,.$$

The axiom c-FAN was introduced in the context of uniform continuity.[1,3]

c-FAN For every $f : \{0,1\}^* \to \{0,1\}$ we have

$$\forall \alpha \in \{0,1\}^{\mathbb{N}}\,\exists n\,\forall w \in \{0,1\}^*\,(f(\overline{\alpha}n * w) = 0) \to$$

$$\exists N\,\forall w \in \{0,1\}^*\,(|w| \geq N \to f(w) = 0)\,.$$

As a third variety of the fan theorem we regard the following axiom.[2,4]

Π^0_1-FAN For every $h : \{0,1\}^* \times \mathbb{N} \to \{0,1\}$ we have

$$\forall \alpha \in \{0,1\}^{\mathbb{N}}\,\exists n\,\forall k\,(h(\overline{\alpha}n, k) = 0) \to$$

$$\exists N\,\forall \alpha \in \{0,1\}^{\mathbb{N}}\,\exists n\,(n \leq N \wedge \forall k\,(h(\overline{\alpha}n, k) = 0))\,.$$

The scope of this paper is to show that

$$\text{FS} \nvdash \text{FAN} \to \text{c-FAN}.$$

First, we show that Π^0_1-FAN implies c-FAN and that c-FAN implies FAN.

Lemma 1.

(1) FS $\vdash \Pi^0_1$-FAN $\to$ c-FAN
(2) FS $\vdash$ c-FAN $\to$ FAN

Proof. *1.* Fix $f : \{0,1\}^* \to \{0,1\}$ and assume that

$$\forall \alpha \in \{0,1\}^{\mathbb{N}}\,\exists n\,\forall w \in \{0,1\}^*\,(f(\overline{\alpha}n * w) = 0)\,. \tag{1}$$

Define $h : \{0,1\}^* \times \{0,1\}^* \to \{0,1\}$ by $h(u,w) = f(u*w)$. Then (1) implies that

$$\forall \alpha \in \{0,1\}^{\mathbb{N}}\,\exists n\,\forall w \in \{0,1\}^*\,(h(\overline{\alpha}n, w) = 0)\,.$$

Thus Π^0_1-FAN yields[a]

$$\exists N\,\forall\alpha\in\{0,1\}^{\mathbb{N}}\,\exists n\,(n\le N\wedge\forall w\in\{0,1\}^*\,(h(\overline{\alpha}n,w)=0))\,.$$

By the definition of h, this implies

$$\exists N\,\forall w\in\{0,1\}^*\;(|w|\ge N\to f(w)=0)\,.$$

2. Fix $f:\{0,1\}^*\to\{0,1\}$ and assume that

$$\forall\alpha\in\{0,1\}^{\mathbb{N}}\,\exists n\,(f(\overline{\alpha}n)=0)\,.$$

Define $g:\{0,1\}^*\to\{0,1\}$ by

$$g(w)=0\overset{def}{\Leftrightarrow}\exists n\,(n\le|w|\wedge f(\overline{w}n)=0)\,.$$

Then we have

$$\forall\alpha\in\{0,1\}^{\mathbb{N}}\,\exists n\,\forall w\in\{0,1\}^*\,(g(\overline{\alpha}n*w)=0)\,.$$

Thus c-FAN implies

$$\exists N\,\forall w\in\{0,1\}^*\;(|w|\ge N\to g(w)=0)\,.$$

We obtain

$$\exists N\,\forall\alpha\in\{0,1\}^{\mathbb{N}}\,\exists n\,(n\le N\wedge f(\overline{\alpha}n)=0)\,.$$ □

The following axiom was investigated by Ishihara.[5]

E_0 For every $h:\mathbb{N}\times\mathbb{N}\to\{0,1\}$ there exists an $\alpha\in\{0,1\}^{\mathbb{N}}$ such that

$$\forall n\,(\alpha(n)=0\leftrightarrow\exists k\,(h(n,k)=0))\,. \tag{2}$$

We will also use the weak König lemma.[b]

WKL_0 Fix $g:\{0,1\}^*\to\{0,1\}$ such that

$$\forall n\,\exists w\in\{0,1\}^*\,(|w|=n\wedge g(w)=0) \tag{3}$$

and

$$\forall u,v\in\{0,1\}^*\,(g(u*w)=0\to g(u)=0) \tag{4}$$

hold. Then there exists an $\alpha\in\{0,1\}^{\mathbb{N}}$ such that $\forall n\,(g(\overline{\alpha}n)=0)$.

---

[a] Here we use the fact that we can identify $\mathbb{N}$ with $\{0,1\}^*$.

[b] Note that WKL_0 is the classical contrapositive form of FAN.

The following relationship between E_0 and WKL_0 follows from results of Ishihara.[5] We give a direct proof.

Lemma 2.

$$\mathrm{FS} \vdash \mathrm{E}_0 \to \mathrm{WKL}_0$$

Proof. Assume E_0 and fix $g : \{0,1\}^* \to \{0,1\}$ such that (3) and (4) hold. Define $h : \{0,1\}^* \times \mathbb{N} \to \{0,1\}$ by

$$h(u,n) = 0 \overset{def}{\Leftrightarrow} \forall w\,(|w| = n \to g(u * w) = 1)\,.$$

By E_0, there is a function $f : \{0,1\}^* \to \{0,1\}$ such that

$$\forall u \in \{0,1\}^*\,(f(u) = 1 \leftrightarrow \exists n\,(h(u,n) = 0))\,.$$

For $u, v \in \{0,1\}^*$ define

$$v \prec u \overset{def}{\Leftrightarrow} |v| = |u| \wedge (\exists k < |v|\,(\overline{v}k = \overline{u}k \wedge v(k) < u(k)))\,.$$

Now define another function $g' : \{0,1\}^* \to \{0,1\}$ by

$$g'(u) = 0 \overset{def}{\Leftrightarrow} f(u) = 0 \wedge \forall w\,(w \prec u \to f(w) = 1)\,.$$

Note that we have

$$\forall n\, \exists_1 u \in \{0,1\}^*\,(|u| = n \wedge g'(u) = 0) \tag{5}$$

and

$$\forall u \in \{0,1\}^*\, \forall i \in \{0,1\}\,(g'(u * (i)) = 0 \to g'(u) = 0)\,. \tag{6}$$

Finally, we define $\alpha \in \{0,1\}^{\mathbb{N}}$ by

$$\alpha(n) = 0 \overset{def}{\Leftrightarrow} \exists u \in \{0,1\}^*\,(|u| = n \wedge g'(u * (0)) = 0)\,.$$

We show that

$$\forall n\,(g'(\overline{\alpha}n)) = 0, \tag{7}$$

by induction on n. For the empty sequence, this property follows from (3) and (4). Now assume that we have shown that $g'(\overline{\alpha}n) = 0$. If $\alpha(n) = 0$, we can conclude that there exists $u \in \{0,1\}^*$ with $|u| = n$ and $g'(u * (0)) = 0$. Now (6) implies that $g'(u) = 0$. Hence, by (5) and the induction hypothesis, $u = \overline{\alpha}n$ and therefore

$$g'(\overline{\alpha}(n+1)) = g'(\overline{\alpha}n * \alpha(n)) = g'(u * (0)) = 0.$$

If $\alpha(n) = 1$, we have

$$\forall u\,(|u| = n \to g'(u * (0)) = 1)\,.$$

In view of (5), there is a u with $|u| = n$ and $g'(u * (1)) = 0$. Again, this implies that that $u = \overline{\alpha}n$ and therefore

$$g'(\overline{\alpha}(n+1)) = g'(\overline{\alpha}n * \alpha(n)) = g'(u * (1)) = 0.$$

This completes the proof of (7), which in particular implies that $g(\overline{\alpha}n) = 0$ for every n. □

Finally, we introduce the axiom c-WKL, which is the classical contrapositive form of c-FAN.

c-WKL For every $f : \{0,1\}^* \to \{0,1\}$ we have

$$\forall n \, \exists w \in \{0,1\}^* \, (|w| \geq n \wedge f(w) = 0) \to$$

$$\exists \alpha \in \{0,1\}^{\mathbb{N}} \, \forall n \, \exists w \in \{0,1\}^* \, (f(\overline{\alpha}n * w) = 0) .$$

Proposition 3.

$$\mathrm{FS} \vdash \text{c-WKL} \leftrightarrow^{\cdot} \mathrm{E}_0$$

Proof. "c-WKL $\Rightarrow$ E_0": Assume c-WKL and fix $h : \mathbb{N} \times \mathbb{N} \to \{0,1\}$. Define $f : \{0,1\}^* \to \{0,1\}$ by

$$f(w) = 0 \overset{def}{\Leftrightarrow} \forall n < |w| \, (w(n) = 0 \leftrightarrow \exists k < |w| \, (h(n,k) = 0)) .$$

Since we have

$$\forall n \, \exists w \in \{0,1\}^* \, (|w| = n \wedge f(w) = 0) ,$$

there exists $\alpha \in \{0,1\}^{\mathbb{N}}$ such that

$$\forall n \, \exists w \in \{0,1\}^* \, (f(\overline{\alpha}n * w) = 0) .$$

We show E_0 by showing that, for arbitrary n, we have

$$\alpha(n) = 0 \leftrightarrow \exists k \, (h(n,k) = 0) . \tag{8}$$

Assume that $\alpha(n) = 0$. Fix a $w \in \{0,1\}^*$ such that $f(\overline{\alpha}(n+1) * w) = 0$. Then

$$(\overline{\alpha}(n+1) * w)(n) = \alpha(n) = 0.$$

Therefore, there exists a k such that $h(n,k) = 0$.

Now assume that there exists a k such that $h(n,k) = 0$. Set

$$N = \max(n,k) + 1.$$

Fix a $w \in \{0,1\}^*$ such that $f(\overline{\alpha}N * w) = 0$. Since $n, k < |\overline{\alpha}N * w|$, we obtain

$$\alpha(n) = (\overline{\alpha}N * w)(n) = 0$$

and the proof of (8) is finished.

"$\mathrm{E}_0 \Rightarrow$ c-WKL": Assume E_0 and fix $f : \{0,1\}^* \to \{0,1\}$ such that

$$\forall n\, \exists w \in \{0,1\}^* \, (|w| \geq n \wedge f(w) = 0)\,.$$

Define a function $h : \{0,1\}^* \times \{0,1\}^* \to \{0,1\}$ by

$$h(v,u) = 0 \overset{def}{\Leftrightarrow} f(v * u) = 0.$$

By E_0, there exists a function $g : \{0,1\}^* \to \{0,1\}$ such that

$$\forall v\, (g(v) = 0 \leftrightarrow \exists u\, (h(v,u)) = 0)\,.$$

Note that the prerequisites (3) and (4) are satisfied by g. Since WKL_0 follows from E_0, there is an $\alpha \in \{0,1\}^{\mathbb{N}}$ such that $\forall n\, (g(\overline{\alpha}n) = 0)$. This implies that

$$\forall n\, \exists w \in \{0,1\}^* \, (f(\overline{\alpha}n * w) = 0)\,. \qquad \square$$

In the sequel we use methods of classical reverse mathematics.[6] This discipline aims at classifying theorems in terms of subsystems of second order arithmetic. Important set-existence-axioms are RCA_0 (the Δ^1_0-comprehension scheme) and ACA_0 (the arithmetical comprehension scheme). In the proof of the next corollary we use the fact that FS can be regarded a subsystem of RCA_0 in the following sense:

Remark 4. Everything which can be expressed in FS can also be expressed in RCA_0. Everything which can be proved in FS can also be proved in RCA_0.

Note that classical reverse mathematics is based on classical logic. Nevertheless, we can use negative results therein to obtain negative results for the constructive formal system FS.

Corollary 5.

$$\mathrm{RCA}_0 \vdash \text{c-FAN} \leftrightarrow \mathrm{ACA}_0$$

Proof. In view of Proposition 3 and Remark 4, it is sufficient to show that ACA_0 and E_0 are equivalent, within RCA_0. First, assume ACA_0 and let $h : \mathbb{N} \times \mathbb{N} \to \{0,1\}$ be given. By arithmetical comprehension, we can define

$$X = \{(n,m) \mid (m = 0 \wedge \exists k\, (h(n,k) = 0)) \vee (m = 1 \wedge \forall k\, (h(n,k) = 1))\}\,.$$

This set defines a function $\alpha \in \{0,1\}^{\mathbb{N}}$ which has the property (2); this completes the proof of E_0.

Now assume E_0. In view of [6, Lemma III.1.3], it is sufficient to show Σ^0_1-comprehension. Therefore, let $\psi(n,k)$ be a quantifier-free formula. We have to show the existence of a subset X of $\mathbb{N}$ such that

$$\forall n\,(n \in X \leftrightarrow \exists k\,\psi(n,k))\,. \tag{9}$$

In RCA_0, we can define a function $h : \mathbb{N} \times \mathbb{N} \to \{0,1\}$ such that

$$\forall n,k\,(h(n,k) = 0 \leftrightarrow \psi(n,k))\,.$$

By E_0, there is a function $\alpha \in \{0,1\}^{\mathbb{N}}$ which has the property (2). Then

$$X = \{n \mid \alpha(n) = 0\}$$

fulfills the property (9). □

Corollary 6.

(1) $\mathrm{FS} \nvdash (\mathrm{FAN} \to \mathrm{c\text{-}FAN})$
(2) $\mathrm{FS} \nvdash (\mathrm{FAN} \to \Pi^0_1\text{-}\mathrm{FAN})$

Proof. *1.* Assume that $\mathrm{FS} \vdash \mathrm{FAN} \to \mathrm{c\text{-}FAN}$. By Remark 4, this implies that

$$\mathrm{RCA}_0 \vdash \mathrm{FAN} \to \mathrm{c\text{-}FAN}.$$

Since RCA_0 is based on classical logic, we obtain that

$$\mathrm{RCA}_0 \vdash \mathrm{WKL}_0 \leftrightarrow \mathrm{c\text{-}FAN}.$$

Now Corollary 5 yields that

$$\mathrm{RCA}_0 \vdash \mathrm{WKL}_0 \leftrightarrow \mathrm{ACA}_0.$$

But we know from [6, Remark I.10.2] that in classical reverse mathematics, the axioms WKL_0 is strictly weaker than the axiom ACA_0. So we have obtained a contradiction.

2. This follows from the first part of the Corollary and Lemma 1. □

Acknowledgments

While writing this paper, the author was being supported by a *Feodor Lynen Return Fellowship*, provided by the *Alexander von Humboldt-Foundation.*

References

1. Josef Berger, 'The Logical Strength of the Uniform Continuity Theorem.' A. Beckmann, U. Berger, B. Löwe, and J. V. Tucker, eds., *Logical Approaches to Computational Barriers*, Lecture Notes in Computer Sciences 3988, (2006), 35–39
2. Josef Berger and Douglas Bridges, 'A bizarre property equivalent to the Π_1^0-fan theorem.' Logic Journal of IGPL 14, no. 6 (2006), 867–871
3. Josef Berger and Douglas Bridges. 'A fan-theoretic equivalent of the antithesis of Specker's theorem.' Indag. Mathem., N.S., 18(2) (2007), 195–202
4. Hannes Diener and Iris Loeb, 'Sequences of real functions on [0,1] in constructive reverse mathematics.' *Annals of Pure and Applied Logic* 157, iss. 1, (2009), 50–61
5. Hajime Ishihara, 'Constructive Reverse Mathematics: Compactness Properties.' In: *From Sets and Types to Analysis and Topology: Towards Practicable Foundations for Constructive Mathematics* (L. Crosilla and P. Schuster, eds), Oxford University Press. Oxford Logic Guides 48 (2005), 245–267
6. S. G. Simpson, *Subsystem of Second Order Arithmetic*, Perspectives in Mathematical Logic, Springer-Verlag, Berlin, 1999.

A TERM-REWRITING CHARACTERIZATION OF PSPACE

N. Eguchi

Graduate School of Engineering, Kobe University,
Rokko-dai, Nada-ku, Kobe 657-8501, Japan
E-mail: eguchi@kurt.scitec.kobe-u.ac.jp

Isabel Oitavem has introduced a term rewriting system (TRS) which captures the class **FPS** of polynomial-space computable functions. We propose an alternative TRS for **FPS**. As a consequence, it is obtained that **FPS** is the smallest class containing certain initial functions and closed under specific operations. It turns out that our characterization is relatively simple and suggests an uniform approach to the space-complexity.

Keywords: Term rewriting; Space complexity; PSPACE; Safe nested recursion.

Introduction

Term rewriting is known as an abstract model of computation, since a term is rewritten by successively replacing subterms by equal terms until no further reduction is possible. It is also known that term rewriting forms a Turing complete model of computation. This paper is an application of term rewriting to small complexity classes which only involve feasibly computable functions.

The immediate motivation has come from a work by I. Oitavem. In 1, a term rewriting characterization of the class **FPS** of functions computed in polynomial space was given. This is based on a recursion-theoretic characterization of **FPS** which was obtained by herself. In 2 Oitavem reformulates **FPS** according to a principle which was initiated by S. Bellantoni and S. Cook. The Bellantoni-Cook principle separates variables in every function by semi-colon as follows: $f(\vec{x};\vec{y})$ – variables $\vec{x}$ occurring to the left of the semi-colon are called normal, while variables $\vec{y}$ to the right are called safe. Roughly speaking, this principle allows recursion only for normal positions (safe recursion), while composition only for safe positions (safe composition). In 3 a recursion-theoretic characterization of the polynomial-time computable functions is given with the use of safe recursion on notation.

Oitavem has shown that it is possible to characterize classes of computational complexity involving space constraints by techniques from the field of term rewriting. Oitavem's function class, however, contains a initial function with large growth rate like the product, and two recursion schemes.

In this paper we introduce a new term rewriting system for **FPS**. As a consequence, an alternative characterization of **FPS** is obtained. This system is also based on the Bellantoni-Cook principle. Nevertheless, our class need only elementary initial functions and one recursion scheme which is called safe nested recursion. Hence our characterization is relatively simple. It also turns out that specific safe recursion schemes capture various space-complexity classes in the presence of the same initial functions (see Remark 1.8). In this sense, our formulation is uniform to the space complexity.

One problem is that translating the safe nested recursion scheme into the rewriting rule results in handling terms of exponential size. This is not admissible to capture **FPS**. We solve this problem by employing the *innermost strategy*. Furthermore, for the converse direction, it is not trivial whether the poly-space computations can be simulated in the present formulation. A main task is to define a pairing and unpairing functions.

The scheme of safe nested recursion has been introduced by T. Arai and the author to characterize the exponential-time computable functions. This paper is not fully self-contained, but owes observations on some properties of safe nested recursion to 4.

In Section 1 we define a function class **N** via the operation of safe nested recursion. The main definition is given in Section 2. Based on **N**, we introduce a rewrite system $R_{\mathbf{N}}$ over a class $\mathcal{F}$ of function symbols. It is shown that for any $f \in \mathcal{F}$, the length of every term occurring in a rewriting sequence which starts with $f(\vec{t})$ is bounded by a polynomial in the lengths of input terms $\vec{t}$ if an innermost strategy is used. As a corollary, we have $\mathbf{N} \subseteq \mathbf{FPS}$. In Section 3, conversely, it is shown that every polyspace computation is simulated in **N**.

1. A function class N with safe nested recursion

The first section is devoted to introduce a class **N** via the operation of safe nested recursion on notation (SNRN). The scheme of SNRN is introduced by Arai and the author.[4]

Notation 1.1. Although the functions in **N** are defined over the natural numbers, numbers are denoted as binary strings. For instance, $2x + i$ is denoted by xi for each $i \in \{0, 1\}$. Similarly, xy denotes the concatenation of numbers x and y in the binary representation.

Fix the signature $\Sigma = \{0, 1, \mathrm{Z}\}$ and put $\Sigma^k := \{\sigma_1 \cdots \sigma_k : \sigma_1, \ldots, \sigma_k \in \Sigma\} \setminus \{\mathrm{Z} \cdots \mathrm{Z}\}$. If $w \in \Sigma^k$ is of the form $\sigma_1 \cdots \sigma_k$, then $w(i)$ denotes σ_i for every $i = 1, \ldots, k$. Numerals are represented by the binary successor $C_0, C_1 \in \mathbf{N}$ such that $C_0(x;) = x0$ and $C_1(x;) = x1$. Extending this notation to $\Sigma = \{0, 1, \mathrm{Z}\}$, we mean $C_{\mathrm{Z}}(x;)$ for 0, and $C_w(\vec{y};)$ abbreviates $(C_{w(1)}(y_1;), \ldots, C_{w(k)}(y_k;))$. For $\vec{y} = (y_1, \ldots, y_k)$, $w \in \Sigma^k$ and $i \in \{k + 1, \ldots, k + k\}$, $C_{w(i)}(y_i;)$ denotes y_{i-k} if $w(i - k) \in \{0, 1\}$, otherwise 0.

Let $f(\vec{x}, z)[g(\vec{y})/z]$ denote $f(\vec{x}, g(\vec{y}))$, the result of a substitution. $|x|$ denotes the length of the binary representation of a number x, i.e., $|x| = \lceil \log_2(x + 1) \rceil$, which is called the binary length of x. Moreover, for $\vec{x} = (x_1, \ldots, x_k)$, let $|\vec{x}| := (|x_1|, \ldots, |x_k|)$ and $\max \vec{x} := \max\{x_i : i = 1, \ldots, k\}$.

The computation for every function defined by nested recursion runs along the lexicographic ordering. The scheme of SNRN is defined via a lexicographic ordering $\prec$ and the $\prec$-functions.

Definition 1.2 ($\prec$-predecessors). Suppose $k \geq 1$ and $w \in \Sigma^k$. For $\vec{v} = (v_1, \ldots, v_k)$ and $\vec{y} = (y_1, \ldots, y_k)$, we define $\vec{v} \prec^k \vec{y}$.

(1) The case $k = 1$ is defined by $v \prec^1 C_i(v;)$ for an $i \in \{0, 1\}$. We write $v \preceq^1 y$ if $v \prec^1 y$ or $v = y$.
(2) For the case $k > 1$, $(v_1, \ldots, v_k) \prec^k (y_1, \ldots, y_k)$ iff there exists $k_0 \in \{1, \ldots, k\}$ such that $\forall i < k_0 (v_i = y_i)$, $v_{k_0} \prec^1 y_{k_0}$ and $\forall i > k_0 \exists j \in \{1, \ldots, k\} (v_i \preceq^1 y_j)$.

If $\vec{v} \prec^k \vec{y}$, then $\vec{v}$ is called a $\prec^k$-predecessor of $\vec{y}$.

Note. The ordering $\prec$ is a natural restriction of the usual lexicographic ordering in the Bellantoni-Cook principle. To see this, for now modify $\prec^2$ as

- $(x, y) \prec^2 (x + 1, 0)$ if $y \in \{x, x + 1, 0, 1\}$,
- $(x + 1, y) \prec^2 (x + 1, y + 1)$, and
- $(x, v) \prec^2 (x + 1, y + 1)$ if $v \in \{x, x + 1, y, y + 1\}$.

Let us consider Ackermann function here. Ackermann function $A(x, y)$ is defined by nested recursion on (x, y): $A(0, y) = y + 1$, $A(x + 1, 0) = A(x, 1)$, $A(x + 1, y + 1) = A(x, A(x + 1, y))$. In the equations, $(x, 1) \prec^2 (x + 1, 0)$ and $(x + 1, y) \prec^2 (x + 1, y + 1)$, but $(x, A(x + 1, y)) \not\prec^2 (x + 1, y + 1)$ due to the presence of $A(x + 1, y)$. Recall that the Bellantoni-Cook principle forbids substituting the recursion terms into the recursion parameters. These observations give rise to the definition of a weaker lexicographic ordering $\prec$.

To determine an operation of SNRN, the $\prec$-functions are introduced. Recall that $C_Z(y;)$ denotes 0. Every $\prec^k$-function $\mathbf{f}$ indicates which $\prec^k$-predecessor of $C_w(\vec{y};) = (C_{w(1)}(y_1;), \ldots, C_{w(k)}(y_k;))$ should be chosen for each $w \in \Sigma^k$.

Definition 1.3 ($\prec$-functions). Suppose that $\mathbf{f}$ is a finite function such that $\mathbf{f} : \{1, \ldots, k\} \times \Sigma^k \to \{1, \ldots, 2k\}$. For $w \in \Sigma^k$, let $C_{w(\mathbf{f}(w))}(\vec{y}_{\mathbf{f}(w)};)$ abbreviate $(C_{w(\mathbf{f}(1,w))}(y_{\mathbf{f}(1,w)});), \ldots, C_{w(\mathbf{f}(k,w))}(y_{\mathbf{f}(k,w)});))$.

Then $\mathbf{f}$ is called a $\prec^k$-function, if $C_{w(\mathbf{f}(w))}(\vec{y}_{\mathbf{f}(w)};) \prec^k C_w(\vec{y};)$ for all $w \in \Sigma^k$ (not depending on choice of $\vec{y} = (y_1, \ldots, y_k)$).

Note. Let us recall a convention in Notation 1.1 that for $k_0 \in \{1, \ldots, k\}$ and $w \in \Sigma^k$, $C_{w(k+k_0)}(y_{k+k_0};)$ denotes y_{k_0} if $w(k_0) \in \{0, 1\}$. By the definition, any $\prec^k$-function $\mathbf{f}$ satisfies the condition: $\forall w \in \Sigma^k$, $\exists k_0 \in \{1, \ldots, k\}$ s.t. $w(k_0) \in \{0, 1\}$ and

$$\begin{cases} \mathbf{f}(i, w) = i & \text{if } i < k_0, \\ \mathbf{f}(k_0, w) = k + k_0, & \\ \mathbf{f}(i, w) \in \{1, \ldots, 2k\} & \text{if } i > k_0. \end{cases}$$

For examples of $\prec^k$-functions, see [4].

Now we introduce the class $\mathbf{N}$, which are defined over the natural numbers, contrary to the characterization of words in Oitavem [2]. Let $a + 1$ stand for the numeric successor of a, whereas the notation n' is used in [2]. The notation $a \dot{-} 1$ is used to denote the corresponding predecessor of a.

Definition 1.4 (The class N). A class $\mathbf{N}^{k,l}$ of functions with k normal and l safe arguments is defined by the following initial functions and operations:

Zero $O^{k,l} \in \mathbf{N}^{k,l}$; $\quad O^{k,l}(x_1, \ldots, x_k; a_1, \ldots, a_l) = 0$
Projections $I_j^{k,l} \in \mathbf{N}^{k,l}$ $(1 \le j \le k + l)$;

$$I_j^{k,l}(x_1, \ldots, x_k; a_1, \ldots, a_l) = \begin{cases} x_j & \text{if } 1 \le j \le k, \\ a_{j-k} & \text{if } k < j \le k + l. \end{cases}$$

Successor $S \in \mathbf{N}^{0,1}$; $\quad S(; a) = a + 1$
Predecessor $P \in \mathbf{N}^{0,1}$; $\quad P(; a) = a \dot{-} 1$

Cases $C \in \mathbf{N}^{0,3}$; $\quad C(;a,b,c) = \begin{cases} b \text{ if } a = 0 \\ c \text{ if } a > 0 \end{cases}$

i-Concatenation $C_i \in \mathbf{N}^{1,0}$ $(i = 0,1)$; $\quad C_i(x;) = 2x + i = xi$

Deletion $D \in \mathbf{N}^{0,1}$; $\quad D(;a) = \lfloor a/2 \rfloor$

Safe composition If $h \in \mathbf{N}^{k',l'}$, $g_1, \ldots, g_{k'} \in \mathbf{N}^{k,0}$ and $\varphi_1, \ldots, \varphi_{l'} \in \mathbf{N}^{k,l}$, then $f \in \mathbf{N}^{k,l}$ is defined by $f(\vec{x};\vec{a}) = h(\vec{g}(\vec{x};);\vec{\varphi}(\vec{x};\vec{a}))$.

Safe nested recursion on notation (SNRN) Suppose that $g \in \mathcal{N}^{k',l+1}$, and $h_w, \varphi_w \in \mathcal{N}^{k+k',l+2}$ for each $w \in \Sigma^k$. Also suppose that $\mathbf{f}_1, \mathbf{f}_2$ are $\prec^k$-functions. Then $f \in \mathcal{N}^{k+k',l+1}$ is defined by

$$\begin{cases} f(\vec{0},\vec{x};\vec{a},b) = g(\vec{x};\vec{a},b), \\ f(C_w(\vec{y};),\vec{x};\vec{a},b) = h_w(\vec{v}_1,\vec{x};\vec{a},b,f(\vec{v}_1,\vec{x};\vec{a},c)) \\ \qquad\qquad [\varphi_w(\vec{v}_2,\vec{x};\vec{a},b,f(\vec{v}_2,\vec{x};\vec{a},b))/c] \quad (w \in \Sigma^k), \end{cases}$$

where, for every $i = 1,2$, $\vec{v}_i$ abbreviates $C_{w(\mathbf{f}_i(w))}(\vec{y}_{\mathbf{f}_i(w)};)$, and hence $\vec{v}_i \prec^k C_w(\vec{y};)$.

Then we define $\mathbf{N} := \bigcup_{k,l \in \mathbb{N}} \mathbf{N}^{k,l}$ and $\mathbf{N}_{normal} := \bigcup_{k \in \mathbb{N}} \mathbf{N}^{k,0}$.

Remark 1.5. In the above scheme of SNRN, the restriction of the nesting is not crucial. Even if we allow any constant number of nestings, arguments in Section 2 work. Further, Arai and the author[4] employ a more general scheme of the form

$$\begin{cases} f(\vec{0},\vec{x};\vec{a}) = g(\vec{x};\vec{a}), \\ f(C_w(\vec{y}),\vec{x};\vec{a}) = h_w(\vec{v}_1,\vec{x};\vec{a},f(\vec{v}_1,\vec{x};\vec{\varphi}_w(\vec{v}_2,\vec{x};\vec{a},f(\vec{v}_2,\vec{x};\vec{a})))) \end{cases} \tag{1}$$

for $\vec{\varphi}_w = \varphi_{w,1}, \ldots, \varphi_{w,l}$ and an $l > 0$. If the projection function $I^{k+k',l}_{k+k'+i}$ is taken as $\varphi_{w,i}$ for each $i = 1, \ldots, l-1$, then this scheme is just the SNRN scheme. If the scheme (1) is contained instead of the one in Definition 1.4, Theorem 2.11 does not hold. Nevertheless, this restriction is not essential either. The same class will be generated even by (1). See also Remark 2.12.

Example 1.6.

(1) $\div(x;a) = D^{2^{|x|}}(;a)$.

$$\div(0;a) = D(;a),$$
$$\div(xi;a) = \div(x;\div(x;a)).$$

(2) $+(x;a) = x + a$.

$$\begin{aligned} +(0;a) &= a, \\ +(x0;a) &= 2x + a = +(x; +(x;a)), \\ +(x1;a) &= 2x + 1 + a = S(; +(x; +(x;a))). \end{aligned}$$

Similarly, $\dot{-}(x;a) = a \dot{-} x$ is defined.

(3) $\times(x, y; a) = y \cdot 2^{|x|} + a$.

$$\begin{aligned} \times(0, y; a) &= y + a, \\ \times(xi, y; a) &= \times(x, y; \times(x, y; a)). \end{aligned}$$

(4) $f(x;a) = 2^{|x|} + a$.

$$\begin{aligned} f(0;a) &= S(;a), \\ f(xi;a) &= 2^{|x|+1} + a \\ &= f(x; f(x;a)). \quad (i = 0, 1) \end{aligned}$$

(5) $f(x, y, z; a) = 2^{|x| \cdot |y| + |z|} + a$.

$$\begin{aligned} f(0,0,0;a) &= S(;a), \\ f(x, y, zi; a) &= 2^{|x| \cdot |y| + |z| + 1} + a = f(x, y, z; f(x, y, z; a)), \\ f(x, yi, 0; a) &= 2^{|x|(|y|+1)} + a = f(x, y, x; a), \\ f(xi, 0, 0; a) &= f(x, 0, 0; a). \quad (i = 0, 1) \end{aligned}$$

Similarly, we can define $f(x, y, z, u, v, w; a) = 2^{|x||y||z| + |u||v| + |w|} + a$ and so on. Hence, a suitable application of safe composition yields $2^{p(|\vec{x}|)} \in \mathbf{N}_{normal}$ for any polynomial $p(\vec{x})$.

(6) $2^{2^{|x|}} \notin \mathbf{N}$, contrary to $f(x;a) = 2^{2^{|x|}} \cdot a \in \mathcal{N}$ in 4. This is due to the choice of initial functions. The binary successors $S_0, S_1 \in \mathcal{N}^{0,1}$ are defined so that $S_i(;a) = 2a + i$ $(i = 0, 1)$. However $C_0, C_1 \in \mathbf{N}^{1,0}$ are defined on the normal argument, and therefore C_i cannot be g in the scheme of SNRN in Definition 1.4.

Recall that **FPS** is the class of functions computed by a deterministic Turing machine with the use of a number of cells bounded by $p(|\vec{x}|)$ for some polynomial $p(\vec{x})$. The following theorem is a direct consequence of results in Sections 2 and 3.

Theorem 1.7. $\mathbf{N}_{normal} = \mathbf{FPS}$.

Remark 1.8. Let us recall briefly the scheme of **safe recursion**:

$$\begin{cases} f(0,\vec{x};\vec{a}) = g(\vec{x};\vec{a}), \\ f(S(;y),\vec{x};\vec{a}) = h(y,\vec{x};\vec{a},f(y,\vec{x};\vec{a})). \end{cases}$$

Via a work of Bellantoni ([5] Chapter 5), it turns out that if the SNRN scheme is replaced by the scheme of safe recursion (and even if both C_i and D are omitted), the resulting class is identical to the class of linear-space computable functions. Moreover, replace SNRN by safe nested recursion (on unary notation), and restrict safe composition suitably, cf. [6] or [4]. Then the resulting class will be identical to the class of functions computable in $2^{O(|x|)}$-space, i.e., EXPSPACE-computable functions. These observations together with Theorem 1.7 suggest an uniform approach to space-complexity classes.

2. Term rewriting system for PSPACE

In this section we introduce a term rewriting system $R_{\mathbf{N}}$ over a class $\mathcal{F}$ of function symbols corresponding to **N**. We show that for any $f \in \mathcal{F}$, the size of every term occurring in a rewriting sequence which starts with $f(\vec{m};\vec{n})$ is bounded by a polynomial in the binary lengths of numerals $\vec{m}$ and $\vec{n}$, whenever an innermost strategy is used. This implies that every $R_{\mathbf{N}}$-reduction (innermost) strategy yields an algorithm for **N** running in polynomial space, i.e., $\mathbf{N}_{normal} \subseteq \mathbf{FPS}$.

Let $\mathcal{V}$ be a countably infinite set of variables. Variables are written as x, y, z. We use $\mathcal{V}ar(t)$ to denote the set of variables occurring in a term t. A finite set R of rewrite rules is called a term rewriting system (TRS for short) if every rewrite rule $l \to r \in R$ satisfies $\mathcal{V}ar(r) \subseteq \mathcal{V}ar(l)$. A TRS R defines a rewriting relation $\to_R$ by $l \to r \in R \Rightarrow r[l\theta] \to_R r[r\theta]$ for any context $r[\cdot]$ and substitution θ. Then $l\theta$ is called a redex. The reflexive and transitive closure of $\to_R$ is denoted by $\to_R^*$.

$\mathcal{F}$ is the smallest class of symbols built up from $O^{k,l}, I_j^{k,l}, S, P, C, C_i, D$ by means of $\mathrm{SUB}^{k,l}$ and $\mathrm{SNRN}^{k,l}$, e.g., if $g \in \mathcal{F}^{k',l+1}$ and $h_w, \varphi_w \in \mathcal{F}^{k+k',l+2}$ for all $w \in \Sigma^k$, then $\mathrm{SNRN}^{k+k',l+1}[g, \{h_w, \varphi_w : w \in \Sigma^k\}] \in \mathcal{F}^{k+k',l+1}$. Superscripts of SUB and SNRN are always omitted if no confusion arises.

Let $\mathcal{T}(\mathcal{F},\mathcal{V})$ be the set of terms over $\mathcal{F}$ and $\mathcal{V}$. Let $\mathcal{T}(\mathcal{F})$ be the set of ground terms in $\mathcal{T}(\mathcal{F},\mathcal{V})$, which are built up from $O^{0,0}$ by means of elements of $\mathcal{F}$. Numerals are built up from $O^{0.0}$ by means of C_0, C_1. We write $n, m, \ldots$ for numerals and 0 instead of $O^{0,0}$. The binary length $|m|$ of a numeral m

is defined by $|0| = 1$, and $C_i(m;) = |m| + 1$. By replacing "=" occurring in the equations in Definition 1.4 by "→", we obtain a schematic TRS for **N**. However, relaxing the definition, we introduce a non-deterministic or non-confluent TRS $R_{\mathbf{N}}$ (according to the referee's suggestion).

Definition 2.1 (Term rewriting system $R_{\mathbf{N}}$).

(1) $O^{k,l}(\vec{x};\vec{y}) \to 0 \ (0 < k+l)$
(2) $I_j^{k,l}(x_1,\ldots,x_k;x_{k+1},\ldots,x_{k+l}) \to x_j \ (1 \le j \le k+l)$
(3) $S(;0) \to C_1(0;)$
(4) $S(;C_0(x;)) \to C_1(x;)$
(5) $S(;C_1(x;)) \to C_0(S(;x);)$
(6) $P(;0) \to 0$
(7) $P(;C_0(x;)) \to C_1(P(;x);)$
(8) $P(;C_1(x;)) \to C_0(x;)$
(9) $C(;0,x,y) \to x$
(10) $C(;C_i(z;),x,y) \to y$
(11) $D(;0) \to 0$
(12) $D(;C_i(x;)) \to x$
(13) $\mathrm{SUB}[h,\vec{g},\vec{\varphi}](\vec{x};\vec{y}) \to h(\vec{g}(\vec{x};);\vec{\varphi}(\vec{x};\vec{y}))$
(14) $\mathrm{SNRN}[g,\{h_w,\varphi_w : w \in \Sigma^k\}](\vec{0},\vec{x};\vec{z},z') \to g(\vec{x};\vec{z},z')$
(15) $\mathrm{SNRN}[g,\{h_w,\varphi_w : w \in \Sigma^k\}](C_{w'}(\vec{y};),\vec{x};\vec{z},z') \to$
$h_{w'}(\vec{v}_1,\vec{x};\vec{z},z',\mathrm{SNRN}[g,\{h_w,\varphi_w : w \in \Sigma^k\}](\vec{v}_1,\vec{x};\vec{z},u))$
$[\varphi_{w'}(\vec{v}_2,\vec{x};\vec{z},z',\mathrm{SNRN}[g,\{h_w,\varphi_w : w \in \Sigma^k\}](\vec{v}_2,\vec{x};\vec{z},z'))/u]$
for some $\vec{v}_1,\vec{v}_2 \prec^k C_{w'}(\vec{y};)$ $(w' \in \Sigma^k)$.

The intended semantic π for $\mathcal{F}$ is obvious. For example, suppose that $f = \mathrm{SUB}[h,\vec{g},\vec{\varphi}]$. Then $\pi(f) \in \mathbf{N}$ is defined by safe composition from $\pi(h)$, $\pi(\vec{g})$ and $\pi(\vec{\varphi})$ in **N**: $\pi(f)(\vec{x};\vec{a}) = \pi(h)(\pi(\vec{g})(\vec{x};);\pi(\vec{\varphi})(\vec{x};\vec{a}))$. However, one should be careful of the case $f = \mathrm{SNRN}[g,\{h_w,\varphi_w : w \in \Sigma^k\}]$. Clearly, the TRS $R_{\mathbf{N}}$ is not confluent due to absence of $\prec^k$-functions $\mathbf{f}_1,\mathbf{f}_2$. Let us recall that a $\prec$-function indicates which $\prec$-predecessor should be chosen. The mapping π is extended to $\mathcal{T}(\mathcal{F}) \to \mathbb{N}$ by $\pi(f(\vec{t};\vec{s})) := \pi(f)(\pi(\vec{t});\pi(\vec{s}))$.

Lemma 2.2. *$R_{\mathbf{N}}$ is terminating.*

Proof. Define a precedence $<_{\mathcal{F}}$ on $\mathcal{F}$ by

- $0 <_{\mathcal{F}} O^{k,l}$ if $k+l > 0$,
- $g <_{\mathcal{F}} f$ for each $g \in \{O^{k,l}, I_j^{k,l}, C_i\}$ and $f \in \{S,P,C,D\}$,
- $f <_{\mathcal{F}} \mathrm{SUB}[h,\vec{g},\vec{\varphi}]$ for each $f \in \{h,\vec{g},\vec{\varphi}\}$, and
- $f <_{\mathcal{F}} \mathrm{SNRN}[g,\{h_w,\varphi_w : w \in \Sigma^k\}]$ for each $f \in \{g,h_w,\varphi_w : w \in \Sigma^k\}$.

Then $R_{\mathbf{N}}$ is reducing under the lexicographic path order LPO induced by $<_{\mathcal{F}}$. Namely, $t \rightarrow_{R_{\mathbf{N}}} s \Rightarrow s <_{\mathrm{LPO}} t$ for each $t, s \in \mathcal{T}(\mathcal{F}, \mathcal{V})$. The well-foundedness of LPO implies the termination of $R_{\mathbf{N}}$. □

By the termination of $R_{\mathbf{N}}$, a normal form always exists, i.e., for any $t \in \mathcal{T}(\mathcal{F}, \mathcal{V})$, there exists a $t' \in \mathcal{T}(\mathcal{F}, \mathcal{V})$ such that $t \rightarrow^{*}_{R_{\mathbf{N}}} t'$ and $\nexists s$ s.t. $t' \rightarrow_{R_{\mathbf{N}}} s$. It is not difficult to see that any normal form of each ground term is a numeral.

Notation 2.3. Since $R_{\mathbf{N}}$ is not confluent, there may be several normal forms of a ground term. At the risk of confusion, let $\underline{t}$ denote a normal form of $t \in \mathcal{T}(\mathcal{F})$. Hence, in paticular, $\underline{m} = m$ for any numeral m.

Definition 2.4. The length $\mathrm{lh}(f)$ of $f \in \mathcal{F}$ is defined by

- $\mathrm{lh}(f) = 1$ if $f \in \{O^{k,l}, I_j^{k,l}, S, P, C, C_i, D\}$,
- $\mathrm{lh}(\mathrm{SUB}[h, \vec{g}, \vec{\varphi}]) = \mathrm{lh}(h) + \sum_{i=1}^{k'} \mathrm{lh}(g_i) + \sum_{i=1}^{l'} \mathrm{lh}(\varphi_i) + 1$, and
- $\mathrm{lh}(\mathrm{SNRN}[g, \{h_w, \varphi_w : w \in \Sigma^k\}]) = \mathrm{lh}(g) + \sum_{w \in \Sigma^k} \mathrm{lh}(h_w) + \sum_{w \in \Sigma^k} \mathrm{lh}(\varphi_w) + 1$

Then, the length $\mathrm{lh}(t)$ of $t \in \mathcal{T}(\mathcal{F})$ is defined by

$$\mathrm{lh}(f(\vec{t}; \vec{s})) = \mathrm{lh}(f) + \sum_{i=1}^{k} \mathrm{lh}(t_i) + \sum_{i=1}^{l} \mathrm{lh}(s_i).$$

Hence, in particular, $|m| = \mathrm{lh}(m)$ for any numeral m. As a corollary, we observe Lemma 2.5.

Lemma 2.5. *If* $t \rightarrow s \in R_{\mathbf{N}}$ *by any of Definition* 2.1(1)—2.1(12), *then* $\mathrm{lh}(s\theta) \leq \mathrm{lh}(t\theta)$ *for any ground substitution* θ.

In general, for $f \in \mathcal{F}$, $\max\{\mathrm{lh}(t) : f(\vec{m}; \vec{n}) \rightarrow^{*}_{R_{\mathbf{N}}} t\}$ is not bounded by a polynomial in $|\vec{m}|, |\vec{n}|$. Consider the rewriting rule $f(C_i(y;); x) \rightarrow f(y; f(y; x))$. For numerals m and n, it is possible to obtain the rewriting sequence $f(m; n) \rightarrow^{*}_{R_{\mathbf{N}}} f(0; f(0; \cdots f(0; n) \cdots))$. However, $\mathrm{lh}(f(0; f(0; \cdots f(0; n) \cdots))) = (\mathrm{lh}(f) + 1)2^{|m|} + |n|$, which is not admissible. Hence we employ the innermost rewriting strategy.

Notation 2.6. Let us use $t \rightarrow_{in} s$ to denote $t \rightarrow_{R_{\mathbf{N}}} s$ according to an innermost strategy. Namely, $t \rightarrow_{in} s$ if $t \rightarrow_{R_{\mathbf{N}}} s$ and the redex r is innermost, i.e., r contains no redex that is a proper subterm of r.

The rest of this section is devoted to show that for every $f \in \mathcal{F}$ there exists a polynomial p such that $\max\{\mathrm{lh}(t) : f(\vec{m};\vec{n}) \to_{in}^{*} t\} \leq p(|\vec{m}|, |\vec{n}|)$. Due to the innermost strategy, $\max\{\mathrm{lh}(t) : f(\vec{t};\vec{s}) \to_{R_{\mathbf{N}}}^{*} t\}$ can be reduced as follows.

Lemma 2.7. *For any* $f \in \mathcal{F}^{k,l}$ *and* $t_1, \dots, t_k, s_1, \dots, s_l \in \mathcal{T}(\mathcal{F})$, $\max\{\mathrm{lh}(t) : f(\vec{t};\vec{s}) \to_{in}^{*} t\}$ *is bounded by either* $\mathrm{lh}(f) + \sum_{i=1}^{k} \max\{\mathrm{lh}(t_i^*) : t_i \to_{in}^{*} t_i^*\} + \sum_{i=1}^{l} \max\{\mathrm{lh}(s_i^*) : s_i \to_{in}^{*} s_i^*\}$, *or* $\max\{\mathrm{lh}(t) : f(\underline{\vec{t}};\underline{\vec{s}}) \to_{in}^{*} t\}$.

In the following proofs, we use $\mathrm{Sp}(t)$ to abbreviate $\max\{\mathrm{lh}(s) : t \to_{in}^{*} s\}$, which intended to denote the *space* required to rewrite $t \in \mathcal{T}(\mathcal{F})$ according to $\to_{in}$.

Proof. Suppose that $f \in \{O^{k,l}, I_j^{k,l}, C_i, D, S, P, C\}$. From Lemma 2.5, we observe that

$$\mathrm{Sp}(f(\vec{t};\vec{s})) \leq \mathrm{lh}(f) + \sum_{i=1}^{k} \mathrm{Sp}(t_i) + \sum_{i=1}^{l} \mathrm{Sp}(s_i). \tag{2}$$

Otherwise, $f \in \{\mathrm{SUB}[h, \vec{g}, \vec{\varphi}], \mathrm{SNRN}[g, \{h_w, \varphi_w : w \in \Sigma^k\}]\}$. Then, due to the innermost strategy, an arbitrary terminating rewriting which starts with $f(\vec{t};\vec{s})$ runs as

$$f(\vec{t};\vec{s}) \to_{in}^{*} f(\underline{\vec{t}};\underline{\vec{s}}) \to_{in}^{*} \underline{f(\underline{\vec{t}};\underline{\vec{s}})}.$$

Thus, $\mathrm{Sp}(f(\vec{t};\vec{s}))$ is bounded by either $\mathrm{lh}(f) + \sum_{i=1}^{k} \mathrm{Sp}(t_i) + \sum_{i=1}^{l} \mathrm{Sp}(s_i)$ or $\mathrm{Sp}(f(\underline{\vec{t}};\underline{\vec{s}}))$. This concludes the lemma. □

Definition 2.8. For $d, b_1, \dots, b_k \in \mathbb{N}$, put

$$\sum(d, \vec{b}) = \sum(d, b_1, \dots, b_k) := \sum_{i=1}^{k} (\max \vec{b} + 1)^{d-i} b_i.$$

Then, as in [4], one can prove a fundamental lemma concerning the descending lengths with respect to the ordering $\prec$.

Lemma 2.9. *Suppose that* $\vec{n}$ *and* $\vec{m}$ *are numerals. If* $\vec{n} \prec^k \vec{m}$ *and* $k \leq d$, *then* $\sum(d, |\vec{n}|) < \sum(d, |\vec{m}|)$.

Lemma 2.10. *For any* $f \in \mathcal{F}$ *there exist some constants* c *and* d *such that for numerals* $\vec{m}, \vec{n}$ *and an arbitrary normal form* $\underline{f(\vec{m};\vec{n})}$ *of* $f(\vec{m};\vec{n})$, $2^{|\underline{f(\vec{m};\vec{n})}|} \leq 2^{c(\max|m|^d+1)} + 2^{\max|n|}$, *and hence* $|\underline{f(\vec{m};\vec{n})}| \leq \max\{c(\max|m|^d + 1), \max|n|\} + 1$.

Proof. We prove the lemma by induction on $\mathrm{lh}(f)$. For the base case, if $\mathrm{lh}(f) = 1$, i.e., $f \in \{O, I, C_i, S, P, D, C\}$, then $2^{|\underline{f(\vec{m};\vec{n})}|} \leq 2^{\max|\vec{m}|+1} + 2^{\max|\vec{n}|}$.

Consider the induction step. Suppose that $f = \mathrm{SUB}[h, \vec{g}, \vec{\varphi}]$ for $h \in \mathbf{N}^{k',l'}$, $g_1, \ldots, g_{k'} \in \mathbf{N}^{k,0}$ and $\varphi_1, \ldots, \varphi_{l'} \in \mathbf{N}^{k,l}$. Then, by the induction hypothesis, there exist some constants c_0, d_0 for h, c_i, d_i for g_i ($i = 1, \ldots, k'$) and c_{k+i}, d_{k+i} for φ_i ($i = 1, \ldots, l'$) which enjoy the condition. Fix a normal form $\underline{g_i(\vec{m};)}$ of $g_i(\vec{m};)$ for each $i = 1, \ldots, k'$ and $\underline{\varphi_i(\vec{m};\vec{n})}$ of $\varphi_i(\vec{m};\vec{n})$ for each $i = 1, \ldots, l'$. Let $c' := \max\{c_0, c_1, \ldots, c_{k'+l'}\}$ and $d' := \max\{d_0, d_1, \ldots, d_{k'+l'}\}$. Then, by IH for $\vec{g}$ and $\vec{\varphi}$,

$$|\underline{g_i(\vec{m};)}| \leq c'(\max|\vec{m}|^{d'} + 1) \ (i = 1, \ldots, k'), \tag{3}$$

$$2^{|\underline{\varphi_i(\vec{m};\vec{n})}|} \leq 2^{c'(\max|\vec{m}|^{d'}+1)} + 2^{\max|\vec{n}|} \ (i = 1, \ldots, l'). \tag{4}$$

Fix a normal form $\underline{f(\vec{m};\vec{n})} = \underline{h(\underline{\vec{g}(\vec{m};)}; \underline{\vec{\varphi}(\vec{m};\vec{n})})}$ of $f(\vec{m};\vec{n})$. By IH for h,

$$\begin{aligned} 2^{|\underline{f(\vec{m};\vec{n})}|} &= 2^{|\underline{h(\vec{g}(\vec{m};\vec{n});\vec{\varphi}(\vec{m};\vec{n}))}|} \\ &\leq 2^{c'(\max|\underline{\vec{g}(\vec{m};)}|^{d'}+1)} + 2^{\max|\underline{\vec{\varphi}(\vec{m};\vec{n})}|}. \end{aligned} \tag{5}$$

Put $c := 2^{d'} c'^{d'+1} + 1$ and $d := d'^2$. Then,

$$\begin{aligned} c'(\max|\underline{\vec{g}(\vec{m};)}|^{d'} + 1) &\leq c'(c'(\max|\vec{m}|^{d'} + 1)^{d'} + 1) \quad \text{by (3)}, \\ &\leq c'(c'^{d'} 2^{d'} \max|\vec{m}|^{d'^2} + 1) \\ &\leq 2^{d'} c'^{d'+1}(\max|\vec{m}|^{d'^2} + 1) \\ &\leq (c-1)(\max|\vec{m}|^{d} + 1) \end{aligned} \tag{6}$$

by the definitions of the constants c and d. Thus, by (5), (6) and (4),

$$\begin{aligned} 2^{|\underline{f(\vec{m};\vec{n})}|} &\leq 2^{(c-1)(\max|\vec{m}|^{d}+1)} + 2^{c'(\max|\vec{m}|^{d'}+1)} + 2^{\max|\vec{n}|} \\ &\leq 2^{(c-1)(\max|\vec{m}|^{d}+1)+1} + 2^{\max|\vec{n}|} \quad \text{by } c' \leq c-1, \\ &\leq 2^{c(\max|\vec{m}|^{d}+1)} + 2^{\max|\vec{n}|}. \end{aligned}$$

For the case for SNRN, it is convenient to consider a more general form (1) in Remark 1.5:

$$\begin{cases} f(\vec{0}, \vec{x}; \vec{a}) = g(\vec{x}; \vec{a}), \\ f(C_w(\vec{y}), \vec{x}; \vec{a}) = h_w(\vec{v}_1, \vec{x}; \vec{a}, f(\vec{v}_1, \vec{x}; \vec{\varphi}_w(\vec{v}_2, \vec{x}; \vec{a}, f(\vec{v}_2, \vec{x}; \vec{a})))) \\ (\vec{\varphi}_w = \varphi_{w,1}, \ldots, \varphi_{w,l}, \ l > 0) \end{cases}$$

Let $f = \mathrm{SNRN}[g, \{h_w, \vec{\varphi}_w : w \in \Sigma^k\}] \in \mathcal{F}^{k+k',l}$ for $g \in \mathcal{F}^{k',l}$ and h_w, $\varphi_{w,1}$, $\ldots$, $\varphi_{w,l} \in \mathcal{F}^{k+k',l+1}$. By IH, there exist some constants c_0, d_0 for g, c_1, d_1

for h_w, and c_2, d_2 for $\vec{\varphi}_w$ which enjoy the condition for all $w \in \Sigma^k$. Put $c := \max\{c_0, c_1, c_2, 2\}$ and $d := \max\{d_0, d_1, d_2, k\}$. Suppose that $m^y_1, \dots, m^y_k$ and $m^x_1, \dots, m^x_{k'}$ are numerals. Then, by side induction on $\sum(d, |\vec{m}^y|)$, we show that for any $\vec{t} = t_1, \dots, t_l \in \mathcal{T}(\mathcal{F})$ and their arbitrary normal forms $\underline{\vec{t}} = \underline{t_1}, \dots, \underline{t_l}$,

$$2^{|\underline{f(\vec{m}^y, \vec{m}^x; \underline{\vec{t}})}|} \le 2^{c((\max\{|\vec{m}^y|, |\vec{m}^x|\}+1)^d + \sum(d, |\vec{m}^y|)+1)} + 2^{\max|\underline{\vec{t}}|}.$$

Put $b := \max\{|\vec{m}^y|, |\vec{m}^x|\} + 1$ and $b_i := |m^y_i|$ for each $i = 1, \dots, k$. First consider the case $\vec{m}^y = \vec{0}$: $f(\vec{0}, \vec{m}^x; \underline{\vec{t}}) \to_{in} g(\vec{m}^x; \underline{\vec{t}})$. This case follows from IH as

$$\begin{aligned} 2^{|\underline{f(\vec{0}, \vec{m}^x; \underline{\vec{t}})}|} &= 2^{|\underline{g(\vec{m}^x; \underline{\vec{t}})}|} \\ &\le 2^{c_0(b^{d_0}+1)} + 2^{\max|\underline{\vec{t}}|} \quad \text{by IH for } g, \\ &\le 2^{c(b^d + \sum(d, \vec{b})+1)} + 2^{\max|\underline{\vec{t}}|} \quad \text{by } c_0 \le c, \text{ and } d_0 \le d. \end{aligned}$$

Next consider the case $\vec{m}^y \neq \vec{0}$:

$$\begin{aligned} f(\vec{m}^y, \vec{m}^x; \underline{\vec{t}}) &\to_{in} h_w(\vec{m}^v_1, \vec{m}^x; \underline{\vec{t}}, f(\vec{m}^v_1, \vec{m}^x; \vec{\varphi}_w(\vec{m}^v_2, \vec{m}^x; \underline{\vec{t}}, f(\vec{m}^v_2, \vec{m}^x; \underline{\vec{t}})))) \\ &\to^*_{in} h_w(\vec{m}^v_1, \vec{m}^x; \underline{\vec{t}}, f(\vec{m}^v_1, \vec{m}^x; \vec{\varphi}_w(\vec{m}^v_2, \vec{m}^x; \underline{\vec{t}}, \underline{f(\vec{m}^v_2, \vec{m}^x; \underline{\vec{t}})}))) \\ &\to^*_{in} h_w(\vec{m}^v_1, \vec{m}^x; \underline{\vec{t}}, f(\vec{m}^v_1, \vec{m}^x; \underline{\vec{\varphi}_w(\vec{m}^v_2, \vec{m}^x; \underline{\vec{t}}, \underline{f(\vec{m}^v_2, \vec{m}^x; \underline{\vec{t}})})})) \\ &\to^*_{in} h_w(\vec{m}^v_1, \vec{m}^x; \underline{\vec{t}}, \underline{f(\vec{m}^v_1, \vec{m}^x; \underline{\vec{\varphi}_w(\vec{m}^v_2, \vec{m}^x; \underline{\vec{t}}, \underline{f(\vec{m}^v_2, \vec{m}^x; \underline{\vec{t}})})})}). \end{aligned}$$

for some $\vec{m}^v_1, \vec{m}^v_2 \prec^k \vec{m}^y$ and a suitable $w \in \Sigma^k$. By Lemma 2.9, for each $j = 1, 2$,

$$\sum(d, |\vec{m}^v_j|) < \sum(d, \vec{b}). \tag{7}$$

Hence, by the side induction hypothesis and the condition that $\max|\vec{m}^v_2| \le \max|\vec{m}^y|$,

$$2^{|\underline{f(\vec{m}^v_2, \vec{m}^x; \underline{\vec{t}})}|} \le 2^{c(b^d + \sum(d, |\vec{m}^v_2|)+1)} + 2^{\max|\underline{\vec{t}}|} \le 2^{c(b^d + \sum(d, \vec{b}))} + 2^{\max|\underline{\vec{t}}|}.$$

From this and IH for $\vec{\varphi}_w$,

$$\begin{aligned} 2^{\max|\underline{\vec{\varphi}_w(\vec{m}^v_2, \vec{m}^x; \underline{\vec{t}}, \underline{f(\vec{m}^v_2, \vec{m}^x; \underline{\vec{t}})})}|} &\le 2^{c_2(b^{d_2}+1)} + 2^{\max\{|\underline{\vec{t}}|, |\underline{f(\vec{m}^v_2, \vec{m}^x; \underline{\vec{t}})}|\}} \\ &\le 2^{c_2(b^{d_2}+1)} + 2^{c(b^d + \sum(d, \vec{b}))} + 2^{\max|\underline{\vec{t}}|} \\ &\le 2 \cdot 2^{c(b^d + \sum(d, \vec{b}))} + 2^{\max|\underline{\vec{t}}|}, \end{aligned} \tag{8}$$

since $c_2 \le c$, $d_2 \le d$, and $1 \le \sum(d, \vec{b})$. Similarly to the case for $\vec{m}_2^v$, by SIH together with (7) for $\vec{m}_1^v$,

$$
\begin{aligned}
& 2^{\underline{|f(\vec{m}_1^v, \vec{m}^x; \underline{\underline{\varphi_w(\vec{m}_2^v, \vec{m}^x; \underline{\vec{t}}, \underline{f(\vec{m}_2^v, \vec{m}^x; \underline{\vec{t}})})}}))|}} \\
& \le 2^{c(b^d + \sum(d, |\vec{m}_1^v|)+1)} + 2^{\max |\underline{\vec{\varphi}_w(\vec{m}_2^v, \vec{m}^x; \underline{\vec{t}}, \underline{f(\vec{m}_2^v, \vec{m}^x; \underline{\vec{t}})})}|} \\
& \le 2^{c(b^d + \sum(d, \vec{b}))} + 2 \cdot 2^{c(b^d + \sum(d, \vec{b}))} + 2^{\max |\underline{\vec{t}}|} \quad \text{by (8)}, \\
& \le 3 \cdot 2^{c(b^d + \sum(d, \vec{b}))} + 2^{\max |\underline{\vec{t}}|}.
\end{aligned}
$$

Therefore,

$$
\begin{aligned}
& 2^{|\underline{f(\vec{m}^y, \vec{m}^x; \underline{\vec{t}})}|} \\
& = 2^{\underline{|h_w(\vec{m}_1^v, \vec{m}^x; \underline{\vec{t}}, \underline{f(\vec{m}_1^v, \vec{m}^x; \underline{\varphi_w(\vec{m}_2^v, \vec{m}^x; \underline{\vec{t}}, \underline{f(\vec{m}_2^v, \vec{m}^x; \underline{\vec{t}})})})})|}} \\
& \le 2^{c_1(b^{d_1}+1)} + 3 \cdot 2^{c(b^d + \sum(d, \vec{b}))} + 2^{\max |\underline{\vec{t}}|} \quad \text{by IH for } h_w, \\
& \le 4 \cdot 2^{c(b^d + \sum(d, \vec{b}))} + 2^{\max |\underline{\vec{t}}|} \quad \text{by } c_1 \le c,\ d_1 \le d \text{ and } 1 \le \sum(d, \vec{b}), \\
& \le 2^{c(b^d + \sum(d, \vec{b})+1)} + 2^{\max |\underline{\vec{t}}|} \quad \text{by } 2 \le c.
\end{aligned}
$$

This completes the proof of the lemma. □

Theorem 2.11. *For any $f \in \mathcal{F}^{k,l}$, there exist c and d such that for numerals $m_1, \ldots, m_k$ and $n_1, \ldots, n_l$,* $\max\{\mathrm{lh}(t) : f(\vec{m}; \vec{n}) \to^*_{in} t\} \le c(\max |\vec{m}|^d + 1)(\max |\vec{n}| + 1)$.

Proof. We prove the theorem again by induction on $\mathrm{lh}(f)$. If $\mathrm{lh}(f) = 1$, then, by (2),

$$
\begin{aligned}
\mathrm{Sp}(f(\vec{m}; \vec{n})) &\le \mathrm{lh}(f) + \sum_{i=1}^{k} |m_i| + \sum_{i=1}^{l} |n_i| \\
&\le (\mathrm{lh}(f) + k + l)(\max |\vec{m}| + 1)(\max |\vec{n}| + 1). \qquad (9)
\end{aligned}
$$

The case $f = \mathrm{SUB}[h, \vec{g}, \vec{\varphi}]$ is seen from IH, Lemma 2.7, and Lemma 2.10.

Finally, consider the case $f = \mathrm{SNRN}[g, \{h_w, \varphi_w : w \in \Sigma^k\}]$. For simplicity, we only consider $f \in \mathcal{F}^{k,1}$ such that $f(\vec{0}; x) \to g(; x)$ and $f(\vec{y}; x) \to f(\vec{v}_1; f(\vec{v}_2; x))$ for some $\vec{v}_1, \vec{v}_2 \prec^k \vec{y}$. By IH, there exists a constant c_g for g enjoying the condition. Furthermore, by Lemma 2.10, there exist constants c', d' for f enjoying the condition in the lemma. Let $c_{\mathrm{lh}} := \mathrm{lh}(f) + k + 1$. Now put $c := \max\{c_{\mathrm{lh}}, c_g, c' + 1\}$ and $d := \max\{k, d'\}$. Suppose that $m_1, \ldots, m_k$ are numerals. Then, by side induction on $\sum(d, |\vec{m}|)$, we show that for any $t \in \mathcal{T}(\mathcal{V})$ and its arbitrary normal form $\underline{t}$,

$$
2^{\mathrm{Sp}(f(\vec{m}; \underline{t}))} \le (2^{c(\sum(d, |\vec{m}|)+1)} + 2^{|\underline{t}|+1})^{c(\sum(d, |\vec{m}|)+1)}. \qquad (10)
$$

This results in $\mathrm{Sp}(f(\vec{m};\underline{t})) \le c^2(\sum(d,|\vec{m}|)+1)^2(|\underline{t}|+1)$.

Case 1. $\mathrm{Sp}(f(\vec{m};\underline{t})) \le \mathrm{lh}(f) + \sum_{i=1}^{k}|m_i| + |\underline{t}|$: In this case, (10) follows, since

$$\begin{aligned}\mathrm{Sp}(f(\vec{m};\underline{t})) &\le \mathrm{lh}(f) + k\max|\vec{m}| + |\underline{t}|\\ &\le c_{\mathrm{lh}}(\max|\vec{m}| + |\underline{t}| + 1)\\ &\le c(\max|\vec{m}|+1)(|\underline{t}|+1) \quad \text{by } c_{\mathrm{lh}} \le c.\end{aligned}$$

Case 2. $\mathrm{Sp}(f(\vec{m};\underline{t})) \le \mathrm{Sp}(g(;\underline{t}))$: In this case, (10) follows, since

$$\begin{aligned}\mathrm{Sp}(g(;\underline{t})) &\le c_g(|\underline{t}|+1) \quad \text{by IH for } g,\\ &\le c(\max|\vec{m}|^d+1)(|\underline{t}|+1) \quad \text{by } c_g \le c.\end{aligned}$$

Case 3. Otherwise: By Lemma 2.7,

$$\mathrm{Sp}(f(\vec{m};\underline{t})) \le \mathrm{Sp}(f(\vec{m}_1;\underline{f(\vec{m}_2;\underline{t})})) \tag{11}$$

for some $\vec{m}_1, \vec{m}_2 \prec^k \vec{m}$ and a normal form $\underline{f(\vec{m}_2;\underline{t})}$ of $f(\vec{m}_2;\underline{t})$. Again, by Lemma 2.9, for each $j = 1, 2$,

$$\sum(d,|\vec{m}_j|) < \sum(d,|\vec{m}|). \tag{12}$$

Hence, by SIH for $\vec{m}_1$,

$$\begin{aligned}2^{\mathrm{Sp}(f(\vec{m}_1;\underline{f(\vec{m}_2;\underline{t})}))} &\le (2^{c(\sum(d,|\vec{m}_1|)+1)} + 2^{|\underline{f(\vec{m}_2;\underline{t})}|+1})^{c(\sum(d,|\vec{m}_1|)+1)}\\ &\le (2^{c\sum(d,|\vec{m}|)} + 2^{|\underline{f(\vec{m}_2;\underline{t})}|+1})^{c(\sum(d,|\vec{m}|)+1)}\end{aligned} \tag{13}$$

by (12). On the other hand, by Lemma 2.10,

$$\begin{aligned}2^{|\underline{f(\vec{m}_2;\underline{t})}|+1} &\le 2^{c'(\max|\vec{m}_2|^{d'}+1)+1} + 2^{|\underline{t}|+1}\\ &\le 2^{c(\max|\vec{m}_2|^{d}+1)} + 2^{|\underline{t}|+1} \quad \text{by } c'+1 \le c \text{ and } d' \le d,\\ &\le 2^{c\sum(d,|\vec{m}|)} + 2^{|\underline{t}|+1},\end{aligned} \tag{14}$$

since $\max|\vec{m}_2|^d + 1 \le \sum(d,|\vec{m}|)$. Combining (13) and (14), we obtain

$$\begin{aligned}2^{\mathrm{Sp}(f(\vec{m}_1;\underline{f(\vec{m}_2;\underline{t})}))} &\le (2^{c\sum(d,|\vec{m}|)} + 2^{c\sum(d,|\vec{m}|)} + 2^{|\underline{t}|+1})^{c(\sum(d,|\vec{m}|)+1)}\\ &= (2^{c\sum(d,|\vec{m}|)+1} + 2^{|\underline{t}|+1})^{c(\sum(d,|\vec{m}|)+1)}\\ &\le (2^{c(\sum(d,|\vec{m}|)+1)} + 2^{|\underline{t}|+1})^{c(\sum(d,|\vec{m}|)+1)}\end{aligned} \tag{15}$$

by $1 \le c$. Now (10) follows from (11) and (15).

Above argument is slightly extended to the case for the general form of SNRN. This completes the case for SNRN and, therefore, the proof of the theorem. □

Remark 2.12. As mentioned in Remark 1.5, if a more general form (1) of SNRN is taken instead, the proof of Theorem 2.11 does not work. However, with the use of a stronger rewriting strategy, e.g. head reduction, a similar argument will work.

Corollary 2.13. *If $f \in \mathbf{N}_{normal}$, then the space required to compute $f(\vec{x};)$ is bounded by $p(|\vec{x}|)$ for some polynomial p, i.e., $\mathbf{N}_{normal} \subseteq$* **FPS**.

Proof. Assume that $f \in \mathbf{N}_{normal}$ and $\vec{m} \in \mathbb{N}$. Let us identify f with the corresponding function symbol in $\mathcal{F}$ and $\vec{m}$ with the corresponding numerals in $\mathcal{T}(\mathcal{F})$. Recall that $\mathrm{Sp}(m) = |m|$ for any numeral m. This together with Theorem 2.11 yields a polynomial p such that $\mathrm{Sp}(f(\vec{m};)) \leq p(|\vec{m}|)$. From this, there exists an algorithm that rewrites an arbitrary ground term into its reduct running in polynomial time and hence in polynomial space. And also, any normal form $\underline{f(\vec{m};)}$ of $f(\vec{m};)$ is a numeral and there is a normal form $\underline{f(\vec{m};)}$ such that $\pi(f(\vec{m};)) = \pi(\underline{f(\vec{m};)})$. Therefore, we can construct a non-deterministic Turing program which computes f in polynomial space. Recall here Savitch's Theorem which states **NPSPACE** $\subseteq$ **PSPACE**. This yields **FNPS** $\subseteq$ **FPS**, and hence $f \in$ **FPS**. □

3. PSPACE-computable functions belong to N

In this section we show that every function from **FPS** is a member of $\mathbf{N}_{normal}$. The proof is divided into two steps.

First we simulate a computation of a $O(|x|^k)$-space-bounded Turing machine by an $2^{O(|x|^l)}$-step-bounded register machine working over the unary representation (for some l). This is done by imitating an argument in Handley and Wainer 6, in which every $O(|x|)$-space Turing computation is simulated by a $2^{O(|x|)}(\approx O(x^k)$ for some $k)$-step-bounded register machine on unary notation.

Then, we simulate an action of the exponentially bounded register machine by functions in $\mathbf{N}_{normal}$. The argument is similar to one in Arai and the author [4], in which an exponential-time Turing computation is arithmetized with the use of SNRN. We notice that the situation is, however, easier. This is due to the fact that the class **PSPACE** is closed under composition whereas **EXP** is not.

To simulate computations within **N**, the closedness under a simultaneous SNRN is necessary:

Lemma 3.1. (Cf. 4 Theorem 3.1) *Suppose that $f_1, \dots, f_l$ are defined from $h_1, \dots, h_l \in \mathbf{N}^{k',l}$ and $\prec^k$-functions $\mathbf{f}_1, \mathbf{f}_2$ by a scheme of simultaneous*

SNRN such that for each $i = 1, \ldots, l$ and $w \in \Sigma^k$,

$$\begin{cases} f_i(\vec{0}, \vec{x}; \vec{a}) = h_i(\vec{x}; \vec{a}), \\ f_i(C_w(\vec{y};), \vec{x}; \vec{a}) = f_i(\vec{v}_1, \vec{x}; f_1(\vec{v}_2, \vec{x}; \vec{a}), \ldots, f_l(\vec{v}_2, \vec{x}; \vec{a})). \end{cases}$$

Then, for any $g_1, \ldots, g_l \in \mathbf{N}^{k',0}$ and for each $i = 1, \ldots, l$,

$$f_i(\vec{y}, \vec{x}; g_1(\vec{x};), \ldots, g_l(\vec{x};)) \in \mathbf{N}^{k+k',0}.$$

Proof. Although the usual simultaneous recursion is easily reduced to a single recursion by a pairing and unpairing functions, it is not a trivial task in the safe representation. Bellantoni[5] defines a pairing function in the Bellantoni-Cook class $\mathcal{B}$. This pairing function is not definable in $\mathbf{N}$, since the i-concatenation C_i is available only on the normal argument. A familiar pairing function $\langle x, y \rangle = \frac{1}{2}(x+y)(x+y+1) + x$ is definable in $\mathbf{N}_{normal}$, while it is not clear for the author whether unpairing functions for $\langle \cdot, \cdot \rangle$ can be defined in $\mathbf{N}_{normal}$. Therefore, we instead employ the following paring function.

For the function $\times$ in Example 1.6(3), $\times(x, y; a) = y \cdot 2^{|x|} + a = y \underbrace{0 \ldots 0}_{|x|-|a|} a$ if $|a| \leq |x|$. Hence we define a pairing function $\pi \in \mathbf{N}^{3,0}$ by

$$\pi(x, y_0, y_1;) = \times(x, y_0; y_1),$$

which works as a pairing function if $|y_1| \leq |x|$. For the function $\dot{-}$ in Example 1.6(1), the corresponding unpairing function π_0 is defined by

$$\pi_0(x, y;) = \dot{-}(x; y),$$

which works as $\pi_0(x, \pi(x, y_0, y_1;);) = y_0$. We mention that the deletion function D is needed only to define π_0. And another one π_1 is defined by

$$\pi_1(x, y;) = y \dot{-} \pi_0(x, y;) \cdot 2^{|x|}.$$

which works as $\pi_1(x, \pi(x, y_0, y_1;);) = \pi(x, y_0, y_1;) \dot{-} y_0 \cdot 2^{|x|} = y_1$.

Let $p(\vec{x}, \vec{y})$ be a polynomial such that

$$\max\{|f_i(\vec{y}, \vec{x}; \vec{g}(\vec{x};))| : i = 1, \ldots, l\} \leq p(|\vec{x}|, |\vec{y}|).$$

The existence of the polynomial $p(\vec{x}, \vec{y})$ is guaranteed by Lemma 2.10. Then, $\pi(2^{p(|\vec{x}|, |\vec{y}|)}, z_0, z_1;)$ works as a pairing function for $f_i(\vec{y}, \vec{x}; \vec{g}(\vec{x};))$ $(i = 1, \ldots, l)$, and $\pi_i(2^{p(|\vec{x}|, |\vec{y}|)}, z;)$ $(i = 0, 1)$ as the corresponding unpairing functions. Recall Example 1.6(5) here. Since $2^{p(|\vec{x}|, |\vec{y}|)}$ is defined in $\mathbf{N}_{normal}$, we thus conclude $f_i(\vec{y}, \vec{x}; \vec{g}(\vec{x};)) \in \mathbf{N}^{k+k',0}$ for each $i = 1, \ldots, l$. □

Theorem 3.2. (Cf. [4] Theorem 4.1) $\mathbf{FPS} \subseteq \mathbf{N}_{normal}$.

Proof. Assume that $f \in \mathbf{FPS}$ and the arity of f is k. Let $\vec{x} = x_1, \ldots, x_k$ be inputs. Then the number of possible configurations is bounded by $2^{p(|\vec{x}|)}$ for some polynomial $p(\vec{x})$. Hence $f(\vec{x})$ is computed within a number of steps bounded by $2^{p(|\vec{x}|)}$, since the same configuration does not repeat in any terminating computations. As in the proof of Theorem 3.9 in [6], actions of the Turing machine can be simulated by a register machine working over the unary representation as follows.

- One register contains the number representing the tape-configuration to the left of the reading head, and
- another register contains the number representing the tape-configuration to the right of the reading head.

We consider the following (unlimited) register machines. A register machine max $|\vec{m}|$ has registers $R_0, R_1, \ldots$ which store natural numbers $r_0, r_1, \ldots$. A register machine program is a finite list $\{I_j : j = 1, \ldots, j_M\}$ of instructions.

- Each instruction has one of four basic types: (Zero) $Z(k)$; $r_k = 0$, (Successor) $S(k)$; $r_k = r_k + 1$, (Predecessor) $P(k)$; $r_k = r_k \dot{-} 1$, (Transfer) $T(k, l)$; $r_k = r_l$, (Jump) $J(k, l, j)$; if $r_k = r_l$ then go to instruction I_j else go to next instruction.
- When a computation starts, the inputs $x_1, \ldots, x_k$ are stored in registers $R_1, \ldots, R_k$, respectively, and 0 in all the other registers.
- When the computation halts, the output is the number in register R_0.

Functions like $x \mapsto 2x$, $x \mapsto 2x + 1$ and $x \mapsto \lfloor x/2 \rfloor$ are computed by a register machine within a number of steps bounded by the exponential of a polynomial in $|x|$. Therefore, $f(\vec{x})$ is computed by a register machine M within $2^{q(|\vec{x}|)}$-steps for some polynomial $q(\vec{x})$.

Now we simulate this computation in $\mathbf{N}$. For a constant $l \geq 3 + k$ depending on M, information on M in step $|y|$ of the computation on inputs $\vec{x}$ is represented by some functions $f'_j(y, \vec{x};)$ $(j = 1, \ldots, l)$, which are defined by simultaneous safe recursion on notation of the form

$$\begin{cases} f'_j(0, \vec{x};) = g_j(\vec{x};), \\ f'_j(C_i(y;), \vec{x};) = h_j(; f_1(y, \vec{x};), \ldots, f_l(y, \vec{x};)) \quad (i = 0, 1) \end{cases}$$

for each $j = 1, \ldots, l$. For the sake of completeness, we give an outline of this arithmetization. Let us assume a (finite) encoding $\lceil I \rceil$ for an instruction I.

We can define the following functions in $\mathbf{N}_{normal}$:

$$\begin{aligned} index(y, \vec{x};) &= \text{the index of the instruction performed next,} \\ inst(y, \vec{x};) &= \text{a code } \lceil I \rceil \text{ of the instruction } I \text{ performed next,} \\ r_j(y, \vec{x};) &= \text{the number } r_j \text{ stored in the register } R_j \text{ in step } |y|. \\ &\quad (j = 0, \ldots, l-3) \end{aligned}$$

Namely, $f'_1 = index$, $f'_2 = inst$, and $f'_j = r_{j-3}$ for each $j = 3, \ldots, l$. For the base case of recursion, $index(0, \vec{x};) = 1$, $inst(0, \vec{x};) = \lceil I_1 \rceil$, and

$$r_j(0, \vec{x};) = \begin{cases} 0 & \text{if } j = 0, \\ x_j & \text{if } 1 \leq j \leq k, \\ 0 & \text{otherwise.} \end{cases}$$

The recursion step is of the form

$$\begin{aligned} index(C_i(y;), \vec{x};) &= \delta_1(; index(y, \vec{x};), inst(y, \vec{x};)), \\ inst(C_i(y;), \vec{x};) &= \delta_2(; index(y, \vec{x};), inst(y, \vec{x};)), \\ r_j(C_i(y;), \vec{x};) &= \delta_{3+j}(; inst(y, \vec{x};), r_0(y, \vec{x};), r_1(y, \vec{x};), \ldots, r_l(y, \vec{x};)), \end{aligned}$$

where $\delta_1, \ldots, \delta_l$ are determined according to the program, which depend also on the encoding $\lceil \cdot \rceil$ for the instructions. Obviously, each δ_j is defined only on safe arguments from O, S, P, I, C by a suitable number of application of safe composition. For the same reason, the above recursion step does not depend on $i = 0, 1$ of $C_i(y;)$.

By the output convention, $f_3(t, \vec{x};) = r_0(t, \vec{x};) = f(\vec{x})$ if $|t| \geq 2^{q(|\vec{x}|)}$. Next, via the functions $h_1, \ldots, h_l$, we define functions $f_j(y, \vec{x}; \vec{a})$ ($j = 1, \ldots, l$) by the following equations of simultaneous SNRN:

$$\begin{cases} f_j(0; \vec{a}) = h_j(; \vec{a}), \\ f_j(C_i(y;); \vec{a}) = f_j(y; f_1(y; \vec{a}), \ldots, f_l(y; \vec{a})) \quad (i = 0, 1) \end{cases}$$

Lemma 3.1 ensures that $F_j(y, \vec{x};) := f_j(y; g_1(\vec{x};), \ldots, g_l(\vec{x};)) \in \mathbf{N}_{normal}$ for each $j = 1, \ldots, l$. By the definition, the functions $F_j(y, \vec{x};)$ ($j = 1, \ldots, l$) represent information on M in step $2^{|y|}$. Therefore, $F_3(t, \vec{x};) = r_0(2^t, \vec{x};) = f(\vec{x})$ whenever $t \geq 2^{q(|\vec{x}|)}$. Thus an application of safe composition yields that $f(\vec{x}) = F_3(2^{q(|\vec{x}|)}, \vec{x};) \in \mathbf{N}_{normal}$. □

We mention that the formation of $F_3(2^{q(|\vec{x}|)}, \vec{x};)$ is now easy contrary to the construction in the proof of Theorem 4.1 in [4]. This is due to the formulation of safe composition. A more restrictive composition of the form $f(\vec{x}; \vec{a}) = h(\vec{I}_j^{k,0}(\vec{x};); \vec{\varphi}(\vec{x}; \vec{a}))$ for projection functions $I_{j_1}^{k,0}, \ldots, I_{j_k}^{k,0}$ is necessary in [4].

Conclusions and final comments

In this paper we give a term-rewriting characterization of the polyspace functions. As a consequence, a recursion-theoretic characterization of the polyspace functions is obtained. This suggests a uniform approach to the space complexity in the sense that adding specific safe recursion schemes to a suitable base class capture various space complexity classes (cf. Remark 1.8):

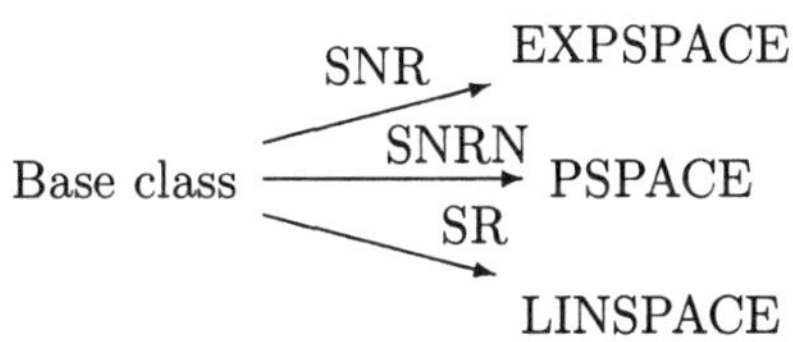

We introduce a two-sorted function class $\mathbf{N}^{k,l}$ such that $\mathbf{N}_{normal} =$ **FPS**. The present work is inspired by Oitavem's works, which also give a term-rewriting and recursion-theoretic characterization of **FPS**. However, it is not clear for the author how to prove the equality of **N** and the Oitavem class. Nor proper inclusion relation between **N** and $\mathcal{N}$ in [4] which captures the exptime functions. For example, it seems quite difficult to simulate safe nested recursion within Oitavem's formulation as well as the converse is. These are future works.

The research in such a framework has been initiated by Cichon and Weiermann,[7] and followed by Beckmann and Weiermann.[8] The underlying idea in [7] and [8] is that term rewriting can be used to prove inclusion relations between complexity classes. In particular, non-trivial closure properties are obtained. An application in this direction will be to prove Savitch's Theorem in the present context (as suggested by the referee).

Clearly, the present result is related to a work by Bonfante, Marion and Moyen.[9] In [9] it is shown that if a program is polynomially quasi-interpretable and its termination is proved via the lexicographic path order LPO, then the program is polyspace-computable. This together with the proof of Lemma 2.2 suggests that the TRS $R_{\mathbf{N}}$ can be polynomially quasi-interpreted (assuming a suitable rewriting strategy).

Acknowledgments

First of all, I would like to acknowledge Georg Moser. He called my attention to Oitavem's works[1,2] and a related work by M. Avanzini.[10] He also pointed out some errors in an earlier draft. I would like to thank the referee for careful reading and invaluable comments. In particular, the introduction of the non-confluent TRS $R_{\mathbf{N}}$ and the use of the innermost strategy are due to him or her. I would also like to thank my thesis advisor Prof. Toshiyasu Arai for his comments on this work.

References

1. I. Oitavem, A term rewriting characterization of the functions computable in polynomial space, *Archive for Mathematical Logic* **41**, 35 (2002).
2. I. Oitavem, New recursive characterizations of the elementary functions and the functions computable in polynomial space, *Revista Mathematica de la Universidad Complutense de Madrid* **10**, 109 (1997).
3. S. Bellantoni and S. Cook, A new recursion-theoretic characterization of the polytime functions, *Computational Complexity* **2**, 97 (1992).
4. T. Arai and N. Eguchi, A new function algebra of EXPTIME functions by safe nested recursion, to appear in *ACM Transactions on Computational Logic* .
5. S. Bellantoni, Predicative recursion and computational complexity, PhD thesis, University of Toronto1992.
6. W. G. Handley and S. S. Wainer, Complexity of primitive recursion, in *Computational Logic, NATO ASI Series F: Computer and Systems Science*, eds. U. Berger and H. Schwichtenberg (Springer, 1999) pp. 273–300.
7. E. A. Cichon and A. Weiermann, Term rewriting theory for the primitive recursive functions, *Annals of Pure and Applied Logic* **83**, 199 (1997).
8. A. Beckmann and A. Weiermann, A term rewriting characterization of the polytime functions and related complexity classes, *Archive for Mathematical Logic* **36**, 11 (1996).
9. G. Bonfante, J.-Y. Marion and J.-Y. Moyen, On lexicographic termination ordering with space bounded certifications, *Perspectives of system informatics, Lecture Notes in Computer Science* **2244**, 482 (2001).
10. M. Avanzini, *Term rewriting characterizations of complexity classes*, seminar report, University of Innsbruck, Institute of Computer Science (July 2007).

$\mathcal{I}$-ultrafilters and summable ideals

J. FLAŠKOVÁ*

Department of Mathematics, University of West Bohemia in Pilsen,
Univerzitní 22, 306 14 Plzeň, Czech Republic
** E-mail: flaskova@kma.zcu.cz*
www.zcu.cz

We study $\mathcal{I}$-ultrafilters and two weaker notions in the setting where $\mathcal{I}$ is a tall summable ideal $\mathcal{I}_g$ on ω. Such ultrafilters are closely related to rapid ultrafilters. We consider various aspects of this relation and show some similarities and differences between the two classes of ultrafilters. Under Martin's axiom for countable posets we construct a rapid ultrafilter which is not an $\mathcal{I}_g$-ultrafilter for arbitrary summable ideal $\mathcal{I}_g$ or an ultrafilter which is an $\mathcal{I}_g$-ultrafilter for every tall summable ideal $\mathcal{I}_g$, but is not a Q-point.

Keywords: $\mathcal{I}$-ultrafilter, summable ideal, Q-point, rapid ultrafilter

1. Introduction

In this paper we study $\mathcal{I}$-ultrafilters and their two modifications — weak $\mathcal{I}$-ultrafilters and $\mathcal{I}$-friendly ultrafilters — for $\mathcal{I}$ a tall summable ideal on ω. In the first section we provide the necessary definitions and present some basic facts about $\mathcal{I}$-ultrafilters and the other notions. We also recall the definition of summable ideals and mention some known results connecting summable ideals and rapid ultrafilters.

The definition of an $\mathcal{I}$-ultrafilter was given by Baumgartner[1]: Let $\mathcal{I}$ be a family of subsets of a set X such that $\mathcal{I}$ contains all singletons and is closed under subsets. Given an ultrafilter $\mathcal{U}$ on ω, we say that $\mathcal{U}$ is an *$\mathcal{I}$-ultrafilter* if for every function $F : \omega \to X$ there exists $A \in \mathcal{U}$ such that $F[A] \in \mathcal{I}$. We modify Baumgartner's definition restricting the class of considered functions to finite-to-one or one-to-one functions: Let $\mathcal{I}$ be a family of subsets of a set X such that $\mathcal{I}$ contains all singletons and is closed under subsets. We say that an ultrafilter $\mathcal{U}$ on ω is

- *a weak $\mathcal{I}$-ultrafilter* if for every finite-to-one function $F : \omega \to X$ there exists $A \in \mathcal{U}$ such that $F[A] \in \mathcal{I}$.

– *an $\mathcal{I}$-friendly ultrafilter* if for every one-to-one function $F : \omega \to X$ there exists $A \in \mathcal{U}$ such that $F[A] \in \mathcal{I}$.

According to the definition every $\mathcal{I}$-ultrafilter is a weak $\mathcal{I}$-ultrafilter, every weak $\mathcal{I}$-ultrafilter is an $\mathcal{I}$-friendly ultrafilter and every $\mathcal{I}$-friendly ultrafilter has nonempty intersection with the ideal $\mathcal{I}$. It is obvious that whenever $\mathcal{I} \subseteq \mathcal{J}$ then every $\mathcal{I}$-ultrafilter is a $\mathcal{J}$-ultrafilter and an analogous result holds for weak $\mathcal{I}$-ultrafilters and $\mathcal{I}$-friendly ultrafilters.

We only consider the situation where $X = \omega$ and $\mathcal{I}$ is an ideal on ω although some of the statements and remarks on $\mathcal{I}$-ultrafilters in the first section are true in other settings as well. Since every principal ultrafilter is obviously an $\mathcal{I}$-ultrafilter for arbitrary $\mathcal{I}$, we restrict all discussions in this paper to nonprincipal ultrafilters. So "ultrafilter" from now on always means "nonprincipal ultrafilter" and we will denote by ω^* the set of all nonprincipal ultrafilters on ω.

An ideal on ω is called *tall* (or *dense*[2]) if for every infinite set $A \subseteq \omega$ there exists infinite $B \subseteq A$ such that $B \in \mathcal{I}$. It is not difficult to show that for ideals which are not tall neither the corresponding $\mathcal{I}$-ultrafilters nor weak $\mathcal{I}$-ultrafilters nor $\mathcal{I}$-friendly ultrafilters exist. However, the existence of $\mathcal{I}$-ultrafilters is consistent with ZFC for every tall ideal $\mathcal{I}$ on ω. This was proved assuming Martin's axiom for σ-centered posets in the author's PhD thesis[3]. Since the existence of $\mathcal{I}$-ultrafilters or weak $\mathcal{I}$-ultrafilters has not been yet established in ZFC for any $\mathcal{I}$, some constructions in this paper make use of instances of Martin's axiom. For $\mathcal{I}$-friendly ultrafilters, however, the situation is slightly different. This is because, for particular ideals, $\mathcal{I}$-friendly ultrafilters have been constructed in ZFC [4,5].

Let us recall the definition of the Rudin-Keisler order $\leq_{RK}$ on ω^*: For $\mathcal{U}$, $\mathcal{V} \in \omega^*$ we write $\mathcal{U} \leq_{RK} \mathcal{V}$ if there is $f \in {}^{\omega}\omega$ such that $(\forall A \in \mathcal{U})$ $f^{-1}[A] \in \mathcal{V}$, which is equivalent to $(\forall B \in \mathcal{V})$ $f[B] \in \mathcal{U}$.

The family of all $\mathcal{I}$-ultrafilters is downwards closed in the Rudin-Keisler order, i.e. whenever $\mathcal{U}$ is an $\mathcal{I}$-ultrafilter and $\mathcal{V} \leq_{RK} \mathcal{U}$ then $\mathcal{V}$ is also an $\mathcal{I}$-ultrafilter. In fact, one can state this observation as an equivalence:

Lemma 1.1. *An ultrafilter $\mathcal{U}$ is an $\mathcal{I}$-ultrafilter if and only if every $\mathcal{V} \in \omega^*$ such that $\mathcal{V} \leq_{RK} \mathcal{U}$ is an $\mathcal{I}$-ultrafilter too.*

For a function $g : \omega \to (0, \infty)$ such that $\sum_{n\in\omega} g(n) = +\infty$ the family

$$\mathcal{I}_g = \{A \subseteq \omega : \sum_{a\in A} g(a) < +\infty\}$$

is an ideal on ω, which we call the *summable ideal determined by function* g. A summable ideal is tall if and only if $\lim_{n\to\infty} g(n) = 0$.

It is important for our purposes that summable ideals are P-ideals, i.e. if $A_n, n \in \omega$ belong to $\mathcal{I}_g$ then there exists $A \in \mathcal{I}_g$ such that $A_n \subseteq^* A$ for every $n \in \omega$ where $A_n \subseteq^* A$ means "A contains all but finitely many elements of A_n".

From now on we will consider only $\mathcal{I}_g$-ultrafilters, weak $\mathcal{I}_g$-ultrafilters and $\mathcal{I}_g$-friendly ultrafilters for a tall summable ideal $\mathcal{I}_g$ if not explicitly stated otherwise. Our aim is to compare such ultrafilters to rapid ultrafilters and Q-points, so let us recall the relevant definitions:

Given f, $g \in {}^\omega\omega$ we write $f \leq^* g$ if the set $\{n \in \omega : f(n) \not\leq g(n)\}$ is finite. A family of functions $\mathcal{S} \subseteq {}^\omega\omega$ is *dominating in* $({}^\omega\omega, \leq^*)$ if for every $f \in {}^\omega\omega$ there exists $g \in \mathcal{S}$ such that $f \leq^* g$.

An ultrafilter $\mathcal{U}$ is called a *rapid ultrafilter* if the collection of the enumeration functions of its sets is a dominating family in $({}^\omega\omega, \leq^*)$, where the enumeration function of a set A is the unique strictly increasing function e_A from ω onto A.

An ultrafilter $\mathcal{U}$ is called a *Q-point* if for every partition $\{Q_n : n \in \omega\}$ of ω into finite sets there is $A \in \mathcal{U}$ such that $|A \cap Q_n| \leq 1$ for every $n \in \omega$.

Clearly, every Q-point is a rapid ultrafilter, but not vice versa[6].

There exist certain connections between weak $\mathcal{I}_g$-ultrafilters and rapid ultrafilters, which we summarize in Theorem 1.1 below. The equivalence (1) $\Leftrightarrow$ (2) is a result of Hrušák[7], who characterized rapid ultrafilters using summable ideals and Katětov-Blass order, stated in terms of weak $\mathcal{I}$-ultrafilters. The equivalence (1) $\Leftrightarrow$ (4) appears in a paper by Vojtáš[8].

Theorem 1.1. *For an ultrafilter $\mathcal{U} \in \omega^*$ the following are equivalent:*

(1) $\mathcal{U}$ is rapid
(2) $\mathcal{U}$ is a weak $\mathcal{I}$-ultrafilter for every tall summable ideal $\mathcal{I}$
(3) $\mathcal{U}$ is an $\mathcal{I}$-friendly ultrafilter for every tall summable ideal $\mathcal{I}$
(4) $\mathcal{U} \cap \mathcal{I} \neq \emptyset$ for every tall summable ideal $\mathcal{I}$

Let us introduce one more notion connected to rapid ultrafilters. We say that an ultrafilter $\mathcal{U}$ is a *hereditarily rapid ultrafilter* if it is a rapid ultrafilter such that for every $\mathcal{V} \leq_{RK} \mathcal{U}$ the ultrafilter $\mathcal{V}$ is again a rapid ultrafilter.

Since every hereditarily rapid ultrafilter is obviously a rapid ultrafilter, the existence of hereditarily rapid ultrafilters is not provable in ZFC. On the other hand, every selective ultrafilter is hereditarily rapid, thus the existence of hereditarily rapid ultrafilters is consistent with ZFC.

As an immediate corollary of Theorem 1.1 and Lemma 1.1 we obtain the following characterization of hereditarily rapid ultrafilters

Theorem 1.2. *For an ultrafilter $\mathcal{U} \in \omega^*$ the following are equivalent:*

(1) $\mathcal{U}$ is hereditarily rapid
(2) $\mathcal{U}$ is an $\mathcal{I}$-ultrafilter for every tall summable ideal $\mathcal{I}$

In Section 2. we construct, assuming Martin's axiom for countable posets, an ultrafilter which is an $\mathcal{I}_g$-ultrafilter for all summable ideals. This provides a new proof of the existence of hereditarily rapid ultrafilters under MA_{ctble} and builds an important prerequisite for the Theorem 4.1.

In Section 3. we will show, assuming Martin's axiom for countable posets, that for an arbitrary summable ideal $\mathcal{I}_g$ there exists a rapid ultrafilter which is not an $\mathcal{I}_g$-ultrafilter. So it is consistent that for every summable ideal $\mathcal{I}_g$ one cannot reverse the obvious implication that every $\mathcal{I}$-ultrafilter is a weak $\mathcal{I}$-ultrafilter. This also implies that consistently there exist rapid ultrafilters which are not hereditarily rapid. In fact, Theorem 3.1 implies even more, namely that under MA_{ctble} there exists a Q-point which is not a hereditarily rapid ultrafilter.

The fourth section of the present paper is concerned with the other possible inclusion between rapid ultrafilters and $\mathcal{I}_g$-ultrafilters. It is not known whether for an arbitrary summable ideal $\mathcal{I}_g$ there exist (at least consistently) $\mathcal{I}_g$-ultrafilters which are not rapid. This has only been proved for $\mathcal{I}_{1/n}$-ultrafilters[3]. Leaving the general problem open we present in section 4. a weaker result that consistently there exists an ultrafilter which is an $\mathcal{I}_g$-ultrafilter for all summable ideals and not a Q-point, i.e. it is consistent that there are hereditarily rapid ultrafilters which are not Q-points.

2. The existence of $\mathcal{I}_g$-ultrafilters

Assuming Martin's axiom for σ-centered posets one can construct $\mathcal{I}$-ultrafilters for an arbitrary tall ideal $\mathcal{I}$. Thus for every tall summable ideal $\mathcal{I}_g$ it is consistent that $\mathcal{I}_g$-ultrafilters (and weak $\mathcal{I}_g$-ultrafilters and $\mathcal{I}_g$-friendly ultrafilters) exist. However, it is sufficient to assume Martin's axiom for countable posets when we consider summable ideals and one can even require more — we will construct an ultrafilter $\mathcal{U} \in \omega^*$ which is an $\mathcal{I}_g$-ultrafilter for all summable ideals $\mathcal{I}_g$.

Theorem 2.1. (MA_{ctble}) *There exists $\mathcal{U} \in \omega^*$ such that $\mathcal{U}$ is an $\mathcal{I}_g$-ultrafilter for every tall summable ideal $\mathcal{I}_g$.*

Proof. Enumerate all pairs $\{\langle f_\alpha, g_\alpha\rangle : \alpha < \mathfrak{c}\}$ where $f_\alpha \in {}^\omega\omega$ and $\mathcal{I}_{g_\alpha}$ is a tall summable ideal. By transfinite induction on $\alpha < \mathfrak{c}$ we will construct filter bases $\mathcal{F}_\alpha$ so that the following conditions are satisfied:

(i) $\mathcal{F}_0$ is the Fréchet filter
(ii) $\mathcal{F}_\alpha \subseteq \mathcal{F}_\beta$ whenever $\alpha \leq \beta$
(iii) $\mathcal{F}_\gamma = \bigcup_{\alpha<\gamma} \mathcal{F}_\alpha$ for γ limit
(iv) $(\forall\alpha)$ $|\mathcal{F}_\alpha| \leq |\alpha| \cdot \omega$
(v) $(\forall\alpha)$ $(\exists A \in \mathcal{F}_{\alpha+1})$ $f_\alpha[A] \in \mathcal{I}_{g_\alpha}$

Conditions (i)–(iii) take care of the beginning the induction and keep it going. Because of (iii) condition (iv) is satisfied at limit stages of the construction and one only has to verify that conditions (iv) and (v) hold at non-limit induction steps.

Induction step: Suppose we already know $\mathcal{F}_\alpha$. If there is $A \in \mathcal{F}_\alpha$ such that $f_\alpha[A] \in \mathcal{I}_{g_\alpha}$ then simply put $\mathcal{F}_{\alpha+1} = \mathcal{F}_\alpha$. If $f_\alpha[A] \notin \mathcal{I}_{g_\alpha}$ for every $A \in \mathcal{F}_\alpha$ (in particular, $f_\alpha[A]$ is infinite in this case) then we will construct a suitable set eventually making use of Martin's axiom.

If there exists $K \in [\omega]^{<\omega}$ such that $A \cap f_\alpha^{-1}[K]$ is infinite for every $A \in \mathcal{F}_\alpha$ then we let $\mathcal{F}_{\alpha+1}$ be the filter base generated by $\mathcal{F}_\alpha$ and $A = f_\alpha^{-1}[K]$. In the following we will assume that no such set exists, i.e.

(♣) for every $K \in [\omega]^{<\omega}$ there is $A_K \in \mathcal{F}_\alpha$ such that $A_K \cap f_\alpha^{-1}[K]$ is finite.

This also means that $A \setminus f_\alpha^{-1}[K]$ is infinite for every $A \in \mathcal{F}_\alpha$.

Consider $P = \{K \in [\omega]^{<\omega} : (\forall u, v \in f_\alpha[K])$ if $u < v$ then $g_\alpha(v) < \frac{1}{2} g_\alpha(u)\}$ and define a partial order $\leq_P$ on P in the following way: $K \leq_P L$ if and only if $K = L$ or $K \supset L$ and $\min K \setminus L > \max L$. For every $A \in \mathcal{F}_\alpha$ and $k \in \omega$ define $D_{A,k} = \{K \in P : |A \cap K| \geq k\}$.

Claim. $D_{A,k} = \{K \in P : |A \cap K| \geq k\}$ *is dense in* P *for every* $A \in \mathcal{F}_\alpha$ *and* $k \in \omega$.
Take arbitrary $L \in P$. Since $\lim_{n\to\infty} g_\alpha(n) = 0$ there exists $n_L > \max f_\alpha[L]$ such that for every $n \geq n_L$ we have $g_\alpha(n) \leq \frac{1}{2} g_\alpha(\max f_\alpha[L])$. According to the assumption (♣) the set $A \setminus f_\alpha^{-1}[0, n_L)$ is infinite and we can inductively pick $m_i \in A \setminus f_\alpha^{-1}[0, n_L)$ for $i = 1, \ldots, k$ so that $m_1 > \max L$, $m_{i+1} > m_i$, $f_\alpha(m_{i+1}) > f_\alpha(m_i)$ and $g_\alpha(f_\alpha(m_{i+1})) < \frac{1}{2} g_\alpha(f_\alpha(m_i))$. Finally, let $K = L \cup \{m_i : i = 1, \ldots, k\}$. It is obvious that $K \leq_P L$ and $K \in D_{A,k}$.

The family $\mathcal{D} = \{D_{A,k} : A \in \mathcal{F}_\alpha, k \in \omega\}$ consists of dense subsets of P and has cardinality less than $\mathfrak{c}$. Thus according to the assumptions there exists a $\mathcal{D}$-generic filter $\mathcal{G}$ on P. Let $C = \bigcup\{K : K \in \mathcal{G}\}$. It remains to verify that C is the set to add to $\mathcal{F}_\alpha$.

(1) $(\forall A \in \mathcal{F}_\alpha)\ C \cap A$ is infinite
For every $k \in \omega$ and every $K \in \mathcal{G} \cap D_{A,k}$ we have $C \supset K$ and $|K \cap A| \geq k$. Thus $C \cap A$ is infinite.

(2) $f_\alpha[C] \in \mathcal{I}_{g_\alpha}$
Enumerate $f_\alpha[C] = \{c_n : n \in \omega\}$. For every $n \in \omega$ there exists $K_n \in \mathcal{G}$ such that $c_n, c_{n+1} \in f_\alpha[K_n]$. The definition of P implies $g_\alpha(c_{n+1}) < \frac{1}{2}\, g_\alpha(c_n)$. Therefore

$$\sum_{n\in\omega} g_\alpha(c_n) \leq \sum_{n\in\omega} \frac{1}{2^n}\, g_\alpha(c_0) = 2\, g_\alpha(c_0)$$

and $f_\alpha[C]$ belongs to the ideal $\mathcal{I}_{g_\alpha}$.

Completing the induction step we let $\mathcal{F}_{\alpha+1}$ be the filter base generated by $\mathcal{F}_\alpha$ and C. By (1) this is a uniform filter base, so all conditions are met.

Finally, let $\mathcal{F} = \bigcup_{\alpha<\mathfrak{c}} \mathcal{F}_\alpha$ and consider an ultrafilter $\mathcal{U} \supseteq \mathcal{F}$. Because of condition (v) in the construction, $\mathcal{U}$ is an $\mathcal{I}_g$-ultrafilter for every tall summable ideal $\mathcal{I}_g$. □

3. Rapid ultrafilters need not be $\mathcal{I}_g$-ultrafilters

It has been proved[3] that, assuming Martin's axiom for countable posets, for every tall ideal $\mathcal{I}$ there exists a Q-point (and hence of course a rapid ultrafilter) which is not an $\mathcal{I}$-ultrafilter. It follows that under $\mathrm{MA}_{\mathrm{ctble}}$ for every summable ideal $\mathcal{I}_g$ there exists a Q-point, which is not an $\mathcal{I}_g$-ultrafilter. In this chapter we strengthen the result and show that assuming Martin's axiom for countable posets there is a Q-point which is not an $\mathcal{I}_g$-ultrafilter for all summable ideals at once. This is a consequence of Theorem 3.1, which is in fact more general. Several lemmas precede the main theorem in this section to make its proof more transparent.

Let us recall that a family $\mathcal{A} = \{A_{\alpha,n} : \alpha \in I, n \in \omega\} \subseteq \mathcal{P}(\omega)$ is called *independent with respect to a filter* (or filter base) $\mathcal{F}$ if $\{A_{\alpha,n} : n \in \omega\}$ is a partition of ω into infinite sets for every $\alpha \in I$ and $(\forall B \in \mathcal{F})\ (\forall M \in [I]^{<\omega})$ $(\forall f : M \to \omega)\ |B \cap \bigcap_{\beta\in M} A_{\beta,f(\beta)}| = \omega$.

Lemma 3.1. (MA_{ctble}) *Assume $\mathcal{F}$ is a filter base on ω with $|\mathcal{F}| < \mathfrak{c}$ and $\mathcal{A} = \{A_{\beta,n} : \beta \leq \alpha, n \in \omega\}$ with $\alpha < \mathfrak{c}$ is an independent family with respect to $\mathcal{F}$.*
Then there exists a partition of ω into infinite sets $\{A_{\alpha+1,n} : n \in \omega\}$ such that $\mathcal{A}' = \mathcal{A} \cup \{A_{\alpha+1,n} : n \in \omega\}$ is independent with respect to $\mathcal{F}$.

Proof. Consider the set of all finite functions $P = ({}^{<\omega}\omega, \leq_P)$ ordered by end extension, i.e. $\sigma_2 \leq_P \sigma_1$ if σ_2 is an end extension of σ_1. For every $B \in \mathcal{F}$, $M \in [\alpha]^{<\omega}$, $f : M \to \omega$ and $m, n \in \omega$ let $D_{B,M,f,m,n} = \{\sigma \in P : (\exists k \geq m)$ $k \in B \cap \bigcap_{\beta \in M} A_{\beta,f(\beta)}$, $k \in \operatorname{dom} \sigma$ and $\sigma(k) = n\}$.

Claim. $D_{B,M,f,m,n}$ *is dense in* $(P, \leq_P)$ *for every* $B \in \mathcal{F}$, $M \in [\alpha]^{<\omega}$, $f : M \to \omega$ *and* $m, n \in \omega$.
Take arbitrary $\rho \in P$. Since $B \cap \bigcap_{\beta \in M} A_{\beta,f(\beta)}$ is infinite there exists $k \in B \cap \bigcap_{\beta \in M} A_{\beta,f(\beta)}$ such that $k > \max \operatorname{dom} \rho$ and $k \geq m$. Put $\sigma = \rho^\frown \langle k, n \rangle$. Obviously, $\sigma \leq_P \rho$ and $\sigma \in D_{B,M,f,m,n}$.

The family $\mathcal{D} = \{D_{B,M,f,m,n} : B \in \mathcal{F}, M \in [\alpha]^{<\omega}, f : M \to \omega, m, n \in \omega\}$ consists of dense subsets of P and has cardinality less than $\mathfrak{c}$. So there exists a $\mathcal{D}$-generic filter $\mathcal{G}$ on P.

Let $g = \bigcup\{\sigma : \sigma \in \mathcal{G}\}$. Observe that there are infinitely many $k \in B \cap \bigcap_{\beta \in M} A_{\beta,f(\beta)}$ such that $g(k) = n$. Thus $\{g^{-1}(n) : n \in \omega\}$ is a partition of ω into infinite sets. Let $A_{\alpha+1,n} = g^{-1}(n)$ for every $n \in \omega$. It follows from the definition of $\{A_{\alpha+1,n} : n \in \omega\}$ that $\mathcal{A}' = \mathcal{A} \cup \{A_{\alpha+1,n} : n \in \omega\}$ is independent with respect to $\mathcal{F}$. □

Lemma 3.2. (MA_{ctble}) *Assume* $\mathcal{F}$ *is a filter base on* ω *with* $|\mathcal{F}| < \mathfrak{c}$, $\mathcal{A} = \{A_{\beta,n} : \beta \leq \alpha, n \in \omega\}$ *with* $\alpha < \mathfrak{c}$ *is an independent family with respect to the filter base* $\mathcal{F}$ *and* $\mathcal{Q} = \{Q_i : i \in \omega\}$ *is a partition of* ω *into finite sets. Then there exists* $C \subseteq \omega$ *such that* $|C \cap Q| \leq 1$ *for every* $Q \in \mathcal{Q}$ *and* $\mathcal{A}$ *is independent with respect to the filter base* $\mathcal{F}'$ *generated by* $\mathcal{F}$ *and* C.

Proof. Consider the set $P = \{K \in [\omega]^{<\omega} : (\forall Q \in \mathcal{Q})\ |K \cap Q| \leq 1\}$ with the partial order defined by $K \leq_P L$ if $K = L$ or $K \supset L$ and $\min(K \setminus L) > \max L$. For every $B \in \mathcal{F}$, $M \in [\alpha]^{<\omega}$, $f : M \to \omega$ and $k \in \omega$ let $D_{B,M,f,k} = \{K \in P : |K \cap B \cap \bigcap_{\beta \in M} A_{\beta,f(\beta)}| \geq k\}$.

Claim: $D_{B,M,f,k}$ *is dense in* $(P, \leq_P)$ *for every* $B \in \mathcal{F}$, $M \in [\alpha]^{<\omega}$, $f : M \to \omega$ *and* $k \in \omega$.
Whenever we take $L \in P$ there is a finite set $S \subseteq \omega$ such that $L \subseteq \bigcup_{i \in S} Q_i$. Since $B \cap \bigcap_{\beta \in M} A_{\beta,f(\beta)}$ is infinite there exists $L' \subseteq B \cap \bigcap_{\beta \in M} A_{\beta,f(\beta)}$ such that $\min L' > \max \bigcup_{i \in S} Q_i$, $|L'| \geq k$ and $|L' \cap Q| \leq 1$ for every $Q \in \mathcal{Q}$. Put $K = L \cup L'$. It is easy to see that $K \leq_P L$ and $K \in D_{B,M,f,k}$. □

The family $\mathcal{D} = \{D_{B,M,f,k} : B \in \mathcal{F}, M \in [\alpha]^{<\omega}, f : M \to \omega, k \in \omega\}$ consists of dense subsets of P and has cardinality less than $\mathfrak{c}$. So there exists a $\mathcal{D}$-generic filter $\mathcal{G}$ on P. Let $C = \bigcup\{K : K \in \mathcal{G}\}$. It remains to verify that C has the required properties:

(1) $(\forall Q \in \mathcal{Q}_\alpha)\ |C \cap Q| \leq 1$
If $u, v \in C$ then there is $K \in \mathcal{G}$ such that $u, v \in K$ and due to the definition of P the elements u, v belong to distinct sets from the partition $\mathcal{Q}$.

(2) $(\forall B \in \mathcal{F})\ (\forall M \in [\alpha]^{<\omega})\ (\forall f : M \to \omega)\ |C \cap B \cap \bigcap_{\beta \in M} A_{\beta,f(\beta)}| = \omega$
For every $k \in \omega$ and every $K \in \mathcal{G} \cap D_{B,M,f,k}$ we have $|K \cap B \cap \bigcap_{\beta \in M} A_{\beta,f(\beta)}| \geq k$ and $C \supset K$. Thus $C \cap B \cap \bigcap_{\beta \in M} A_{\beta,f(\beta)}$ is infinite and the family $\mathcal{A}$ is independent with respect to the filter base $\mathcal{F}'$ generated by $\mathcal{F}$ and C. □

Theorem 3.1. (MA_{ctble}) *Assume $\{\mathcal{I}_\alpha : \alpha < \mathfrak{c}\}$ is a family of tall ideals. There is a Q-point which is not an $\mathcal{I}_\alpha$-ultrafilter for any $\alpha < \mathfrak{c}$.*

Proof. Enumerate all partitions of ω into finite sets as $\{\mathcal{Q}_\alpha : \alpha < \mathfrak{c}\}$. By transfinite induction on $\alpha < \mathfrak{c}$ we will construct filter bases $\mathcal{F}_\alpha$ and families $\mathcal{A}_\alpha = \{A_{\beta,n} : \beta \leq \alpha, n \in \omega\}$ such that $\{A_{\alpha,n} : n \in \omega\}$ is a partition of ω into infinite sets for every $\alpha < \mathfrak{c}$ and the following conditions are satisfied:

(i) $\mathcal{F}_0$ is the Fréchet filter, $\mathcal{A}_0$ is partition of ω into infinite sets
(ii) $\mathcal{F}_\alpha \supseteq \mathcal{F}_\beta$, $\mathcal{A}_\alpha \supseteq \mathcal{A}_\beta$ whenever $\alpha \geq \beta$
(iii) $\mathcal{F}_\gamma = \bigcup_{\alpha<\gamma} \mathcal{F}_\alpha$, $\mathcal{A}_\gamma = \bigcup_{\alpha<\gamma} \mathcal{A}_\alpha$ for γ limit
(iv) $(\forall \alpha)\ |\mathcal{F}_\alpha| \leq |\alpha| \cdot \omega$ and $|\mathcal{A}_\alpha| \leq |\alpha| \cdot \omega$
(v) $(\forall \alpha)\ \mathcal{A}_\alpha$ is independent with respect to $\mathcal{F}_\alpha$
(vi) $(\forall \alpha)\ (\exists B \in \mathcal{F}_{\alpha+1})\ (\forall Q \in \mathcal{Q}_\alpha)\ |B \cap Q| \leq 1$

Conditions (i)–(iii) allow us to start the induction and keep it going. Moreover (iii) ensures that (iv)–(vi) are satisfied at limit stages of the construction, so it is necessary to verify conditions (iv)–(vi) only at non-limit steps.

Induction step: Suppose we already know $\mathcal{F}_\alpha$ and $\mathcal{A}_\alpha$. Due to (iv) and (v) we may apply Lemma 3.1 to $\mathcal{F}_\alpha$ and $\mathcal{A}_\alpha$ and put $\mathcal{A}_{\alpha+1} = \mathcal{A}'$. The family $\mathcal{A}_{\alpha+1}$ satisfies condition (iv) and is independent with respect to $\mathcal{F}_\alpha$. Now, we may apply Lemma 3.2 to $\mathcal{F}_\alpha$, $\mathcal{A}_{\alpha+1}$ and $\mathcal{Q}_\alpha$ and put $\mathcal{F}_{\alpha+1} = \mathcal{F}'$. The filter base $\mathcal{F}_{\alpha+1}$ satisfies (iv)–(vi).

Finally, let $\mathcal{F} = \bigcup_{\alpha<\mathfrak{c}} \mathcal{F}_\alpha$. Every ultrafilter which extends $\mathcal{F}$ is a Q-point because of condition (vi). Let us show that $\mathcal{F}$ can be extended to an ultrafilter which is not an I_α-ultrafilter for any $\alpha < \mathfrak{c}$. For every $C \in \mathcal{I}_\alpha$ let $G_\alpha^C = \bigcup\{A_{\alpha,n} : n \notin C\}$. The family $\mathcal{G}_\alpha = \{G_\alpha^C : C \in \mathcal{I}_\alpha\}$ is a filter base for every $\alpha < \mathfrak{c}$ because it is closed under finite intersections.

Claim 1. $\mathcal{G} = \mathcal{F} \cup \bigcup\{\mathcal{G}_\alpha : \alpha < \mathfrak{c}\}$ *has the finite intersection property.*
We have to verify that $G_{\alpha_1}^{C_1} \cap G_{\alpha_2}^{C_2} \cap \cdots \cap G_{\alpha_k}^{C_k} \cap B$ is infinite whenever $B \in \mathcal{F}$, $G_{\alpha_i}^{C_i} \in \mathcal{G}_{\alpha_i}$. We may assume $\alpha_1 < \alpha_2 < \cdots < \alpha_k < \mathfrak{c}$ and $B \in \mathcal{F}_\beta$.

Let $\alpha = \max\{\beta, \alpha_i : i = 1, 2, \ldots, k\}$ and for every $i = 1, 2, \ldots, k$ choose $n_i \notin C_i$.
The set $B \cap \bigcap_{i=1}^{k} A_{\alpha_i, n_i}$ is infinite because the family $\mathcal{A}_\alpha$ is independent with respect to $\mathcal{F}_\alpha \supseteq \mathcal{F}_\beta$. It follows from $G_{\alpha_i}^{C_i} \supseteq A_{\alpha_i, n_i}$ that $B \cap \bigcap_{i=1}^{k} G_{\alpha_i}^{C_i}$ is also infinite.

Let $\mathcal{U}$ be an ultrafilter which extends $\mathcal{G}$. The ultrafilter $\mathcal{U}$ is obviously a Q-point because it extends $\mathcal{F}$. To complete the proof it remains to verify that $\mathcal{U}$ is not an $\mathcal{I}_\alpha$-ultrafilter for any $\alpha < \mathfrak{c}$.

Claim 2. $(\forall \alpha < \mathfrak{c})$ *$\mathcal{U}$ is not an $\mathcal{I}_\alpha$-ultrafilter.*
Given $\alpha < \mathfrak{c}$ define $f_\alpha : \omega \to \omega$ by $f_\alpha[A_{\alpha,n}] = \{n\}$. Assume for the contrary that there exists $H \in \mathcal{U}$ such that $f_\alpha[H] = C \in \mathcal{I}_\alpha$. Then $G_\alpha^C = f_\alpha^{-1}[\omega \setminus C]$ is disjoint from H and $\mathcal{U}$ is not an ultrafilter — a contradiction. □

Corollary 3.1. (MA_{ctble}) *There is a Q-point which is not an $\mathcal{I}_g$-ultrafilter for an arbitrary summable ideal $\mathcal{I}_g$.*

Proof. This is an immediate consequence of Theorem 3.1 because there are only $\mathfrak{c}$-many summable ideals. □

Corollary 3.2. (MA_{ctble}) *For every summable ideal $\mathcal{I}_g$ there exist a rapid ultrafilter which is not an $\mathcal{I}_g$-ultrafilter.* □

4. $\mathcal{I}_g$-ultrafilters need not be Q-points

We will prove that for an arbitrary summable ideal $\mathcal{I}_g$ consistently there exist an $\mathcal{I}_g$-ultrafilter which is not a Q-point using the fact that, unlike Q-points, $\mathcal{I}_g$-ultrafilters are closed under sums and products. So let us first recall the definition of ultrafilter sum and product (for details consult e.g. Comfort, Negrepontis[9]).

Assume $\mathcal{U}$ and $\mathcal{V}_n$, $n \in \omega$, are ultrafilters on ω. The ultrafilter on $\omega \times \omega$, denoted as $\sum_{\mathcal{U}} \langle \mathcal{V}_n : n \in \omega \rangle$, is defined by $A \in \sum_{\mathcal{U}} \langle \mathcal{V}_n : n \in \omega \rangle$ if and only if $\{n : \{m : \langle n, m \rangle \in A\} \in \mathcal{V}_n\} \in \mathcal{U}$. The ultrafilter $\sum_{\mathcal{U}} \langle \mathcal{V}_n : n \in \omega \rangle$ is called the *$\mathcal{U}$-sum of ultrafilters* $\mathcal{V}_n$, $n \in \omega$. If $\mathcal{V}_n = \mathcal{V}$ for every $n \in \omega$ then we write $\sum_{\mathcal{U}} \langle \mathcal{V}_n : n \in \omega \rangle = \mathcal{U} \cdot \mathcal{V}$ and the ultrafilter $\mathcal{U} \cdot \mathcal{V}$ is called the *product of ultrafilters $\mathcal{U}$ and $\mathcal{V}$.*

Notice that the product of two ultrafilters is never a Q-point. To see this consider for example the partition of $\omega \times \omega$ into finite sets $A_k = \{\langle m, n \rangle : \max\{m, n\} = k\}$, $k \in \omega$. On the other hand, the theorem below implies that the product of two $\mathcal{I}$-ultrafilters is always an $\mathcal{I}$-ultrafilter if $\mathcal{I}$ is a P-ideal.

Proposition 4.1. *Let $\mathcal{I}$ be an arbitrary P-ideal on ω. If $\mathcal{U}$ is an $\mathcal{I}$-ultrafilter and $\{n : \mathcal{V}_n$ is $\mathcal{I}$-ultrafilter$\} \in \mathcal{U}$ then $\sum_{\mathcal{U}}\langle \mathcal{V}_n : n \in \omega\rangle$ is an $\mathcal{I}$-ultrafilter.*

Proof. Suppose $\mathcal{U}$ is an $\mathcal{I}$-ultrafilter and for some $B_0 \in \mathcal{U}$ the ultrafilters $\mathcal{V}_n$, $n \in B_0$, are also $\mathcal{I}$-ultrafilters. Let $f : \omega \times \omega \to \omega$ be an arbitrary function. We want to find $A \in \sum_{\mathcal{U}}\langle \mathcal{V}_n : n \in \omega\rangle$ such that $f[A] \in \mathcal{I}$.

For every $n \in B_0$ define $f_n : \omega \to \omega$ by $f_n(m) = f(\langle n, m\rangle)$ for every $m \in \omega$. According to the assumption for every $n \in B_0$ there exists $C_n \in \mathcal{V}_n$ such that $f_n[C_n] \in \mathcal{I}$. Since $\mathcal{I}$ is a P-ideal there is a set $C \in \mathcal{I}$ such that $f_n[C_n] \subseteq^* C$ for every $n \in B_0$.

The set $f_n^{-1}[f_n[C_n]]$ obviously belongs to $\mathcal{V}_n$. Therefore either $f_n^{-1}[f_n[C_n] \cap C]$ or $f_n^{-1}[f_n[C_n] \setminus C]$ belongs to $\mathcal{V}_n$ for every $n \in B_0$. Define $B_1 = \{n \in B_0 : f_n^{-1}[f_n[C_n] \cap C] \in \mathcal{V}_n\}$ and $B_2 = \{n \in B_0 : f_n^{-1}[f_n[C_n] \setminus C] \in \mathcal{V}_n\}$. Since $\mathcal{U}$ is an ultrafilter one of the sets B_1, B_2 belongs to the ultrafilter $\mathcal{U}$.

Case A. $B_1 \in \mathcal{U}$

Put $A = \{\{n\} \times f_n^{-1}[f_n[C_n] \cap C] : n \in B_1\}$. It is easy to see that $A \in \sum_{\mathcal{U}}\langle \mathcal{V}_n : n \in \omega\rangle$ and $f[A] = \bigcup_{n \in B_1} f_n[C_n] \cap C \subseteq C \in \mathcal{I}$.

Case B. $B_2 \in \mathcal{U}$

Since $f_n[C_n] \setminus C$ is finite and $\mathcal{V}_n$ is an ultrafilter, there exists $k_n \in f_n[C_n] \setminus C$ such that $f_n^{-1}\{k_n\} \in \mathcal{V}_n$ for every $n \in B_2$. Define $g : \omega \to \omega$ by $g(n) = k_n$ if $n \in B_2$ and arbitrarily otherwise. Since $\mathcal{U}$ is an $\mathcal{I}$-ultrafilter there exists $B_3 \in \mathcal{U}$ such that $g[B_3] \in \mathcal{I}$. It remains to put $A = \{\{n\} \times f_n^{-1}\{k_n\} : n \in B_2 \cap B_3\}$. It is easy to verify that $A \in \sum_{\mathcal{U}}\langle \mathcal{V}_n : n \in \omega\rangle$ and $f[A] \subseteq g[B_3] \in \mathcal{I}$. □

Theorem 4.1. (MA_{ctble}) *There exists $\mathcal{U} \in \omega^*$ such that $\mathcal{U}$ is an $\mathcal{I}_g$-ultrafilter for every tall summable ideal $\mathcal{I}_g$ and $\mathcal{U}$ is not a Q-point.*

Proof. According to Theorem 2.1 there exists an ultrafilter $\mathcal{U} \in \omega^*$ such that $\mathcal{U}$ is an $\mathcal{I}_g$-ultrafilter for every tall summable ideal $\mathcal{I}_g$. Since every summable ideal $\mathcal{I}_g$ is a P-ideal by Proposition 4.1 the ultrafilter $\mathcal{U} \times \mathcal{U}$ is again an $\mathcal{I}_g$-ultrafilter for every tall summable ideal $\mathcal{I}_g$, but it is not a Q-point because no product of two ultrafilters is a Q-point. □

Unfortunately, ultrafilter products cannot be used for construction of $\mathcal{I}_g$-ultrafilters which are not rapid because rapid ultrafilters are closed under products[6]. Thus it remains an open problem whether for every tall summable ideal $\mathcal{I}_g$ there exists an $\mathcal{I}_g$-ultrafilter which is not rapid.

Acknowledgement

The author wishes to thank Jörg Brendle for his helpful suggestions concerning Theorem 3.1.

References

1. Baumgartner, J., Ultrafilters on ω, *J. Symbolic Logic* **60**, no. 2, 624–639, 1995.
2. Farah, I., Analytic Quotients: Theory of Liftings for Quotients over Analytic Ideals on the Integers, *Mem. Amer. Math. Soc.* **148**, no. 702, 2000.
3. Flašková, J., Ultrafilters and small sets, Ph.D. Thesis, Charles University, Prague, 2006.
4. Gryzlov, A., Some types of points in N^*, in: *Proceedings of the 12th Winter School on Abstract Analysis (Srní, 1984)*, Rend. Circ. Mat. Palermo (2) **1984**, Suppl. No. 6, 137–138.
5. Flašková, J., More than a 0-point, *Comment. Math. Univ. Carolinae* **47**, no. 4, 617–621, 2006.
6. Miller, A. W., There are no Q-points in Laver's model for the Borel conjecture, *Proc. Amer. Math. Soc.* **78**, no. 1, 103–106, 1980.
7. Hrušák, M., *personal communication.*
8. Vojtáš, P., On ω^* and absolutely divergent series, *Topology Proceedings* **19**, 335 - 348, 1994.
9. Comfort, W. W., Negrepontis, S., *The Theory of Ultrafilters*, Springer, Berlin, New York, 1974

Lowness and highness properties for randomness notions

Johanna N.Y. Franklin

Department of Mathematics, National University of Singapore, Singapore, 117543, Singapore
** E-mail: franklin@math.nus.edu.sg*
www.nus.edu.sg

Given two relativizable classes R and P and a real A, we say that A is in Low(R,P) if $\mathsf{R} \subseteq \mathsf{P}^A$ and that A is in High(R,P) if $\mathsf{R}^A \subseteq \mathsf{P}$. In this paper, we survey the current results on highness and lowness for Kurtz, Schnorr, recursive, Martin-Löf, and weak 2-randomness.

1. Introduction

Lowness and highness are notions of relative computational strength. A real is said to be low for a relativizable class if the class generated using this real as an oracle is no different from the class generated using no oracle at all, and a real is said to be high for a relativizable class if using it as an oracle is equivalent to using the strongest possible oracle for some particular function. In this paper, we will discuss lowness and highness notions for algorithmic randomness and provide an overview of the existing results.

The most familiar and perhaps oldest example of these notions is based on the Turing jump: a real A is said to be low if $A' \equiv_T 0'$ and high if $A' \equiv_T 0''$. These terms were first published in a paper by Soare in 1972,[1] but applied only to the Δ_2^0 degrees with these properties. Later, the concept of lowness was developed in other contexts, such as lowness for EX-learning.[2] We formalize a more general notion of lowness below.

Definition 1.1. Let R be a relativizable class. We say that a real A is *low for R* if $\mathsf{R} = \mathsf{R}^A$. The class of reals that is low for R will be denoted here by Low(R).

We note that while lowness can clearly be defined for any relativizable class, it is more difficult to define highness in any sort of generality. The definition of A as a high Turing degree does not involve a class relativized to A, but rather the equality of the jump operator relativized to A and the maximal element of the Δ_2^0 reals, $0'$. However, in other cases, there may not be a natural choice for such a maximal oracle and function.

These definitions were extended to lowness and highness for pairs of related classes instead of simply single classes as below in Refs. 3 and 4, respectively.

Definition 1.2. Suppose that R and P are relativizable classes. Given a real A, we say that A is in Low(R,P) if $\mathsf{R} \subseteq \mathsf{P}^A$ and that A is in High(R,P) if $\mathsf{R}^A \subseteq \mathsf{P}$.

In short, P^A will always be a subclass of P. A real is in Low(R,P) if this subclass still contains R. On the other hand, we say that a real is in High(R,P) if its use as an oracle with respect to R generates a class of reals contained entirely within P. Note that for any notion R, Low(R,R) is precisely the class of reals that are low for R and that for any notions $\mathsf{R} \subseteq \bar{\mathsf{R}} \subseteq \bar{\mathsf{P}} \subseteq \mathsf{P}$, Low($\bar{\mathsf{R}}$,$\bar{\mathsf{P}}$) $\subseteq$ Low(R,P) and High(R,P) $\subseteq$ High($\bar{\mathsf{R}}$,$\bar{\mathsf{P}}$).

Although we have defined Low(R,P) and High(R,P) in full generality above, we will restrict our attention to Low(R,P) for R and P such that $\mathsf{R} \subseteq \mathsf{P}$ and High(R,P) for R and P such that $\mathsf{R} \supset \mathsf{P}$. The reason for this is that if $\mathsf{P} \subset \mathsf{R}$, we will have Low($\mathsf{R}$,$\mathsf{P}$) $= \emptyset$ and that if $\mathsf{R} \subseteq \mathsf{P}$, we will have High($\mathsf{R}$,$\mathsf{P}$) $= 2^\omega$.

We note that Low(P) can often be computed easily from Low(P,R) for some other notion R.

In Sections 3 through 6, we present results concerning lowness for pairs of randomness notions, and in Section 7, we present results concerning highness for pairs of randomness notions. Finally, in Section 8, we summarize and comment on these results and present some currently open questions. In general, we will not give detailed, technical proofs, but rather thorough outlines and references to the appropriate sources for the full proofs.

1.1. *Definitions and notation*

Our notation is standard and generally follows Soare (Ref. 5) and Odifreddi (Ref. 6,7). We refer to the elements of the Cantor space, 2^ω, as reals, and μ will always denote the standard Lebesgue measure on this space. We write $\tau \preceq \sigma$ to indicate that τ is an initial segment of σ, $\tau\sigma$ to denote the concatenation of τ and σ, and $[\sigma]$ to denote the set of reals extending σ. If S is a subset of $2^{<\omega}$, we define $[S]$ similarly. Finally, for any string σ and any measurable set R, we will use $\mu_\sigma(R)$ to represent the fraction of R that extends σ: $2^{|\sigma|}\mu(R \cap [\sigma])$.

For a general reference on randomness, please see Downey and Hirschfeldt (Ref. 8), Nies (Ref. 9), or Downey, Hirschfeldt, Nies, and Terwijn (Ref. 10); however, we will remind the reader of the most important definitions below. There are three primary approaches one can take when formalizing notions of randomness, and each of these has advantages and disadvantages when it comes to particular randomness notions. In this paper, we will only present the approaches that are most convenient for the theorems we are interested in. Any of the references above will provide equivalent characterizations using the other approaches for a given randomness notion.

We begin by recalling the definition of the most commonly studied randomness notion, Martin-Löf randomness. All three characterizations will be used at some point in this paper: that of prefix-free Kolmogorov complexity relative to a universal Turing machine, the unpredictability definition, and the test definition. If $M : 2^{<\omega} \to 2^{<\omega}$ is a prefix-free Turing machine, the complexity of a string σ with respect to M is $K_M(\sigma) = \min\{\tau \mid M(\tau) = \sigma\}$. When a universal Turing machine is used, the subscript will be omitted. For more background on Kolmogorov complexity, please see Ref. 11. Also recall that a Martin-Löf test is a uniformly r.e. sequence $\langle V_i \rangle_{i\in\omega}$ of Σ^0_1 classes such that $\mu(V_i) \leq 2^{-i}$ for each i.

A martingale is a function $d : 2^{<\omega} \to \mathbb{R}^{\geq 0}$ that satisfies the equation

$$d(\sigma) = \frac{d(\sigma 0) + d(\sigma 1)}{2}$$

for all $\sigma \in 2^{<\omega}$. We may think of a martingale d as a function that tells us, for each string σ, how much capital we would have if we started with $d(\langle\rangle)$ and bet on the bits of σ in order using an associated betting strategy. An r.e. (recursive) martingale is simply a martingale whose values are uniformly r.e. (recursive) reals. We say that a martingale d *succeeds* on a real A if $\limsup_n d(A\restriction n) = \infty$; i.e., if there is no bound on the amount of capital d can have after betting on some initial segment of A.

Definition 1.3.[12,13] A real A is Martin-Löf random if the following three equivalent conditions hold. We will denote the class of such reals by ML.

(1) $(\exists c)(\forall n)[K(A\restriction n) \geq n - c]$.
(2) No r.e. martingale d succeeds on A.
(3) For every Martin-Löf test $\langle V_i \rangle_{i\in\omega}$, $A \notin \cap_i V_i$.

Note that there is a universal Martin-Löf test; i.e., a Martin-Löf test $\langle U_i \rangle_{i\in\omega}$ such that for all Martin-Löf tests $\langle V_i \rangle_{i\in\omega}$, $\cap_i V_i \subseteq \cap_i U_i$.[12] There is also a universal r.e. martingale.

To define recursive randomness, we will only use the martingale notion.

Definition 1.4.[14] A real is recursively random if no recursive martingale succeeds on it. We will denote this class of reals by Rec.

When we discuss Schnorr randomness together with Martin-Löf randomness, the most natural characterization is in terms of tests, but when we discuss it together with recursive randomness, the most natural characterization is the martingale characterization. We present both below.

Definition 1.5.[14] The following two statements are equivalent and characterize Schnorr randomness for a real A. We will denote the class of Schnorr random reals by Schnorr.

(1) If $\langle V_i \rangle_{i \in \omega}$ is a Martin-Löf test such that $\mu(V_i) = 2^{-i}$ for each i, then $A \notin \cap_i V_i$.
(2) There is no recursive martingale d such that $d(A \upharpoonright n) \geq h(n)$ infinitely often for some unbounded, nondecreasing, recursive function h.

We will refer to an unbounded, nondecreasing, recursive function as an order function from this point onwards.

The three notions presented above are the most commonly considered randomness notions we will discuss in this paper. However, we will also consider two other notions: weak 2-randomness, which is stronger than all of these, and Kurtz randomness, which is weaker. Both were introduced in Ref. 15. Kurtz randomness was originally called weak 1-randomness, and it should be noted that weak 2-randomness is sometimes called Kurtz 2-randomness, as in Refs. 8 and 10 or strong randomness, as in Ref. 4.

Kurtz's original definition of weak 2-randomness is listed first in the definition below.

Definition 1.6.[15–17] The following two statements are equivalent to weak 2-randomness for a real A. We will denote the class of weakly 2-random reals by W2R.

(1) For every Σ^0_2 set U of measure 1, $A \in U$.
(2) $A \notin \cap_i U_i$ for any recursive sequence of r.e. open sets $\langle U_i \rangle_{i \in \omega}$ such that $U_i \supseteq U_{i+1}$ for all i and $\lim_i \mu(U_i) = 0$. Such a sequence is called a generalized Martin-Löf test.

Finally, we present Kurtz randomness.

Definition 1.7.[15] A real A is Kurtz random if for every r.e. open set U of measure 1, $A \in U$. Such a set U is called a (positive) Kurtz test. We will denote this class of reals by Kurtz.

Finally, we describe the relations between these classes of reals.

Theorem 1.1.[14,15,17] *The following chain of inclusions holds, and none of them are reversible.*

$$\mathsf{W2R} \subset \mathsf{ML} \subset \mathsf{Rec} \subset \mathsf{Schnorr} \subset \mathsf{Kurtz}$$

We conclude this section with the Kraft-Chaitin Theorem, which will be used frequently throughout the paper.

Theorem 1.2 (Kraft-Chaitin Theorem[18]). *Let $\langle d_i, \sigma_i \rangle_{i \in \omega}$ be a recursive sequence with $d_i \in \omega$ and $\sigma_i \in 2^{<\omega}$ for all i such that $\sum_i \frac{1}{2^{d_i}} \leq 1$. (Such a sequence is called a Kraft-Chaitin set, and each element of the sequence is called a Kraft-Chaitin axiom.) Then there are strings τ_i and a prefix-free machine M such that $dom(M) = \{\tau_i \mid i \in \omega\}$ and for all i and j in ω,*

(1) if $i \neq j$, then $\tau_i \neq \tau_j$,
(2) $|\tau_i| = d_i$,
(3) and $M(\tau_i) = \sigma_i$.

The Kraft-Chaitin Theorem allows us to construct a prefix-free Turing machine by specifying only the lengths of the strings in the domain rather than the strings themselves. We will therefore occasionally identify $\langle \tau, \sigma \rangle$ with $\langle d, \sigma \rangle$, where $d = |\tau|$.

2. Lowness for Martin-Löf randomness

The class Low(ML) is a particularly interesting one. It is unusual in that its only known characterizations involve other randomness notions (and, as we shall see, every other class we discuss with this property is identical to Low(ML)). We present some of these equivalent notions here. We begin by defining two other ways in which a real can be said to be "far from Martin-Löf random."

Definition 2.1. A real A is *K-trivial* if there is some c such that for all n, $K(A\restriction n) \leq K(0^n) + c$, and A is *low for K* if there is some c such that for all $\sigma \in 2^{<\omega}$, $K(\sigma) \leq K^A(\sigma) + c$.

Informally, A is K-trivial if every initial segment of A can be described in a way that is no more complicated than a description of a string of 0s of the same length (up to a constant), and A is low for K if using A as an oracle does not materially alter the complexity of any string σ. Since Martin-Löf randomness can be defined in terms of prefix-free complexity as in Definition 1.3, every real that is low for K is also low for Martin-Löf randomness and vice versa.

Here, we show that the K-trivial reals are precisely those that are low for K, so as a corollary, we can see that the K-trivial reals are precisely those that are low for Martin-Löf randomness. Theorem 2.1 was proven by Nies and Hirschfeldt and first appeared in Ref. 19.

Prop 2.1. Every real that is low for K (and thus every real that is low for Martin-Löf randomness) is K-trivial.

Proof. Suppose that c_0 witnesses A's lowness for K. Given a universal prefix-free machine U, we can find another universal prefix-free machine M such that $M^X(\sigma) = X\restriction|U(\sigma)|$ for every X and σ whenever either of these terms is defined. Then there is a constant c_1 such that for all reals X, $K^X(X\restriction n) \leq K(0^n) + c_1$ for every n, and we can see that

$$K(A\restriction n) \leq K^A(A\restriction n) + c_0 \leq K(0^n) + c_0 + c_1$$

for every n. Therefore, A is K-trivial. □

Theorem 2.1.[19] *Every K-trivial real is low for K.*

Proof. This proof utilizes Nies's "golden run" method. Due to the complicated nature of this construction, we will provide very few technical details and simply give a general outline of the proof and a description of the main ideas involved.

Let A be K-trivial, and let b witness this fact. We begin by noting that A is Δ^0_2: A is a path on the tree $T = \{\sigma \mid (\forall \rho \subseteq \sigma)[K(\rho) \leq K(0^{|\rho|})+b]\}$, and all paths on this tree are isolated.[18] Therefore, we can choose a recursive approximation $\langle A_s\rangle_{s\in\omega}$ of A.

Two Kraft-Chaitin sets will appear in this proof. The first, L, will allow us to make use of the fact that A is K-trivial. The second, W, will witness A's lowness for K.

To construct the set W, we use a tree of runs of procedures. At each branching node, we will try to ensure that W is a witness to A's lowness for K. However, problems may arise if our approximation to A changes. The automatic inclination would be to add more elements to L to compensate for the change; however, we would risk increasing the measure of each set to the extent that neither would be a Kraft-Chaitin set. Therefore, we must develop another system to handle these changes. We will say that each axiom $\langle r, n \rangle$ that enters L must either reach an acceptable level of the tree or be garbage. The measure of each of these categories will be bounded above by 2^{-1}, which will guarantee that the measure requirement, at least, of the definition of a Kraft-Chaitin set is met for L.

We may assume that a machine M_d and its index d are given, and we will build L to be a Kraft-Chaitin set for M_d. For the rest of the proof, we will let $c = b + d$ and $k = 2^{c+1}$. If $\langle r, n \rangle$ enters L, we will ensure that $K(A \upharpoonright n) \leq K(0^n) + b \leq r + d$.

The first technical definition we need is that of a j-set for $1 \leq j \leq k$. We say that a finite subset E of ω is a j-set at a stage t if for all $n \in E$, some axiom $\langle r_n, n \rangle$ entered L at a stage $u < t$ and there are j different strings of the form $A_v \upharpoonright n$ at stages $u \leq v \leq t$ such that $K_v(A_v \upharpoonright n) \leq r_n + c$. An r.e. set $E = \cup_t E_t$ is a j-set if for every t, E_t is a j-set at stage t. We note here that if E is a k-set, we can bound its measure as follows. If we put a description of 0^n into L, any matching description that might also go in must be within c of n, so for any k-set E, $\mu(E) \leq 2^{-1}$.

We now describe our tree of runs of procedures. At each stage, our tree will have $2k - 2$ levels. The root procedure will be P_k, which will call several different procedures of type Q_{k-1}. Each of these will call the procedure P_{k-1}, and each of those will call several different procedures of type Q_{k-1}, and so on until we reach the procedures of type Q_1. Each of these procedures has an associated goal, which is some given amount of measure. The runs of each of these procedures will enumerate a set, and the enumeration will only stop if it reaches its goal or if the run is cancelled by runs of procedures that appear above it on the tree. The procedure P_k has 1 as its goal, and the set we will try to enumerate is a Kraft-Chaitin set of measure 1 that we will call C_k. If C_k never achieves a measure of 1, we will see that there must be a run of some procedure P_i that never returns, even though all of its subprocedures will either return or be cancelled. This "golden run" will enable us to define our set W.

In order to reach C_k, a number must get through j-sets C_j and D_j for all $1 \leq j < k$ first in the order $C_1, D_1, \ldots, D_{k-1}, C_k$. A procedure P_i will move a number n from D_{i-1} to C_i when the approximation $A \upharpoonright n$ changes. This has the effect of adding a new string of the form $A_u \upharpoonright n$ to an $(i-1)$-set and creating an i-set, guaranteeing us that C_i will be an i-set for any i. The procedure Q_1 will enumerate C_1, and the procedures of type Q_i for larger i will move numbers from C_i to D_i.

A procedure of type Q_i will be indexed by a 4-tuple $\langle i, \sigma, \tau, w \rangle$, where σ is a description, τ is the object being described, and w is a particular A-use. The procedure P_i will call procedures $Q_{i,\sigma,\tau,w}$ such that $U^A(\sigma) = \tau$, where U is a particular universal prefix-free Turing machine.

The goal of a procedure P_i is the amount of measure it wants to move from D_{i-1} to C_i, and the goal of a procedure of type Q_i is the amount of measure it wants to move from C_i to D_i. If a procedure of type Q_i returns a set D, P_i waits for a change in $A \restriction w$ for the appropriate w. If one occurs, it will add D to C_i. If $A \restriction w$ changes before the appropriate Q_i returns, that change will be enough to convert Q_i's current set D to an i-set, so P_i can add D to C_i immediately and cancel the run of Q_i. The additions of these weights are accompanied by the additions of approximations to initial segments of A, so we will be able to keep track of the behavior of U relativized to A.

It is the weights of the cancelled runs and the weights of the procedures that return but for which $A \restriction w$ never changes that make up the "garbage" component of the set L. To control the size of this garbage component, we associate a garbage quota with each run of a procedure. For a procedure P_i, this is the amount it is allowed to leave in $D_{i-1} - C_i$; for a procedure of type Q_i, this is the amount it is allowed to leave in $C_i - D_i$. We will choose the goal parameter of each run to be small enough that the garbage quota of the run immediately above it is not threatened.

The verification proceeds as follows. We can see by induction that each C_i is an i-set. The set L must be a Kraft-Chaitin set based on our careful balancing of the garbage quotas and goals. Furthermore, there is a "golden run" of some procedure P_i which is not cancelled, does not return itself, and for which every run of a procedure of the form Q_i that it starts is either cancelled or returns. This can be seen by assuming that every run of every procedure is either returned or cancelled and obtaining a contradiction.

Finally, we show that A is low for K. We choose a golden run of a procedure P_i and enumerate a Kraft-Chaitin set W as follows. If a run of the type Q_j returns, we add an axiom to W of the form $\langle |\sigma| + g + 1, \tau \rangle$, where the components of this axiom are determined by the elements of the 4-tuple indexing the Q_j run and the goal and garbage quota of P_i. Since we have carefully assigned the garbage quotas and goals for each procedure to be sufficiently small, the total measure of W will be no more than 1, and W will be a Kraft-Chaitin set. Let M_e be the machine for W obtained by the Kraft-Chaitin Theorem. Then we can show that for all τ, $K(\tau) \leq K^A(\tau) + g + e + 1$. If $K_s^A(\sigma) = \tau$, where σ is the shortest description of τ and s is the least stage at which this computation is stable, $Q_{i-1,\sigma,\tau,w}$ must be called. Since the run of P_i is not cancelled, it must return, and at that stage, $\langle|\sigma| + g + 1, \tau\rangle$ will enter W. This gives us $K(\tau) \leq |\sigma| + g + 1$, and since $K^A(\tau) \leq |\sigma| + e$, we have

$$K(\tau) \leq K^A(\tau) + g + e + 1$$

and we are done. □

Corollary 2.1. *A real is K-trivial if and only if it is low for Martin-Löf randomness.*

3. Lowness for recursive randomness

In this section, we present the characterizations of Low(ML,Rec) and Low(Rec) originally obtained by Nies in Ref. 19. In this case, the latter can be obtained easily from the former. The proofs in this section make heavy use of martingales, since this is the simplest way to characterize recursive randomness. In the first proof, we also use one of the characterizations of Low(ML) given in the previous section.

Theorem 3.1.[19] *A real A is in Low(ML,Rec) if and only if A is low for K and thus in Low(ML) (and K-trivial).*

Proof. As previously noted, if A is low for K, then $A \in$ Low(ML). Since Low(ML) $\subseteq$ Low(ML,Rec), we need only show that if A is in Low(ML,Rec), then A is low for K. We will build a martingale functional L such that if L^A only succeeds on reals that are not Martin-Löf random, then A must be low for K. Since such a real A must be in Low(ML,Rec), we will have the desired inclusion.

The general outline of the proof is as follows. We will begin with an r.e. open set R in the Cantor space with measure strictly less than 1 that contains all reals that are not Martin-Löf random. We will then construct a martingale functional L (a Turing functional such that L^X is a martingale for every oracle X) such that if L^A only succeeds on reals that are not Martin-Löf random, then A must be low for K. To demonstrate that A will be low for K in such a case, we will define a Kraft-Chaitin set W that witnesses this fact by ensuring that for some $c \in \omega$, if $K^{A\upharpoonright n}(\sigma) = m$, then $\langle m + c, \sigma\rangle \in W$. As we build W, we will ensure that it is a Kraft-Chaitin set by balancing the axioms that we allow to enter W against the measure that is enumerated into R at any given stage.

We begin by noting that it is not difficult to find an open set R as described above: simply let $R = \{\sigma \mid (\exists\tau \preceq \sigma)[K(\tau) \leq |\tau| - 1]\}$. In this case, the measure of R is actually less than 2^{-1}, and we can see that any real that is not Martin-Löf random will be contained in R. We now note the following lemma.

Lemma 3.1. *Let d be a martingale that does not succeed on any Martin-Löf random real. Then there are $\sigma \in 2^{<\omega}$ and $c \in \omega$ such that $\sigma \notin R$ and for all $\tau \succeq \sigma$, τ is in R only if $d(\tau) \geq 2^c$.*

This lemma is justified as follows. If it failed, we could define a sequence of strings $\langle\sigma_n\rangle_{n\in\omega}$ starting with the empty string such that σ_{n+1} is a proper extension of σ_n, $d(\sigma_{n+1}) \geq 2^n$, and $\sigma_{n+1} \notin R$. Then d would succeed on $X = \lim \sigma_n$, but, since X cannot be in R, X would have to be Martin-Löf random.

Now we turn to our martingale functional L. We will independently construct a martingale functional L_n for each n that has the value 2^{-n} on any input of length $\leq n$. If we define L to be $\sum_{n\geq 1} L_n$, this will ensure that L will be a rational-valued martingale functional as well.

We enumerate all triples $\delta_n = \langle\tau, a, u\rangle$, where $\tau \in 2^{<\omega}$ and $a, u \in \omega$. If τ and a are witnesses for the above Lemma and $0 < 2^{-u} < 1 - \mu_\tau(R)$, we will define a Kraft-Chaitin set witnessing A's lowness for K. However, since we cannot identify the witnesses τ, a, and u in advance, we must consider all possible triples δ_n. For each δ_n, we will build a sequence of finite trees $\langle T_s\rangle_{s\in\omega}$ in $2^{<\omega}$ such that $T = \lim_s T_s$ exists and such that if δ_n is actually a witness, A is a path in T. As we build T, we build an accompanying Kraft-Chaitin set W such that if $\gamma \in T$ and $K^\gamma(\rho) = r$, then $\langle r + c, \rho\rangle$ will enter W, where $c = n + a + u + 3$.

At each stage s for each triple δ_n, after we determine T_s, we may carry out a *procedure* α to construct our W. Each procedure is a triple of strings $\langle\sigma, \rho, \gamma\rangle$ with certain length conditions. We start α at the least stage s such that $\gamma \in T_s$ and $U_s^\gamma(\sigma) = \rho$, where U is the universal Turing machine we are using to compute prefix-free Kolmogorov complexity. At this point, α will try to make $\langle|\sigma| + c, \rho\rangle$ enter W. However, this must be done with care. For W to be a Kraft-Chaitin set, it must have measure no greater than 1. To ensure this, we will wait to add $\langle|\sigma| + c, \rho\rangle$ to W until a particular clopen set of measure $2^{-(|\sigma|+c)}$ enters R. Once that happens, α will enumerate this tuple into W. We must also take care that the clopen sets associated with different procedures are disjoint so we do not overcount the measure and make $\mu(W)$ greater than 1. To do this, we will build the clopen set for α in pieces that do not overlap with any such set previously chosen and whose measure will be a fixed fraction of $2^{-(|\sigma|+c)}$. When one such set enters R, we assign a new set of a fixed measure less than the remaining part of the allotted $2^{-(|\sigma|+c)}$ and wait until α appears again. If α reappears infinitely often, the clopen set will have the appropriate measure and $\langle|\sigma|+c, \rho\rangle$ will enter W. Otherwise, it will keep away a set of very small measure.

This procedure α will also build the martingale functional L_n by choosing certain strings υ and ensuring that for all reals X extending γ, $L_n^X(\upsilon) \geq a$. It does so by claiming a certain amount of L_n's initial capital, betting it on these chosen strings, and withdrawing it from other strings to ensure that L_n remains a martingale functional. Here, care must be taken to choose strings υ that are not chosen by other procedures and that have not previously been used by α. This will be possible if we choose the υs to be sufficiently long that they only consume a small fraction of the available measure. □

This result gives us a quick proof that the reals that are low for recursive randomness are precisely those that are recursive.

Theorem 3.2.[19] *A real is low for recursive randomness if and only if it is recursive.*

Proof. We begin by observing that Low(Rec) $\subseteq$ Low(ML,Rec) $\subseteq \Delta^0_2$. In Ref. 20, Bedregal and Nies showed that for every hyperimmune real A, there is an A-recursive martingale d^A that succeeds on some recursively random real R, so every real that is low for recursive randomness must be hyperimmune free. Since Martin and Miller proved in Ref. 21 that the only Δ^0_2 reals that are hyperimmune free are the recursive reals, Low(Rec) is simply the set of recursive reals. □

4. Lowness for Schnorr randomness

In this section, we present characterizations of Low(ML,Schnorr), Low(Rec,Schnorr), and Low(Schnorr). The theorems whose proofs are presented in this section were originally proved by Kjos-Hanssen, Nies, and Stephan;[3] however, several of the techniques they used originated in Terwijn and Zambella's work.[22]

First, we mention a notion related to lowness for Schnorr randomness: lowness for Schnorr tests. A real A is said to be low for Schnorr tests if for every Schnorr test relative to A, $\langle V_i^A\rangle_{i\in\omega}$, there is a Schnorr test $\langle U_i\rangle_{i\in\omega}$

such that $\cap_i V_i^A \subseteq \cap_i U_i$. Any real that is low for Schnorr tests will clearly be low for Schnorr, but the converse is not obviously true since there is no universal Schnorr test.[14] Ambos-Spies and Kučera asked in Ref. 23 whether these notions were equivalent.

In Ref. 22, Terwijn and Zambella characterized the reals that were low for Schnorr tests using the notion of traceability, which is a way of placing bounds on the functions that can be obtained when a particular real is used as an oracle. Later, Kjos-Hanssen, Nies, and Stephan answered Ambos-Spies and Kučera's question affirmatively by showing that these are precisely the reals that are low for Schnorr, and they were able to use the techniques appearing in Ref. 22 to describe Low(ML,Schnorr) as well.[3] The characterizations of Low(Schnorr) and Low(Rec,Schnorr) can be obtained easily from the characterization of Low(ML,Schnorr) and a result of Bedregal and Nies.[20]

Definition 4.1. A real A is said to be r.e. traceable if there is an order function p, called a bound, such that for all $f \leq_T A$, there is a recursive function r such that for all n, $|W_{r(n)}| \leq p(n)$ and $f(n) \in W_{r(n)}$. A real is said to be recursively traceable if the same statement holds when $W_{r(n)}$ is replaced by $D_{r(n)}$, where $\langle D_m \rangle_{m \in \omega}$ is a canonical ordering of the finite sets (for instance, we may say that D_m contains precisely the natural numbers which are positions in which a 1 occurs in the binary expansion of m).

We can see that a recursively traceable real may be considered to be uniformly hyperimmune free. This notion provides us with a characterization of lowness for Schnorr tests.

Theorem 4.1.[22] *A real is recursively traceable if and only if it is low for Schnorr tests.*

The following proposition will be used in the proof of Theorem 4.1.

Prop 4.1.[22] Suppose that A is r.e. (recursively) traceable with respect to a bound function p. Then A is r.e. (recursively) traceable with respect to any bound function q.

To see this, we note that we can simply make our bound function grow more slowly by considering $g \restriction f(i) \leq_T A$ rather than $g \leq_T A$ for some suitably fast-growing recursive function f.

Theorem 4.2.[3] *A real A is in Low(ML,Schnorr) if and only if it is r.e. traceable.*

Proof. Suppose that A is in Low(ML,Schnorr). To prove that A is r.e. traceable, we must show that we can find an r.e. trace for an arbitrary $f \leq_T A$.

We begin by coding f into a Schnorr test relative to A as follows: we define $B_{k,f(k)}$ to be $\cup_k\{[\tau 1^k] \mid |\tau| = f(k)\}$ and let $V_i^f = \cup_{k>i} B_{k,f(k)}$. Without loss of generality, we can assume that this is a Schnorr test relative to A. Since A is in Low(ML,Schnorr), we can further assume that $\cap_i V_i^f \subseteq \cap_i U_i$, where $\langle U_i \rangle_{i\in\omega}$ is a universal Martin-Löf test.

This will allow us to calculate a trace r for f using only the fact that $\cap_i V_i^f \subseteq U_3$. To ensure that each $W_{r(k)}$ has a small enough cardinality, we will only let n enter $W_{r(k)}$ if $B_{k,n} - U_3$ is sufficiently small. Since $\mu(U_3) \leq 4^{-1}$, it will normally be the case that $B_{k,n} - U_3$ has relatively large measure. Therefore, there will not be many n such that $B_{k,n} - U_3$ is small enough, so each $W_{r(k)}$ will contain only a few elements. We must balance this requirement on the size of $W_{r(k)}$ against our need to ensure that r is actually a trace for f. We will be able to do this based on our method of encoding the initial segments $f\lceil n$ into our V_i^fs. The formal proof requires several long measure-theoretic calculations that we omit here.

Now suppose that A is r.e. traceable, and let $\langle V_i^A \rangle_{i\in\omega}$ be a Schnorr test relative to A. For each i, we will let $\langle V_{i,s}^A \rangle_{s\in\omega}$ be a sequence of clopen sets that is uniform in s and i such that $V_i^A = \cup_s V_{i,s}^A$ and $\mu(V_{i,s}^A) > 2^{-i}(1-2^{-s})$ for all i and s. We now let $f \leq_T A$ be a function such that $[S_{f(\langle i,s \rangle)}] = V_{i,s}^A$, where the S_is are the canonical finite sets of elements of $2^{<\omega}$. Since A is r.e. traceable via some function r, we can apply Proposition 4.1 and choose $p(n) = n$ as our bound function.

We will now construct a function $\widehat{r}$ such that $W_{\widehat{r}(\langle i,s \rangle)} \subseteq W_{r(\langle i,s \rangle)}$. Our goal is to make each $W_{\widehat{r}(\langle i,s \rangle)}$ so small that we can use them to build a Martin-Löf test whose intersection contains $\cap_i V_i^A$ and still avoid losing any information in $W_{r(\langle i,s \rangle)}$. To do this, we consider only the elements e of $W_{r(\langle i,s \rangle)}$ such that $2^{-i}(1-2^{-s}) < \mu([S_e]) < 2^{-i}$ and $[S_e] \supseteq [S_d]$ for some $d \in W_{\widehat{r}(\langle i,s \rangle)}$. Then we let V_i be the union of all $[S_e]$s obtained by considering pairs of the form $\langle i,s \rangle$. The measures of these V_is will be bounded, and we can recursively find a subsequence of them that forms a Martin-Löf test. The intersection of the elements of this Martin-Löf test will contain $\cap_i V_i^A$, so every real that is Martin-Löf random will be Schnorr random relative to A, and $A \in$ Low(ML,Schnorr). □

Theorem 4.3.[3] *The following are equivalent for a real A.*

(1) A is in Low(Rec,Schnorr).

(2) A is in Low(Schnorr).
(3) A is recursively traceable.

Proof. By Theorem 4.1, a recursively traceable real is low for Schnorr tests, so it must be low for Schnorr and thus in Low(Rec,Schnorr). It will therefore be enough to show that any real in Low(Rec,Schnorr) is recursively traceable.

In Ref. 20, Bedregal and Nies showed that Low(Rec,Schnorr) is actually a subset of the hyperimmune-free reals. It is clear that any real A that is hyperimmune free and r.e. traceable is actually recursively traceable, since, given a function $f \leq_T A$ and an r.e. trace r, we can compute the least stage at which $f(n)$ appears in $W_{r(n)}$ recursively and obtain a recursive trace. Finally, since Low(Rec,Schnorr) $\subseteq$ Low(ML,Schnorr), we can use Theorem 4.2 to see that every element of Low(Rec,Schnorr) is recursively traceable.

□

5. Lowness for Kurtz randomness

In contrast to the previous section, the characterization of Low(Kurtz) is not presented as a corollary of the characterizations of Low(R,Kurtz) for other classes R. Instead, we present the characterization of Low(Kurtz) and then characterize the potentially larger classes Low(ML,Kurtz), Low(Rec,Kurtz), and Low(Schnorr,Kurtz).

In Ref. 24, Downey, Griffiths, and Reid showed that any Schnorr low real is low for Kurtz tests and that every real that is low for Kurtz tests is hyperimmune free using similar techniques to those in Ref. 22. It was hypothesized that Low(Kurtz)=Low(Schnorr); however, Stephan and Yu showed that this was not the case and demonstrated that any real that was hyperimmune free and not DNR was low for Kurtz randomness.[25] Greenberg and Miller have recently shown that the converse is true, completing the characterization.[26] Recall that a function f is *diagonally nonrecursive* (DNR) if for any e such that $\varphi_e(e)$ converges, $f(e) \neq \varphi_e(e)$, and that a real is said to be DNR if it computes such a function. We begin with Stephan and Yu's result.

Theorem 5.1.[25] *If a real is hyperimmune free and not DNR, then it is low for Kurtz randomness.*

Proof. Suppose that A is hyperimmune free and not DNR, and let U^A be a Σ_1^A class of measure 1. We will build a Σ_1^0 class T of measure 1 such that $T \subseteq U^A$.

We begin by observing that U^A must be dense. We can therefore find a function F recursive in A such that, for every n, every string of length n has a proper extension of length $F(n)$ that determines a neighborhood that is entirely contained in U^A and such that the sum of the measures of these neighborhoods exceeds $1 - 2^{-n}$. We can then use the fact that A is hyperimmune free to find a recursive function f such that $f(n+1) > F(f(n))$ for all n. This allows us to compute an A-recursive function g such that $g(n)$ codes a set of finite binary strings of length $f(n+1)$ extending every string of length $f(n)$ whose neighborhoods are contained in U^A and have measure within 2^{-n} of 1. We now make use of the following result of Kjos-Hanssen, Merkle, and Stephan.

Prop 5.1.[27] If A is hyperimmune free and not DNR, then for every $g \leq_T A$, there are recursive functions h and $\widehat{h}$ such that

$$(\forall n)(\exists m \in \{n, n+1, \ldots, \widehat{h}(n)\})[h(m) = g(m)].$$

We can now define T to be

$$\{X \mid (\exists n)(\forall m \in \{n, n+1, \ldots, \widehat{h}(n)\})[X \restriction f(m+1) \in h(m)]\}.$$

Our definition of g lets us see easily that this set is dense, and it is clearly defined in a Σ^0_1 way. Finally, if $X \in T$, there are n and m such that $m \in \{n, n+1, \ldots, \widehat{h}(n)\}$ and $h(m) = g(m)$, so X extends $g(m)$. Therefore, $T \subseteq U^A$. □

We now turn our attention to the converse implication, first proven by Greenberg and Miller. Kurtz showed in his thesis that every hyperimmune degree contains a Kurtz random real, so every real that is low for Kurtz randomness is necessarily hyperimmune free.[15] Therefore, it is sufficient to show that a real that is low for Kurtz randomness cannot be DNR. Recently, Miller has obtained a shorter proof of this result which involves an argument similar to the proof given here of Theorem 7.2.[28]

Greenberg and Miller began by proving the corresponding result for the notion of lowness for Kurtz tests. Lowness for Kurtz tests is analogous to lowness for Schnorr tests, which we mentioned in Section 4.

Theorem 5.2.[26] *If a real is low for Kurtz tests, then it is not DNR.*

Proof. This proof is heavily computational. Greenberg and Miller first showed that any real that is low for Kurtz tests is not DNR. To do this, they introduced the notion of svelte trees.

Definition 5.1. A finite subtree T of $\omega^{<\omega}$ is said to be k-svelte if there is a sequence $\langle S_{k+i}\rangle_{i\in\omega}$ of subsets of T such that the following three conditions hold for a particular fixed increasing sequence of natural numbers $\langle n_m\rangle_{m\in\omega}$.

(1) $S_m \in T \cap \omega^{n_m}$.
(2) $|S_m| \leq 2^{m-(k+1)}$.
(3) Every leaf of T extends a string in $\cup_m S_m$.

Such a tree can be covered by so few basic clopen sets that none of its paths can be made to be DNR.

To show that a real that is low for Kurtz tests is not DNR, we show that if a function f is DNR, there must be a Π_1^f class of measure 0 that is not contained in any Π_1^0 class of measure 0. We define an operator that takes an $f \in \omega^{\leq\omega}$ to a closed set $P^f \subseteq 2^\omega$ in such a way that $\mu(P^f) = 0$ for all f and that P^f and P^g will be sufficiently independent for $f \neq g$. This allows us to define P^T as the union of P^σ for all leaves σ of T if T is finite and as the union of P^f for all branches f in T if T has no dead ends.

We then show that any finite tree T such that $\mu(P^T) \leq 2^{-(k+1)}$ is k-svelte and use this information to work with infinite trees $T \subseteq \omega^{<\omega}$. Such a tree is defined to be full-by-finite if there is a finite tree S with each leaf at some level n_m such that $T = S \cup \{\sigma \in \omega^{<\omega} \mid \sigma \text{ extends a leaf of } S\}$. We now consider a full-by-finite tree T such that $\mu(P^T) \leq 2^{-(k+1)}$ and suppose that its full-by-finiteness is witnessed by S. Such an S can be seen to be k-svelte. We can then show that every clopen $C \subseteq 2^\omega$ has an associated full-by-finite tree T such that $[T] = \{f \in \omega^\omega \mid P^f \subseteq C\}$.

Finally, we build a partial recursive function ψ such that for the kth Π_1^0 class Q_k, if we can see that $\mu(Q_k, s) < 2^{-(k+1)}$ at some stage s, we will compute a finite tree S that is k-svelte and whose upwards closure is the tree of paths f such that $P^f \subseteq Q_{k,s}$. This gives us witnesses $\langle S_{k+1}, \ldots\rangle$ to S's k-svelteness. We then define ψ in such a way that for any string $\sigma \in S_m$, σ is not a DNR string by using the Recursion Theorem. Since every f in this tree of paths extends such a σ, no such f can be DNR itself. □

This leads easily to the following theorem, which finishes the characterization of Low(Kurtz).

Theorem 5.3.[26] *Any real that is DNR is not low for Kurtz randomness.*

Proof. First, we show that if $f \in \omega^\omega$ and a nonempty clopen subclass of P^f is covered by a null Π_1^0 class, then P^f itself is, too. We can then see from the previous theorem that if f is DNR, P^f is not contained in a null

Π^0_1 class, so neither are any of its clopen subsets. We can then build a Kurtz random real contained in P^f by extensions that avoid all null Π^0_1 classes.□

Finally, we present characterizations of Low(ML,Kurtz), Low(Rec,Kurtz), and Low(Schnorr,Kurtz).

Theorem 5.4.[26,29] *A real is in Low(ML,Kurtz) if and only if it is not DNR.*

Proof. Kjos-Hanssen showed in Ref. 29 that if A is a real that does not compute a DNR function, every Π^A_1 null class P is contained in the intersection of a Martin-Löf test. This is enough since lowness for Martin-Löf tests coincides with lowness for Martin-Löf randomness. To do this, he used an A-recursive, nested sequence $\langle C_i \rangle_{i\in\omega}$ of clopen sets such that $\mu(C_i) = 2^{-i}$ and $P = \cap_i C_i$. Since A is not DNR, there are infinitely many n such that $\varphi_n(n)$ is a code for C_n. We can then define a Martin-Löf test $\langle V_i \rangle_{i\in\omega}$ such that V_i is the union of all C_n such that $n > i$ and $\varphi_n(n)$ codes C_n. Clearly, $P \subseteq \cap_i V_i$.

Greenberg and Miller proved the other direction of this theorem by using the machinery from the previous theorems mentioned here. Given a universal Martin-Löf test $\langle U_i \rangle_{i\in\omega}$, we can build a finite k-svelte tree whose upward closure is the tree of paths f such that $P^f \subseteq U_{k+1,s}$ when k is enumerated into $0'$ at stage s and ensured that none of the paths through the tree are DNR. Then we can see that if f is DNR and $P^f \subseteq \cap_k U_k$, it must be the case that $0' \leq_T f$. Since there is a Martin-Löf random set $R \leq_T 0'$, no degree $\geq_T 0'$ can be in Low(ML,Kurtz), and the theorem is proved. □

The classes Low(Rec,Kurtz) and Low(Schnorr,Kurtz) coincide and strictly contain Low(ML,Kurtz).

Theorem 5.5.[26] *The classes Low(Rec,Kurtz) and Low(Schnorr,Kurtz) are both precisely the class of reals that are not high or DNR.*

Proof. Since Low(Schnorr,Kurtz) $\subseteq$ Low(Rec,Kurtz) $\subseteq$ Low(ML,Kurtz), by the previous theorem, all of these classes are disjoint from DNR. Since every high degree contains a recursively random real,[30] we can also see that Low(Schnorr,Kurtz) and Low(Rec,Kurtz) are disjoint from the high degrees, so we simply need to show that every real A that is not high or DNR is in Low(Schnorr,Kurtz). To do this, we need Proposition 5.1 again.

Given a null Π_1^A class, we consider the sequence $\langle C_i \rangle_{i \in \omega}$ of the previous proof. We then use the recursive function h given by Kjos-Hanssen, Merkle, and Stephan's result to "pad" the Martin-Löf test built in the previous proof and convert it to a Schnorr test, thus proving the theorem. □

6. Lowness for weak 2-randomness

We now turn our attention to weak 2-randomness, the strongest of the randomness notions discussed in this paper. This is the only notion for which two classes remain uncharacterized: Low(W2R,Schnorr) and Low(W2R,Kurtz). In this section, we will present characterizations of Low(W2R), Low(W2R,ML), and Low(W2R,Rec).

The first results in this area come from Downey, Nies, Weber, and Yu, who showed that Low(W2R,ML)=Low(ML).[31] This provides a partial characterization of Low(W2R), since Low(W2R) $\subseteq$ Low(W2R,ML). Later, Kjos-Hanssen, Miller, and Solomon showed that Low(ML) $\subseteq$ Low(W2R), which completed the characterization.[32] There is also a proof of this result by Nies using the "golden run" argument described in Section 2 in Ref. 9. Finally, a variation by Nies on the proof of Theorem 3.1 provides a characterization of Low(W2R,Rec).[9] We begin with the results of Downey, Nies, Weber, and Yu.

Theorem 6.1.[31] *The reals that are low for Martin-Löf randomness are precisely the elements of Low(W2R,ML).*

Proof. We can see immediately that Low(ML) $\subseteq$ Low(W2R,ML). We now suppose that a real A is not low for Martin-Löf randomness. This allows us to apply the following result based on a theorem by Stephan appearing in Ref. 19.

Prop 6.1. Suppose that A is not low for Martin-Löf randomness and that β and γ are rationals such that $\beta < \gamma < 1$. For every r.e. open set V and string σ, if $\mu_\sigma(V) \leq \beta$, then there is a string τ such that $K^A(\tau) \leq |\tau| - 1$ and $\mu_{\sigma\tau}(V) \leq \gamma$.

We define a sequence of strings $\langle \sigma_i \rangle_{i \in \omega}$ such that for each i, $K^A(\sigma_i) \leq |\sigma_i| - 1$. By a result of Merkle that appears in Ref. 9, the concatenation of these σ_is will not be Martin-Löf random relative to A.

To build this sequence, we take an enumeration of all potential generalized Martin-Löf tests. If the eth test, $\langle V_i^e \rangle_{i \in \omega}$, is actually a generalized

Martin-Löf test, we will define a number n_e and ensure that the concatenation of the σ_is is not in $V^e_{n_e}$. At each stage e, we will define the string σ_e. We let U_e be the union of those $V^i_{n_i}$ such that n_i has been defined and $i < e$. We can ensure that $\mu_{\sigma_0 \dots \sigma_{e-1}}(U_e) \leq 1 - 2^{-e}$ inductively, and this will allow us to find an appropriate σ_e via Proposition 6.1. □

Corollary 6.1. *Every real that is low for weak 2-randomness is low for Martin-Löf randomness.*

Theorem 6.2.[9,32] *Every real that is low for Martin-Löf randomness is low for weak 2-randomness.*

Proof. Kjos-Hanssen, Miller, and Solomon's proof uses a reducibility originally defined in Ref. 19, $\leq_{LR}$, which is based on lowness for Martin-Löf randomness. Given two reals A and B, we say that $A \leq_{LR} B$ if every real that is Martin-Löf random relative to B is Martin-Löf random relative to A. Clearly, if A is low for Martin-Löf randomness, then $A \leq_{LR} 0^\omega$. We recall that every element of Low(ML) is Δ^0_2 and thus $\leq_T 0'$, which allows us to apply the following theorem from the same paper.

Theorem 6.3. *The following are equivalent for any two reals A and B.*

(1) $A \leq_T B'$ *and* $A \leq_{LR} B$.
(2) *Every* Π^A_1 *class has a* Σ^B_2 *subclass of the same measure.*
(3) *Every* Σ^A_2 *class has a* Σ^B_2 *subclass of the same measure.*

This theorem lets us see that every Σ^A_2 class has a Σ^0_2 subclass of the same measure, so every Π^A_2 class of measure 0 is a subclass of some Π^0_2 class of measure 0. Therefore, every generalized Martin-Löf test relative to A is contained in an unrelativized generalized Martin-Löf test. □

Corollary 6.1 and Theorem 6.2 allow us to see that the reals that are low for weak 2-randomness are precisely those that are low for Martin-Löf randomness.

Finally, we discuss the characterization of Low(W2R,Rec).

Theorem 6.4.[9] *The class Low(W2R,Rec) is precisely Low(ML).*

Proof. As previously mentioned, this is a modification of the proof of Theorem 3.1. Instead of triples of the form $\delta_m = \langle \tau, a, u \rangle$, we use 5-tuples of the form $\delta_m = \langle \tau, a, V, q, \epsilon \rangle$, where $\tau \in 2^{<\omega}$, $a \in \omega$, V is an r.e. open set, and q and ϵ are positive dyadic rationals such that $q + \epsilon \leq 1$. For each witness

δ_m, we define R to be the set $\{\sigma \succeq \tau \mid \mu_\sigma(V) \geq q + \epsilon\}$, and we let u be the least natural number such that $\frac{q}{q+\epsilon} < 1 - 2^{-u}$. Now a procedure for δ_m will be active only at stages s such that $\mu_\tau(R_s) < 1 - 2^{-u}$. Finally, instead of Lemma 3.1, we use the following lemma.

Lemma 6.1. *Suppose that d is a martingale that only succeeds on reals that are not weakly 2-random, and list the quintuples $\delta_m = \langle \sigma, c, V, q, \epsilon \rangle$ where $\sigma \in 2^{<\omega}$, $c \in \omega$, V is an r.e. open set, and q and ϵ are positive dyadic rationals such that $q + \epsilon \leq 1$. Then there is a δ_m such that $\mu_\sigma(V) \leq q$ and for all $\tau \succeq \sigma$, if $d(\tau) \geq 2^c$, then $\mu_\tau(V) \geq q + \epsilon$.*

Given these changes, the argument proceeds as before. □

7. Highness for randomness notions

We now turn our attention to highness for pairs of randomness notions. This concept was originally introduced by Franklin, Stephan, and Yu in Ref. 4 and has been further explored by Barmpalias, Miller, and Nies in Ref. 28 for relativized versions of several of the randomness notions discussed here. Almost all of the results presented in this section appear in Ref. 4; the exception is the characterization of High(ML,W2R), which was determined by Miller[33] and appears in Ref. 28.

High(Schnorr,W2R), High(Schnorr,ML), and High(Schnorr,Rec) all turn out to be the same class of reals, and the proofs are essentially identical.

Theorem 7.1.[4] *The classes High(Schnorr,W2R), High(Schnorr,ML), and High(Schnorr,Rec) are all equal to the class of reals that compute the halting problem: $\{X \in 2^\omega \mid X \geq_T 0'\}$.*

Proof. One direction of the proof of this theorem is reasonably simple. If $A \geq_T 0'$, then any real in Schnorr^A is actually weakly 2-random, so it is certainly recursively random and Martin-Löf random. The other direction of the proof of this theorem is based on the following result, which also appears in Ref. 4.

Prop 7.1. If A and B are reals and $A \not\geq_T 0'$, then there is a real R such that the following three conditions hold.

(1) $B \leq_T R$.
(2) R is not recursively random.
(3) R is Schnorr random relative to A.

Given this proposition, the theorem will follow easily. If $A \not\geq_T 0'$ and we let $B = A$, then the proposition tells us that there is a real that is Schnorr random relative to A that is not recursively random. Therefore, it will not be Martin-Löf random or weakly 2-random, either.

Proof of Proposition 7.1. We will construct R explicitly in intervals. Some of these intervals will be used to code B into R, and others will be used to ensure that no A-recursive martingale is too successful on R.

We fix an enumeration of the halting problem, K. The lengths of the intervals will be defined by a recursive, injective enumeration of all pairs $\langle a_m, b_m \rangle$ such that $a_m > 0$ and either $b_m = 0$ or some element below a_m is enumerated into K at stage b_m. We then divide ω into consecutive intervals I_m of length $3a_m + 1$ and let d be a weighted sum of all total A-recursive martingales. We now define an infinite A-recursive set E of elements using a K-recursive function maximizing c_K and use this set to divide the intervals I_m into two infinite classes. For all intervals I_m of the first type, we define R on I_m so it is not 0 on all of the least $2a_m$ elements of I_m and so d will grow on I_m by a very small factor. Otherwise, we define R to be 0 on all of the least $2a_m$ elements of I_m and let the next $a_m + 1$ elements of R be the first $a_m + 1$ elements of B.

It is clear that $B \leq_T R$: to compute the nth bit of B, we find some $a_m \geq n+1$ such that the first $2a_m$ bits of R on I_m are 0 and then read off the nth bit of B. To see that R is recursively random, we build a recursive martingale that succeeds on it. To do this, we make our martingale bet that R will be 0 on the first $2a_m$ bits of every interval and then bet nothing on the other a_m+1 bits. The losses of this martingale will be bounded, while it will gain a certain fixed amount infinitely often, ensuring that R will not be recursively random. Finally, we have chosen R so that d does not gain too much on enough intervals I_m, so we can see that R is not Schnorr random relative to A. □

As argued above, this is enough to prove our theorem.

We now consider the class High(ML,W2R). The following proof appears in Ref. 28.

Theorem 7.2.[28] *The class High(ML,W2R) is the class of all reals A such that there is no function recursive in $0'$ that is DNR relative to A; i.e., there is no $f \leq_T 0'$ such that $\varphi_e^A(e) \neq f(e)$ whenever $\varphi_e^A(e)$ is defined.*

Proof. We will first show that if A is in High(ML,W2R), then no function recursive in $0'$ is DNR relative to A. This proof involves another class of reals: High(ML,Kurtz[$0'$]), where Kurtz[$0'$] is the class of reals that are Kurtz random relative to $0'$. We begin by observing that since every $\Pi_1^{0,0'}$ class is a Π_2^0 class, High(ML,W2R) is a subset of High(ML,Kurtz[$0'$]). Now we must show that for each element A of High(ML,Kurtz[$0'$]), there is no function f recursive in $0'$ that is DNR relative to A.

We suppose that there is a function f as described above. We will show that $0'$ computes an infinite subset D of an set that is Martin-Löf random with respect to A. This set D will be a member of a null $\Pi_1^{0,0'}$ class and thus not Kurtz random relative to $0'$, but it will also be Martin-Löf random with respect to A, giving us a contradiction.

We begin by taking a nonempty $\Pi_1^{0,A}$ class Q of sets that are Martin-Löf random relative to A and observing that, by a result of Kučera, if we are given a nonempty $\Pi_1^{0,A}$ class $P \subseteq Q$, we can compute a k such that $2^{-k} < \mu(P)$ uniformly from an index for P.[34] Then we use f to compute an increasing sequence $\langle d_n \rangle_{n\in\omega}$ such that the $\Pi_1^{0,A}$ class P_n of all elements of Q containing $\{d_0, \ldots, d_n\}$ is nonempty for every n. At each stage, we will choose d_{n+1} to avoid the set $G_n = \{m \mid \forall Z \in P_n(Z(m) = 0)\}$. We then use the fact that f is DNR relative to A and recursive in $0'$ to show that we can compute the sequence $\langle d_n \rangle_{n\in\omega}$ from $0'$. We can then see that the intersection of the P_ns is a nonempty $\Pi_1^{0,0'}$ class of measure 0, and it is clearly contained in Q. Therefore, any element of their intersection must be Martin-Löf random relative to A but not Kurtz random relative to $0'$.

Now suppose that there is no function recursive in $0'$ that is DNR relative to A, and let $\langle V_n \rangle_{n\in\omega}$ be a sequence of Σ_1^0 classes whose measure converges to 0 as n increases. It will be enough to build a Martin-Löf test relative to A, $\langle U_n^A \rangle_{n\in\omega}$, such that $\cap_n V_n \subseteq \cap_n U_n^A$.

For any Σ_1^0 class V and positive rational ϵ, we will let $(V)_\epsilon$ denote the Σ_1^0 class obtained from V by enumerating it just as long as its measure is less than or equal to ϵ. We also note that $0'$ can compute a function f such that $\mu(V_{f(k)}) \leq 2^{-k}$ for all $k \in \omega$. We now define U_n^A to be $\cup_{k>n}(V_{\varphi_k^A(k)})_{2^{-k}}$ for each n. (We will say that $V_{\varphi_k^A(k)} = \emptyset$ if $\varphi_k^A(k)$ diverges.) Since, by our assumption, $0'$ does not compute a function that is DNR relative to A, there are infinitely many k such that $f(k) = \varphi_k^A(k)$. Therefore, U_n^A covers $\cap_n V_n$ for all n. By definition, $\mu(U_n^A) \leq 2^{-n}$, so $\langle U_n^A \rangle_{n\in\omega}$ is a Martin-Löf test relative to A that covers $\cap_n V_n$, and A must be an element of High(ML,W2R). $\square$

The classes High(Kurtz,R) are identical for all other R discussed in this article.

Theorem 7.3.[4] *The classes High(Kurtz,Schnorr), High(Kurtz,Rec), High(Kurtz,ML), and High(Kurtz,W2R) are all equal to the empty set.*

Proof. Given any A, we can construct a real $R \leq_T A'$ that is Kurtz random relative to A and not Schnorr random. This will, of course, ensure that R is not random with respect to any stronger randomness notion, either.

To do this, we note that we can choose an A'-recursive sequence of numbers $\langle a_i \rangle_{i \in \omega}$ in such a way that R will always be Kurtz random relative to A regardless of the values of R on the intervals $2^{a_m} \leq n < 2^{a_n+1}$. If we let R be constant on each of these intervals, R cannot be Schnorr random, and we are done. □

Finally, we present the partial results obtained on the class High(Rec,ML). We first observe that a natural class of Turing degrees, the PA-complete degrees, is contained in High(Rec,ML). Recall that the PA-complete degrees are those Turing degrees that compute a complete extension of Peano Arithmetic.

Prop 7.2.[4] If a real A is PA-complete, then $A \in$ High(Rec, ML).

Proof. We begin by recalling that $A \in$ High(Schnorr,Rec) if and only if $A \geq_T 0'$. We also note that if A is PA-complete, the universal r.e. martingale is majorized by an A-recursive martingale. Therefore, if a real R is recursively random relative to A, neither the A-recursive martingale above nor the universal r.e. martingale succeeds on R, so R must be Martin-Löf random. □

While the above provides a partial characterization, a reversal is not known. However, the following partial result appears in Ref. 4.

Theorem 7.4. *If $A \in$ High(Rec, ML), then there is a Martin-Löf random real $R \leq_T A$.*

Proof. We prove the contrapositive and begin by supposing that A does not compute any Martin-Löf random real. We will show that A is not in High(Rec,ML) by constructing a real that is not Martin-Löf random that is still recursively random relative to A.

To do this, we will build a function $F \leq_T A'$ such that $A \oplus F$ has high Turing degree relative to A and computes no Martin-Löf random real. By relativizing a result of Nies, Stephan, and Terwijn,[30] we can see that there will be a real $Q \leq_T A \oplus F$ that is recursively random relative to A, but Q will not be Martin-Löf random.

To build F, we use a universal r.e. martingale and an injective enumeration of the indices of partial A-recursive functions $\langle e_i \rangle_{i \in \omega}$ such that for all k, there is an $n \leq k$ such that $\varphi^A_{e_k}(n)$ is undefined. This enumeration will be recursive in A'. The function F is built in intervals in such a way that the universal martingale will take on larger and larger values on each interval, while information about A'' is coded into $(A \oplus F)'$. We can use the fact that A does not compute a Martin-Löf random real to ensure that the algorithm does not terminate and F is total. The latter will guarantee the highness of $A \oplus F$ relative to A, while the former will ensure that no Martin-Löf random real will be computable from $A \oplus F$. □

On the other hand, we can see from the following theorem that this result does not lead to a full characterization.

Theorem 7.5.[4] *There is a real A that is Martin-Löf random that is not in High(Rec,ML).*

Proof. This proof relies on a result in Ref. 35. In this paper, Cholak, Greenberg, and Miller proved that there is an incomplete r.e. set B and an almost everywhere dominating (a.e.d.) function f such that $f \leq_T B$. Recall that an a.e.d. function is one that, for a subclass of the Cantor space of measure 1, dominates every function that is recursive relative to a member of that subclass. We use this function to construct a martingale d recursive in B. To do so, we construct a martingale d^E for every real E by using f as a bound on the convergence speed and use of the associated betting strategy and then integrate over all reals E. We can then build a real R recursively in B on which this martingale is not successful. Since B is r.e. and Turing incomplete, R cannot be Martin-Löf random.

Finally, we show that R is recursively random relative to every member of a class of measure 1. To do so, we assume that this is not the case and note that there must be a martingale d_1 such that d_1^A succeeds on R for some set of oracles A that does not have measure 0. Since f dominates all A-recursive functions, we can show that our original martingale, d, must succeed on R, which gives us a contradiction.

Since R is recursively random relative to every member of a class of measure 1, it must be recursively random relative to a Martin-Löf random real. This Martin-Löf random real will not be in High(Rec,ML), and the theorem is proven. □

Therefore, the question remains open for the reals that compute a Martin-Löf random real but not a complete extension of PA.

8. Conclusions

Summaries of the known characterizations of the classes Low(R,P) and High(R,P) appear in Tables 1 and 2 below. The entry in row R and column P represents the characterization of Low(R,P) in Table 1 and High(R,P) in Table 2. A blank box in the table indicates that the characterization of lowness (or highness) for that particular pair of randomness notions is trivially $\emptyset$ (or 2^ω), while a question mark indicates that no full characterization of that set of reals exists.

In Table 2, $\mathcal{D}$ represents the class of reals A such that there is no function f recursive in K that is DNR relative to A.

It is interesting to note that while these five randomness notions give rise to the chain of classes of Theorem 1.1, the basic lowness notions Low(R) do not. In fact, Low(ML), Low(Rec), and Low(Schnorr) only overlap pairwise on the recursive reals. However, we can create such chains by considering the classes Low(R,P) or High(R,P) for a fixed R or P. Note that in Table 1, the classes form increasing chains going up the columns and from left to right and that in Table 2, the classes form increasing chains going down the columns and from right to left.

It should be noted that similar work has been done on lowness for genericity notions and highness for pairs of genericity notions. While one can argue that some results on this subject have been presented here, since Kurtz randomness can reasonably be considered to be a genericity notion, there are further results in this area. The reader is referred to Refs. 36 and 25 for work on lowness for genericity notions and to Ref. 4 for work on highness for pairs of genericity notions.

We conclude with some open questions.

Question 8.1.[26] Characterize Low(W2R,Schnorr) and Low(W2R,Kurtz).

Question 8.2.[4] (Fully) characterize High(Rec,ML) and High(Rec,W2R).

Question 8.3.[23] Can Low(ML) be characterized in a recursion-theoretic way that is not directly related to a randomness notion?

Table 1. Low(R,P)

	W2R	ML	Rec	Schnorr	Kurtz
W2R	K-trivial	K-trivial	K-trivial	?	?
ML		K-trivial	K-trivial	r.e. traceable	non-DNR
Rec			recursive	recursively traceable	nonhigh, non-DNR
Schnorr				recursively traceable	nonhigh, non-DNR
Kurtz					hyperimmune-free, non-DNR

Table 2. High(R,P)

	Kurtz	Schnorr	Rec	ML	W2R
Kurtz		$\emptyset$	$\emptyset$	$\emptyset$	$\emptyset$
Schnorr			$A \geq_T 0'$	$A \geq_T 0'$	$A \geq_T 0'$
Rec				?	?
ML					$\mathcal{D}$

Acknowledgements

The author thanks Robert Soare for his correspondence and Yu Liang for his helpful comments.

References

1. R. Soare, *Canadian J. of Math.* **24**, 1070 (1972).
2. T. A. Slaman and R. M. Solovay, When oracles do not help, in *Fourth Annual Workshop on Computational Learning Theory*, (Morgan Kaufman., Los Altos, CA, 1991) pp. 379–383.
3. B. Kjos-Hanssen, A. Nies and F. Stephan, *SIAM J. Comput.* **35**, 647 (2005).
4. J. N. Y. Franklin, F. Stephan and L. Yu, *Relativizations of randomness and genericity notions*, Tech. Rep. TRA2/09, School of Computing, National University of Singapore (2009).
5. R. I. Soare, *Recursively Enumerable Sets and Degrees*, Perspectives in Mathematical Logic (Springer-Verlag, 1987).
6. P. Odifreddi, *Classical Recursion Theory*, Studies in Logic and the Foundations of Mathematics, no. 125 (North-Holland, 1989).
7. P. Odifreddi, *Classical Recursion Theory, Volume II*, Studies in Logic and the Foundations of Mathematics, no. 143 (North-Holland, 1999).
8. R. Downey and D. R. Hirschfeldt, *Algorithmic Randomness and Complexity.* To appear.
9. A. Nies, *Computability and randomness* (Clarendon Press, Oxford, 2008). To appear.
10. R. Downey, D. R. Hirschfeldt, A. Nies and S. A. Terwijn, *Bull. Symbolic Logic* **12**, 411 (2006).
11. M. Li and P. Vitányi, *An introduction to Kolmogorov complexity and its applications*, Second edn. (Springer-Verlag, 1997).
12. P. Martin-Löf, *Information and Control* **9**, 602 (1966).
13. C.-P. Schnorr, *Math. Systems Theory* **5**, 246 (1971).
14. C.-P. Schnorr, *Zufälligkeit und Wahrscheinlichkeit*, Lecture Notes in Mathematics, Vol. 218 (Springer-Verlag, Heidelberg, 1971).
15. S. A. Kurtz, Randomness and genericity in the degrees of unsolvability, PhD thesis, University of Illinois, 1981.
16. S. M. Kautz, Degrees of random sets, PhD thesis, Cornell University, 1991.
17. Y. Wang, Randomness and complexity, PhD thesis, University of Heidelberg, 1996.
18. G. J. Chaitin, *J. Assoc. Comput. Mach.* **22**, 329 (1975).

19. A. Nies, *Adv. Math.* **197**, 274 (2005).
20. B. R. C. Bedregal and A. Nies, *Electronic Notes in Theoretical Computer Science* **84**, 73 (2003).
21. W. Miller and D. A. Martin, *Z. Math. Logik Grundlagen Math.* **14**, 159 (1968).
22. S. A. Terwijn and D. Zambella, *J. Symbolic Logic* **66**, 1199 (2001).
23. K. Ambos-Spies and A. Kučera, Randomness in computability theory, in *Computability theory and its applications (Boulder, CO, 1999)*, Contemp. Math. Vol. 257 (Amer. Math. Soc., Providence, RI, 2000) pp. 1–14.
24. R. G. Downey, E. J. Griffiths and S. Reid, *Theoret. Comput. Sci.* **321**, 249 (2004).
25. F. Stephan and L. Yu, Lowness for weakly 1-generic and Kurtz-random, in *Theory and applications of models of computation*, Lecture Notes in Comput. Sci. Vol. 3959 (Springer, Berlin, 2006) pp. 756–764.
26. N. Greenberg and J. S. Miller, *J. Symbolic Logic* (2009), To appear.
27. B. Kjos-Hanssen, W. Merkle and F. Stephan, Kolmogorov complexity and the recursion theorem, in *STACS 2006*, , Lecture Notes in Comput. Sci. Vol. 3884 (Springer, Berlin, 2006) pp. 149–161.
28. G. Barmpalias, J. S. Miller and A. Nies, Randomness notions and partial relativization, Submitted.
29. D. Diamondstone and B. Kjos-Hanssen, Members of random closed sets, In progress.
30. A. Nies, F. Stephan and S. A. Terwijn, *J. Symbolic Logic* **70**, 515 (2005).
31. R. Downey, A. Nies, R. Weber and L. Yu, *J. Symbolic Logic* **71**, 1044 (2006).
32. B. Kjos-Hanssen, J. S. Miller and R. Solomon, Lowness notions, measure, and domination, In progress.
33. J. S. Miller(July, 2008), Private communication.
34. A. Kučera, Measure, Π^0_1-classes and complete extensions of PA, in *Recursion theory week (Oberwolfach, 1984)*, Lecture Notes in Math. Vol. 1141 (Springer, Berlin, 1985) pp. 245–259.
35. P. Cholak, N. Greenberg and J. S. Miller, *J. Symbolic Logic* **71**, 1057 (2006).
36. L. Yu, *Arch. Math. Logic* **45**, 233 (2006).

Countable Borel equivalence relations, Borel reducibility, and orbit equivalence

Greg Hjorth

Department of Mathematics and Statistics
University of Melbourne
Parkville, 3010
Victoria
Australia
E-email: greg.hjorth@gmail.com

This is based on a sequence of four lectures given at Kobe during the 10th Asian Logic Conference.

I will discuss a little of the theory of Borel equivalence relations, from the point of view of Borel reducibility, and then try to indicate why countable Borel equivalence relations forced set theorists in this area to become acquainted with the research in *orbit equivalence.* This is an eclectic and tangly area which falls outside the usual neat divisions into the various clubs and sub-specialties which dominate so much of modern pure mathematics. Among other areas, it makes connections with functional analysis, geometric group theory, Lie groups and *super rigidity* in the sense of Margulis and Zimmer, operator algebras, ergodic theory, and only most recently descriptive set theory.

There are several peculiarities of this exposition. Let me state them explicitly, just so it is clear the extent to which they are intentional.

First of all, I am explicitly writing from the view point of a set theorist. Calculations which are routine given a perfect working knowledge of say the first eighteen chapters of[16] are omitted – thus the various assertions that certain functions are Borel or measurable or that certain spaces can naturally be viewed as standard Borel are all passed over without serious discussion or proof.

On the other hand, many well know results from outside set theory, in some cases quite minor, are often addressed. Although most of the proofs are at the level of sketches, alerting the reader to key ideas without unravelling all the details, I have tried to provide at least some indications of all the main ideas. The only real black box which is being used below is the *spectral theorem* for infinite dimensional unitary operators – in one place, in the discussion of *property (T)*, we will not only use the theorem but take certain steps which can only be justified after considerable reflection on the theorem and how it can it can be used in the context of group representations.

Many numerical calculations are skipped. Instead of fixing a $\epsilon > 0$ and writing $\|f - g\| < \epsilon$, I will just note $f \sim g$, leaving it as understood that with sufficient care we could go back and find the precise ϵ which would make the various inequalities unfold in the way we desire.

It should be understood that these notes travel through a lot of very deep work quickly, with all sorts of simplifications. Very often I am proving a theorem in a special case. Indeed very often I am taking one tiny bite from a much larger body of work and illustrating a key idea. The exposition here *does not* do justice to the very deep theory which has been developed in the area by mathematicians whose theorems I discuss below.

This is part of an introductory sequence of lectures. If you like, it is an "invitation". It is neither a text book nor a survey paper.

In these notes nothing at all is said about about the motivation for studying equivalence relations or why this has become a fashionable area in descriptive set theory. There are already a number of survey and expository papers, such as,[9],[17] and,[26] which do that job.

Finally, I should make it clear that not every theorem is attributed. In some cases this is laziness on my behalf, but more often because the results are folklore or have multiple and hazy origins. Only the theorems which I explicitly attribute to myself are due to me.

Acknowledgments: The author thanks the Australian Research Council for support in the form of an Australian Professorial Fellowship. The author also thanks the referee for an exceptionally acute and penetrating reading, which has led to the correction of a multitude of typos and ambiguities.

1. Lecture I

1.1. *Borel equivalence relations*

Definition 1.1. X is a *Polish space* if it is separable and it accepts a complete metric. The *Borel sets* are those appearing in the σ-algebra generated by the open sets.

Examples (i) $\mathbb{R}$

(ii) $C([0,1])$, the collection of continuous functions from the unit interval to the reals, with distance

$$d(f,g) = \sup_{z\in[0,1]}|f(z) - g(z)|$$

inducing the topology.

(iii) For S a countable set, $\mathcal{P}(S)$, the collection of all subsets is a Polish space, under the natural identification of this space with

$$2^S = \{0,1\}^S = \prod_{\mathbb{N}}\{0,1\},$$

with product topology.

More generally, the countable product of Polish spaces is Polish in the product topology.

(iv) Any separable Banach space – such as ℓ^1, ℓ^2, c_0.

(v) The isometry group of a separable, complete metric space is Polish – take a dense subset of the space, $\{x_n : n \in \mathbb{N}\}$ and let

$$d(\phi,\psi) = \sum_{n\in\mathbb{N}} \min(2^{-n}, d(\phi(x_n), \psi(x_n))).$$

Viewing the Hilbert space ℓ^2 as a metric space in the norm metric, we obtain that the metric isomorphisms which are *linear* on ℓ^2 form a closed subspace of the group of isometries. This group is sometimes denoted as

$$U(\ell^2)$$

and referred to as the *unitary group* on ℓ^2.

Note that in general a closed subspace of a Polish space is again Polish

(vi) $M([0,1],\lambda)$, the measurable subsets of the unit interval, with the identification of sets agreeing almost everywhere, forms a Polish space. Its isometry group is alternately denoted by

$$\mathrm{Aut}([0,1],\lambda)$$

or

$$M_\infty([0,1],\lambda),$$

and consists of the measure preserving bijections.

Definition 1.2. A set X equipped with a σ-algebra is said to be a *standard Borel space* if there is some choice of a Polish topology on X which gives rise to that σ-algebra as the collection of Borel subsets of X. A function between two standard Borel spaces,

$$f : X \to Y,$$

is said to be *Borel* if for any Borel $B \subset Y$ we have $f^{-1}[B]$ Borel in X.

The next two theorems are classical and can be found in [16].

Theorem 1.1. *A Borel subset A of a Polish space X is a standard Borel space – with the σ-algebra of all $B \subset A$ which are Borel in X.*

Theorem 1.2. *Any standard Borel space has cardinality one of:*

$$0, 1, 2, 3, ..., \aleph_0, 2^{\aleph_0}.$$

Moreover, any two standard Borel spaces of the same cardinality are Borel isomorphic.

The first part of this theorem holds for the cardinality of the number of equivalence classes of a Borel equivalence relation, as was shown in [24].

Definition 1.3. An equivalence relation E on a standard Borel space X is said to be *Borel* if it is in the σ-algebra generated by the Borel rectangles, $A \times B$, where A and B are Borel subsets of X.

Theorem 1.3. *(Silver,[24]) If an equivalence relation E is Borel and it has uncountably many equivalence classes, then it has $2^{\aleph_0}$ many classes.*

However the *moreover* part of the classical theorem 1.2 fails. Among the Borel equivalence relations with uncountably many classes there is no simple or natural catalogue, and under any reasonable notion of isomorphism there are many, many non-isomorphic Borel equivalence relations with uncountably many classes.

1.2. *Countable Borel equivalence relations*

Definition 1.4. A Borel equivalence relation is said to be *countable* if all its equivalence classes are countable.

Examples (i) $\mathrm{id}(\mathbb{R}) = \{(x,y) \in \mathbb{R}^2 : x = y\}$. More generally, given a Polish space X we can form $\mathrm{id}(X)$, the identity relation on X.

(ii) E_0, eventual agreement on infinite binary sequences – thus for $x, y \in 2^{\mathbb{N}}$ we set xE_0y if there is an N such that for all $n > N$

$$x(n) = y(n).$$

(iii) Given a countable group Γ, we can let it act on 2^Γ, the set of all functions $f : \Gamma \to \{0,1\}$ in the product topological and measure theoretic structure, by

$$\gamma \cdot f(\sigma) = f(\gamma^{-1}\sigma),$$

and form the induced orbit equivalence relation E_Γ. This is sometimes called the *shift* action.

More generally, if we have any Borel action of a countable group Γ on a Polish space, we can form the induced orbit equivalence relation E_Γ.

Definition 1.5. For E, F Borel equivalence relations on X, Y, we say that *E is Borel reducible to F*,

$$E \leq_B F,$$

if there is a Borel function $\theta : X \to Y$ such that for all $x_1, x_2 \in X$

$$x_1 E x_2 \Leftrightarrow \theta(x_1) F \theta(x_2).$$

By the classical isomorphism theorem for uncountable standard Borel spaces, whenever X, Y are uncountable we have $\mathrm{id}(X) \leq_B \mathrm{id}(Y)$.

Theorem 1.4. *E_0 is not Borel reducible to id($\mathbb{R}$).*

Proof. Let $\mathbb{Z}(2) = \{0,1\}$ with addition mod 2. Let Γ be the group

$$\mathbb{Z}(2)^{<\infty},$$

consisting of all infinite sequences from $\mathbb{Z}(2)$ which are eventually zero. Let Γ act on $2^{\mathbb{N}}$ by

$$\gamma \cdot f(n) = \gamma(n) + f(n) \bmod 2,$$

and note that $E_\Gamma = E_0$.

Suppose $\theta : 2^{\mathbb{N}} \to \mathbb{R}$ is Borel with

$$x_1 E_0 x_2 \Rightarrow \theta(x_1) = \theta(x_2).$$

Let C be a comeager set on which θ is continuous. Take

$$C^* = \bigcap_{\gamma \in \Gamma} \gamma \cdot C$$

to obtain an invariant comeager set on which it is continuous.

For any $x \in C^*$ we will have that $[x]_{E_0}$, the equivalence class of x, is dense in C^*. Since θ is constant on this dense set, θ is constant on C^*. But C^* must be uncountable, so we can find an inequivalent $y \in C^*$ with $\theta(y) = \theta(x)$. □

Definition 1.6. An equivalence relation E is *smooth* if $E \leq id(X)$ for some standard Borel space X.

Thus we have seen in 1.4 that E_0 is not smooth. Generalizations of the argument can be used to obtain corollaries of the proof. For instance:

Theorem 1.5. *Let Γ be a countable group acting by homeomorphisms on an uncountable Polish space X. If every orbit is dense, then E_Γ is not smooth.*

Now it gets hard.

Definition 1.7. An equivalence relation E is *hyperfinite* if there is a increasing sequence of equivalence relations $E_1 \subset E_2 \subset E_3...$ such that each E_n is Borel with finite equivalence classes and

$$E = \bigcup_{n \in \mathbb{N}} E_n.$$

As shown in [15], if E is a countable Borel equivalence relation, then $E \leq_B E_0$ if and only if it is hyperfinite.

Theorem 1.6. *(Sullivan, Weiss, Wright) Let X be a Polish space and E a countable Borel equivalence relation on X. Then there is a comeager set $C \subset X$ on which E is hyperfinite.*

The original reference for this is [25], though a cleaner exposition is presented in [18]. It in effect tells us that beyond the first simplest step, categoricity arguments, working on comeager sets, will be of no use.

All the known proofs to obtain countable Borel equivalence relations beyond E_0 use measure theory.

1.3. *Amenability and hyperfiniteness*

Lemma 1.1. *Let $\mathbb{Z}$ act by Borel automorphisms on a standard Borel space X. Then the resulting orbit equivalence relation is hyperfinite.*

Proof. For conceptual simplicity I will assume that X is Polish and the action is continuous.

Now if we are really, really lucky, the resulting orbits, $[x] = \{T^\ell(x) : \ell \in \mathbb{Z}\}$, will all be dense and so will each of the *forward orbits*,

$$\{T^\ell(x) : \ell \geq 0\},$$

and the *backwards orbits*,

$$\{T^\ell(x) : \ell \leq 0\}.$$

Now choose a sequence of decreasing non-empty open sets, $U_0 \supset U_1 \supset U_2 ...$ such that

$$\bigcap_{n \in \mathbb{N}} U_n = \emptyset.$$

We then set $xE_nT^\ell(x)$ if either

(i) $\ell = 0$; or

(ii) $\ell > 0$ and $x, T(x), T^2(x), ...T^\ell(x)$ are all *outside* U_n; or

(ii) $\ell < 0$ and $x, T^{-1}(x), T^{-2}(x), ...T^\ell(x)$ are all *outside* U_n.

A more complicated case is when the orbits are dense, but not necessarily in both directions. Then we choose for each x some basic open W_x so that $[x]$ has a last or first moment when it meets that open set. With some care we can do this so that $xEy \Rightarrow W_x = W_y$ and if we let $s(x)$ be that special last or first point, then

$$x \mapsto s(x)$$

is not only $\mathbb{Z}$-invariant but Borel.[a] We then let xE_ky if $x = y$ or for some $i, j \in \{-k, -k+1, ..., 0, 1, ...k\}$ we have $T^j(x) = s(x)$, and

$$T^i(y) = s(x) = s(y).$$

In general there is no guarantee, of course, that the orbits will be dense. The argument in this more typical case involves decomposing X into Borel subsets on which all orbits have the same closure and working on each of

[a]Here is another way of looking at the argument: We obtain an invariant Borel set on which $E_{\mathbb{Z}}$ is smooth; the remainder of the proof consists in verifying that if a countable Borel equivalence relation is smooth, then it is hyperfinite.

these components separately. Suffice to say there are technicalities, but they are not deep. □

Definition 1.8. $\mathbb{F}_2 = \langle a, b\rangle$ is the free group on generators a and b. An element of $\mathbb{F}_2$ consists of *reduced* words in the letters $\{a, a^{-1}, b, b^{-1}\}$ – where *reduced* means there should be no adjacent appearances of a and a^{-1} or b and b^{-1}. We multiply elements of $\mathbb{F}_2$ by concatenating and then reducing.

For instance let $\sigma = ab^2a^{-1}$ and $\tau = ab^3aba^{-1}$. (The usual notational shortcut: I write ab^2a^{-1} instead of $abba^{-1}$.) We multiply by

$$\sigma\tau = ab^5aba^{-1}.$$

Technically speaking, the identity of $\mathbb{F}_2$ is the empty string. We usually denote this by e – as against lyrically just leaving an empty space and hoping it is recognized for its role as the identity of the group.

Definition 1.9. For Γ a countable group, we let $\ell^1(\Gamma)$ be the space of functions

$$f : \Gamma \to \mathbb{R}$$

with

$$\sum_{\sigma\in\Gamma} |f(\sigma)| < \infty.$$

For $f \in \ell^1(\Gamma)$ we let

$$\|f\| = \sum_{\sigma\in\Gamma} |f(\sigma)|.$$

We let Γ act on $\ell^1(\Gamma)$ by

$$\sigma \cdot f(\tau) = f(\sigma^{-1}\tau).$$

We then say that the group Γ is *amenable* if for any finite $F \subset \Gamma$ and $\epsilon > 0$ there is some $f \in \ell^1(\Gamma)$ with

$$\|f\| = 1$$

and

$$\|f - \sigma \cdot f\| < \epsilon$$

all $\sigma \in F$.

The choice taken here for defining amenability, in terms of *almost invariant vectors* in $\ell^1(\Gamma)$, is technical but convenient to our goals. A discussion of other characterizations and more can be found in [8], but it is worth giving an alternative characterization which is combinatorially more transparent and probably more appealing to a set theorist.

Lemma 1.2. *A countable group* Γ *is amenable if and only if there are "almost invariant finite subsets of* Γ*". That is to say, for all* $\epsilon > 0, F \subset \Gamma$ *finite, there is some finite non-empty* $A \subset \Gamma$ *such that for all* $\gamma \in F$

$$\frac{|A\Delta\gamma A|}{|A|} < \epsilon.$$

Proof. Just in regards to notation, $|A|$ is the size of the set A; γA is the set $\{\gamma\sigma : \sigma \in A\}$; and $A\Delta\gamma A$ is the symmetric difference of the sets – the set of points in one but not the other.

The main issue in the proof is to go from the almost invariant functions to the almost invariant sets. So let us suppose $F \subset \Gamma, \epsilon > 0$ are given, and we have $f \in \ell^1(\Gamma)$ with

$$\|f - \gamma \cdot f\| < \frac{\epsilon}{|F|}\|f\|$$

all $\gamma \in F$. Without loss, $f \geq 0$.

At each $r \in \mathbb{R}$, let f_r be the characteristic function for the set on which f assumes value greater than r – that is to say, $f_r(h) = 1$ if $f(h) \geq r$, and $= 0$ otherwise.

Then at all $h \in \Gamma$

$$f(h) = \int_{\mathbb{R}^{\geq 0}} f_r(h) d\lambda(r).$$

Putting this into the inequality

$$\sum_{\gamma \in F} \|\gamma \cdot f - f\| < \epsilon \|f\|$$

we obtain

$$\sum_{\gamma \in F} \sum_{h \in \Gamma} \int_{\mathbb{R}^{\geq 0}} |\gamma \cdot f_r(h) - f_r(h)| d\lambda(r) < \epsilon \sum_{h \in \Gamma} \int_{\mathbb{R}^{\geq 0}} f_r(h) d\lambda(r).$$

Interchanging the order of summation, a legitimate step given that all the quantities involved are ≥ 0, we obtain

$$\int_{\mathbb{R}^{\geq 0}} \sum_{\gamma \in F} \sum_{h \in \Gamma} |\gamma \cdot f_r(h) - f_r(h)| d\lambda(r) < \epsilon \int_{\mathbb{R}^{\geq 0}} \sum_{h \in \Gamma} |f_r(h)| d\lambda(r).$$

Thus in particular there must be some $r > 0$ with

$$\sum_{\gamma \in F} \sum_{h \in \Gamma} |\gamma \cdot f_r(h) - f_r(h)| < \epsilon \sum_{h \in \Gamma} |f_r(h)|.$$

Take A to be the set of points at which $f(h) \geq r$. □

Lemma 1.3. $\mathbb{F}_2$ *is not amenable.*

Proof. For $u \in \{a, a^{-1}, b, b^{-1}\}$ we let A_u be the reduced words beginning with u. For $f \in \ell^1(\mathbb{F}_2)$ we define f_u by

$$f_u(\sigma) = f(\sigma)$$

if $\sigma \in A_u$, and $f_u(\sigma) = 0$ otherwise.

Thus f is the sum of $f_a, f_{a^{-1}}, f_b, f_{b^{-1}}$, as well as its value on e. Note moreover that if $w \neq u^{-1}$, $w, u \in \{a, a^{-1}, b, b^{-1}\}$, then

$$w \cdot A_u \subset A_w.$$

We will take as our F the set $\{a, a^{-1}, b, b^{-1}\}$ and as our ϵ the value $\frac{1}{4}$. Let $f \in \ell_1(\Gamma)$ with $\|f\| = 1$. We will show there is some $\sigma \in F$ with

$$\|f - \sigma \cdot f\| \geq \frac{1}{4}.$$

First choose $u \in F$ with

$$\|f_u + f_{u^{-1}}\| \leq \frac{1}{2}.$$

Note that this implies

$$\|f_u\| + \|f_{u^{-1}}\| = \|f_u + f_{u^{-1}}\| \leq \frac{1}{2}.$$

Thus either $\|f_u\| \leq 1/4$ or $\|f_{u^{-1}}\| \leq 1/4$. For simplicity assume $\|f_{u^{-1}}\| \leq 1/4$. For $\sigma \notin A_{u^{-1}}$ we have $u \cdot \sigma \in A_u$ and $(u \cdot f)(u \cdot \sigma) = f(\sigma)$. This yields

$$\|(u \cdot f)_u\| \geq \|f - f_{u^{-1}}\| \geq \|f\| - \|f_{u^{-1}}\| \geq \frac{3}{4}.$$

Thus

$$\|u \cdot f - f\| \geq \|(u \cdot f)_u - f_u\| \geq \|(u \cdot f)_u\| - \|f_u\| \geq \frac{1}{4}.$$ □

Lemma 1.4. $\mathbb{Z}$ *is amenable.*

Proof. Let

$$A_n = \{-n, -n+1, ..., 0, 1, ..., n-1, n\}.$$

For $\ell \in Z$ and $A \subset \mathbb{Z}$ we let $\ell \cdot A = \{\ell + k : k \in A\}$. It is then easily seen that

$$\frac{|A_n \Delta \ell \cdot A_n|}{|A_n|} \to 0$$

as $n \to \infty$. Thus if we let

$$f_n = \frac{1}{|A_n|} \chi_{A_n}$$

then each $\|f_n\| = 1$ and

$$\|f_n - \ell \cdot f_n\| \to 0$$

as $n \to \infty$. Thus given any finite $F \subset \mathbb{Z}$ and $\epsilon > 0$ we have

$$\|f_n - \ell \cdot f_n\| < \epsilon$$

all $\ell \in F$ and n sufficiently large. □

Definition 1.10. A measure space (X, μ) is *standard* if X is a standard Borel space and μ is a σ-finite, atomless, Borel measure. It is *finite* if $\mu(X) < \infty$ and it is a *standard Borel probability* space if in addition $\mu(X) = 1$.

Any standard Borel probability space is measurable isomorphic to $([0, 1], \lambda)$, the unit interval equipped with Lebesgue measure.

Theorem 1.7. *Let $\mathbb{F}_2$ act freely and by measure preserving Borel automorphisms on a standard Borel probability space (X, μ). Then the resulting orbit equivalence relation*

$$E_{\mathbb{F}_2} = \{(x, \sigma \cdot x) : \sigma \in \mathbb{F}_2\}$$

is not hyperfinite.

Proof. Suppose instead for a contradiction

$$E_{\mathbb{F}_2} = \bigcup_{i \in \mathbb{N}} E_i$$

where the E_i's are finite Borel equivalence relations with $E_i \subset E_{i+1}$. Then at each $i \in \mathbb{N}, x \in X$ we let

$$f_{i,x}(\sigma) = \frac{1}{|[x]_{E_i}|}$$

if $xE_i\sigma^{-1} \cdot x$, and $= 0$ otherwise. (Here $|[x]_{E_i}|$ denotes the number of points E_i-equivalent to x.) Let

$$f_i(\sigma) = \int f_{i,x}(\sigma) d\mu(x).$$

Since the action is free we obtain each

$$f_{i,x} \in \ell^1(\mathbb{F}_2)$$

with $\|f_{i,x}\| = 1$. Interchanging integration with summation yields

$$\|f_i\| = \sum_{\sigma \in \mathbb{F}_2} \int f_{i,x}(\sigma) d\mu(x)$$

$$= \int \sum_{\sigma \in \mathbb{F}_2} f_{i,x}(\sigma) d\mu(x)$$

$$= \int \|f_{i,x}\| d\mu(x) = 1.$$

Claim: For each $\sigma \in \mathbb{F}_2$

$$\lim_{i\to\infty} \|f_i - \sigma \cdot f_i\| \to 0.$$

Proof of Claim: Let $A_i = \{x : xE_i\sigma \cdot x\}$. Since $\bigcup_{i \in \mathbb{N}} E_i = E_{\mathbb{F}_2}$ we obtain

$$\mu(A_i) \to 1$$

as $i \to \infty$. Moreover if $x \in A_i$ then for any $\tau \in \mathbb{F}_2$

$$\tau^{-1} \cdot xE_i x \Leftrightarrow \tau^{-1} \cdot xE_i\sigma \cdot x$$

$$\Leftrightarrow (\sigma\tau)^{-1} \cdot (\sigma \cdot x)E_i\sigma \cdot x,$$

and thus

$$f_{i,\sigma \cdot x}(\sigma\tau) = f_{i,x}(\tau);$$

equivalently we can state this as saying that for any $\tau \in \mathbb{F}_2$

$$f_{i,\sigma \cdot x}(\tau) = f_{i,x}(\sigma^{-1}\tau).$$

Thus

$$\int_{A_i} \sigma \cdot f_{i,x}(\tau) d\mu = \int_{A_i} f_{i,x}(\sigma^{-1}\tau) d\mu$$

$$= \int_{A_i} f_{i,\sigma\cdot x}(\tau)d\mu$$

$$= \int_{\sigma\cdot A_i} f_{i,x}(\tau)d\mu,$$

since σ acts in a measure preserving manner. Thus

$$\|f_i - \sigma\cdot f_i\| = \sum_{\tau\in\mathbb{F}_2}(|\int f_{i,x}(\tau)d\mu - \int f_{i,x}(\sigma^{-1}\tau)d\mu|)$$

$$\leq \sum_{\tau\in\mathbb{F}_2}(|\int_{\sigma\cdot A_i} f_{i,x}(\tau)d\mu - \int_{A_i} f_{i,x}(\sigma^{-1}\tau)d\mu|) + \sum_{\tau\in\mathbb{F}_2}(\int_{X\setminus\sigma\cdot A_i} f_{i,x}(\tau)d\mu$$

$$+ \int_{X\setminus A_i} f_{i,x}(\sigma^{-1}\tau)d\mu)$$

$$= \int_{X\setminus\sigma\cdot A_i}\sum_{\tau\in\mathbb{F}_2} f_{i,x}(\tau)d\mu + \int_{X\setminus A_i}\sum_{\tau\in\mathbb{F}_2} f_{i,x}(\sigma^{-1}\tau)d\mu$$

$$= \mu(X\setminus\sigma\cdot A_i) + \mu(X\setminus A_i).$$

Since σ acts in a measure preserving manner, this in turn equals

$$2\mu(X\setminus A_i),$$

which goes to 0 as $i\to\infty$. (Claim□)

But now for any finite $F\subset\mathbb{F}_2$ and $\epsilon>0$ we will have at all sufficiently large i

$$\forall\sigma\in F(\|f_i-\sigma\cdot f_i\|<\epsilon),$$

with a contradiction to non-amenability of $\mathbb{F}_2$. □

All we used about $\mathbb{F}_2$ is its non-amenability. Thus the proof shows:

Theorem 1.8. *Let Γ be a non-amenable group acting freely and by measure preserving transformations on a standard Borel probability space (X,μ). Then the resulting orbit equivalence relation is not hyperfinite.*

As a corollary to this theorem we obtain another proof that $\mathbb{Z}$ is amenable.

Theorem 1.9. *(Gao-Jackson) Let Γ be a countable abelian group acting by Borel automorphisms on a standard Borel probability space X. Then the resulting orbit equivalence relation E_Γ is hyperfinite.*

Their long proof of this deep theorem is still yet to be published.

While we do not know whether this theorem can be extended to amenable groups, in the measure theoretic context the situation is settled.

Theorem 1.10. *(Ornstein-Weiss,[222]) If Γ is an amenable group acting by measure preserving transformations on a standard Borel probability space (X, μ), then there is a conull set $M \subset X$ with*

$$E_\Gamma|_M$$

hyperfinite.

2. Lecture II

2.1. *Orbit equivalence*

Since all the known proofs to distinguish countable Borel equivalence relations $\geq_B E_0$ rely on measure theory, one is led to consider a definition which predates any of the work by descriptive set theorists on equivalence relations and arises by considering an equivalence relation up to its measure theoretic properties.

Definition 2.1. An equivalence relation E on a standard measure space (X, μ) is *measure preserving* if for any two measurable $A, B \subset X$, and Borel bijection

$$\phi : A \to B,$$

with $\phi(x)Ex$ all $x \in A$, we have

$$\mu(A) = \mu(B).$$

(This arises for instance if E is induced by the measure preserving action of a countable group – for any such $\phi : A \to B$ we can decompose A into pieces $(A_\gamma)_{\gamma \in \Gamma}$ with $\phi(x) = \gamma \cdot x$ all $x \in A_\gamma$; then ϕ will be measure preserving on each A_γ, and hence on A.)

E is *ergodic* if for any invariant, Borel A we have that A is either null or conull. (Here *invariant* means that if $x \in A$ and yEx, then as well we must have $y \in A$.)

Two ergodic, measure preserving, countable, Borel equivalence relations E_1, E_2 on standard Borel measure spaces $(X_1, \mu_1), (X_2, \mu_2)$ are *orbit equivalent* if there is a measure preserving, Borel bijection

$$\phi : X_1 \to X_2$$

such that for μ_1-a.e. $x \in X_1$ we have exact preservation of its orbit under ϕ – namely,

$$\{\phi(y) : yE_1x\} = \{z : zE_2\phi(x)\}.$$

Traditionally orbit equivalence has focused on the case of ergodic equivalence relation. An ergodic equivalence relation should be thought of as irreducible – we cannot divide the space up into two non-trivial subspaces which are E-unrelated. An analogue of *simple group* or *indecomposable module.*

Building on earlier results in the area, the following was proved in early 2002.

Theorem 2.1. *(Hjorth,[10]) Any countable non-amenable group has at least two free, ergodic, measure preserving actions on standard Borel probability spaces up to orbit equivalence.*

I may have been happy with the result at the time, but it raised more questions than it answered. The proof showed for the class of *property (T)* groups we have continuum many actions up to orbit equivalence. Connes, Schmidt, and Weiss had previously shown that for non-amenable, non-(T) groups we have at least 2. Hence in general we have at least two, but no hint how to obtain larger numbers for groups such as $\mathbb{F}_2$.

We will start working towards a result of Gaboriau and Popa a couple of years later which showed the non-abelian free groups to have continuum many orbit inequivalent actions. First we need some entirely new concepts.

2.2. *The spectral theorem and direct integrals of representations*

The purpose of this section is to provide a very fast introduction to how the spectral theorem for abelian operators can be levered to analyze the representations of a group with a normal abelian subgroup. The key idea is to reduce to the case of irreducible representations.

In general given a finite dimensional unitary representation we can write it as a *direct sum* of irreducible representations. Given an invariant closed subspace, its orthogonal complement will be invariant as well, and we can decompose the space into two invariant orthogonal pieces. Finite dimensionality ensures the process will terminate after a finite number of steps.

The situation with infinite dimensional representations is more subtle. There is no guarantee of termination after finitely many steps, and indeed we can not generally write a unitary representation as a direct sum of irreducible representations. Rather one obtains that the representation can be written as a kind of *direct integral* of irreducible representations.

We will want to apply this kind of analysis to the case of a countable group Γ with a normal abelian subgroup Δ. I will try to describe the process informally, in loose general terms, before going on to set up the formal details.

Irreducible representations of abelian groups are one dimensional, and thus the irreducible representations of Δ correspond to elements of the dual group, $\hat{\Delta}$, consisting of homomorphisms from Δ to the circle group S^1 of complex numbers of norm one. In other words, if $\Delta \curvearrowright \mathcal{H}$ is an irreducible representation, then $\mathcal{H}$ will be one dimensional and there will be a homomorphism $\chi : \Delta \to S^1$ with $\delta \cdot v = \chi(\delta)v$ all $\delta \in \Delta$, $v \in \mathcal{H}$.

We can lever this in the situation that $\Gamma \to U(\mathcal{H})$ is a unitary representation. In very rough terms – and the details are cleaned up in the passages below – we can first of all write the unitary representation so that at the level of Δ it is a direct integral of irreducible Δ representations. After suitably rearranging the space notationally, we thereby obtain

$$\mathcal{H} = \int_{\hat{\Delta}} \mathcal{H}_\chi \nu(\chi),$$

where each $\mathcal{H}_\chi$ is a direct sum of one dimensional representations of Δ corresponding to the character χ and ν is a probability measure on the dual $\hat{\Delta}$. This in place, we can think of any element in $\mathcal{H}$ as being a ν-measurable function which assigns each to element $\chi \in \hat{\Delta}$ a corresponding $w_\chi \in \mathcal{H}_\chi$. In the notation above we are thinking of $w \in \mathcal{H}$ in exactly that way – as a function. The key point here is that the normality of Δ means that we can think of Γ/Δ as acting in a non-singular manner on the space $(\hat{\Delta}, \nu)$.

In the literature, this whole process, of analyzing a group representation in terms of a direct integral of irreducible representations for some normal abelian subgroup, is sometimes referred to as the "Mackey machine." The various mathematical assertions which I make in the text below are standard for the area. They can be found in [20] or [30].

Definition 2.2. Let $\mathcal{H}$ be a separable Hilbert space. $B(\mathcal{H})$ is the collection linear operators

$$\phi : \mathcal{H} \to \mathcal{H}$$

which are bounded in the sense that there exists a single bound $c > 0$ such that for all $v \in \mathcal{H}$

$$\|\phi(v)\| \leq c\|v\|.$$

$U(\mathcal{H}) \subset B(\mathcal{H})$ is the collection of *unitary operators*; in other words, the collection of linear bijections $\phi : \mathcal{H} \to \mathcal{H}$ with

$$\langle \phi(v), \phi(w) \rangle = \langle v, w \rangle$$

for all $v, w \in \mathcal{H}$. For $\phi \in B(\mathcal{H})$ we let ϕ^*, the adjoint, be defined by

$$\langle \phi(v), w \rangle = \langle v, \phi^*(w) \rangle.$$

In the case that ϕ is unitary one has

$$\phi^{-1} = \phi^*.$$

Definition 2.3. Let (X, μ) be a standard Borel probability space. Let $(\mathcal{H}_n)_{n\in\mathbb{N}}$ be a collection of separable Hilbert spaces and let $(X_n)_{n\in\mathbb{N}}$ be a partition of X into measurable sets. We then define

$$\mathcal{H}((\mathcal{H}_n)_n, (X_n)_n, \mu)$$

to be the collection of all functions

$$f : X \to \bigcup_n \mathcal{H}_n$$

such that f is *measurable*, in the sense that for each $n \in \mathbb{N}$, $\epsilon > 0$, and $v \in \mathcal{H}_n$ the set

$$\{x \in X_n : \|f(x) - v\| < \epsilon\}$$

is measurable, and

$$\int_X \langle f(x), f(x)\rangle d\mu(x) < \infty,$$

and $f(x) \in \mathcal{H}_n$ all $x \in X_n$. We then view $\mathcal{H}((\mathcal{H}_n)_n, (X_n)_n, \mu)$ as a Hilbert space in its own right: For $f, g \in \mathcal{H}((\mathcal{H}_n)_n, (X_n)_n, \mu)$, $c \in \mathbb{C}$

(1) $\|f\| = \int_X \langle f(x), f(x)\rangle d\mu(x)$
(2) $(f + g)(x) = f(x) + g(x)$
(3) $(cf)(x) = c(f(x))$
(4) $\langle f, g\rangle = \int_X \langle f(x), g(x)\rangle d\mu(x)$.

Given $\phi \in B(\mathcal{H}((\mathcal{H}_n)_n, (X_n)_n, \mu))$ we say that ϕ is a *multiplication operator* if there is some $g : X \to \mathbb{C}$ with the property that for any $f \in \mathcal{H}((\mathcal{H}_n)_n, (X_n)_n, \mu)$

$$\phi(f)(x) = g(x) \cdot f(x)$$

a.e. $x \in X$.

Definition 2.4. For Γ a countable group and $\mathcal{H}$ a Hilbert space, a *unitary representation* of Γ is a homomorphism

$$\varphi : \Gamma \to U(\mathcal{H}),$$

$$\gamma \mapsto \varphi_\gamma.$$

The representation is said to be *irreducible* if there is no closed non-trivial proper subspace

$$\mathcal{H}_0 \subset \mathcal{H}$$

which is invariant under all the operators $(\varphi_\gamma)_{\gamma\in\Gamma}$.

The only kinds of group representations we will want to consider are unitary representations. Thus we will usually just say *representation* and leave it understood that it is unitary representation.

With the definitions in hand, we can at least start the relevant form of the spectral theorem.

Theorem 2.2. *Let $\mathcal{H}$ be a separable Hilbert space. Let $A \subset B(\mathcal{H})$ be a subset which is commutative in the sense that*

$$\phi\psi(v) = \psi\phi(v)$$

all $v \in \mathcal{H}, \phi, \psi \in B(\mathcal{H})$ and closed under adjoints in the sense that $\phi^* \in A$ whenever $\phi \in A$. Then there is a standard Borel probability space (X, μ) and an isomorphism of Hilbert spaces

$$\pi : \mathcal{H} \cong \mathcal{H}((\mathcal{H}_n)_n, (X_n)_n, \mu)$$

such that for each $a \in A$ there exists measurable bounded $g_a : X \to \mathbb{C}$ such that

$$\pi(a \cdot v)$$

always equals the function

$$x \mapsto g_a(x)(\pi(v))(x).$$

In other words, every element of A is conjugate by π to a multiplication operator on $\mathcal{H}((\mathcal{H}_n)_n, (X_n)_n, \mu)$. Note that in particular this theorem implies that every irreducible representation of an abelian group is one dimensional – in other words, it corresponds to a character.

Definition 2.5. Let $S^1 = \{\xi \in \mathbb{C} : |\xi| = 1\}$, equipped with the group operation of complex multiplication. For Δ an abelian group, the *dual group* or *character group* of Δ, written $\hat{\Delta}$, consists of all homomorphisms

$$\chi : \Delta \to S^1.$$

Theorem 2.3. *Let Δ be a countable group. Let*

$$\varphi : \Delta \to U(\mathcal{H})$$

$$\delta \mapsto \varphi_\delta$$

be a unitary representation of Δ on a separable Hilbert space $\mathcal{H}$. Then we can find a standard Borel probability space (X, μ), along with $(\mathcal{H}_n)_{n\in\mathbb{N}}$ a *collection of separable Hilbert spaces, and* $(X_n)_{n\in\mathbb{N}}$ *a partition of* X *into measurable sets, and an isomorphism*

$$\pi : \mathcal{H} \cong \mathcal{H}((\mathcal{H}_n)_n, (X_n)_n, \mu),$$

and measurable assignments

$$\psi_\delta : X \to \bigcup_{n\in\mathbb{N}} U(\mathcal{H}_n)$$

for each $\delta \in \Delta$ *such that:*

(1) if $x \in X_n$ *then* $\psi_\delta(x) \in U(\mathcal{H}_n)$*;*
(2) if $x \in X_n$ *then as representation of* Δ

$$\delta \mapsto \psi_\delta(x)$$

is an irreducible representation;
(3) there is an isomorphism of Hilbert spaces

$$\pi : \mathcal{H} \cong \mathcal{H}((\mathcal{H}_n)_n, (X_n)_n, \mu)$$

such that for any $\delta \in \Delta$ *and* $v \in \mathcal{H}$

$$\pi(\varphi_\delta(v))(x) = \psi_\delta(x)(\pi(v)(x)).$$

The proof of this theorem derives from 2.2. In essence the idea is to look at all the operators

$$P \in B(\mathcal{H})$$

which are *projections* in the sense that $P^2 = P$ and commute with the representation, in the sense that for each $\delta \in \Delta$ and $v \in \mathcal{H}$

$$P(\varphi_\delta(v)) = \varphi_\delta(P(v)).$$

These can be thought of as measuring the failure of the representation to be irreducible, in the sense that if this collection of projections is trivial then the representation is already irreducible.

We then form a maximal abelian A in the collection of projections which commute with the representation φ. We apply 2.2 to write

$$\pi : \mathcal{H} \cong \mathcal{H}((\mathcal{H}_n)_n, (X_n)_n, \mu)$$

with each $P \in A$ conjugate to a multiplication operator. It follows from the P being a projection that at a.e. $x \in X$ we have

$$\pi(P(v))(x) \in \{\vec{0}, v(x)\}.$$

It is then a tedious, though routine, calculation to verify that at each $\delta \in \Delta$ there will be an assignment

$$\psi_\delta : X \to \bigcup_{n \in \mathbb{N}} U(\mathcal{H}_n)$$

such that at $v \in \mathcal{H}$

$$\pi(\varphi_\delta(v))(x) = \psi_\delta(x)(\pi(v)(x)).$$

Our interest in this decomposition of representations into direct integrals of irreducibles will be in particular directed to the case that the group Δ is a normal abelian subgroup of some larger group Γ which comes equipped with some representation

$$\varphi : \Gamma \to U(\mathcal{H}),$$

$$\gamma \mapsto \varphi_\gamma.$$

In this special case, we obtain the following:

Theorem 2.4. *Let Γ be a countable group and $\Delta \lhd \Gamma$ be a normal abelian subgroup. Let*

$$\varphi : \Gamma \to U(\mathcal{H})$$

$$\gamma \mapsto \varphi_\gamma$$

be a unitary representation of Γ on a separable Hilbert space $\mathcal{H}$. Then we can find a Borel probability measure μ on $\hat{\Delta}$ along with $(\mathcal{H}_n)_{n \in \mathbb{N}}$ a collection of separable Hilbert spaces, and $(X_n)_{n \in \mathbb{N}}$ a partition of $\hat{\Delta}$ into measurable sets, an isomorphism

$$\pi : \mathcal{H} \cong \mathcal{H}((\mathcal{H}_n)_n, (X_n)_n, \mu),$$

and a non-singular action of Γ on $(\hat{\Delta}, \mu)$,

$$a : \Gamma \times \hat{\Delta} \to \hat{\Delta},$$

such that for each $\delta \in \Delta$:

(1) there is an isomorphism of Hilbert spaces

$$\pi : \mathcal{H} \cong \mathcal{H}((\mathcal{H}_n)_n, (X_n)_n, \mu)$$

such that for any $\delta \in \Delta$, $\chi \in \hat{\Delta}$, $v \in \mathcal{H}$

$$\pi(\varphi_\delta(v))(\chi) = \chi(\delta)(\pi(v)(\chi));$$

(2) for each $\gamma \in \Gamma$, $\chi \in \hat{\Delta}$, $\delta \in \Delta$

$$a(\gamma, \chi)(\delta) = \chi(\gamma^{-1}\delta\gamma);$$

(3) for each $\gamma \in \Gamma$, $v \in \mathcal{H}$, *and measurable* $B \subset \hat{\Delta}$

$$\int_B \|(\pi(v))(x)\| d\mu(x) = \int_{\gamma \cdot B} \|(\pi(\varphi_\gamma(v)))(x)\| d\mu.$$

There is more that can be said with regards to point 3, in fact more that probably should be said. First of all when I write $\gamma \cdot B$ in the subscript of the second integral, I mean this with respect to the indicated action a of Γ on $\hat{\Delta}$; more formally we should probably write $\{a(\gamma, x) : x \in B\}$ instead of $\gamma \cdot B$. There is also a kind of naturalness to the action of Γ on the dual space $\hat{\Delta}$, which arises from the induced action of Γ on the representations of Δ. This is discussed thoroughly in [20], and while the ideas are straightforward and very intuitive, the effort to make them precise notationally would take us too far afield. Indeed the statement at 3 is really a crude approximation to those ideas, justified only by its sufficiency in the course of the measure theoretic arguments underlying the proof of relative property (T) in the next section.

2.3. *Relative property (T)*

Definition 2.6. Let $\mathrm{SL}_2(\mathbb{Z})$ be the collection of matrices of the form

$$\begin{bmatrix} a_{11} & a_{12} \\ a_{21} & a_{22} \end{bmatrix}$$

with integer entries and determinant, $a_{11}a_{22} - a_{21}a_{12} = 1$.

This forms a group under matrix multiplication. The main issue is existence of inverses – but we can row reduce every element of $\mathrm{SL}_2(\mathbb{Z})$ down to the identity using elementary row operations and verify that each of those will have an inverse in $\mathrm{SL}_2(\mathbb{Z})$.

Definition 2.7. Think of $\mathbb{Z}^2$ as the collection of 2×1 column matrices with addition conducted pointwise on each coordinate. Let $\mathrm{SL}_2(\mathbb{Z})$ act it in the obvious way.

$$\begin{bmatrix} a_{11} & a_{12} \\ a_{21} & a_{22} \end{bmatrix} \cdot \begin{bmatrix} m \\ n \end{bmatrix} = \begin{bmatrix} a_{11}m + a_{12}n \\ a_{21}m + a_{22}n \end{bmatrix}.$$

In this way each element of $\mathrm{SL}_2(\mathbb{Z})$ induces an automorphism of $\mathbb{Z}^2$, and we can form the semi-direct product

$$\Gamma = \mathrm{SL}_2(\mathbb{Z}) \ltimes \mathbb{Z}^2$$

with the group operation

$$(M, \vec{a})(N, \vec{b}) = (MN, N^{-1} \cdot \vec{a} + \vec{b}).$$

Definition 2.8. Let $\hat{\mathbb{Z}^2}$ be the *character group* of $\mathbb{Z}^2$ – that is to say the collection of all homomorphism

$$\chi : \mathbb{Z}^2 \to S^1,$$

where S^1 is the collection of complex numbers with norm one considered as a group under complex multiplication.

Facts 2.5. (i) Every $\chi \in \hat{\mathbb{Z}^2}$ has the form

$$\chi : \begin{bmatrix} m \\ n \end{bmatrix} \mapsto e^{2(sm+tn)\pi i},$$

for some $s, t \in \mathbb{R}$.

Definition 2.9. I will write $\chi_{(s,t)}$ for the character

$$\chi_{(s,t)} : \begin{bmatrix} m \\ n \end{bmatrix} \mapsto e^{2(sm+tn)\pi i}.$$

Later on we will need to consider $\mathrm{SL}_2(\mathbb{Z})$ acting on these characters using its canonical action on $\mathbb{R}^2$ and hence $\mathbb{R}^2/\mathbb{Z}^2$. I will write $M \cdot (s,t)$ to indicate $(sa_{11} + ta_{12}, ta_{21} + ta_{22})$, where

$$M = \begin{bmatrix} a_{11} & a_{12} \\ a_{21} & a_{22} \end{bmatrix}.$$

I will use I to denote the identity matrix

$$I = \begin{bmatrix} 1 & 0 \\ 0 & 1 \end{bmatrix}.$$

Facts 2.6. (Continued) (ii) From (i), there is a natural isomorphism between $\hat{\mathbb{Z}^2}$ and $\mathbb{R}^2/\mathbb{Z}^2$.

(iii) If $\mathbb{Z}^2$ acts in a unitary (i.e. a linear action with $\langle \vec{a} \cdot v, \vec{a} \cdot u \rangle$ is equal to $\langle \vec{v}, \vec{u} \rangle$ for all $u, v \in \mathcal{H}$, $\vec{a}, \vec{b} \in \mathbb{Z}^2$) manner on a one dimensional complex Hilbert space $\mathcal{H} = \mathbb{C} \cdot u$, then there are $s, t \in \mathbb{R}$ with

$$\begin{bmatrix} m \\ n \end{bmatrix} \cdot u = e^{2(sm+tn)\pi i} u.$$

Thus there is an exact correspondence between irreducible unitary representations of $\mathbb{Z}^2$ and elements of the dual group $\hat{\mathbb{Z}^2}$.

(iv) If $\mathcal{W}$ is a Hilbert space on which Γ acts unitarily, and $\mathcal{V}$ a $\mathbb{Z}^2$ invariant subspace, with

$$\begin{bmatrix} m \\ n \end{bmatrix} \cdot u = e^{2(sm+tn)\pi i} u,$$

all $u \in \mathcal{V}$. Then for any $u \in \mathcal{V}$, $\begin{bmatrix} m \\ n \end{bmatrix} \in \mathbb{Z}^2$ and

$$M = \begin{bmatrix} a_{11} & a_{12} \\ a_{21} & a_{22} \end{bmatrix}$$

we have

$$(M, \vec{0})(I, \begin{bmatrix} m \\ n \end{bmatrix})(M^{-1}, \vec{0}) = (I, M \cdot \begin{bmatrix} m \\ n \end{bmatrix}),$$

and hence for $v = (M^{-1}, \vec{0}) \cdot u$

$$(I, \begin{bmatrix} m \\ n \end{bmatrix}) \cdot v = (M^{-1}, \vec{0}) \cdot (I, M \cdot \begin{bmatrix} m \\ n \end{bmatrix})) \cdot u$$

$$= (M^{-1}, \vec{0}) \cdot e^{2(sma_{11}+sna_{12}+ta_{21}m+ta_{22}n)\pi i} u = e^{2(sma_{11}+sna_{12}+ta_{21}m+ta_{22}n)\pi i} v$$

$$= e^{2((a_{11}s+a_{21}t)m+(a_{12}s+a_{22}t)n)\pi i} v = \chi_{M^t(s,t)}(\begin{bmatrix} m \\ n \end{bmatrix})v.$$

Thus if $\mathbb{Z}^2$ acts on $\mathbb{C}u$ as indicated by

$$\chi_{(s,t)}$$

then it acts on $\mathbb{C}v = \mathbb{C}M^{-1} \cdot u$ as indicated by

$$\chi_{M^t(s,t)},$$

where M^t is the transpose of M.

Notation 2.1. Since $\hat{\mathbb{Z}^2}$ is canonically isomorphic as a space to $\mathbb{R}^2/\mathbb{Z}^2$, we can take the induced $\mathrm{SL}_2(\mathbb{Z})$ invariant measure μ which arises from the usual Lebesgue measure on $\mathbb{R}^2/\mathbb{Z}^2$.

This is in fact equal to the *Haar measure* on $\hat{\mathbb{Z}^2}$ – the unique two sided invariant Borel probability measure which arises from viewing it as a compact group in point wise multiplication.

Theorem 2.7. *(Kazhdan)* $\Gamma = \mathrm{SL}_2(\mathbb{Z}) \ltimes \mathbb{Z}^2$ *has "relative property (T)" over* $\mathbb{Z}^2$ *– in other words, if*

$$\Gamma \to U(\mathcal{H})$$

is a unitary representation of Γ *on a Hilbert space* $\mathcal{H}$ *with "almost invariant vectors", then there is a "non-trivial"* $\mathbb{Z}^2$*-invariant vector.*

Definition 2.10. Here when we say that a representation

$$\varphi : \Gamma \to U(\mathcal{H})$$

$$\gamma \mapsto \varphi_\gamma$$

has *almost invariant vectors* we mean that for all $\epsilon > 0$ and finite $F \subset \Gamma$ then we can find some $v \in \mathcal{H}, v \neq 0$ with

$$\|\varphi_\gamma(v) - v\| < \epsilon \|v\|$$

all $\gamma \in F$. Then $\mathbb{Z}^2 < \Gamma$ has a *non-trivial invariant vector* if there is some $v \in \mathcal{H}$ with $v \neq 0$ with

$$\varphi_{\vec{a}}(v) = v$$

all $\vec{a} \in \mathbb{Z}^2$.

Proof. I will sketch a proof which was discovered independently by Burger and Shalom.

Let us start with some $v \in \mathcal{H}$ which has

$$\gamma \cdot v \sim v$$

all

$$\gamma \in \{(I, \begin{bmatrix} 1 \\ 0 \end{bmatrix}), (I, \begin{bmatrix} 0 \\ 1 \end{bmatrix}), (\begin{bmatrix} 1 & \pm 3 \\ 0 & 1 \end{bmatrix}, \vec{0}), (\begin{bmatrix} 1 & 0 \\ \pm 3 & 1 \end{bmatrix}, \vec{0})\}.$$

Now for the non-trivial appeal to the spectral theorem. Appropriately understood, and applied to the abelian group $\mathbb{Z}^2$, this enables us to write the Hilbert space $\mathcal{H}$ as a direct integral

$$\mathcal{H} = \int_{\hat{\mathbb{Z}^2}} \mathcal{H}_\chi d\nu(\chi),$$

where ν is a Borel probability measure on $\hat{\mathbb{Z}}^2$, and where each $\mathcal{H}_\chi$ is a Hilbert space on which $\mathbb{Z}^2$ acts unitarily in accordance with χ – that is to say, for any $\vec{a} \in \mathbb{Z}^2$ and $w \in \mathcal{H}$

$$(I, \vec{a}) \cdot w = \int (\vec{a} \cdot w)_\chi d\nu(\chi) = \int \chi(\vec{a}) w_\chi d\nu(\chi).$$

Appealing to this decomposition, we can also write

$$v = \int_{\hat{\mathbb{Z}^2}} v_\chi d\nu(\chi).$$

Note then that by almost invariance of v we get that $\|v_\chi\|$ is very small for χ outside of some small neighborhood of $\vec{0}$ in $\mathbb{R}^2/\mathbb{Z}^2$. In fact, we may as well assume there is a tiny $\delta > 0$ such that

$$v_\chi = 0$$

unless

$$\chi = \chi_{(s,t)}$$

for some $s, t \in \mathbb{R}, |s|, |t| < \delta$. This is setting up the key combinatorial idea for the argument: Around this small neighborhood of the identity in $\hat{\mathbb{Z}^2}$, $\mathrm{SL}_2(\mathbb{Z})$ acts in a nearly linear fashion, resembling its canonical action on $\mathbb{R}^2$. The ultimate punch line of the proof is that the only "almost" $\mathrm{SL}_2(\mathbb{Z})$-invariant probability measures on $\mathbb{R}^2$ are those that largely concentrate with point mass at $\vec{0}$.

Then we can also think of $\mathrm{SL}_2(\mathbb{Z})$ as acting on the base space $\hat{\mathbb{Z}^2}$. By 2.6 (iv) we get at a.e. χ

$$\|((M^{-1}, \vec{0}) \cdot v)_{M^t\chi}\| = \|v_\chi\|.$$

Now forget everything else in the space $\mathcal{H}$ and just consider the direct integral presentation of v. From this we can define a measure

$$m(A) = \int_A \|v_\chi\| d\nu(\chi).$$

The almost invariance of v implies almost invariance of m.

m concentrates on the block

$$B = \{\chi_{(s,t)} : |s|, |t| < \delta\}.$$

B in turn can be broken up into four pieces, B_1, B_2, B_3, and B_4, depending on their quadrant in the plane. At each $i \leq 4$ let m_i be the restriction of m to the set B_i.

Consider the first of these.

$$B_1 = \{\chi_{(s,t)} : 0 \leq s, t < \delta\}.$$

Note then that

$$\begin{bmatrix} 1 & 0 \\ 3 & 1 \end{bmatrix} \cdot B_1 \cap \begin{bmatrix} 1 & 3 \\ 0 & 1 \end{bmatrix} \cdot B_1 = \{\vec{0}\},$$

and hence, since these two matrices both map the first quadrant of $\mathbb{R}^2$ into itself, almost invariance of m implies that $m|_{B_1}$ can only have a very small amount of mass away from the point $\vec{0}$. With different uses of the matrices in our finite set, we obtain the same result for $m|_{B_i}$ all $i \leq 4$.

Thus m concentrates largely on $\vec{0}$. This in turn implies that $v(\{\vec{0}\}) \neq 0$, and we obtain a $\mathbb{Z}^2$-invariant vector in $\mathcal{H}$ with the measurable assignment

$$\chi_{(0,0)} \mapsto v_{\chi_{(0,0)}}$$

and

$$\chi \mapsto 0$$

for

$$\chi \neq \chi_{(0,0)}. \qquad \square$$

2.4. *Gaboriau-Popa*

We are now in a position to prove Gaboriau-Popa, and I am going to present it in a way which disguises its roots in another area.

There is an essential point of view which motivates and inspires the construction; and while this point of view is thought of as absolutely trivial and not worthy of comment or even conscious recognition in the area of *operator algebras*, it is initially remarkable and surprising to mathematicians from vastly different fields.

From this perspective, every space is thought of as being an abelian object in a much more general kind of category. For instance, if (X, μ) is a measure space, then we associate it to $L^\infty(X, \mu)$, viewed as a group with the operation of pointwise addition.

And as a kind of converse, every abelian group is thought of as a space in waiting – given a countable group Δ we can form $\hat{\Delta}$, the compact group of all homomorphisms from Δ to $\{\xi \in \mathbb{C} : |\xi| = 1\}$ with group operation

$$\chi_1 \cdot \chi_2(\delta) = \chi_1(\delta)\chi_2(\delta).$$

We can in turn equip that dual group $\hat{\Delta}$ with its Haar measure ν, and view Δ as sitting inside $L^\infty(\hat{\Delta}, \nu)$ identifying each $\delta \in \Delta$ with the function

$$\chi \mapsto \chi(\delta).$$

It is that point of view which makes it possible to link the *rigidity*, the relative property (T), of $\mathbb{Z}^2$ inside $\Gamma = \mathrm{SL}_2(\mathbb{Z}) \ltimes \mathbb{Z}^2$ to the rigidity of certain kinds of measure preserving actions of $\mathrm{SL}_2(\mathbb{Z})$, and then for certiain copies of the free group lying inside.

Theorem 2.8. *(Gaboriau-Popa;[5]) For $n \geq 2$, $\mathbb{F}_n$ has continuum many orbit inequivalent measure preserving, free, ergodic actions on standard Borel probability spaces.*

Proof. I will sketch an argument for $n = 3$ based around Asger Törnquist's thesis – see [27].

First of all, we need to represent $\mathbb{F}_2 = \langle a, b \rangle$ as a subgroup of $\mathrm{SL}_2(\mathbb{Z})$ such that $\mathbb{F}_2 \ltimes \mathbb{Z}^2$ still has relative property (T) over $\mathbb{Z}^2$.

One way to do this is as follows. We take

$$a = \begin{bmatrix} 1 & 3 \\ 0 & 1 \end{bmatrix},$$

$$b = \begin{bmatrix} 1 & 0 \\ 3 & 1 \end{bmatrix}.$$

The argument from the last section still gives relative property (T), since we only used these elements and their inverses in $\mathrm{SL}_2(\mathbb{Z})$, but we need to see that they freely generate the subgroup $\langle a, b \rangle$. For this it suffices to see that reduced words of the form

$$W = a^{k_{2n-1}} b^{k_{2n-2}} .. b^{k_2} a^{k_1}$$

or

$$W = b^{k_{2n}} a^{k_{2n-1}} b^{k_{2n-2}} .. b^{k_2} a^{k_1}$$

are never equal to the identity. This can be shown by induction on n. Writing

$$W = \begin{bmatrix} w_{11} & w_{12} \\ w_{21} & w_{22} \end{bmatrix}$$

we obtain that in the first case $|w_{12}|$ is larger than all the other $|w_{ij}|$'s, while in the second case $|w_{22}|$ is the dominant entry.

From now on I will identify $\mathbb{F}_2$ with the collection of matrices we generate in $\mathrm{SL}_2(\mathbb{Z})$ from

$$\begin{bmatrix} 1 & 3 \\ 0 & 1 \end{bmatrix}, \begin{bmatrix} 1 & 0 \\ 3 & 1 \end{bmatrix}.$$

Now let $X = \hat{\mathbb{Z}^2} \cong \mathbb{R}^2/\mathbb{Z}^2 = \mathbb{T}^2$ and equip it with the canonical probability measure μ from before. Let $L(\hat{\mathbb{Z}^2}, S^1, \mu)$ be the collection of measurable functions from $\hat{\mathbb{Z}^2} = X$ to complex numbers of norm 1. This becomes a Polish group with the metric

$$d(\phi, \psi) = \int |\phi(\chi) - \psi(\chi)| d\mu(\chi),$$

provided we identify functions agreeing a.e. We have a natural embedding

$$\mathbb{Z}^2 < L(\hat{\mathbb{Z}^2}, S^1, \mu),$$

where we think of $\vec{a} \in \mathbb{Z}^2$ as being identified with the function

$$\vec{a} : \chi \mapsto \chi(\vec{a}).$$

Let $\mathbb{F}_2 < \mathrm{SL}_2(\mathbb{Z})$ act on $\hat{\mathbb{Z}^2}$ in the manner suggested previously,

$$M : \chi_{(s,t)} \mapsto \chi_{M(s,t)},$$

and recall we have a certain intertwining of the action of $\mathbb{F}_2$ and $\mathbb{Z}^2$:

$$(M \cdot \vec{a})(\chi_{(s,t)}) = \vec{a}(\chi_{M^t \cdot (s,t)}).$$

Some properties of the action of $\mathbb{F}_2$ on $\hat{\mathbb{Z}^2}$:

(a) *It is free a.e.* This follows because the fixed points for some $M \in \mathbb{F}_2$ would correspond to eigenvectors for M, and only the identity element in $\mathbb{F}_2$ has a two dimensional collection of eigenvalues.

(b) *The action is ergodic.* This is a little bit more subtle. The main point here is that the functions arising from $\mathbb{Z}^2 \subset L(\hat{\mathbb{Z}^2}, S^1, \mu)$,

$$\vec{a} : \chi \mapsto \chi(\vec{a}),$$

form an orthonormal basis of the Hilbert space of $L^2(\hat{\mathbb{Z}^2}, \mu)$. It is a simple calculation to see that they orthonormal, and then to see they are a basis we appeal to Stone-Weierstrass: They separate points and thus the algebra they generate will be dense in the collection of continuous functions on $\hat{\mathbb{Z}^2}$, and hence dense in $L^2(\hat{\mathbb{Z}^2}, \mu)$.

Now given any non-constant $\phi \in L^2(\hat{\mathbb{Z}^2}, \mu)$ we want to show it is not invariant under the action of $\mathbb{F}_2$. This will suffice to prove ergodicity.

We write it in terms of the orthonormal basis:

$$\phi = \sum_{\mathbb{Z}^2} c_{\vec{a}} \vec{a},$$

for some choice of $c_{\vec{a}}$'s in $\mathbb{C}$. Invariance would imply invariance of the sets $A_c = \{\vec{a} \in \mathbb{Z}^2 : c_{\vec{a}} = c\}$ under the action of $\mathbb{F}_2$.

Now note that each A_c must be finite, since the Hilbert space norm of $\sum_{A_c} c\vec{a}$ equals $\sqrt{(|A_c|c^2)}$. The only finite subset of $\mathbb{Z}^2$ fixed under the action of $\mathbb{F}_2$ is $\{\vec{0}\}$, and thus we have $f = c\vec{0}$ for some $c \in \mathbb{C}$.

Now consider the Polish space $M_\infty(X,\mu)$, consisting of all measure preserving bijections from X to itself, with subbasic open sets consisting of $\{\phi \in M_\infty(X,\mu) : \mu(A\Delta\phi[B]) < \epsilon\}$, for A, B measurable, $\epsilon > 0$. We will need a result due to Törnquist. As a word of notation, for groups $G_1, G_2 < H$, we say that G_1 is *freely joined with* G_2, if for any sequence $a_1, a_2, ..., a_n$ of non-identity elements in G_1, and $b_1, ...b_n$ of non-identity elements in G_2, the elements

$$a_n b_n a_{n-1} .... a_1 b_1,$$

$$a_n b_n a_{n-1} .... a_1,$$

$$b_n a_{n-1} .... b_2 a_1,$$

$$b_n a_{n-1} .... b_2 a_1 b_1,$$

are all unequal to the identity in H. In other words, the subgroup generated by G_1 and G_2 is naturally isomorphic to $G_1 * G_2$.

Theorem 2.9. *(Törnquist,[27]) For any countable $\Delta < M_\infty(X,\mu)$, the collection of $\varphi \in M_\infty(X,\mu)$ such that $\{\varphi^\ell : \ell \in \mathbb{Z}\}$ is freely joined with Δ is comeager.*

Applying that here, we in particular have that the collection of $\varphi \in M_\infty(X,\mu)$ such that

$$\langle a, b, \varphi\rangle$$

forms a free group in the indicated generators is comeager. Since the collection of ergodic transformations[b] is also comeager, we can obtain an ergodic $\varphi \in M_\infty(X,\mu)$ with $\langle a, b, \varphi\rangle$ generating a free group.

Now we will define a parameterized of measurable subsets $(B_t)_{0<t<1}$ of X such that

$$B_s \subset B_t, \ \ \mu(B_s) < \mu(B_t)$$

[b] $\psi \in M_\infty(X,\mu)$ is *ergodic* if every measurable ψ-invariant set is either null or conull.

for $s < t$.

First fix a measurable partition $(A_n)_{n\in\mathbb{N}}$ of $X = \hat{\mathbb{Z}^2}$ with each $\mu(A_n) > 0$. Let $\{q_n : n \in \mathbb{N}\}$ be an enumeration of the rationals lying between 0 and 1. For each $t \in \mathbb{R}$ let

$$B_t = \bigcup\{A_n : q_n < t\}.$$

Note then that for $s < t$ we have $B_s \subset B_t$ and $B_t \setminus B_s$ non-null.

For each $t \in \mathbb{R}$ and $\chi \in \hat{\mathbb{Z}^2}$ we define $N_t(\chi)$ in one of two ways, depending on whether χ is in B_t: If $\chi \in B_t$ then $N_t(\chi)$ is the least $N > 0$ such that

$$\varphi^N(\chi) \in B_t;$$

if $\chi \in A_n$ for some n with $q_n \geq t$, then $N_t(\chi)$ is the least $N > 0$ such that

$$\varphi^N(\chi) \in A_n.$$

By the Poincare recurrence lemma, N_t is defined a.e.[c] We then define

$$\phi_t : X \to X$$

$$\chi \mapsto \varphi^{N_t(\chi)}(\chi).$$

ϕ_t is defined a.e. and the assignment

$$\mathbb{R} \to M_\infty(X,\mu),$$

$$t \mapsto \phi_t$$

is measurable. By considering the decomposition of the space as

$$X = B_t \dot{\cup} \dot{\bigcup_{q_n \geq t}} A_n$$

it can be shown that each ϕ_t is measure preserving and invertible.

We then let $\mathbb{F}_3$ be the group freely generated by $\{a, b, c\}$ and let $a_t : \mathbb{F}_3 \times X \to X$ be the action defined by

$$a_t(a,\chi) = \begin{bmatrix} 1 & 3 \\ 0 & 1 \end{bmatrix} \cdot \chi,$$

[c]The Poincare recurrence lemma states that if (Y, m) is a standard Borel probability space, $T : Y \to Y$ measure preserving, and $B \subset Y$ non-null, then almost every point in B meets it again under the forward orbits of T. This is easily seen in the case T is invertible, since the failure of the lemma would enable us to find a positive measure set C whose forward orbits never meet C; that in turn gives $\{T^n[C] : n \geq 0\}$ as a sequence of disjoint sets with equal measure – contradicting the finiteness of the measure.

$$a_t(b, \chi) = \begin{bmatrix} 1 & 0 \\ 3 & 1 \end{bmatrix} \cdot \chi,$$

$$a_t(c, \chi) = \phi_t(\chi).$$

Some properties of *these* morphisms

(a) At each t we still have that $\langle a, b, \varphi_t \rangle$ generates a copy of $\mathbb{F}_3$ in the indicated generators. The induced action of $\mathbb{F}_3$ is free a.e., measure preserving, and ergodic.

(b) If for $0 < t < 1$ we let E_t be the orbit equivalence relation induced by the action of $\langle a, b, \varphi_t \rangle$, then we obtain that for $0 < s_1 < s_2 < 1$, the equivalence relation E_{s_1} has *infinite index* in E_{s_2} – each E_{s_2} equivalence class contains infinitely many E_{s_1} equivalence classes.

To see (b), Let C be the set of y for which $\{\phi^\ell_{s_1}(y) : \ell \in \mathbb{Z}\}$ is strictly included in $\{\phi^\ell_{s_2}(y) : \ell \in \mathbb{Z}\}$. It is non-null by construction. The saturation of C under a and b will be co-null by the ergodicity of their action. For a.e. x we can find an infinite set $\{y_n : n \in \mathbb{N}\} \subset C$ included in the orbit under $\langle a, b \rangle$ of x. Each of these y_n's in C will have its orbit under ϕ_{s_1} strictly included in its orbit under ϕ_{s_2}. Freeness of the action of $\langle a, b, \varphi \rangle$ then implies that for $y \in [y_n]_{\phi_{s_2}} \setminus [y_n]_{\phi_{s_1}}$ we have y not an element of $[y_m]_{\langle a,b,\phi_{s_1} \rangle}$ for any $m \neq n$.

Claim: Each E_s is orbit equivalent to only countably many other E_t's.

Proof of Claim: If not, let $W \subset [0,1]$ be uncountable such that for some fixed $t \in (0,1)$ we have at each $s \in W$ that E_s orbit equivalent to E_t via

$$\theta_s : X \to X,$$

a measurable bijection.

At each $s \in W$ define

$$\alpha_s : \mathbb{F}_3 \times X \to \mathbb{F}_3$$

by the requirement

$$\alpha(\gamma, \chi) \cdot \theta_s(\chi) = \theta_s(\gamma \cdot \chi).$$

By the separability of the space of measurable functions from $\mathbb{F}_3 \times X$ to $\mathbb{F}_3$ and the separability of $L(\hat{\mathbb{Z}}^2, S^1, \mu)$ we obtain $s_1 \neq s_2$ in W such that:

(a) $\alpha_{s_1}(\gamma, \chi) = \alpha_{s_2}(\gamma, \chi)$ for a large measure set of $\chi \in X$ and all $\gamma \in \mathbb{F}_2 = \langle a, b \rangle$ corresponding to a matrix of the form

$$\begin{bmatrix} 1 & \pm 3 \\ 0 & 1 \end{bmatrix}, \begin{bmatrix} 1 & 0 \\ \pm 3 & 1 \end{bmatrix};$$

(b) for $\vec{a} \in \mathbb{Z}^2$ of the form

$$\begin{bmatrix} 0 \\ 1 \end{bmatrix}, \begin{bmatrix} 1 \\ 0 \end{bmatrix}$$

we have $\vec{a}(\theta_{s_1}(\chi)) \sim \vec{a}(\theta_{s_2}(\chi))$ for a large measure set of $\chi \in X$.[d]

Let $\theta = \theta_{s_2}^{-1}\theta_{s_1}$. This witnesses E_{s_1} orbit equivalent to E_{s_2}.

Now let $\mathcal{H}$ be the collection of all functions ψ from X to $\ell^2(X)$ with the following properties:

(1) for each $\chi \in X$

$$\psi(\chi) \in \ell^2([\theta(\chi)]_{E_{s_2}});$$

(2) ψ is *measurable* in the sense that for any $w \in \langle a, b, \phi_{s_2}\rangle$ and open $O \subset \mathbb{C}$, the set

$$\{\chi \in X : \langle \psi(\chi), w \cdot \chi\rangle \in O\}$$

is measurable in X;

(3) $\int \|\psi(x)\| d\mu(x) < \infty$.

Provided we identify functions agreeing a.e, this becomes a Hilbert space with inner product

$$\langle \psi, \phi\rangle = \int_X \langle \psi(\chi), \phi(\chi)\rangle d\mu(\chi)$$

Let $\mathbb{Z}^2$ act on $\mathcal{H}$ by

$$(\vec{a} \cdot \psi)(\chi)(\xi) = \vec{a}(\chi)\vec{a}(\xi)^{-1}\psi(\chi)(\xi).$$

Let $\mathbb{F}_2 < \mathrm{SL}_2(\mathbb{Z})$ act on $\mathcal{H}$ by

$$(M \cdot \psi)(\chi)(\xi) = \psi(M^t \cdot \chi)(M^t \cdot \xi).$$

It follows from the intertwining property

$$(M \cdot \vec{a})(\chi_{(s,t)}) = \vec{a}(\chi_{M^t \cdot (s,t)})$$

that this indeed defines an action of $\mathbb{F}_2 \ltimes \mathbb{Z}^2$.

Let $\phi_\Delta \in \mathcal{H}$ be defined by

$$\phi_\Delta(\chi)(\theta(\chi)) = 1$$

but

$$\phi_\Delta(\chi)(\xi) = 0$$

[d] As a word on unpacking the notation, $\theta_s(\chi)$ will be a character, and then we use $\vec{a}(\theta_s(\chi))$ to denote $\theta_s(\chi)(\vec{a}) \in S^1$, the result of applying the character $\theta_s(\chi)$ to $\vec{a}$.

for $\xi \neq \theta(\chi)$.

Our assumptions on θ give that ϕ_Δ is almost invariant. By relative property (T), we obtain a non-trivial $\mathbb{Z}^2$-invariant vector.

Now we need an extra little trick. I claim we can choose the invariant vector so that it is close to ϕ_Δ.

One way to see this is to go back through the proof of theorem 2.7 and see that it indeed gives this sharper result. Another approach is to observe that whenever $H \lhd G$, with G having relative property (T) over H, and with G acting unitarily on some Hilbert space $\mathcal{W}$, then we can write $\mathcal{W} = \mathcal{W}_0 \oplus \mathcal{W}_1$, where $\mathcal{W}_0$ is the collection of H-invariant vectors and $\mathcal{W}_1$ is its orthogonal complement. By relative property (T), no vector in $\mathcal{W}_1$ can be almost invariant, and thus every almost invariant vector is close to some vector in $\mathcal{W}_0$.

Granting this, we obtain $\mathbb{Z}^2$-invariant $\psi \in \mathcal{H}$ with $\psi \sim \phi_\Delta$. Thus on a positive measure set A we have for all $\chi \in A$

$$\exists! \zeta_\chi \in [\theta(\chi)]_{E_{s_2}}$$

such that

$$|\psi(\chi)(\zeta_\chi)| > \frac{3}{4}.$$

Then $\mathbb{Z}^2$ invariance gives for all $\chi \in A$,

$$(\vec{a} \cdot \chi)(\vec{a} \cdot \zeta_\chi)^{-1} = 1,$$

$$\therefore \vec{a} \cdot \chi = \vec{a} \cdot \zeta_\chi.$$

But $\mathbb{Z}^2$ separates points in $X = \hat{\mathbb{Z}^2}$. Thus on A, $\chi = \zeta_\chi$. Since $\zeta_\chi \in \theta[[\chi]_{E_{s_1}}]$ and θ effects an orbit equivalence,

$$E_{s_1}|_A = E_{s_2}|_A.$$

Since every E_{s_2} class contains infinitely many E_{s_1} classes and any non-null measurable set, such as A, is a selector for the ergodic equivalence relation, E_{s_1}, this gives a contradiction. (□Claim)

2.5. *But are we really using the spectral theorem?*

No, not really.

Here is another way to think about the entire argument.

Definition 2.11. Let Γ be a countable group acting on metric space (X, d). We say that the action is *expansive* if there is some finite $F \subset \Gamma$, and $\epsilon > 0$, and a finite collection of subsets

$$A_1, A_2, \ldots A_n \subset X \times X$$

such that:
(a) $\bigcup_{i \leq n} A_i \supset \{(x, y) \in X^2 : d(x, y) < \epsilon\}$;
(b) for each A_i there is a finite subset $F_i \subset F$ such that

$$\forall j \neq i (F_i \cdot A_i \cap A_j) \subset \Delta(X) =_{df} \{(x, x) : x \in X\}$$

and

$$\bigcap_{\gamma \in F_i} \gamma \cdot A_i \subset \Delta(X).$$

It is easily verified that the action of

$$\langle \begin{bmatrix} 1 & \pm 3 \\ 0 & 1 \end{bmatrix}, \begin{bmatrix} 1 & 0 \\ \pm 3 & 1 \end{bmatrix} \rangle \cong \mathbb{F}_2$$

on $\mathbb{R}^2/\mathbb{Z}^2$ is expansive in any choice of a compatible metric. This was basically the proof we went through in the course of 2.7 – we take $n = 4$ and let

$$A_1 = \{((s, t), (s', t') : 0 \leq s - s' \leq \frac{1}{100}, 0 \leq t - t' \leq \frac{1}{100}\},$$

$$A_2 = \{((s, t), (s', t') : -\frac{1}{100} \leq s - s' \leq 0, 0 \leq t - t' \leq \frac{1}{100}\},$$

$$A_3 = \{((s, t), (s', t') : 0 \leq s - s' \leq \frac{1}{100}, -\frac{1}{100} \leq t - t' \leq 0\},$$

$$A_4 = \{((s, t), (s', t') : -\frac{1}{100} \leq s - s' \leq 0, -\frac{1}{100} \leq t - t' \leq 0\}$$

and the argument devolves in to an exercise in linear algebra for F the set of matrices of the form

$$\begin{bmatrix} 1 & \pm 3 \\ 0 & 1 \end{bmatrix}, \begin{bmatrix} 1 & 0 \\ \pm 3 & 1 \end{bmatrix}.$$

Abstracting from the proof given above for Gaboriau-Popa we obtain theorems such as:

Theorem 2.10. *Let (X, d) be a complete, separable metric space equipped with an atomless Borel probability measure μ. Suppose Γ acts ergodically by* *measure preserving transformations on* (X,μ) *and the action on* (X,d) *is expansive. Let* $(E_t)_{0<t<1}$ *be a collection of distinct countable Borel equivalence relations on* X *with:*

(a) on any non-null measurable $A \subset X$ *we have for all* $s \neq t$ *and a.e.* $x \in A$

$$\{y \in A : yE_s x\} \neq \{y \in A : yE_t x\};$$

(b) each $E_s \supset E_\Gamma$.

Then each E_s *is orbit equivalent to only countably many* E_t*'s.*

I mention this not as an exercise in proof mining, but rather because it might be interesting to explore dynamical properties such as expansiveness. Conceivably there could be other applications to the theory of countable Borel equivalence relations.

For instance, just for example, I do not know which countable groups admit measure preserving expansive actions.

2.6. *Countable Borel equivalence relations up to Borel reducibility*

For a long while we only had finitely many countable Borel equivalence relations which were known to be distinct in the $\leq_B$ ordering. This was finally settled by Adams and Kechris:

Theorem 2.11. *(Adams-Kechris,[1]) There are continuum many* $\leq_B$*-incomparable countable Borel equivalence relations.*

Their argument ultimately relied on super rigidity results for certain classes of algebraic groups, as developed in.[29] For the free groups there is no hint that such a rigidity theory should exist, and by implication there is not even the slightest suggestion that there is a parallel theory for treeable equivalence relations.

Definition 2.12. A countable Borel equivalence relation E on a standard Borel space X is *treeable* if there is a symmetric, irreflexive Borel relation $R \subset X \times X$ such that

(a) R is acyclic. (If $x_0, x_1, ...x_n$ is a path with each $(x_i, x_{i+1}) \in R$ with no $x_i = x_j$ for $1 \leq i < j < n$, then $x_1 \neq x_n$).

(b) The R-connected components form the E-classes.

Definition 2.13. For E, F Borel equivalence relations on Polish X, Y, E is *strongly Borel reducible to* F, written

$$E \leq_B^s F,$$

if there is a Borel function

$$\theta : X \to Y$$

such that

$$\{\theta(z) : zEx\} = [\theta(x)]_F;$$

in other words, we can witness Borel reducibility with a function whose image is F-invariant.

With minor adjustment, we can obtain from Gaboriau-Popa style arguments that:

Theorem 2.12. *There are continuum many treeable, countable Borel equivalence relations up to* $\leq_B^s$.

Going back to the proof we had before, it suffices to show that each E_t has only countably many E_s's with

$$E_s \leq_B^s E_t.$$

For a contradiction, suppose otherwise. The argument for Gaboriau-Popa allows us then to assume that there is an uncountable collection of s's with

$$\theta_s : X \to X$$

witnessing $E_s \leq_B^s E_t$ and the images are essentially disjoint – that is, for $s \neq s'$ we have conull $A \subset X$ with

$$\theta_s[A] \cap \theta_{s'}[A] = 0.$$

For convenience, let's assume outright that the images are always disjoint.

Then we can find $s_1 < s_2$ and witnesses

$$\theta_1, \theta_2 : X \to X$$

to

$$E_{s_1} \leq_B^s E_t, \quad E_{s_2} \leq_B^s E_t,$$

such that on a large measure set of $x \in X$ and for a large finite collection of $M \in \mathrm{SL}_2(\mathbb{Z})$:

(a) There will be a common $\gamma \in \langle a, b, \varphi \rangle$ such that

$$\theta_1(\gamma \cdot x) = M \cdot \theta_1(x),$$

$$\theta_2(\gamma \cdot x) = M \cdot \theta_2(x).$$

For future reference, if $\gamma \in \langle a, b, \varphi \rangle$ has the property that

$$\theta_i(\gamma \cdot x) = M \cdot \theta_i(x),$$

then we will denote by it $\gamma_{M,i,x}$.

We admittedly haven't set this up so the choice of $\gamma_{M,i,x}$ will be unique, but standard selection theorems enable us to choose such a group element for each $\theta_i(x)$ and M, and let us go forward assuming this has been put in to place and we obtain the indicated agreement on a large set.

(b) For $\vec{a} \in \mathbb{Z}^2$ of the form

$$\begin{bmatrix} 0 \\ 1 \end{bmatrix}, \begin{bmatrix} 1 \\ 0 \end{bmatrix}$$

we obtain

$$\vec{a} \cdot (\theta_1(x)) \sim \vec{a}(\theta_2(x)).$$

Now we take our Hilbert space to be the collection of functions

$$\phi : X \to \ell^2(X) \times \ell^2(X)$$

with $\phi(x) \in \ell^2([\theta_1(x)]_{E_{s_1}}) \times \ell^2([\theta_2(x)]_{E_{s_2}})$ all $x \in X$. We let $\Gamma = \mathrm{SL}_2(\mathbb{Z}) \ltimes \mathbb{Z}$ act with

$$(M, \vec{0}) \cdot \phi(\gamma_{(M^t)^{-1},1,x} \cdot x)((M^t)^{-1} \cdot y_1)((M^t)^{-1} \cdot y_2) = \phi(x)(y_1)(y_2)$$

and

$$\vec{a} \cdot \phi(x)(y_1)(y_2) = \vec{a}(y_1)\vec{a}(y_2)^{-1}\phi(x)(y_1)(y_2).$$

Since $\theta_1(\gamma \cdot x) = M \cdot \theta_1(x) = \theta_2(\gamma \cdot x) = M \cdot \theta_1(x)$ for a large finite collection of M and for a large measure set of $x \in X$, we can obtain our almost invariant vector with

$$\phi_\Delta^+(x)(y_1)(y_2) = 1$$

if $y_i = \theta_i(x)$, and

$$\phi_\Delta^+(x)(y_1)(y_2) = 0$$

otherwise. Then as before we obtain a $\mathbb{Z}^2$-invariant vector which is close to ϕ^+_Δ. As before, the invariance will imply agreement on ranges, with the same contradiction as previously.

Question 2.1. Does every countable non-amenable group give rise to continuum many free, measure preserving, ergodic actions up to strong Borel reducibility?

2.7. *Further results on orbit equivalence for non-amenable groups*

From about 2004 until about 2007 there were a sequence of partial results towards trying to prove that all non-amenable groups have continuum many orbit inequivalent measure preserving, free, ergodic actions on standard Borel probability spaces.

Theorem 2.13. *(Popa,[23]) If Γ is a countable group having a relative property (T) over an infinite normal subgroup, then it has continuum many orbit inequivalent measure preserving, free, ergodic actions on standard Borel probability spaces.*

Theorem 2.14. *(Monod-Shalom,[21]) If Γ is a countable product of non-amenable hyperbolic[e] groups, then it has has continuum many orbit inequivalent measure preserving, free, ergodic actions on standard Borel probability spaces.*

Theorem 2.15. *(Ioana,[12]) If Γ is a countable product of a non-amenable group and an infinite amenable group, then it has continuum many orbit inequivalent measure preserving, free, ergodic actions on standard Borel probability spaces.*

Theorem 2.16. *(Ioana,[13]) If Γ is a countable group containing $\mathbb{F}_2$ as a subgroup, then it has continuum many orbit inequivalent measure preserving, free, ergodic actions on standard Borel probability spaces.*

The most general result possible was finally obtained by Inessa Epstein in 2007. Before presenting a sketch of her proof, we need to spend some time discussing the notion of cost.

[e] A group is said to be *hyperbolic* or *word hyperbolic* if there is a finite generating set and the induced Cayley graph with the induced path metric is hyperbolic as a space – that is to say there is some $N \in \mathbb{N}$ such that in any geodesic triangle, each side is contained within the ball of radius N around the other two.

3. Lecture III

3.1. *Introduction to the theory of cost*

For technical reasons we will want to spend some time working with measure spaces which are finite without necessarily having total measure 1. Many of the unattributed results in this section can be found in [19].

Definition 3.1. Let E be a countable Borel, measure preserving equivalence relations on a standard finite measure space, (X, μ). A *graphing* of E is a countable collection Φ such that:

(a) each $\phi \in \Phi$ is a bijection of the form

$$\phi : A_\phi \to B_\phi$$

for A_ϕ, B_ϕ measurable sets, with

$$xE\phi(x)$$

all $x \in A_\phi$;

(b) for any $(x, y) \in E$ there is a sequence $\phi_0, \phi_1, ...\phi_n \in \Phi \cup \Phi^{-1}$ with $y = \phi_n \circ \phi_{n-1} \circ ...\phi_0(x)$. (Here Φ^{-1} is the collection of morphisms $\phi^{-1} : B_\phi \to A_\phi$.)

Φ is a *treeing* of E if for μ almost every $x \in X$ we have that for all yEx there is a unique choice of $\phi_0, \phi_1, ...\phi_n \in \Phi \cup \Phi^{-1}$ with $y = \phi_n \circ \phi_{n-1} \circ ... \phi_0(x)$ and

$$\phi_i \circ \phi_{i-1} \circ ...\phi_0(x) \neq \phi_j \circ \phi_{j-1} \circ ...\phi_0(x)$$

all $i < j \leq n$.

We let $C_\mu(\Phi)$, the *cost of* Φ, be

$$\sum_{\phi \in \Phi} \mu(A_\phi).$$

We let $C_\mu(E)$, the *cost of* E, be the infimum over $C_\mu(\Phi)$ as Φ ranges over graphings of E.

Technically this will be the most useful definition of cost for our purposes. There is however another definition which is arguably more elegant.

Definition 3.2. A Borel graph $R \subset X \times X$ is said to *graph* E if its connected components are the E-classes. We then define $C_\mu(R)$ to be

$$\int_X \frac{1}{2}\deg(x) d\mu(x),$$

where $\deg(x)$ denotes the number of edges incident to x. (If this $\deg(x)$ is infinite on a non-null set, then $C_\mu(R)$ is infinite.)

The $\frac{1}{2}$ in this definition is to avoid counting a morphism twice when we use an R that graphs E to generate a graphing in the previous sense.

Lemma 3.1. *$C_\mu(E)$ equals the infimum of $C_\mu(R)$ as R ranges over Borel graphs which graph E.*

Lemma 3.2. *If the cost is attained by a specific graphing, $C_\mu(E) = C_\mu(\Phi)$ for Φ some graphing Φ of E, then Φ is a treeing.*

In a nutshell, here is why: If Φ fails to be a treeing, then on a non-null set there will be a specific reason for this failure. Namely, there will be some $A \subset X$ non-null and $\phi_0, \phi_1, ..., \phi_n$ such that every $x \in A$ has $\phi_n \circ \phi_{n-1} \circ ...\phi_0(x) = x$ as a witness to it not being a treeing. Then we can reduce A such that the sets of the form

$$A_i = \phi_i \circ \phi_{i-1} \circ ...\phi_0[A]$$

are disjoint. Then we can replace ϕ_0 by

$$\phi_0|_{A_{\phi_0} \setminus A}$$

and obtain a graphing with lower cost.

Lemma 3.3. *If E has every equivalence class of size N for some fixed $N \in \mathbb{N}$, then*

$$C_\mu(E) = \frac{N-1}{N}\mu(X).$$

Roughly speaking this holds because we can graph a set of size N with $N-1$ edges, but any choice of less than $N-1$ edges will leave it disconnected.

Lemma 3.4. *If E has finite classes, then we can choose a measurable selector*[f] *$A \subset X$ and for any such choice of A we have*

$$C_\mu(E) = \mu(X \setminus A).$$

This is a variation on the last lemma.

Lemma 3.5. *If all E classes are infinite, then*

$$C_\mu(E) \geq \mu(X).$$

---

[f]That is to say, a set which meets each equivalence class in exactly one point.

In very rough terms, this lemma is true because an infinite connected graph must have infinitely many edges; however it does take some care and effort to turn that combinatorial fact in to a proof of the measure theoretic statement at lemma 3.5.

Lemma 3.6. *If E is hyperfinite, then $C_\mu(E) \leq \mu(X)$.*

The idea of the proof of this lemma is based around an iterative construction: We build a sequence of partial morphisms

$$\phi_1 \subset \phi_2 \subset ... \phi_n \subset ...$$

such that at each n the set consisting of the nth morphism, $\{\phi_n\}$, provides a treeing for E_n and each ϕ_n is injective. Given ϕ_n, then for each x we let

$$[y_{0,x}]_{E_n}, [y_{1,x}]_{E_n}, ..., [y_{m(n,x),x}]_{E_n}$$

enumerate the E_n-classes included in $[x]_{E_{n+1}}$. We can do this so that each $y_{i,x}$ is *not* in the domain of ϕ. We can also make the process sufficiently uniform that the sets of the form

$$B_k = \{x \in X : m(x,n) \geq k\}$$

are measurable and at each $i \leq k$ the function

$$B_k \to B_k$$

$$x \mapsto y_{i,x}$$

is measurable. We then let ϕ_{n+1} extend ϕ_n with

$$\phi_{n+1}(y_{i,x}) = y_{i+1,x}.$$

There is one more basic fact about cost which is somewhat more involved, and I mention it now only for later reference.

Theorem 3.1. *(The Cost Restriction Formula; Gaboriau,[4]) If E is a measure preserving, ergodic, countable equivalence relation on a standard finite measure space (X, μ), then for any measurable non-null $A \subset X$*

$$C_\mu(E|_A) = C_\mu(E) - \mu(X \setminus A).$$

The proof of this theorem resists easy summary, but I will say something about the rough idea. We start with a graphing Φ of E and we want to reorganize it into a new graphing of the same cost containing morphisms which spread A out across X.

In the simplest case, let us suppose $\mu(A) = \frac{1}{2}\mu(X)$. We will reorganize Φ so it has the form $\Phi = \{\phi\} \cup \Psi$ with $\phi : A \to X \setminus A$. Then we suitably post compose and or pre compose each element of Ψ by ϕ or its inverse to obtain Φ' of the same cost with $\Phi' \setminus \{\phi\}$ graphing $E|_A$. For instance if $\psi : A_\psi \to B_\psi$ is in Φ we can assume, after appropriately subdividing, that both of sets A_ψ and B_ψ are either disjoint from or included in A. If they are both disjoint from A, then we replace ψ by $\phi^{-1} \circ \psi \circ \phi$. If just A_ψ is disjoint from A, then we replace by $\psi \circ \phi$, and so on.

The hard part is to find ϕ. We build it by gluing together morphisms in Φ.

First of all, by ergodicity, there must be some $\psi \in \Phi$ and non-null $B \subset A_\psi \cap A$ such that

$$\psi[B] \subset X \setminus A.$$

Then we adjust Φ to obtain $\Phi_0 = \Phi \setminus \{\psi\} \cup \{\psi|_{A_\psi \setminus B}, \psi|_B\}$.

The next step will be more complicated, since there is a split in cases. A typical case might find some $C \subset A_{\phi_1}$ and ϕ_1, ϕ_2, ϕ_3 such that

$$C \subset A \setminus B,$$

$$\phi_1[C] \subset B,$$

$$\phi_2 \circ \phi_1[C] \subset \psi[A],$$

$$\phi_3 \circ \phi_2 \circ \phi_1[C] \subset X \setminus (A \cup \psi[A]).$$

Then we replace ϕ_1 in Φ_0 with

$$\phi_1|_{A_{\phi_1} \setminus C}$$

and $\psi|_B$ with

$$\psi|_B \cup (\phi_3 \circ \phi_2 \circ \phi_1)|_C.$$

And on from there with many slightly differing cases all to be accounted for with suitable variations of the procedure described here.

3.2. *Gaboriau's breakthrough result*

We will go through a sketch of a proof of Gaboriau's converse to 3.2, but before doing so I want to emphasize the dramatic originality of his argument.

The notion of cost was not first defined by Gaboriau, but it stood there as an isolated definition with no apparent applications. Specific calculations of cost appeared to be intractable – except in the relatively trivial cases of hyperfiniteness or finite classes.

In the neighborhood of this invariant many problems remained apparently impervious to any known method of attack. It had been an embarrassment for 30 years that no one knew whether the shift action of $\mathbb{F}_2$ on $2^{\mathbb{F}_2}$ was orbit equivalent to the shift action of $\mathbb{F}_3$ on $2^{\mathbb{F}_3}$.

All this was settled by Gaboriau's theorem:

Theorem 3.2. *(Gaboriau,[4]) Let E be a countable, measure preserving, equivalence relation on a standard finite measure space (X, μ). Let Φ be a treeing of E. Then*

$$C_\mu(E) = C_\mu(\Phi).$$

Proof. I am going to sketch the proof in a special but representative case.

Let us first make the harmless assumption that $\mu(X) = 1$. I will also assume the action is ergodic; the more general argument would require us to first pass to the ergodic components of E in order to apply the cost restriction formula. Now let us also make the much more specific and drastic assumption that $E = E_{\mathbb{F}_2}$ is induced by a free measure preserving action of the free group, $\mathbb{F}_2 = \langle a, b \rangle$. The action provides a natural treeing of E, with $\{\phi_a, \phi_b\}$, where

$$\phi_a : x \mapsto a \cdot x$$

and

$$\phi_b : x \mapsto b \cdot x,$$

both defined on all of X. That graphing has cost 2. It suffices to see that no graphing can have lower cost.

Let Φ be some other graphing of E. We may assume without loss of generality that for each $\phi \in \Phi$ there is a specific $w_\phi \in \mathbb{F}_2$ such that on all $x \in A_\phi$ we have

$$\phi(x) = w_\phi \cdot x.$$

We may also assume that Φ is finite. The point is that if Φ provides a counterexample, with $C_\mu(\Phi) < 2$, then we can take some other graphing which is finite, arising from replacing infinitely many of the morphisms of Φ with ϕ_a and ϕ_b restricted to sufficiently small sets, which will still have cost less than 2.

We are going to expand X, obtaining a kind of covering space.

We let $\hat{X}$ consist of all triples (x, u, ϕ) where $\phi \in \Phi$ and u is a *right segment* of w_ϕ. We identify x with (x, e, ϕ) and $\phi(x)$ with (x, w_ϕ, ϕ). Thus between A_ϕ and B_ϕ we create a kind of bridge which passes through the copies of the points we should have traversed if we had instead used w_ϕ and the graphing provided by the action of $\mathbb{F}_2$.

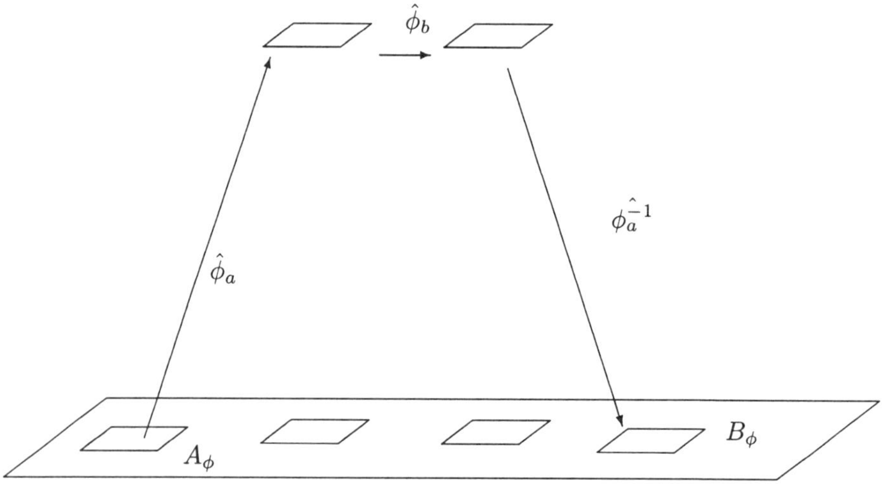

Fig. 1. Here is the situation we would have in the event that $w_\phi = a^{-1}ba$.

Now let $\hat{\Phi}$ consist of the morphisms of the form

$$\{(x, u, \phi) : x \in A_\phi\} \to \{(x, cu, \phi) : x \in A_\phi\}$$

$$(x, u, \phi) \mapsto (x, cu, \phi),$$

for u and cu right segments of w_ϕ, $c \in \{a^{\pm 1}, b^{\pm 1}\}$. We define

$$p : \hat{X} \setminus X \to X$$

by

$$(x, u, \phi) \mapsto u \cdot x.$$

We let $\hat{E}_a$ be the equivalence relation on $\hat{X}$ generated by p and the action of the generator $a \in \mathbb{F}_2$ – thus, $\hat{x}\hat{E}_a\hat{y}$ if for some $\ell \in \mathbb{Z}$

$$a^\ell \cdot p(\hat{x}) = p(\hat{y}).$$

Similarly $\hat{E}_b$ is generated from p and $b \in \mathbb{F}_2$.

We let $\hat{\mu}$ be the measure we obtain by extending μ to $\hat{X}$ with

$$\hat{\mu}(\{(x,u,\phi) : x \in A_\phi\}) = \mu(A_\phi).$$

Then by the cost restriction formula we have

$$C_{\hat{\mu}}(\hat{\Phi}) = C_\mu(\Phi) + \hat{\mu}(\hat{X} \setminus X).$$

Thus, it suffices to show $C_{\hat{\mu}}(\hat{\Phi}) \geq 2 + \hat{\mu}(\hat{X} \setminus X)$.

Since $\hat{E}_a$ and $\hat{E}_b$ have infinite classes, they both have cost at least equal to

$$1 + \hat{\mu}(\hat{X} \setminus X) = \hat{\mu}(\hat{X}).$$

Thus it will suffice for us to prove that

$$C_{\hat{\mu}}(\hat{\Phi}) \geq C_{\hat{\mu}}(\hat{E}_a) + C_{\hat{\mu}}(\hat{E}_b) - \hat{\mu}(\hat{X} \setminus X).$$

We let E_p be the equivalence relation generated by p. Note then that it has cost equal to $\hat{\mu}(\hat{X} \setminus X)$ by 3.4. We let $\hat{\Phi}_a$ be the morphisms in $\hat{\Phi}$ of the form

$$(x,u,\phi) \mapsto (x,cu,\phi),$$

for $c = a^{\pm 1}$, and $\hat{\Phi}_b$ be the morphisms in $\hat{\Phi}$ of the form

$$(x,u,\phi) \mapsto (x,cu,\phi),$$

for $c = b^{\pm 1}$.

Claim: There exist Ψ_a, Ψ_b such that:

(a) Ψ_a, Ψ_b are disjoint;

(b) $\Psi_a \cup \Psi_b$ is a treeing of a subequivalence relation of E_p.

(c) If we let $F_a = E_{(\hat{\Phi}_a \cup \Psi_a)}$ be the equivalence relation generated by the collection or morphisms $\hat{\Phi}_a \cup \Psi_a$, and similarly set $F_b = E_{(\hat{\Phi}_b \cup \Psi_b)}$, then

$$F_a \cap E_p = F_b \cap E_p.$$

Proof of Claim: We build a sequence of collections of morphisms

$$\Psi_a^0 \subset \Psi_a^1 \subset \Psi_a^2 \subset ...$$

$$\Psi_b^0 \subset \Psi_b^1 \subset \Psi_b^2 \subset ...,$$

commencing with $\Psi_a^0 = \Psi_b^0 = \emptyset$, both empty, and maintaining at each stage:

(i) $\Psi_a^n \cap \Psi_0^b = \emptyset$;

(ii) every morphism in $\Psi_a^{n+1} \setminus \Psi_a^n$ connects points which are equivalent under the equivalence relation

$$E_{(\hat{\Phi}_b \cup \Psi_b^n)} \cap E_p;$$

and moreover

$$E_{(\hat{\Phi}_b \cup \Psi_b^n)} \cap E_p \subset E_{(\hat{\Phi}_a \cup \Psi_a^{n+1})};$$

(iii) every morphism in $\Psi_b^{n+1} \setminus \Psi_b^n$ connects points which are equivalent under the equivalence relation

$$E_{(\hat{\Phi}_a \cup \Psi_a^{n+1})} \cap E_p;$$

moreover

$$E_{(\hat{\Phi}_a \cup \Psi_a^{n+1})} \cap E_p \subset E_{(\hat{\Phi}_b \cup \Psi_b^{n+1})};$$

(iv) $\Psi_a^n \cup \Psi_b^n$ is a treeing of a subequivalence relation of E_p;
(v) $\Psi_a^{n+1} \cup \Psi_b^n$ is a treeing of a subequivalence relation of E_p.

Clearly if we maintain this, then taking $\Psi_a = \bigcup_{n \in \mathbb{N}} \Psi_a^n$ and $\Psi_b = \bigcup_{n \in \mathbb{N}} \Psi_b^n$ completes the proof of the claim.

We first start working on Ψ_a^1: The equivalence relation

$$E_{(\hat{\Phi}_a \cup \Psi_a^0)} \cap E_p = E_{\hat{\Phi}_a} \cap E_p$$

obviously has finite classes. If no two disjoint equivalence classes contain elements which are

$$E_{(\hat{\Phi}_b \cup \Psi_b^0)} \cap E_p = E_{\hat{\Phi}_b} \cap E_p$$

equivalent, then there is no further work to do and we let $\Psi_a^1 = \Psi_a^0 = \emptyset$.

In the case that there are distinct classes, say

$$[x_0]_{E_{\hat{\Phi}_a} \cap E_p},$$

$$[x_1]_{E_{\hat{\Phi}_a} \cap E_p},$$

$$\ldots$$

$$[x_m]_{E_{\hat{\Phi}_a} \cap E_p},$$

which are linked in $E_{\hat{\Phi}_b} \cap E_p$, we choose exactly one point

$$y_0 \in [x_0]_{E_{(\hat{\Phi}_a \cap E_p)}},$$

$$y_1 \in [x_1]_{E_{(\hat{\Phi}_a \cap E_p)}},$$

$$\dots$$

$$y_m \in [x_m]_{E_{\hat{\Phi}_a} \cap E_p},$$

which are $E_{\hat{\Phi}_b} \cap E_p$-equivalent and require that some morphisms link them in Ψ^1_a. We can do this sparingly and in a way which does not introduce a cyclic amongst the various equivalence classes, for instance setting $\phi(y_i) = y_{i+1}$, and only being defined on $y_0, ..., y_{m-1}$ in this E_p-equivalence class.

(Of course there are various selection problems, and we need to argue that all these tasks can be completed in a Borel manner, giving only countably many new morphisms in Ψ^1_a. These steps use nothing beyond the classical theory of Borel sets and measurable functions, and I omit them. The key point here is that E_p has finite classes.)

Now for the other side. We consider the equivalence relation,

$$E_{(\hat{\Phi}_b \cup \Psi^0_b)} \cap E_p = E_{\hat{\Phi}_b} \cap E_p,$$

which again has finite classes. Again we consider a sequence of classes

$$[x_0]_{E_{\hat{\Phi}_b} \cap E_p},$$

$$[x_1]_{E_{\hat{\Phi}_b} \cap E_p},$$

$$\dots$$

$$[x_m]_{E_{\hat{\Phi}_b} \cap E_p},$$

which contain elements which are linked in

$$E_{(\hat{\Phi}_a \cup \Psi^1_a)} \cap E_p;$$

we choose elements

$$y_0 \in [x_0]_{E_{(\hat{\Phi}_b \cap E_p)}},$$

$$y_1 \in [x_1]_{E_{(\hat{\Phi}_b \cap E_p)}},$$

$$\dots$$

$$y_m \in [x_m]_{E_{\hat{\Phi}_b} \cap E_p},$$

and require one of the morphisms in Ψ^1_b to provide links without introducing cycles.

Here we do need to check that $\Psi_a^1 \cup \Psi_b^1$ is still a treeing. This amounts to checking that there is no morphism in Ψ_a^1 linking any of the

$$[x_i]_{E_{\hat{\Phi}_b} \cap E_p}$$

to

$$[x_j]_{E_{\hat{\Phi}_b} \cap E_p}$$

for $i \neq j$ in the above situation. But if there was such a link it would have to have been in $E_{\hat{\Phi}_b} = E_{\hat{\Phi}_b \cup \Phi_n^0}$ by the assumption that every morphism in $\Psi_a^1 \setminus \Psi_a^0$ connects points which are equivalent under the equivalence relation

$$E_{(\hat{\Phi}_b \cup \Psi_b^0)} \cap E_p.$$

Now we go back to the Ψ_a^1 side. We look for disjoint equivalence classes which are linked in

$$E_{(\hat{\Phi}_b \cup \Psi_b^1)} \cap E_p.$$

We add links as needed. Again we do not introduce cycles, since the the equivalence classes which demand work cannot be connected under Ψ_b^1 since every morphism in $\Psi_b^1 \setminus \Psi_b^0$ already appeared in $E_{(\hat{\Phi}_a \cup \Psi_a^1)} \cap E_p$.

And so on. (□Claim)

Claim: $\Psi_a \cap \hat{\Phi}_a$ is a graphing of $\hat{E}_a$.

Proof of Claim: Suppose $z\hat{E}_a z'$. Then we can certainly link them in $\hat{\Phi}$, and in particular we can find a chain

$$z_0 = z, z_1, \ldots z_n = z'$$

where each $(z_i, z_{i+1}) \in E_{(\hat{\Phi}_a \cup \Psi_a)}$ or in $E_{(\hat{\Phi}_b \cup \Psi_b)}$.

Let us assume n is minimal. That would involve strictly alternating between the two equivalence relations.

By considering $x_i = p(z_i)$ as a path through X, we note that either $n = 1$ or there is some pair $x_i = x_{i+1}$ – this goes back to the fact that $\mathbb{F}_2 = \langle a, b \rangle$ is acting freely. This gives then that (z_i, z_{i+1}) is in either $E_{(\hat{\Phi}_a \cup \Psi_a)} \cap E_p$ or in $E_{(\hat{\Phi}_b \cup \Psi_b)} \cap E_p$. But the assumptions on Ψ_a and Ψ_b imply that once it is one then it is in both, and we can shorten the length of the chain.

So now we have $n = 1$. But then we can again use $\hat{E}_a \cap \hat{E}_b = E_p$ to obtain $zE_{\Psi_a \cap \hat{\Phi}_a} z'$. (□Claim)

By exactly the same argument:

Claim: $\Psi_b \cap \hat{\Phi}_b$ is a graphing of $\hat{E}_b$.

$\Psi_a \cup \Psi_b$ is a treeing of a subequivalence relation of E_p. By the considerations raised at 3.3 and 3.4, we have that

$$C_{\hat{\mu}}(\Psi_a \cup \Psi_b) \leq \hat{\mu}(\hat{X} \setminus X).$$

We obtain

$$2 + 2\hat{\mu}(\hat{X} \setminus X) = 2\hat{\mu}(\hat{X}) \leq C_{\hat{\mu}}(\hat{E}_a) + C_{\hat{\mu}}(\hat{E}_b),$$

since $\hat{E}_a$ and $\hat{E}_b$ have infinite classes. Since $\hat{\Phi}_a \cup \Psi_a$ graphs $\hat{E}_a$ and $\hat{\Phi}_b \cup \Psi_b$ graphs $\hat{E}_b$ we obtain

$$C_{\hat{\mu}}(\hat{E}_a) + C_{\hat{\mu}}(\hat{E}_a) \leq C_{\hat{\mu}}(\hat{\Phi}_a \cup \Psi_a) + C_{\hat{\mu}}(\hat{\Phi}_b \cup \Psi_b) \leq C_{\hat{\mu}}(\hat{\Phi}) + C_{\hat{\mu}}(\Psi_a \cup \Psi_b)$$

which in turn is bounded by

$$C_{\hat{\mu}}(\hat{\Phi}) + \hat{\mu}(\hat{X} \setminus X),$$

exactly as required. □

4. Lecture IV

4.1. *More ramifications of cost*

Before Gaboriau's result we knew almost nothing about cost. After, there was a flood of results.

From now on I will go back to the "real world", so to speak, of all our measures spaces having $\mu(X) = 1$.

Theorem 4.1. *(Kechris-Miller,[18]) Let E be a countable, measure preserving equivalence relation on a standard Borel probability space (X, μ) with*

$$C_\mu(E) > \alpha.$$

Then E contains a treeable sub-equivalence relation $F \subset E$ also with

$$C_\mu(F) > \alpha.$$

Theorem 4.2. *(Hjorth,[11]) Let E be a countable, measure preserving, equivalence relation on a standard Borel probability space (X, μ). Assume that E is treeable, ergodic, and $C_\mu(E) = n$, for some $n \in \{1, 2, 3, ..., \infty\}$.*

Then there is a free, measure preserving action of $\mathbb{F}_n$ on X such that on some conull set $A \subset X$

$$E|_A = E_{\mathbb{F}_2}|_A.$$

Corollary 4.1. *Let E be a measure preserving equivalence relation on a standard Borel probability space (X, μ) with $C_\mu(E) > 1$. Then there is a non-null $A \subset X$ and a free, Borel, measure preserving action of $\mathbb{F}_2$ on A with*

$$E_{\mathbb{F}_2} \subset E.$$

Proof. Applying 4.1 we may assume that E is treeable. After passing to to the ergodic decomposition[g] it suffices to consider the case that E is ergodic. Applying the cost restriction formula we may find a set A with

$$C_{\mu|_A}(E|_A) = n\mu(A),$$

for some $n \in \{2, 3, ..., \infty\}$. We renormalize the measure with

$$\nu = \frac{\mu}{\mu(A)}$$

[g]The ergodic decomposition states that any equivalence relation may be written as a *direct integral* or ergodic measure preserving transformations – in a way very parallel to the direct integral decomposition of a unitary representation in terms of irreducible representations. See for instance [28].

to obtain that (A, ν) is a standard Borel probability space.

Finishing with 4.2 we obtain a free action, measure preserving action of $\mathbb{F}_n$ on A with $E_{\mathbb{F}_n} = E|_A$. $\square$

In fact, it is something of an overkill to apply 4.2 in obtaining this consequence of 4.1. The real point of 4.2 is not that treeable equivalence relations of high cost include orbit equivalence relations induced by free actions of non-abelian groups; it is rather that under certain circumstances they are exactly equal. As noted in [6], a careful analysis of the proof of 4.1 given at [18] will show that if $C_\mu(E) \geq 2$ then we obtain a free action of $\mathbb{F}_2$ with $E_{\mathbb{F}_2}$ included in the F they provide.

Notation 4.1. For A a set, $\mathcal{P}(A)$ denotes the collection of all its subsets and $\mathcal{P}_{<\aleph_0}(A)$ denotes the collection of all its finite subsets.

Definition 4.1. Let R be a connected graph with underlying set V. An *end* of R is a function

$$\varphi : \mathcal{P}_{<\aleph_0}(V) \to \mathcal{P}(V)$$

such that:

(a) $\varphi(S)$ is always a connected component of the remainder, $(V \setminus S, R|_{(V\setminus S)^2})$.

(b) $\forall S_1, S_2 \in \mathcal{P}_{<\aleph_0}(V)(\varphi(S_1) \cap \varphi(S_2) \neq \emptyset)$.

Thus an end is a consistent way of choosing connected components whenever we remove a finite subset.

Definition 4.2. An equivalence relation E on a standard Borel space X is said to *have infnitely many ends* if there is a Borel graphing $R \subset X \times X$ with every vertex having finite degree and such that R graphs E and on each equivalence class the induced graph has infinitely many ends.

Theorem 4.3. *(Gaboriau,[4]) If E is a measure preserving equivalence relation on a standard Borel probability space (X, μ) with infinitely many ends, then $C_\mu(E) > 1$.*

This result is a based on analysis by Ghys. I am not going to give a proof, but I will give a plausibility argument.

Using standard Borel techniques, we can assume that there is a Borel collection $\mathcal{C} \subset \mathcal{P}_{<\aleph_0}(X)$ such that:

(a) Each $C \in \mathcal{C}$ is a connected subgraph inside one of the equivalence classes.

(b) Any two elements of $\mathcal{C}$ are either in different equivalence classes, or they have distance at least 3 apart in the graphing.
(c) Every equivalence class includes some element of $\mathcal{C}$.
(d) And finally, $[x]_E \setminus C$ always has more than three component when $C \subset [x]_E$.

At each $C \in \mathcal{C}$, with say

$$C \subset [x]_E,$$

$$[x]_E \setminus C$$

having $n(C)$ many components, we choose $x_{C,0}, ..., x_{C,n(C)}$ in the distinct components outside C, each adjacent to C; it is possible to do this so that each assignment of the form

$$C \mapsto x_{C,i}$$

is Borel. We now restriction our attention the set B of all points of the form $x_{C,i}$ some $i \leq n(C)$.

Let us make the simplifying assumption that $n(C)$ is always 3. Then we let E_1 be the equivalence relation generated by the morphisms of the form

$$x_{C,0} \mapsto x_{C,1}$$

and E_2 the equivalence relation generated by morphisms of the form

$$x_{C,0} \mapsto x_{C,2}.$$

Finally we set xFy if they are E-equivalent and in the same connected component of $[x]_E \setminus \bigcup \mathcal{C}$. Since the elements in $\mathcal{C}$ are always at least a certain set distance apart, every element in B is F-equivalent to some other point.

These equivalence relations will be freely joined, in the sense that there is no cycle of the form

$$x_0, x_1, .., x_n = x_0$$

with each (x_i, x_{i+1}) in one of the three equivalence relations, and then (x_{i+1}, x_{i+2}) in one of the other two.

E_1 and E_2 both have cost on $(B, \mu|_B)$ equal to

$$\frac{\mu(B)}{3}.$$

F has cost at least equal to

$$\frac{\mu(B)}{2}.$$

Thus with an appropriate appeal to a suitable generalization of theorem 3.2 to the cost of freely joined equivalence relations, which is indeed presently explicitly in [4], we obtain $\mu(E|_B) > \mu(B)$, and now we can derive $\mu(E) > 1$ by the cost restriction formula.

Question 4.1. Is there a more direct proof of 4.3?

It remains very much open at the time of writing whether every countable, measure preserving, ergodic equivalence relation which is not hyperfinite contains an equivalence relation induced by a free action of $\mathbb{F}_2$. Several years ago Gaboriau and Lyons showed that for every non-amenable countable Γ there is *some* free measure preserving action which has this property. At the time they viewed it as simply a partial result towards this more general question, and did not bother to publish or even write it up as a preprint.

Theorem 4.4. *(Gaboriau-Lyons,[6]) Let Γ be a countable non-amenable group. Then there is a free measure preserving, ergodic, action of Γ on a standard Borel probability space such that E_Γ includes an equivalence relation with infinitely many ends.*

By theorem 4.3, we then obtain some $E \subset E_\Gamma$ such that E has cost greater than 1. Applying corollary 4.1 we can then get a free action of $\mathbb{F}_2$ on some non-null $A \subset X$ with $E_{\mathbb{F}_2} \subset E|_A$; in fact, the free group on two generators contains the free group on infinitely many generators, so we can as well get some $E_{\mathbb{F}_\infty} \subset E|_A$. With careful reorganization we can then get a free action of $\mathbb{F}_2$, spread out from A to the rest of the space, again with orbit equivalence relation included in $E \subset E_\Gamma$.

Corollary 4.2. *(Gaboriau-Lyons) Let Γ be a countable non-amenable group. Then there is a free measure preserving action of Γ on a standard Borel probability space (X, μ) and a free measure preserving action of $\mathbb{F}_2$ on (X, μ) with*

$$E_{\mathbb{F}_2} \subset E_\Gamma.$$

This is the corollary, which at the time seemed so innocuous, which underpins the final chapter of this story.

4.2. *Open problems in the theory of cost and treeable equivalence relations*

A quick rest at the side of the road.

Although it is not relevant to the main story, there are many open problems in the theory of cost. Most of these have been circulated in the literature for several years, and in some cases have been subjected to repeated attacks.

Definition 4.3. For Γ a group, we define the *price* of Γ to be the infimum over all

$$C_\mu(E),$$

where E arises as the orbit equivalence relation induced by a free, measure preserving, ergodic action of Γ on a standard Borel probability space (X, μ). Γ is said to have *fixed price* if every free measure, preserving, ergodic action of Γ on a standard Borel probability space gives rise to an orbit equivalence relation with the same cost – namely, equal to that of the price of Γ.

Question 4.2. Does every countable group have fixed price?

Definition 4.4. A countable group Γ is said to have *property (T)* if it has property (T) over itself in the earlier sense – that is to say, whenever it acts unitarily on a Hilbert space with almost invariant vectors, there is a non-trivial Γ-invariant vector.

Question 4.3. Does every property (T) group have price 1? For that matter, does every property (T) group have fixed price 1?

In general there are non-amenable groups with fixed price equal to one.[4] shows that whenever Γ and Δ are countable infinite groups, then $\Gamma \times \Delta$ is fixed price equal to 1. Thus, if one of the groups were to be non-amenable, which in turn implies $\Gamma \times \Delta$ non-amenable, we will have a non-amenable group with fixed price equal to one.

Definition 4.5. Let E, F be countable Borel equivalence relations on a standard Borel space X. We say that *E has finite index in F* if $E \subset F$ and every F class contains only finitely many E classes. We say that *E has index N in F* if every F-class contains exactly N many E-classes.

Question 4.4. If E and F are measure preserving, countable Borel equivalence relations on a standard Borel probability space (X, μ), and E has finite index in F, and E is treeable, must F be treeable a.e.?

Finally, an innocent looking question which has unexpectedly been shown by Abert and Nikolov to have important consequences in geometry:

Question 4.5. Let Γ be a countable group and $\Delta \lhd \Gamma$ a normal subgroup of finite index N. Suppose Γ acts freely, ergodically, and in a measure preserving manner on a standard Borel probability space (X, μ), then do we have necessarily have

$$C_\mu(E_\Delta) = C_\mu(E_\Gamma) + \frac{N-1}{N}?$$

Indeed, one can put this in a more general setting, which is also open:

Question 4.6. If E and F are measure preserving, countable Borel equivalence relations on a standard Borel probability space (X, μ) and E has index N in F, must we have

$$C_\mu(E) = C_\mu(F) + \frac{N-1}{N}?$$

4.3. *Epstein's theorem*

This result should be put in to proper context. It appeared about a year or so after [13] and certain aspects of the proof were inspired by the argument given there. I am going to organize the proof somewhat differently, admittedly taking numerous shortcuts, leaving key claims unproved, and making minor simplifying assumptions. The presentation I am giving will tend to obscure the links with [13].

Theorem 4.5. *(Epstein,[3]) Let Γ be a countable, non-amenable group. Then Γ has continuum many orbit inequivalent free, measure preserving, ergodic, actions on standard Borel probability spaces.*

Proof. (Sketch only)

By Gaboriau-Lyons we can find a free, measure preserving, ergodic action of Γ on a standard Borel probability space (X, ν) which includes an equivalence relation $E_{\mathbb{F}_3}$ induced by a free measure preserving action of $\mathbb{F}_3$. For each t between 0 and 1, we recall the constructions given in the proof of Gaboriau-Popa and obtain an action of

$$\mathbb{F}_3 \cong \langle a, b, \varphi_t \rangle$$

on $\hat{\mathbb{Z}^2} \cong \mathbb{R}^2/\mathbb{Z}^2$.

Now let X_t be the collection of

$$\{(x, f) : x \in X, f : [x]_\Gamma \to \hat{\mathbb{Z}^2}, \forall \sigma \in \langle a, b, \varphi_t \rangle, \forall y \in [x]_\Gamma (f(\sigma \cdot y) = \sigma \cdot (f(y)))\}.$$

We let Γ act on X_t by

$$\gamma \cdot (x, f) = (\gamma \cdot x, f)$$

and we let $\langle a, b, \varphi_t \rangle$ act by

$$\sigma \cdot (x, f) = (\sigma \cdot x, f),$$

where here $\sigma \cdot x$ refers to the action of $\mathbb{F}_3 \cong \langle a, b, \varphi_t \rangle$ on X established as a result of Gaboriau-Lyons. We let F_t be the orbit equivalence relation of Γ on X_t and we let E_t be the orbit equivalence relation of $\langle a, b, \varphi_t \rangle$ on X.

There is only one natural Borel structure on X_t, though there are a few different ways this can be described. Let $\mathcal{B}_t$ be the σ-algebra on X_t generated by sets of the form

$$\{(x, f) : f(\gamma \cdot x) \in B\}$$

for $\gamma \in \Gamma$ and $B \subset \hat{\mathbb{Z}}^2$ Borel. It can then be shown that $(X_t, \mathcal{B}_t)$ forms a standard Borel space.

The measure for the space X_t involves taking the product measure along the separate $\mathbb{F}_3$ orbits: Given $\{y_i : i \in \mathbb{N}\}$ a sequence of representatives of the orbits of $\langle a, b, \varphi_t \rangle$ included in $[x]_\Gamma$, we let

$$\mu_x^t = \prod_{\mathbb{N}} \mu$$

where the ith copy of μ, our canonical measure on $\hat{\mathbb{Z}}^2$, is applied to y_i and obtain a measure on

$$X_{t,x} = \{(y, f) \in X_t : y = x\}.$$

Thus for $B_1, B_2, \ldots, B_N \subset \hat{\mathbb{Z}}^2$ Borel, $y_1, y_2, \ldots, y_N \in [x]_\Gamma$ with each $[y_i]_{E_t} \neq [y_j]_{E_t}$ for $i \neq j$, we have

$$\mu_x^t(\{(x, f) : \forall i \leq N(f(y_i) \in B_i)\}) = \prod_{i \leq N} \mu(B_i).$$

Using $\mathbb{F}_3$ invariance of μ, it turns out that this measure is independent of the choice of representatives $\{y_i : i \in \mathbb{N}\}$ we choose. From this it follows that if we define

$$m_t = \int_X \mu_x^t d\nu(x)$$

we obtain an E_Γ-*invariant* probability measure on X_t.

Thus we obtain an induced free, measure preserving action of Γ on X_t with

$$\gamma \cdot (x, f) = (\gamma \cdot x, f).$$

Let F_t denote the orbit equivalence relation arising from this action of Γ.

At this stage we do not know the new action of Γ is ergodic, but for conceptual simplicity, towards seeing the main idea, let us just assume the action is ergodic. (At this point Epstein took the ad hoc solution of going to an ergodic component – though later Ioana, Kechris, and Tsankov, in [14], performed a more general analysis of this peculiar kind of skew product action.)

For $\vec{a} \in \mathbb{Z}^2$ we can again define a function

$$\vec{a} : X_t \to \mathbb{C}$$

$$(x, f) \mapsto \vec{a}(f(x)),$$

where $\vec{a}(f(x))$ now carries the meaning for us it had during the proof of Gaboriau-Popa.

Claim: Each F_t is orbit equivalent to only countably many other F_s's.

Proof of Claim: Otherwise, adapting the earlier argument, we obtain $s_1 > s_2$ with

$$\theta : X_{s_1} \to X_{s_2}$$

witnessing orbit equivalence such that on a large finite set $F \subset \mathbb{F}_2 = \langle a, b \rangle$ and for a large measure of

$$(x, f) \in X_{s_1}$$

(a) $\theta(\gamma \cdot (x, f)) = \gamma \cdot \theta(x, f)$ all $\gamma \in F$;
(b) $\vec{a}(x, f) \sim \vec{a}(\theta(x, f))$ for $\vec{a}$ one of our two basis elements for $\mathbb{Z}^2$.

Note that I am taking $s_1 > s_2$ – this will make the argument slightly cleaner in a few steps time.

Arguing in analogy with the previous proof of Gaboriau-Popa, there is some positive measure

$$B \subset X_{s_1}$$

such that

$$\vec{a}(x, f) = \vec{a}(\theta(x, f))$$

for all $(x, f) \in B$.

Now we do not know that this yields $\theta(x, f) = (x, f)$, since $\mathbb{Z}^2$ is sufficient to separate points in $\hat{\mathbb{Z}^2}$ but retreats to agnosticism when faced with questions on X.

Subclaim: For m_{s_1} a.e. $(x,f) \in B$

$$\theta[[(x,f)]_{\langle a,b,\varphi_{s_1}\rangle}]$$

is included in a single $\langle a,b,\varphi_{s_1}\rangle$ orbit.

Proof of Subclaim: The nature of the measure m_{s_2} gives that for a.e. $(y,g) \in X_{s_2}$ the different E_{s_2}-equivalence classes included in $[y]_\Gamma$ are unrelated under the entire action of $\langle a,b,\varphi\rangle$; in particular, if $[y_1]_{\langle a,b,\varphi_{s_2}\rangle}$ and $[y_2]_{\langle a,b,\varphi_{s_2}\rangle}$ are distinct orbits with

$$y_1 E_\Gamma y_2,$$

then $g(y_1)$ is not an element of

$$[y_2]_{\langle a,b,\varphi_{s_1}\rangle}.$$

(□Subclaim)

We are still not done. Let us define

$$\hat{F} \subset F_{s_1}$$

by setting $(x,f)\hat{F}(y,g)$ if

$$(x,f)F_{s_1}(y,g)$$

and

$$f(x)E_{s_2}g(y).$$

It is clear that $\hat{F}$ has infinite index in F_{s_1}, but that does not suffice to finish the proof. It is the next subclaim which completes the task.

Subclaim: $\hat{F}|_B$ has index greater than 1 in $F_{s_1}|_B$.

Proof of Subclaim: I will assume otherwise and head for a contradiction.

For notational convenience, let us assume $\mu_{s_1}(B) > \frac{1}{2}$. In fact this is minor assumption, because an appropriately tight form of (a) and (b) will go through the Gaboriau-Popa argument and yield this conclusion.

Given the similarity of the action of $\mathbb{F}_2 = \langle a,b\rangle$ on the spaces X_{s_1} and X_{s_2} we may assume without loss of generality that B is $\mathbb{F}_2$-invariant. It can be then shown that there exists some $C \subset B$ of positive measure such that:

(a) $\varphi_{s_1}|_C$ sends every point in C to an $\hat{F}$-*inequivalent* point.
(b) $\mathbb{F}_2 \cdot C = B$.

(a) in particular implies $\varphi_{s_1}(z) \notin B$ all $z \in B$.

For each $z \in B$ we choose $s(z) \in C \cap \mathbb{F}_2 \cdot z$ in a uniformly Borel manner. We can also then choose $\sigma_z \in \mathbb{F}_2$, again in a Borel manner, such that

$$\sigma_z \cdot z = s(z).$$

We then define

$$\rho : B \to X \setminus B$$

$$z \mapsto \sigma_z \cdot \varphi_{s_1}(\sigma_z \cdot z).$$

This is a Borel map included in the graph of F_{s_1} and hence is measure preserving. Now I claim it is one-to-one, which will give us a contradiction to $m_{s_1}(B) > \frac{1}{2}$.

The point is that if

$$\sigma_z \cdot \varphi_{s_1}(\sigma_z \cdot z) = \sigma_w \cdot \varphi_{s_1}(\sigma_w \cdot w)$$

for $z, w \in B$ then we would have that $z\hat{F}w$ by the assumption on B, which means they can be connected in the action of $\mathbb{F}_3 \cong \langle a, b, \varphi_{s_1} \rangle$ without passing through any applications of $\varphi_{s_1}|_C$.

Thus, since $\langle a, b, \varphi_{s_1} \rangle$ is acting freely, we have must have

$$\varphi_{s_1}(\sigma_z \cdot z) = \varphi_{s_1}(\sigma_w \cdot w),$$

which implies $\sigma_z = \sigma_w$ and then that $z = w$. (□Subclaim)

(□Claim)

References

1. S. Adams, A.S. Kechris, *Linear algebraic groups and countable Borel equivalence relations,* **Journal of the American Mathematical Society,** 13 (2000), no. 4, 909–943
2. A. Connes, J. Feldman, B. Weiss, *An amenable equivalence relation is generated by a single automorphism,* **Journal of Ergodic Theory and Dynamical Systems,** vo. 1(1981), pp. 431-450.
3. I. Epstein, *Orbit inequivalent actions of non-amenable groups,* preprint, 2007.
4. D. Gaboriau, *Coût des relations d'équivalence et des groupes,* (French. English summary) [Cost of equivalence relations and of groups] **Inventiones Mathematicae,**139 (2000), no. 1, 41–98.
5. D. Gaboriau, S. Popa, *An uncountable family of nonorbit equivalent actions of* $\mathbb{F}_n$, **Journal of the American Mathematical Society,** 18 (2005), no. 3, 547–559
6. D. Gaboriau, R. Lyons, *A Measurable-Group-Theoretic Solution to von Neumann's Problem,* preprint, 2007.
7. E. Ghys, *Topologie des feuilles genriques. (French) [Topology of generic leaves],* **Annals of Mathematics,** (2) 141 (1995), no. 2, 387–422.
8. F.P. Greenleaf, **Invariant means on topological groups and their applications,** Van Nostrand Mathematical Studies, No. 16 Van Nostrand Reinhold Co., New York-Toronto, London 1969.
9. G. Hjorth, *When is an equivalence relation classifiable?,* **Proceedings of the International Congress of Mathematicians, Vol. II (Berlin, 1998),** Documenta Mathematica, 1998, Extra Vol. II, 23–32
10. G. Hjorth, *A converse to Dye's theorem,* **Transactions of the American Mathematical Society,** 357 (2005), no. 8, 3083–3103
11. G. Hjorth, *A lemma for cost attained,* **Annals of Pure and Applied Logic,** 143 (2006), no. 1-3, 87–102.
12. A. Ioana, *A relative version of Connes'* $\chi(M)$ *invariant and existence of orbit inequivalent actions,* preprint, 2004.
13. A. Ioana, *Orbit inequivalent actions for groups containing a copy of* $\mathbb{F}_2$, preprint, 2006.
14. A. Ioana, A.S. Kechris, T. Tsankov, *Subequivalence relations and positive-definite functions,* to appear in **Groups, Geometry, and Dynamical Systems.**
15. S. Jackson, A.S. Kechris, A. Louveau, *Countable Borel equivalence relations,* **Journal Mathematical Logic,** vol. 2 (2002), no. 1, 1–80
16. A.S. Kechris, **Classical descriptive set theory,** Graduate Texts in Mathematics 156, Springer-Verlag, New York, 1995.
17. A.S. Kechris, *Descriptive dynamics,* **Descriptive set theory and dynamical systems (Marseille-Luminy, 1996),** 231–258, London Mathematical Society Lecture Note Series, 277, Cambridge Univ. Press, Cambridge, 2000
18. A.S. Kechris, B. Miller, **Topics in Orbit Equivalence,** Lecture Notes in Mathematics, Springer, Vol. 1852, 2004.
19. G. Levitt, *On the cost of generating an equivalence relation,* **Journal of Ergodic Theory and Dynamical Systems,** 15 (1995), no. 6, 1173–1181

20. G.W. Mackey, **The theory of unitary group representations,** University of Chicago Press, Chicago, 1976.
21. N. Monod, Y. Shalom, *Orbit equivalence rigidity and bounded cohomology,* **Annals of Mathematics,** (2) 164 (2006), no. 3, 825–878.
22. D.S. Ornstein, B. Weiss, *Ergodic theory of amenable group actions, I. The Rohlin lemma,* **Bulletin of the American Mathemtaical Society,** (N.S.) 2 (1980), no. 1, 161–164.
23. S. Popa, *Strong rigidity of* II_1 *factors arising from malleable actions of w-rigid groups. II,* **Inventiones Mathematicae,** 165 (2006), no. 2, 409–451.
24. J. Silver, *Counting the number of equivalence classes of Borel and coanalytic equivalence relations,* **Annals of Mathematical Logic,** 18 (1980), no. 1, 1–28
25. D. Sullivan, B. Weiss. J.D. Wright, *Generic dynamics and monotone complete* C^**-algebras,* **Transactions of the American Mathematical Society,** 295 (1986), no. 2, 795–809.
26. S. Thomas, *On the complexity of the classification problem for torsion-free abelian groups of finite rank,* **Bulletin of Symbolic Logic,** 7 (2001), no. 3, 329–344
27. A. Törnquist, *Orbit equivalence and actions of* $\mathbb{F}_n$*,* **Journal of Symbolic Logic,** 71 (2006), no. 1, 265–282.
28. V.S. Varadarajan, *On a problem in ergodic theory,* **Bulletin of the American Mathematical Society,** vol. 68 (1962), pp. 345–350
29. R.J. Zimmer, **Ergodic theory and semisimple groups,** Monographs in Mathematics, 81. Birkhäuser Verlag, Basel, 1984.
30. R.J. Zimmer, **Essential results of functional analysis,** Chicago Lectures in Mathematics, University of Chicago Press, Chicago, IL, 1990.

A REMARK ON GENERIC STRUCTURES AND THE NUMBER OF COUNTABLE MODELS

KOICHIRO IKEDA *

Faculty of Business Administration, Hosei University, Tokyo 102-8160, Japan
E-mail: ikeda@hosei.ac.jp

Let T be a theory of a generic structure in a finite relational language. Under some conditions we show that the number of countable models of T is one or infinite.

Keywords: Generic structure, Countable model, Free amalgamation property

1. Introduction

A notion of a $(\mathbf{K}, \leq)$-*generic structure* was introduced by Hrushovski as an adaptation of the Ehrenfeucht-Fraïssé homogeneous universal relational structures.[8,9] He defined a class $\mathbf{K}$ of finite structures which is derived by a predimension, and constructed new infinite structures by amalgamating the elements of $\mathbf{K}$.

By modifying Hrushovski's method, Herwig gave a small stable theory with a type of infinite preweight with respect to itself.[7] His example is very nice one, because if there exists a stable theory with more than one but finitely many countable models then it must be a small theory with a type of infinite preweight.

In this paper, we discuss a $(\mathbf{K}, \leq)$-generic structure in a finite relational language L, where $\mathbf{K}$ is a class of finite L-structures defined by a predimension $\delta(A) = |A| - \sum_{R\in L} \alpha_R|R^A|$. Then we assume that $\mathbf{K}$ satisfies the following two assumptions: One is the assumption that $\mathbf{K}$ is *closed under subgraphs* with respect to each relation in L (see Definition 2.6). In addition to Hrushovski's examples,[8,9] almost all of known examples satisfy this assumption (for example, Refs. 2,7,10). Another is the assumption that

*Research partially supported by Grants-in-Aid for Scientific Research (no.19540150), Ministry of Education, Science and Culture.

$\mathbf{K}$ has the *strong free amalgamation property* (see Definition 2.5). This is stronger than the assumption that $\mathbf{K}$ has the free amalgamation property, and implies the saturation of a generic structure. From this it follows that a theory of a generic structure is stable (see Remark 2.4). This assumption is satisfied by Hrushovski's generic pseudoplane[8] and Herwig's generic structure.[7]

Our aim is to show that if $\mathbf{K}$ is closed under subgraphs and has the strong free amalgamation property, then a (stable) theory of a $(\mathbf{K}, \leq)$-generic structure has one or infinitely many countable models (Theorem 4.1). This means that the Lachlan conjecture is true for a theory of a generic structure under our assumptions.

2. Preliminaries

In addition to original work of Hrushovski,[8,9] there are many papers which deal with generic structures in a similar setting to ours.[1–7,10–12,15,16] In particular, Refs. 4–7,13,16 have laid out the basics of generic structures in various situations. So we do not explain all of those details here.

Our notations and definitions are standard (see, e.g., Refs. 4,16). Let $L = \{R_i\}_{i\in I}$ be a finite relational language, where each R_i is a relation of (n-hyper)graph where $n \geq 2$, i.e., if $\models R_i(\bar{a})$ then the elements of $\bar{a}$ are without repetition and $\models R_i(\sigma(\bar{a}))$ for any permutation σ. Thus, for any L-structure A and R_i with arity n, R_i^A can be thought of as a set of n-element subsets of A. For a finite L-structure A, a *predimension* of A is defined by

$$\delta(A) = |A| - \sum_{i\in I} \alpha_i |R_i^A|$$

where $0 < \alpha_i \leq 1$. Let $\delta(B/A)$ denote $\delta(BA) - \delta(A)$. For $A \subset B$ and $n \in \omega$, A is *n-closed* in B (denoted $A \leq_n B$), if $\delta(X/A \cap X) \geq 0$ for any finite $X \subset B$ with $|X - A| \leq n$. In addition, A is *closed* in B (denoted $A \leq B$), if $A \leq_n B$ for any $n \in \omega$. The *closure* $\mathrm{cl}_B(A)$ of A in B is defined by $\bigcap\{C : A \leq C \leq B\}$. Let $\mathbf{K}_{\bar{\alpha}}$ be a class of the finite L-structures A satisfying $\delta(B) \geq 0$ for any $B \subset A$.

Assumption 2.1. *$\mathbf{K}$ is a subclass of $\mathbf{K}_{\bar{\alpha}}$ which is closed under substructures and isomorphisms.*

Definition 2.1. $\mathbf{K}$ has the *amalgamation property* (in short, AP), if for any $A, B, C \in \mathbf{K}$ with $A \leq B$ and $A \leq C$ there is $C' \cong_A C$ with $B \leq BC', C' \leq BC'$ and $BC' \in \mathbf{K}$.

Definition 2.2. A countable L-structure M is said to be $(\mathbf{K}, \leq)$-*generic*, if

(1) any finite $A \subset M$ belongs to $\mathbf{K}$;
(2) whenever $A \leq B \in \mathbf{K}$ and $A \leq M$, then there is $B' \cong_A B$ with $B' \leq M$;
(3) for any finite $A \subset M$, $|\mathrm{cl}_M(A)|$ is finite.

Remark 2.1. If $\mathbf{K}$ has AP then there is a unique $(\mathbf{K}, \leq)$-generic structure. (The proof is folklore.)

Assumption 2.2. *M is the $(\mathbf{K}, \leq)$-generic structure and $\mathcal{M}$ a big model of* $\mathrm{Th}(M)$.

Remark 2.2.

(1) If M is saturated then $\mathcal{M}$ also satisfies (1)–(3) of Definition 2.2.
(2) By (1) and a back-and-forth argument, if $A \leq B, B'(\leq \mathcal{M})$ are finite with $B \cong B'$ then $\mathrm{tp}(B/A) = \mathrm{tp}(B'/A)$.

Definition 2.3.

(1) Let ABC be an L-structure. Then B and C are *free* over A (denoted $B \perp_A C$), if $R_i^{ABC} = R_i^{AB} \cup R_i^{AC}$ for each $i \in I$.
(2) $\mathbf{K}$ has the *free amalgamation property* (in short, FAP), if for any $A, B, C \in \mathbf{K}$ with $A \leq B$ and $A \leq C$ there is $C' \cong_A C$ with $B \perp_A C'$ and $BC' \in \mathbf{K}$.

We abbreviate $\mathrm{cl}_{\mathcal{M}}(*)$ to $\mathrm{cl}(*)$. Let $\mathrm{acl}(A)$ denote the algebraic closure of A in $\mathcal{M}$.

Lemma 2.1.

(1) FAP implies AP.
(2) $\mathrm{cl}(A) \subset \mathrm{acl}(A)$ *for any* $A \subset \mathcal{M}$.
(3) If M is saturated and $\mathbf{K}$ has FAP, then $\mathrm{cl}(A) = \mathrm{acl}(A)$ *for any* $A \subset M$.

Proof. (1) and (2) are clear. For (3), suppose by way of contradiction that $\mathrm{acl}(A) \neq \mathrm{cl}(A)$. By 2 there is a finite $A_0 \leq \mathrm{cl}(A)$ with $\mathrm{acl}(A_0) - A_0 \neq \emptyset$. Take $\bar{e} \in \mathrm{acl}(A_0)$ with $\bar{e} \notin A_0$. Let $E = \mathrm{cl}(\bar{e}A_0)$ and $E_1, ..., E_n$ be the all conjugates of E over A_0. Let $D = \mathrm{cl}(E_1...E_n)$. By FAP there is $E^* \cong_{A_0} E$ with $D \perp_{A_0} E^*$ and $DE^* \in \mathbf{K}$. By genericity we can assume that $E^* \leq DE^* \leq \mathcal{M}$. By Remark 2.2 we have $\mathrm{tp}(E^*/A_0) = \mathrm{tp}(E/A_0)$. Clearly $E^* \not\subset D$. This contradicts to the assumption that $E_1, ..., E_n$ are all the conjugates of E over A_0. □

The following facts (Fact 2.1 and Fact 2.2) can be essentially found in Refs. 4,16.

For a finite $A \subset \mathcal{M}$, a *dimension* $d(A)$ of A is defined by $\min\{\delta(B) : A \subset B \subset \mathcal{M}, |B| < \aleph_0\}$. In Ref. 16, Wagner introduced the condition DS on a dimension d, and proved that under DS, any theory of a saturated generic structure is stable. Note that his dimension d is abstractly defined, but our d is defined by a concrete predimension $\delta(A) = |A| - \sum_i \alpha_i |R_i^A|$. Therefore it can be checked that our d satisfies DS (see Lemma 1.7 of Ref. 1). So we have the following fact.

Fact 2.1. *Assume that M is saturated. Then* $\mathrm{Th}(M)$ *is stable. In particular, if $\alpha_i = 1$ for each $i \in \omega$ then* $\mathrm{Th}(M)$ *is ω-stable.*

By Fact 2.1, the notion of *forking* can be defined in a theory of saturated generic structures. The following fact gives a characterization of forking in generic structures.

Fact 2.2. *Assume that M is saturated and $\mathbf{K}$ has FAP. Then for any $B, C \leq \mathcal{M}$ with $A = B \cap C$, the following are equivalent:*

(1) $\mathrm{tp}(B/C)$ *does not fork over A.*
(2) $B \perp_A C$ and $BC \leq \mathcal{M}$.

Remark 2.3. Since L is finite, for any $n \in \omega$ and finite $A \leq_n \mathcal{M}$, the property $A \leq_n \mathcal{M}$ is definable, i.e., there is $\varphi \in \mathrm{tp}(A)$ such that if φ satisfies A' then $A' \leq_n \mathcal{M}$.

Using Remark 2.3, we have Fact 2.3 below. This fact gives a characterization of the saturation of generic structures. It can be found in Refs. 4,7,16 and the detailed proof is in Ref. 7.

Fact 2.3. *The following are equivalent.*

(1) M is saturated.
(2) Whenever $A \leq B \in \mathbf{K}$ and $m \in \omega$, then there is $n \in \omega$ such that if $A \leq_n C \in \mathbf{K}$ then there is $C' \cong_A C$ with $B \leq_m BC'$, $C' \leq BC'$ and $B'C \in \mathbf{K}$.

Definition 2.4. For $A \subset B \in \mathbf{K}$, B is said to be A-*minimal*, if $A \leq_{|B-A|-1} B$ and $A \not\leq B$.

Lemma 2.2. *Assume that M is saturated. Let $A \leq B \leq \mathcal{M}$ be finite. Then the following are equivalent.*

(1) $\mathrm{tp}(B/A)$ *is isolated.*
(2) There is $n \in \omega$ *such that if* $B \cong_A B' \leq_n \mathcal{M}$ *then* $B' \leq \mathcal{M}$.
(3) There is $m \in \omega$ *such that if* $D \in \mathbf{K}$ *is* B*-minimal then* $|D| \leq m$.

Proof. The equivalence of (2) and (3) follows immediately from the definition of B-minimality. So we show the equivalence of (1) and (2). Since L is finite, clearly $X \cong_A B$ can be expressed by some formula $\varphi(X)$. By Remark 2.3, $X \leq_n \mathcal{M}$ can be also expressed by some formula $\psi_n(X)$. By Remark 2.2, we have $\{\varphi(x)\} \cup \{\psi_n(X)\}_{n\in\omega} \vdash \mathrm{tp}(B/A)$. It follows that the equivalence of (1) and (2) can be proved. □

Definition 2.5. $\mathbf{K}$ is said to have the *strong free amalgamation property* (in short, SFAP), if whenever $A \leq B \in \mathbf{K}$, then there is $n \in \omega$ such that if $A \leq_n C \in \mathbf{K}$ then there is $C' \cong_A C$ with $B \perp_A C'$ and $BC' \in \mathbf{K}$.

Remark 2.4.

(1) If $\mathbf{K}$ has SFAP then it has FAP.
(2) If $\mathbf{K}$ has SFAP then it satisfies (2) of Fact 2.3. Therefore if $\mathbf{K}$ has SFAP then M is saturated.
(3) By (2) and Fact 2.1, if $\mathbf{K}$ has SFAP then $\mathrm{Th}(M)$ is stable.

Remark 2.5. It is easily checked that Hrushovski's pseudoplane[8] and Herwig's generic structure[7] have the following common properties.

- $\mathbf{K}$ has FAP.
- For any $A \in \mathbf{K}$ there is $n \in \omega$ such that if $A \leq_n C \in \mathbf{K}$ then $A \leq C$.

From this, it follows that these examples satisfy SFAP.

Definition 2.6. $\mathbf{K}$ is said to be *closed under subgraphs (with respect to each relation in* L*)*, if whenever $B \in \mathbf{K}$ and $A|_R$ (the restriction of A to R) is a sub(hyper)graph of $B|_R$ for every $R \in L$, then $A \in \mathbf{K}$.

Remark 2.6. It is easy to construct a class $\mathbf{K}$ which is not closed under subgraphs (see Example 2.3 of Ref. 1), but almost all of known examples are closed under subgraphs.[2,7–10]

3. Proposition

Assumption 3.1. *M is a $(\mathbf{K}, \leq)$-generic structure. T is a complete theory of M and $\mathcal{M}$ a big model of T. $\mathbf{K}$ has SFAP and is closed under subgraphs.*

Notation 3.1. For $i \in I$ and $A \in \mathbf{K}$, we denote $r_i(A) = |R_i^A|$ and $r(A) = \sum r_i(A)$. For $i \in I$ and $AB \in \mathbf{K}$ with $A \cap B = \emptyset$, we denote $r_i(A, B) = r_i(AB) - r_i(A) - r_i(B)$, $r(A, B) = \sum r_i(A, B)$ and $\bar{\alpha}r(A, B) = \sum \alpha_i r_i(A, B)$. An L-structure A is said to be *discrete*, if it has no relations.

Lemma 3.1. *Assume that T is not ω-stable and that there is j with $\alpha_j = 1$ and $R_j^M \neq \emptyset$. Then there is a discrete $B \in \mathbf{K}$ and a t with $0 < t < 1$ such that for any $n \in \omega$ there is a B-minimal $D \in \mathbf{K}$ with $|D - B| \geq n$ and $\delta(D/B) \geq -1 + t$.*

Proof. Without loss of generality, we can assume that $R_i^M \neq \emptyset$ for each $i \in I$. Since T is not ω-stable, by Fact 2.1 there is k with $\alpha_k \neq 1$. So we can take $e\bar{e}$ with $\models R_k(e\bar{e})$ and $r(e\bar{e}) = 1$. Clearly $\bar{e} \leq \bar{e}e$. Since $R_k^M \neq \emptyset$ and $\mathbf{K}$ is closed under subgraphs, we have $e\bar{e} \in \mathbf{K}$. By SFAP there is $u \in \omega$ such that if $\bar{e} \leq_u C \in \mathbf{K}$ then there is $C' \cong_{\bar{e}} C$ with $e \perp_{\bar{e}} C'$ and $eC' \in \mathbf{K}$. Take t with $0 < t \leq 1 - \alpha_k$. Let $j_0 = \text{arity}(R_j)$ and $k_0 = \text{arity}(R_k)$. Fix any $n \in \omega$. Take a sufficiently large $m \in \omega$ for n, j_0, k_0 and u.

Claim 1: There are sequences $b_1, ..., b_m$ and $B_0, ..., B_m$ of elements of $\mathbf{K}$ with the following:

(1) $B_0 = \emptyset$;
(2) $B_{i+1} = B_i b_{i+1}$ for $i = 0, ..., m-1$;
(3) $b_{i-j_0+2}...b_i$ is discrete for $i = j_0 - 1, ..., m-1$;
(4) $B_{i+1} \models R_j(b_{i-j_0+2}...b_{i+1})$ and $r(b_{i+1}, B_i) = 1$ for $i = j_0 - 1, ..., m-2$;
(5) $B_m \models R_k(\bar{b}b_{m-1}b_m)$ for some $\bar{b} \in B_{m-2}$ and $r(b_m, B_{m-1}) = 1$.

Proof: We prove by induction on i. Suppose that $b_1, ..., b_i$ and $B_1, ..., B_i$ have been constructed. Since $\mathbf{K}$ is closed under subgraphs, every finite discrete graph belongs to $\mathbf{K}$. So we can assume that $i \geq j_0 - 1$.
Case 1: $i < m - 1$.

Let $\bar{b} = b_{i-j_0+2}...b_i$. By induction hypothesis, $\bar{b}$ is discrete and $\bar{b} \leq B_i \in \mathbf{K}$. Take b with $\models R_j(b\bar{b})$ and $r(b, \bar{b}) = 1$. Then $\bar{b} \leq b\bar{b} \in \mathbf{K}$. By FAP we can assume that $b \perp_{\bar{b}} B_i$ and $B_i b \in \mathbf{K}$. Let $b_{i+1} = b$ and $B_{i+1} = B_i b$. Then (2), (3) and (4) hold.
Case 2: $i = m - 1$.

Since m is sufficiently large for n, j_0, k_0 and u, we can take $\bar{b} \in B_{m-2}$ with $\text{length}(\bar{b}) = k_0 - 2, \bar{b}b_{m-1} \leq_u B_{m-1}(\in \mathbf{K})$ and $\bar{b}b_{m-1}$ discrete. Take b with $\models R_k(\bar{b}b_{m-1}b)$ and $r(b, B_{m-1}) = 1$. Clearly $\bar{b}b_{m-1} \leq \bar{b}b_{m-1}b$. Since $\bar{b}b_{m-1}$ is discrete, we have $\bar{b}b_{m-1}b \in \mathbf{K}$. By our choice of u and SFAP, we can assume that $b \perp_{\bar{b}b_{m-1}} B_{m-1}$ and $B_{m-1}b \in \mathbf{K}$. Let $b_m = b$ and $B_m = B_{m-1}b$. Then (2) and (5) hold.

Let $B = B_{j_0-1}b_m$ and $D = B_m$. By our construction, B is discrete. Since m is sufficiently large, we can assume that $|D - B| = m - j_0 + 1 \geq n$.

Claim 2: $0 > \delta(D/B) \geq -1 + t$.

Proof: By claim 1, we have

$$\begin{aligned}
\delta(D/B) &= \sum_{i=j_0}^{m-1} \delta(b_i/B_{i-1}b_m) \\
&= \sum_{i=j_0}^{m-2} \delta(b_i/b_{i-j_0+1}...b_{i-1}) + \delta(b_{m-1}/B_{m-2}b_m) \\
&= (m - j_0 - 1)(1 - \alpha_j) + \delta(b_{m-1}) - \bar{\alpha}r(b_{m-1}, B_{m-2}b_m) \\
&= 1 - \alpha_j - \alpha_k = -\alpha_k.
\end{aligned}$$

Hence $0 > \delta(D/B) \geq -1 + t$.

Claim 3: D is B-minimal.

Proof: It is enough to see that $\delta(X/B) \geq 0$ for any proper subset X of $D - B$. Note that $D - B$ can be considered as a sequence $b_{j_0}...b_{m-1}$. First assume that X is a segment of $b_{j_0}...b_{m-1}$. Let b_f be the first term of X and b_l the last term of X. We divide into two cases.

Case 1: $b_l = b_{m-1}$.

Clearly $b_f \neq b_{j_0}$, and hence $\delta(b_f/B) = 1$. Then

$$\begin{aligned}
\delta(X/B) &= \delta(b_f...b_{m-1}/B) \\
&= \delta(b_f/B) + \delta(b_{f+1}/b_fB) + \cdots + \delta(b_{m-1}/b_f...b_{m-2}B) \\
&\geq \delta(b_f/B) + \sum_{i=f+1}^{m-2} \delta(b_i/B_{i-1}) + \delta(b_{m-1}/B_{m-2}b_m) \\
&= 1 + (m - j_0 - 2)(1 - \alpha_j) + (1 - \alpha_j - \alpha_k) \\
&= 1 - \alpha_k > 0.
\end{aligned}$$

Case 2: $b_l \neq b_{m-1}$.

For every $i < m-1$, we have $\delta(b_i/B_{i-1}b_m) = \delta(b_i/B_{i-1})$. Then

$$\begin{aligned}\delta(X/B) &= \delta(b_f...b_l/B)\\ &\geq \delta(b_f...b_l/B_{f-1}b_m)\\ &= \delta(b_f/B_{f-1}b_m) + \delta(b_{f+1}/B_f b_m) + \cdots + \delta(b_l/B_{l-1}b_m)\\ &= \delta(b_f/B_{f-1}) + \delta(b_{f+1}/B_f) + \cdots + \delta(b_l/B_{l-1})\\ &= (l-f+1)(1-\alpha_j) = 0.\end{aligned}$$

Therefore we have $\delta(X/B) \geq 0$ for any case.

Next assume that X is not a segment of $b_{j_0}...b_{m-1}$. Let $X_0, ..., X_s$ be the segments of $b_{j_0}...b_{m-1}$ with $X = X_0 \cup \cdots \cup X_s$. Then we have $\delta(X/B) = \delta(X_0/B) + ... + \delta(X_s/B) \geq 0$. Hence D is B-minimal. □

For graphs $G = (V, E), G' = (V', E')$, G' is said to be a spanning subgraph of G, if $V' = V$ and $E' \subset E$. For L-structures D, E, we call E a *spanning subgraph* of D (with respect to each relation in L), if $E|_{R_i}$ is a spanning sub(hyper)graph of $D|_{R_i}$ for every $i \in I$.

Lemma 3.2. *Assume that T is not ω-categorical and that there is no j with $\alpha_j = 1$ and $R_j^M \neq \emptyset$. Then there is a discrete $B \in \mathbf{K}$ and a t with $0 < t < 1$ such that for any $n \in \omega$ there is a B-minimal $D \in \mathbf{K}$ with $|D - B| \geq n$ and $\delta(D/B) \geq -1 + t$.*

Proof. Without loss of generality, we can assume that $R_i^M \neq \emptyset$ for each $i \in I$. So we assume that there is no j with $\alpha_j = 1$. Since T is not ω-categorical, there is a finite $B \leq \mathcal{M}$ such that $\mathrm{tp}(B)$ is nonisolated. By Lemma 2.2, for any $n \in \omega$ there is a B-minimal $D_n \in \mathbf{K}$ with $|D_n - B| \geq n$. Since $\mathbf{K}$ is closed under subgraphs, we can assume that B is discrete, and moreover that

(*) for each n, if E_n is a spanning subgraph of D_n then $\delta(E_n/B) > 0$.

Let $t = 1 - \max\{\alpha_i : i \in I\}$. By our assumption, we have $t > 0$. We show that $\delta(D_n/B) \geq -1 + t$ for each $n \in \omega$. Suppose that there is $n \in \omega$ with $\delta(D_n/B) < -1 + t$. Let $E_n (\supset B)$ be a spanning subgraph of D_n which is obtained by removing one (hyper)edge from D_n. Then $\delta(E_n/B) \leq \delta(D_n/B) + \max\{\alpha_i : i \in I\} < -1 + t + \max\{\alpha_i : i \in I\} = 0$. This contradicts (*). □

Proposition 3.1. *Assume that T has more than one but finitely many countable models. Then there is a discrete $B \in \mathbf{K}$ and a t with $0 < t < 1$*

such that for any $n \in \omega$ there is a B-minimal $D \in \mathbf{K}$ with $|D - B| \geq n$ and $\delta(D/B) \geq -1 + t$.

Proof. It is known that if T has more than one but finitely many countable models, then it is not superstable.[14] So T is neither ω-categorical nor ω-stable. By Lemma 3.1 and 3.2 we have our proposition. □

4. Theorem

We summarize our situation: $L = \{R_i\}_{i \in I}$ is a finite relational language, where each R_i is a relation of a hypergraph. A predimension is $\delta(A) = |A| - \sum_i \alpha_i |R_i^A|$, where $0 < \alpha_i \leq 1$. $\mathbf{K}_{\bar{\alpha}}$ is a class of all finite structures whose predimension is hereditarily nonnegative. $\mathbf{K}$ is a subclass of $\mathbf{K}_{\bar{\alpha}}$ which is closed under substructures and isomorphisms.

Theorem 4.1. *Assume that $\mathbf{K}$ has SFAP and is closed under subgraphs. Let T be a theory of a $(\mathbf{K}, \leq)$-generic structure M. Then the number of countable models of T is one or infinite.*

Proof. Let $\mathcal{M}$ be a big model of T. Since $\mathbf{K}$ has SFAP, by Remark 2.4, T is stable. Suppose by way of contradiction that T has more than one but finitely many countable models. Then there is a *powerful type* $p \in S(T)$, i. e., any model prime over a realization of p realizes every type over $\emptyset$. Let A be a realization of p. We can assume that $A \leq \mathcal{M}$. By Proposition 3.1, there is a discrete $B \in \mathbf{K}$ and a t with $0 < t < 1$ such that for any $n \in \omega$ there is a B-minimal $D \in \mathbf{K}$ with $|D - B| \geq n$ and $\delta(D/B) \geq -1 + t$. We can assume that $B \leq \mathcal{M}$. Take $k \geq |A| + 4\delta(A)/t$. Let $B_1, ..., B_k$ be a Morley sequence of $\mathrm{tp}(B)$. Since p is powerful, we can assume that each $\mathrm{tp}(B_i/A)$ is isolated. In other words, $A, B_1, ..., B_k$ satisfies the following properties:

- $\mathrm{tp}(B_i) = \mathrm{tp}(B)$ for each $i = 1, ..., k$;
- $\{B_1, ..., B_k\}$ is independent (in a sense of forking);
- $\mathrm{tp}(B_i/A)$ is isolated for each $i = 1, ..., k$.

Claim 1: $B_i \cap B_j = \emptyset$ for each $i \neq j$.

Proof: Since B_i and B_j are independent, we have $B_i \cap B_j \subset \mathrm{acl}(\emptyset)$. By Lemma 2.1, we have $\mathrm{acl}(\emptyset) = \emptyset$. Hence $B_i \cap B_j = \emptyset$.

Claim 2: There is $I_0 \subset \{1, ..., k\}$ with $|I_0| \geq 4\delta(A)/t$ such that $A \cap B_i = \emptyset$ for each $i \in I_0$.

Proof: Let $J_0 = |\{i : A \cap B_i \neq \emptyset\}|$. By Claim 1 we have $|J_0| \leq |A|$. Let $I_0 = \{1, ..., k\} - J_0$. Then we have $|I_0| \geq 4\delta(A)/t$.

Claim 3: There is $I_1 \subset I_0$ with $|I_1| \geq 2\delta(A)/t$ such that $\bar{\alpha}r(A, B_i) < t/2$ for each $i \in I_1$.

Proof: Let $J_1 = \{i \in I_0 : \bar{\alpha}r(A, B_i) \geq t/2\}$. Suppose that $|J_1| > 2\delta(A)/t$. Then $\delta(A/\bigcup_{i\in J_1} B_i) \leq \delta(A) - \sum_{i\in J_1} \bar{\alpha}r(A, B_i) \leq \delta(A) - |J_1|t/2 < 0$. On the other hand, since $\{B_i\}_{i\in J_1}$ is independent, by Fact 2.2, we have $\bigcup_{i\in J_1} B_i \leq M$. A contradiction. Hence $|J_1| \leq 2\delta(A)/t$. Let $I_1 = I_0 - J_1$. Then we have $|I_1| \geq 2\delta(A)/t$.

For each $i \in I_1$ let $C_i = \mathrm{cl}(AB_i)$. Since $A \leq C_i \in \mathbf{K}$, by FAP, for each $i \in I_1$ there is $C_i' \cong_A C_i$ such that $\{C_i'\}_{i\in I_1}$ is free over A and $\bigcup_{i\in I_1} C_i' \in \mathbf{K}$. We can assume that $\bigcup_{i\in I_1} C_i' \leq \mathcal{M}$. For each $i \in I_1$ let B_i' be a copy of B_i with $B_i'C_i' \cong_A B_iC_i$.

Claim 4: $\{B_i'\}_{i\in I_1}$ satisfies the following.

(1) $\mathrm{tp}(B_i') = \mathrm{tp}(B)$ for each $i \in I_1$;
(2) $\{B_i'\}_{i\in I_1}$ is independent;
(3) $\mathrm{tp}(B_i'/A)$ is isolated for each $i \in I_1$.

Proof: Since $A \leq C_i' \leq (\bigcup_{i\in I_1} C_i' \leq)\mathcal{M}, A \leq C_i \leq \mathcal{M}$ and $C_i' \cong_A C_i$, we have $\mathrm{tp}(C_i'/A) = \mathrm{tp}(C_i/A)$, and therefore $\mathrm{tp}(B_i'/A) = \mathrm{tp}(B_i/A)$. Hence (1) and (3) hold. We show (2) by induction. Let $i_0, ..., i_m$ be an enumeration of I_1. Assume that $\{B_{i_0}', ..., B_{i_s}'\}$ has been independent. Let

- $E_0 = C_{i_0} \cup ... \cup C_{i_s}$, $E_1 = C_{i_{s+1}}$, $E_0' = C_{i_0}' \cup ... \cup C_{i_s}'$, $E_1' = C_{i_{s+1}}'$,
- $F_0 = B_{i_0} \cup ... \cup B_{i_s}$, $F_1 = B_{i_{s+1}}$, $F_0' = B_{i_0}' \cup ... \cup B_{i_s}'$, $F_1' = B_{i_{s+1}}'$.

Then we show that F_0' and F_1' are independent. Note that $F_0' \perp F_1'$. By Fact 2.2, it is enough to see that $F_0'F_1' \leq E_0'E_1'$. Take any $X' \subset E_0'E_1' - F_0'F_1'$. Let $f : E_0'E_1' \to E_0E_1$ be a map such that $f|_{C_{i_0}'}, ..., f|_{C_{i_s}'}$ are A-isomorphisms. Let

- $X_0' = X' \cap (E_0' - f^{-1}(E_0 \cap E_1))$, $Y_0' = X' \cap f^{-1}(E_0 \cap E_1) \cap (E_0' - A)$,
- $X_1' = X' \cap (E_1' - f^{-1}(E_0 \cap E_1))$, $Y_1' = X' \cap f^{-1}(E_0 \cap E_1) \cap (E_1' - A)$,
- $X_0 = f(X_0')$, $Y_0 = f(Y_0')$, $X_1 = f(X_1')$, $Y_1 = f(Y_1')$,
- $Z = X' \cap A$, $Y = Y_0 \cap Y_1$, $Y_0^* = Y_0 - Y$.

Then

$$\begin{aligned}
&\delta(X') - \delta(f(X'))\\
&= \delta(X_0'Y_0'ZX_1'Y_1') - \delta(X_0Y_0ZX_1Y_1)\\
&= \delta(X_0'Y_0'/ZX_1'Y_1') + \delta(ZX_1'Y_1') - \delta(X_0Y_0ZX_1Y_1)\\
&= \delta(X_0'Y_0'/Z) + \delta(ZX_1'Y_1') - \delta(X_0Y_0ZX_1Y_1)\\
&= \delta(X_0Y_0/Z) + \delta(ZX_1Y_1) - \delta(X_0Y_0ZX_1Y_1)\\
&= \delta(X_0Y_0/Z) + \delta(ZX_1Y_1) - \delta(ZX_1Y_1) - \delta(X_0Y_0/ZX_1Y_1)\\
&= \delta(X_0Y_0/Z) - \delta(X_0Y_0/ZX_1Y_1)\\
&= \delta(X_0YY_0^*/Z) - \delta(X_0Y_0^*/ZX_1Y_1)\\
&= \delta(X_0Y_0^*/YZ) + \delta(Y/Z) - \delta(X_0Y_0^*/ZX_1Y_1)\\
&\geq \delta(Y/Z) \geq 0.
\end{aligned}$$

On the other hand, it is clear that $r_j(X', F_0'F_1') = r_j(f(X'), F_0F_1)$ for each $j \in I$. Since $\{B_i\}_{i\in I_1}$ is independent, we have $F_0F_1 \leq \mathcal{M}$. Then $\delta(X'/F_0'F_1') = \delta(X') - \bar{\alpha}r(X', F_0'F_1') \geq \delta(f(X')) - \bar{\alpha}r(f(X'), F_0F_1) = \delta(f(X')/F_0F_1) \geq 0$. Hence $F_0'F_1' \leq E_0'E_1'$.

By Claim 4, without loss of generality, we can assume that $B_i = B_i'$ and $C_i = C_i'$ for each $i \in I_1$.

Claim 5: There is $i \in I_1$ with $-t/2 \leq \delta(C_i/AB_i)$.

Proof: Suppose that $\delta(C_i/AB_i) < -t/2$ for all $i \in I_1$. Then

$$\begin{aligned}
&\delta(C_{i_0}...C_{i_m}/B_{i_0}...B_{i_m})\\
&= \delta(C_{i_0}...C_{i_m}/B_{i_0}...B_{i_m}A) + \delta(A/B_{i_0}...B_{i_m})\\
&= \sum_{j=0}^{m} \delta(C_{i_j}/C_{i_0}...C_{i_{j-1}}B_{i_0}...B_{i_m}A) + \delta(A/B_{i_0}...B_{i_m})\\
&\leq \sum_{j=0}^{m} \delta(C_{i_j}/B_{i_j}A) + \delta(A)\\
&< -|I_1|t/2 + \delta(A) \leq 0
\end{aligned}$$

On the other hand, since $\{B_i\}_{i\in I_1}$ is independent, by Fact 2.2 we have $\bigcup_{i\in I_1} B_i \leq \mathcal{M}$. A contradiction.

Fix $i \in I_1$ satisfying Claim 5, and write $B = B_i$ and $C = C_i$. Since $\mathrm{tp}(C/A)$ is isolated, by Lemma 2.2 there is $n \in \omega$ such that $C \cong_A C' \leq_n M$ implies $C' \leq \mathcal{M}$ (and hence $\mathrm{tp}(C'/A) = \mathrm{tp}(C/A)$). On the other hand, since $B \leq C \in \mathbf{K}$, by SFAP there is $m \in \omega$ such that if $B \leq_m F \in \mathbf{K}$ then there is $F' \cong_B F$ with $CF' \in \mathbf{K}$ and $C \perp_B F'$. By our choice of B and t, there is B-minimal $D \in \mathbf{K}$ with $|D-B| > \max(n,m)$ and $\delta(D/B) \geq -1+t$. By SFAP we can assume that $C \perp_B D$ and $CD \in \mathbf{K}$.

Claim 6: $A \leq CD$.

Proof: Take any $X \subset CD-A$. Let $X_C = X \cap (C-AB), X_B = X \cap B, X_D = X \cap (D-C)$. Assume $X_B = \emptyset$. Then

$$\begin{aligned}&\delta(X/A)\\&= \delta(X_D/X_CA) + \delta(X_C/A)\\&= \delta(X_D) + \delta(X_C/A) \geq 0.\end{aligned}$$

Assume $X_B \neq \emptyset$. Since B is discrete, by Claim 3 we have $\delta(X_B/A) = \delta(X_B) - \bar{\alpha}r(X_B, A) \geq 1 - t/2$. So

$$\begin{aligned}&\delta(X/A)\\&= \delta(X_D/X_CX_BA) + \delta(X_C/X_BA) + \delta(X_B/A)\\&= \delta(X_D/X_B) + \delta(X_C/X_BA) + \delta(X_B/A)\\&\geq \delta(X_D/B) + \delta(X_C/BA) + \delta(X_B/A)\\&\geq \delta(D/B) + \delta(C/BA) + \delta(X_B/A)\\&\geq (-1+t) - t/2 + (1-t/2) = 0.\end{aligned}$$

Hence we have $A \leq CD$.

By Claim 6 we can take A-copy $C'D'$ of CD with $C'D' \leq \mathcal{M}$. In particular $C' \leq_n \mathcal{M}$ and $C' \not\leq \mathcal{M}$. Then $\mathrm{tp}(C'/A) \neq \mathrm{tp}(C/A)$. This contradicts our choice of n. □

References

1. Y. Anbo and K. Ikeda, A Note on stability spectrum of generic structures, submitted
2. J. T. Baldwin, An almost strongly minimal non-Desarguesian projective plane, Transactions of American Mathematical Society 342 (1994) 695–711
3. J. T. Baldwin, Rank and homogeneous structures, *Tits buildings and the model theory of groups (Wurzburg, 2000)*, Cambridge University Press (2002) 215–233
4. J. T. Baldwin and N. Shi, Stable generic structures, Annals of Pure and Applied Logic 79 (1996) 1–35
5. D. Evans, $\aleph_0$-categorical structures with a predimension, Annals of Pure and Applied Logic 116 (2002) 157–186

6. J. Goode, Hrushovski's geometries, In Helmut Wolter Bernd Dahn, editor, *Proceedings of 7th Easter Conference on Model Theory* (1989) 106–118
7. B. Herwig, Weight ω in stable theories with few types, Journal of Symbolic Logic 60 (1995) 353–373
8. E. Hrushovski, A stable $\aleph_0$-categorical pseudoplane, preprint, 1988
9. E. Hrushovski, A new strongly minimal set, Annals of Pure and Applied Logic 62 (1993) 147–166
10. K. Ikeda, Minimal but not strongly minimal structures with arbitrary finite dimension, Journal of Symbolic Logic 66 (2001) 117–126
11. K. Ikeda, A note on generic projective planes, Notre Dame Journal of Formal Logic 43(2002) 249–254
12. K. Ikeda, A remark on the stability of saturated generic graphs, Journal of the Mathematical Society of Japan 57 (2005) 1229–1234
13. D. W. Kueker and C. Laskowski, On generic structures, Notre Dame Journal of Formal Logic 33(1992) 147–166
14. A. H. Lachlan, On the number of countable models of a countable superstable theory, *Logic, methodology and philosophy of science, IV (Proc. Fourth Internat. Congr. (Bucharest, 1971)*, Studies in Logic and Foundations of Math., Vol. 74, North-Holland, Amsterdam (1973) 45–56
15. Bruno Poizat, Amalgames de Hrushovski, *Tits buildings and the model theory of groups (Wurzburg, 2000)*, Cambridge University Press (2002) 195–214
16. F. O. Wagner, Relational structures and dimensions, In *Automorphisms of first-order structures*, Clarendon Press, Oxford (1994) 153–181

GEOMETRIC SIMPLICITY THEORY

BYUNGHAN KIM*

Department of Mathematics
Yonsei University
134 Shinchon-dong, Seodaemun-gu
Seoul 120-749, South Korea
E-mail: bkim@yonsei.ac.kr

The class of *simple* structures, properly containing that of *stable* structures is introduced by Shelah.[51] Specific algebraic simple unstable structures are studied by Hrushovski and others.[8,9,24,27,29] After the author establishes in mid 90s the symmetry and transitivity of forking,[30,31] and with Pillay,[38] type-amalgamation over models, an enormous amount of works has been made on *simplicity theory*, the study of simple structures, by generalizing the results on stability theory(=the study of stable structures), discovering more complicated nature of simple structures, and producing the interplay between stability theory and simple structures.

The initial developments of simplicity theory, mostly focused on general foundational issues, are surveyed in[39] and also an expository book[56] by F. O. Wagner is published in 2000.

During the last several years, the research trend of simplicity theory has moved toward so called *geometric* simplicity theory, primarily concerned about combinatorial geometric or algebraic sides of simplicity theory, parallel to *geometric stability theory*.[45] It is well-known geometric stability theory is a major technical bridge connecting pure model theory and its applications to algebraic geometry and number theory. Hrushovski's full resolution of the function field version Mordell-Lang conjecture in number theory[21] is a spectacular example of such.

In this note, we try to summarize (a part of) the recent works on geometric simplicity theory. It should be noted that we have absolutely no intention to make an encyclopedic presentation, which by now is an al-

*The author was supported by KRF grant 2009-0075931.

most impossible task as too much are developed to completely absorb (even present) all. The results in this paper are selected (within the author's narrow knowledge) mainly to expedite reaching the explanations of current works on geometric simplicity theory. The content is a skimmed skeleton of the author's book which will be published soon.[36] This is also a summary of the author's tutoring lectures at Asian Logic Conference 10, Kobe University, Japan, September 1-6, 2008. We thank the conference organizers, in particular Toshiyasu Arai and Hirotaka Kikyo, for their hospitality, and encouragements for writing this note.

We take standard terminology. T is a complete theory in the language $\mathcal{L}$. As usual we fix a saturated model $\mathcal{M}$ of T having cardinality $\bar{\kappa}$. We indeed work in $\mathcal{M}^{\text{eq}}$. Unless mentioned otherwise all tuples and sets $A, B, C, ...,$ are from $\mathcal{M}^{\text{eq}}$ of sizes $< \bar{\kappa}$. Models M, N are also elementary submodels of small sizes. We will deal with infinite tuples but mostly $a, b, x, y, ..$ are finite tuples. For a type p, $\text{dom}(p)$ is the parameter set for p. Abusive notation may be used such as AB, or Ab for $b = (b_1, ..., b_n)$, which indeed means $A \cup B$ or $A \cup \{b_1, ..., b_n\}$ respectively. When we say *bounded* sets, or *bounded* indexes, we mean they are small, i.e. $< \bar{\kappa}$. We write $a \equiv_A b$ to denote $tp(a/A) = tp(b/A)$. Recall that

$$\text{acl}(A) := \{b \in \mathcal{M}^{\text{eq}} \mid b \text{ is algebraic over } A, \text{ i.e. there are only finitely many solutions to } \text{tp}(b/A)\}, \text{ and}$$

$$\text{dcl}(A) := \{b \in \mathcal{M}^{\text{eq}} \mid b \text{ is definable over } A, \text{ i.e. } b \text{ is the unique solution to } \text{tp}(b/A)\}.$$

When we write a formula as $\psi(x)$, it may contain parameters, but if written $\varphi(x; y)$ it usually is an $\mathcal{L}$-formula, and $\varphi(x, c)$ with some parameters c is said to be an *instance* of φ. Recall that the *name* of $\psi(x)$ is the canonical code $e \in \mathcal{M}^{\text{eq}}$ such that for any automorphism f of $\mathcal{M}^{eq}$, f fixes the solution set of $\psi(x)$ iff it fixes e. For a rank R, by $R(a/A, ...)$ we mean $R(\text{tp}(a/A), ...)$.

1. Forking and Simplicity

In this section we present general basic facts on forking and simple theories. Most of the results here are mentioned in,[38] and their proofs can be found in.[56]

Definition 1.1.

(1) For an integer $k \geq 1$, and a formula $\varphi(x, c)$, we say $\varphi(x, c)$ *k-divides over* a set A if there is a sequence $\langle c_i | i \in \omega \rangle$ such that

(a) $\text{tp}(c/A) = \text{tp}(c_i/A)$ for all i, and

(b) $\{\varphi(x, c_i) | i \in \omega\}$ is k-inconsistent, i.e. any finite subset of size k is inconsistent.

(2) A type $p(x)$, *k-divides over A* if there is a formula $\varphi(x, a)$ such that $p \vdash \varphi(x, a)$, and $\varphi(x, a)$ k-divides over A. The type p *divides over A* if it k-divides over A for some k. (Conventionally, any inconsistent formula divides over any set.)

(3) p *forks over A* if there are formulas $\varphi_1(x, a_1), ..., \varphi_n(x, a_n)$ such that
   (a) $p \vdash \bigvee_{1 \leq i \leq n} \varphi_i(x, a_i)$, and
   (b) $\varphi_i(x, a_i)$ divides over A for each i.

(4) For $A \subseteq B$, if a complete type p over B does not fork (divide, resp.) over A, then we say p is a *nonforking (nondividing, resp.) extension* of $p \lceil A$, and conversely the latter an *unforked (undivided, resp.) restriction* of the former.

(5) Given $p \in S(A)$, we say $I = \langle a_i | \; i < \omega \rangle$ is a *Morley sequence* of p if I is an A-indiscernible sequence, and for $i \in \omega$, $\mathrm{tp}(a_{i+1}/Aa_0...a_i)$ is a nonforking extension of p.

From the definition, the following easily follows.

Fact 1.2. (Finite character) *Let $A \subseteq B$. Then* $\mathrm{tp}(a/B)$ *does not fork/divide (resp.) over A iff for any finite $b \in B$,* $\mathrm{tp}(a/Ab)$ *does not fork/divide (resp.) over A.*

The prototypical example of dividing appears in the theory of equivalence relation E having infinitely many infinite equivalence classes. There, xEa 2-divides over the empty set. There is an example of k-dividing for $k > 2$. Consider $(\mathbb{Q}^2, E)$ where, for any $a = (a_1, a_2), b = (b_1, b_2) \in \mathbb{Q}^2$, $E(a, b)$ iff ($a_1 = a_2$ or $b_1 = b_2$). Here xEb 3-divides (but does not 2-divide) over $\emptyset$.

Intuitively dividing is the right notion for dependence: For type-definable non-disjoint sets $X, Y \subseteq \mathcal{M}^n$, we may temporarily say that Y *k-divides* X if there are arbitrarily many automorphisms f_i fixing X set-wise such that $\{Y_i = f_i(Y)\}_i$ is k-inconsistent. Then $X_i = X \cap Y_i \neq \emptyset$ as well. It intuitively means that the set Y chops up X into unboundedly many pieces X_i by conjugates Y_i. It follows that $\mathrm{tp}(c/Ab)$ divides over A iff the set Y of realizations of $\mathrm{tp}(c/Ab)$ *divides* X, that of $\mathrm{tp}(c/A)$. This, in a certain sense, means that c is satisfying more relations (or dependent) with Ab than it is with A.

Why, then is forking introduced over dividing? With forking, we have the following extension axiom which is one of crucial factors establishing further arguments. From now on we write $A \mathop{\smile\!\!\!\!\!\!\,|}_B C$ if for any finite $a \in A$, $tp(a/BC)$ does not divide over B.

Fact 1.3.[52] (Extension) *Let $A \subseteq B \subseteq C$. If $p \in S(B)$ does not fork over A, then there is an extension $q \in S(C)$ of p, which does not fork over A; equivalently, if $c \mathop{\smile\!\!\!\!\!\!\,|}_A B$, then there is $c' \equiv_B c$ such that $c' \mathop{\smile\!\!\!\!\!\!\,|}_A C$.*

The following is the core fact of forking.

Theorem 1.4.[31,34] (Fundamental theorem of forking) *Let T be arbitrary. The following are equivalent.*

(1) Forking[a] *satisfies symmetry: For any A, B, C, $A \mathop{\smile\!\!\!\!\!\!\,|}_B C$ iff $C \mathop{\smile\!\!\!\!\!\!\,|}_B A$.*
(2) Forking satisfies transitivity: For any $A \subseteq B \subseteq C$ and D, $D \mathop{\smile\!\!\!\!\!\!\,|}_A C$ iff $D \mathop{\smile\!\!\!\!\!\!\,|}_A B$ and $D \mathop{\smile\!\!\!\!\!\!\,|}_B C$.
(3) Forking satisfies local character: For any set A and finite d, there is $A_d (\subseteq A)$ of size $\leq |\mathcal{L}|$ such that $d \mathop{\smile\!\!\!\!\!\!\,|}_{A_d} A$.

Definition 1.5.

(1) T is said to be *simple* if one of the above equivalent properties of forking holds.
(2) T is *unstable* if there is $\varphi(x,y)$ and $a_i, b_i \in \mathcal{M}$ $(i < \omega)$ such that $\mathcal{M} \models \varphi(a_i, b_j)$ iff $i < j$.
(3) T is *stable* if it is not unstable.

Fact 1.6.

(1) If T is stable, then it is simple.
(2) [31] *Assume T is simple. Then any complete type has a Morley sequence. Moreover, the two notions of forking and dividing coincide. In fact the following are equivalent.*

(a) $\varphi(x,a)$ divides over A.
(b) For any Morley sequence I of tp(a), $\{\varphi(x,a') | \ a' \in I\}$ *is inconsistent.*
(c) For some Morley sequence I of tp(a), $\{\varphi(x,a') | \ a' \in I\}$ *is inconsistent.*
(d) $\varphi(x,a)$ forks over A.

---

[a] That forking satisfies the property X is the same amount of saying nonforking satisfies X.

(3) [51] *T is simple iff dividing satisfies local character for 1-types, i.e. for any $p \in S_1(A)$, there is $A_d \subseteq A$ with $|A_d| \le |\mathcal{L}|$ such that p does not divides over A_d.*

Now in simple T, we say a family of sets $\mathcal{C} = \{C_i |\ i \in I\}$ is *A-independent* (equivalently saying, *independent over A*) if

$$C_i \underset{A}{\perp} \bigcup \{C_j |\ j (\neq i) \in I\}$$

holds.

It is a good exercise, using symmetry and transitivity, to check that $\mathcal{C} = \{C_i |\ i \in \omega\}$ is *A-independent* iff $C_{i+1} \perp_A C_0 ... C_i$ for all $i \in \omega$. In particular, I is a Morley sequence of p iff I, a sequence in p, is indiscernible and independent over $\mathrm{dom}(p)$.

Definition 1.7. (with Fact)

(1) We say T is *supersimple* if for any set A and finite d, there is finite $A_d (\subseteq A)$ such that $d \perp_{A_d} A$.
(2) We say T is *superstable* if it is supersimple and stable.
(3) Recall that countable T is said to be *ω-stable* if there are only countably many complete types over a countable model.
(4) Fact: T is supersimple iff for any $p \in S_1(A)$, p does not divide over some finite $A_d (\subseteq A)$.

We present some examples of simple structures.

Example 1.8.
Classification chart of simple structures:

$$\begin{array}{llll} & & \subset \text{supersimple} & \\ \aleph_1\text{-categorical} \subset \omega\text{-stable} \subset \text{superstable} & & & \subset \text{simple} \\ & & \subset \text{stable} & \end{array}$$

Examples in (n) below is not in the category of (m) for $m < n$. Applying quantifier elimination to 1.6.3, 1.7.4 is the key idea verifying the membership of the example in the given category. For algebraic examples, analysis of definable sets (in the case of pseudo-finite fields[9]) or giving an independence notion to check type-amalgamation over a model (in the case of ACFA, smoothly approximated structures[8,24]) is the due course to verify (super)simplicity.

(1) $\aleph_1$-categorical structures: An infinite set. A vector space $(V, +, 0, \{r : V \to V\} |\ r \in F\})$ (each r is a scalar multiplication) over a field F. Algebraically closed fields.

(2) ω-stable structures: Differentially closed fields. A theory of equivalence relations E_i $(i \in \omega)$ such that E_i refines each E_{i+1}-class into infinitely many infinite E_i-classes.
(3) Superstable structures: $(\mathbb{N}, +, 0)$. A theory of equivalence relations E_i $(i \in \omega)$ such that E_{i+1} refines each E_i-class into two infinite E_{i+1}-classes.
(4) Stable structures: Separably (but not algebraically) closed fields. A theory of equivalence relations E_i $(i \in \omega)$ such that E_{i+1} refines each E_i-class into infinitely many infinite E_i-classes.
(5) Supersimple structures: The theory of the random graph. The theory of the bipartite random graph. Pseudo-finite fields(= infinite models of the theory of all finite fields). ACFA(=the model companion of the theory of fields equipped with an automorphism). (Unstable) smoothly approximated structures such as a vector space over a finite field (as in (1)) equipped with a non-degenerated bilinear form. Any (non-separably closed) pseudo-algebraically closed fields (see 8.1.1) is perfect and bounded.
(6) Simple structures: (Imperfect bounded) Pseudo-algebraically closed fields.

In section 8, we will get back to those field examples more in detail.

We now introduce various notions of ranks for simple theories.

Definition 1.9. Let $\varphi(x; y)$ be an $\mathcal{L}$-formula, and let k be an integer ≥ 2. For any type $p(x)$, $D(p(x), \varphi(x, y), k) = D_{\varphi,k}(p)$ (either a natural number or ∞) is recursively defined as follows:

(1) $D(p, \varphi, k) \geq 0$ for any consistent type p.
(2) $D(p, \varphi, k) \geq n + 1$ if for some a, $\varphi(x, a)$ k-divides over $\mathrm{dom}(p)$, and $D(p \cup \{\varphi(x; a)\}, \varphi, k) \geq n$.
(3) $D(p, \varphi, k) = n$ if $D(p, \varphi, k) \geq n$ and $D(p, \varphi, k) \not\geq n + 1$.
(4) $D(p, \varphi, k) = \infty$ if $D(p, \varphi, k) \geq n$ for all $n \in \omega$.

Definition 1.10. For a type $p(x)$, $D(p(x), \varphi(x, y)) = D_\varphi(p)$ (either an ordinal or ∞) is defined as:

$D(p, \varphi) \geq \alpha + 1$ if there is a such that $\varphi(x, a)$ divides over $\mathrm{dom}(p)$ and $D(p \cup \{\varphi(x, a)\}, \varphi) \geq \alpha$.

For limit δ, $D(p, \varphi) \geq \delta$ if $D(p, \varphi) \geq \alpha$ for all $\alpha < \delta$. The other part of the definition is similar to above.

Definition 1.11. For a formula $\psi(x)$ (with parameters), we define $D(\psi(x)) \geq \alpha + 1$ if there is a formula $\varphi(x)$ dividing over $\mathrm{dom}(\psi(x))$, and $D(\psi(x) \wedge \varphi(x)) \geq \alpha$; The other part of the definition is similar.

For a type, we let

$$D(p(x)) := \min\{D(\psi_1(x) \wedge ... \wedge \psi_n(x)) |\ \psi_i(x) \in p(x)\}.$$

Definition 1.12. For $p \in S(A)$, the crucial clause $SU(p) \geq \alpha + 1$ holds if p has a forking complete extension q such that $SU(q) \geq \alpha$.

Fact 1.13. *The following are equivalent.*

(1) $\varphi(x, y)$ *has the k-*tree property ($k \geq 2$), *i.e. there exist tuples* c_α *with* $\alpha \in \omega^{<\omega}$ *such that*
  (a) *for any* $\alpha \in \omega^{<\omega}$, $\{\varphi(x, c_{\alpha\frown n}) | n \in \omega\}$ *is k-inconsistent,*
  (b) *for all* $\beta \in \omega^{\omega}$, $\{\varphi(x, c_{\beta\lceil n}) | n \in \omega\}$ *is consistent.*

(2) *For* φ, k, $D(x = x, \varphi, k) = \infty$.

Each of $D_{\varphi,k}/D$ or SU-rank characterizes simplicity/supersimplicity, respectively.

Fact 1.14.

(1) *The following are equivalent.*
  (a) T *is simple.*
  (b) $D(x = x, \varphi, k) < \infty$ *for any* φ, k.
  (c) $D(x = x, \varphi, k) < \omega$ *for any* φ, k.
  (d) T *does not have the tree property i.e. there does not exist* φ *satisfying some k-tree property.*

(2) $D(x = x, \varphi, k) \leq D(x = x, \varphi) \leq D(x = x)$.

(3) $SU(p) \leq D(p)$.

Fact 1.15. *The following are equivalent.*

(1) T *is supersimple.*

(2) $D(x = x) < \infty$.

(3) $SU(p) < \infty$ *for any complete* p.

The rank $D_{\varphi,k}$ (SU, resp.) represents forking too for simple (supersimple, resp.) theories.

Fact 1.16.

(1) T *simple.* $c \mathop{\smile\!\!\!\!\!\mid}_A B$ *iff* $D(c/A, \varphi, k) = D(c/AB, \varphi, k)$ *for any* φ, k.

(2) T supersimple. $c \mathop{\smile\!\!\!\!\!\!\,|}_A B$ iff $SU(c/AB) = SU(c/A)$.

D-rank reflects forking for superstable T, but not for supersimple T in general. Namely, if T is supersimple, then that $SU(c/AB) = SU(c/A)$ implies $c \mathop{\smile\!\!\!\!\!\!\,|}_A B$. If T is superstable, then the converse holds as well.

By 1.14.2, if T is supersimple, then $D_\varphi(x = x) < \infty$. But there is a simple theory where $D_\psi(x = x) = \infty$ for some ψ.[6] Interesting class defined by D_ψ-rank is the class of *low* theories.

Definition 1.17. (with Fact)

(1) T is said to be *low* if $D(x = x, \varphi) < \omega$ for any $\varphi(x, y)$. There even exists a non-low supersimple theory.[5]
(2) Fact: T is low iff T is simple, and for any $\varphi(x, y)$, there is k such that whenever $\varphi(x, a)$ divides over $\emptyset$, it k-divides. Hence if T is low then for any $\varphi(x, y)$ and a set A, there is a set of formulas $\Sigma(y)$ over A such that $\models \Sigma(a)$ iff $\varphi(x, a)$ divides over A.

We finish this section by stating the extension of Lachlan's classical result on superstable theories to the context of supersimple theories. Recall that $I(T, \kappa)$ is the number of nonisomorphic models of T having size κ.

Theorem 1.18.

(1) (Lachlan[41]) If T is countable superstable, then either $I(T, \omega) = 1$ or $\geq \omega$.
(2) [33] If T is countable supersimple, then either $I(T, \omega) = 1$ or $\geq \omega$.

2. Type Amalgamation and Canonical Bases

Before Theorem 1.4 is established, historically Shelah already has proved in 1970s[52] that various forking axioms mentioned in section 1 hold for stable theories, and which has initiated stability theory, the study of stable structures.

Fact 2.1. *Let T be stable. Then two notions of forking and dividing are equivalent (see 1.6.2). Nonforking also satisfies the following.*

(1) (Finite character) *1.2.*
(2) (Symmetry) *1.4.1.*
(3) (Transitivity) *1.4.2.*
(4) (Extension) *1.3.*
(5) (Local character) *1.4.3.*

(6) (Uniqueness over a model) *Assume* $p(x) \in S(M)$. *For* $M \subseteq A$, *there is a unique nonforking extension* $q(x) \in S(A)$ *of* p.

Moreover the six axioms characterize stability and forking. Namely if, in a theory T, *an automorphism-invariant relation between complete types and sets satisfies all the axioms (1) to (6), then the theory is stable and the relation has to be nonforking.*

It is then natural to ask, for simple theories, what axiom can substitute for (Uniqueness over a model). The following are responded in [38].

Theorem 2.2.

(1) (Type amalgamation[b] over a model) T *is simple. Given a model and sets* $M \subseteq A_0, A_1$, *if* $c_0 \equiv_M c_1$, $A_0 \downarrow_M A_1$, *and* $c_i \downarrow_M A_i$ $(i = 0, 1)$, *then there is* $c \equiv_{A_i} c_i$ *such that* $c \downarrow_M A_0 A_1$.

(2) The five basic axioms in 2.1 together with the above axiom characterize simplicity and forking.

Now we explain the notions of various strong types introduced by Shelah and Lascar. Recall that an equivalence relation $E(x, y)$ is said to be *finite* if it has finitely many classes, *bounded* if it has $< \bar{\kappa}$ many classes. E is said to be *invariant* over A if for any A-automorphism f, $E(a, b)$ holds iff $E(f(a), f(b))$ holds.

Definition 2.3.

(1) We say finite tuples a, b have the same *strong type* over A (write $a \equiv^s_A b$, or $\mathrm{stp}(a/A) = \mathrm{stp}(b/A)$) if $E(a, b)$ holds, for any finite definable equivalence relation $E(x, y)$ over A. When a, b are infinite, $a \equiv^s_A b$ holds if $\mathrm{stp}(a'/A) = \mathrm{stp}(b'/A)$ for any finite subtuples $a' \subseteq a$, $b' \subseteq b$ of corresponding indexes.

(2) We say a, b possibly infinite, have the same KP (Kim-Pillay) type over A (write $a \equiv^{KP}_A b$), if $E(a, b)$ holds, for any bounded type-definable equivalence relation $E(x, y)$ over A.

(3) We say a, b possibly infinite, have the same Lascar (strong) type over A (write $a \equiv^L_A b$, or $\mathrm{Ltp}(a/A) = \mathrm{Ltp}(b/A)$ if $E(a, b)$ holds, for any bounded invariant equivalence relation $E(x, y)$ over A.

---

[b]It is originally called *the Independence Theorem.*

The notion of Lascar type is introduced by Lascar.[42] Note that $a \equiv_A b \Rightarrow a \equiv^s_A b \Rightarrow a \equiv^{KP}_A b \Rightarrow a \equiv^L_A b$ holds. In general, $a \equiv^s_A b$ iff $a \equiv_{\mathrm{acl}(A)} b$ (where acl is taken in $\mathcal{M}^{eq}$). If T is simple, Lascar type is KP type. Indeed more is true.

Proposition 2.4.[32] *Suppose that T is simple. Then T is G-compact, i.e. for tuples a, b of arbitrary length, $a \equiv^L_A b$ holds iff $a' \equiv^{KP}_A b'$ holds for any finite subtuples $a' \subseteq a$, $b' \subseteq b$ of corresponding indexes.*

Recall now that in stable theories, the uniqueness axiom holds over algebraically closed sets too.

Fact 2.5. (with Definition) *Let T be stable.*

(1) (Uniqueness for strong types) *For $B \subseteq C$, if $a \equiv^s_B a'$ and $a \mathop{\smile\!\!\!\!\!\mid}_B C$, $a' \mathop{\smile\!\!\!\!\!\mid}_B C$, then $a \equiv_C a'$.*
(2) *A complete type* $\mathrm{tp}(c/A)$ *is said to be* stationary *if it satisfies the uniqueness axiom, equivalently* $\mathrm{tp}(c/A)$ *and* $\mathrm{tp}(c/\mathrm{acl}(A))$ *have the same set of realizations.*

Now in simple theories, the following substitute the uniqueness axiom.

Theorem 2.6. (Type amalgamation for Lascar types) *For $B \subseteq C_0, C_1$, if $a_0 \equiv^L_B a_1$, $C_0 \mathop{\smile\!\!\!\!\!\mid}_B C_1$, and $a_i \mathop{\smile\!\!\!\!\!\mid}_B C_i$ $(i = 0, 1)$, then there is $a \equiv_{C_i} a_i$ such that $a \mathop{\smile\!\!\!\!\!\mid}_B C_0C_1$.*

Then again it is natural to question whether type amalgamation holds for strong types too. This is still an open question. But Buechler affirmatively answer for low theories.

Theorem 2.7.[3] *If T is low, then Lascar types are strong types. Equivalently, type amalgamation holds for strong types.*

For supersimple T, Buechler, Pillay, and Wagner also give a positive answer. Indeed they prove more, namely elimination of hyperimaginaries for supersimple theories. This important issue comes next.

Definition 2.8.

(1) T has *elimination of imaginaries* if for any definable equivalence class a/E, there is finite $b \in \mathcal{M}$ such that they are interdefinable (i.e. for any automorphism fixes the class a/E setwise iff it fixes the tuple b).

(2) T has *elimination of hyperimaginaries* if for any type-definable equivalence class a/E (a possibly infinite), there are definable equivalence classes $\{a_i/E_i\}_i$ such that they are interdefinable. In other words, for any type-definable equivalence relation $E(x,y)$ over $\emptyset$, and any complete type $p(x)$ over $\emptyset$, there are $\emptyset$-definable equivalence relations $E_i(x_i, y_i)$ (finite $x_i (\subseteq x)$, $y_i (\subseteq x)$ have corresponding subindexes), such that

$$p(x) \wedge p(y) \models E(x,y) \leftrightarrow \bigwedge_i E_i(x_i, y_i).$$

By working in $\mathcal{M}^{eq}$, there is no harm to assume that any theory has elimination of imaginaries.

Theorem 2.9 (Buechler, Pillay, Wagner[4]). *If T is supersimple, then T has elimination of hyperimaginaries.*

Therefore if T is supersimple, then $a \equiv^L_A b$ implies $a \equiv^s_A b$ (so $a \equiv_{\mathrm{acl}(A)} b$) even for infinite tuples, and we can identify $\mathrm{Ltp}(a/A) = \mathrm{stp}(a/A) = \mathrm{tp}(a/\mathrm{acl}(A))$.

We do not know still whether every simple T has elimination of hyperimaginaries. Hence in general, to go further we need to introduce the notion of *hyperimaginary* types and *the bounded closures of hyperimaginaries.* But for the rest of this note we assume, for convenience, T **is simple having elimination of hyperimaginaries**, unless stated otherwise. Hence we freely use, in simple T, type amalgamation over an algebraically closed set. Most results stated afterward indeed hold for arbitrary simple T as hyperimaginary versions.

The next issue is about canonical base, which is immensely important notion in developing further deeper works in simplicity theory.

Definition 2.10.

(1) We say $\mathrm{tp}(a/A)$ is an *amalgamation base* if type amalgamation holds for $\mathrm{tp}(a/A)$, equivalently both $\mathrm{tp}(a/A)$ and $\mathrm{tp}(a/\mathrm{acl}(A))$ have the same set of realizations.

(2) Let $p = \mathrm{tp}(a/A)$ be an amalgamation base. We say $A_0 (\subseteq A)$ is the *canonical base* of p, written $\mathrm{Cb}(p) = \mathrm{Cb}(a/A) = \mathrm{Cb}(a/\mathrm{acl}(A))$, if
   (a) p does not fork over A_0,
   (b) $p \lceil A_0$ is an amalgamation base, and
   (c) whenever $A' (\subseteq A)$ satisfies (a),(b), then $A_0 \subseteq \mathrm{dcl}(A)$.

In stable T, p is stationary iff it is amalgamation base.

Theorem 2.11.[18] *Any amalgamation base has the canonical base.*

Due to the canonical base notion, forking satisfies more axiom as such. Let $B_i = \mathrm{acl}(B_i) \subseteq B$ $(i = 0, 1)$. If $a \perp_{B_0} B$ and $a \perp_{B_1} B$, then $a \perp_{B_0 \cap B_1} B$: Let $C = \mathrm{Cb}(a/B)$. Then $C \subseteq B_i$, Therefore by transitivity, it follows $a \perp_C B_0 \cap B_1$ and $a \perp_{B_0 \cap B_1} B$.

There now is a definability of type issue to do with the canonical base. Different from the stable case, amalgamation base p need not satisfy the uniqueness axiom. But it does if we restrict our attention to p-stable formulas. Recall that given a complete type $p(x)$ and $\varphi(x, y) \in \mathcal{L}$, φ-type $p\lceil\varphi$ is the set of formulas in p, each of which is equivalent in T to some Boolean combination of instances of φ.

Definition 2.12. Let $p(x) \in S(A)$ be an amalgamation base. We call $\varphi(x, y) \in \mathcal{L}$, *p-stable* if the φ-type $p\lceil\varphi$ has a unique nonforking φ-type extension over any $B \supseteq A$.

Fact 2.13.

(1) *$\varphi(x, y)$ is stable iff $\varphi(x, y)$ is p-stable for any amalgamation base $p(x)$.*
(2) *If $\varphi(x, y)$ is p-stable, then there is $\psi_\varphi(y)$ over* Cb(p), *called the φ-*definition *of p, such that $\models \psi_\varphi(b)$ holds iff $\varphi(x, b)$ is in some nonforking extension of p.*

Theorem 2.14.[34] *Assume T is supersimple, and $p(x)$ is an amalgamation base. Then*

$$\mathrm{Cb}(p) = \{\textit{names of } \psi_\varphi(y) |\ \varphi(x, y) \textit{ is p-stable}\}.$$

Lastly we state the relationship between Morley sequences and canonical bases.

Fact 2.15. *Given amalgamation base p,* $\mathrm{Cb}(p) \subseteq I$ *where I is a Morley sequence of p.*

3. Modularity

In this section we introduce another important subclass, called *modular*. Studying such class is one of main themes of geometric simplicity theory.

Consider the example in 1.8 of vector space $(V, +, 0, \{r : V \to V\}|\ r \in F\})$ over a field F. Here the nonforking independence $\perp$ is exactly linear independence, i.e. $A \perp_B C$ iff for any finite $a \in A$, $\dim(a/B) = \dim(a/BC)$, and SU-rank measures the basis dimension. Moreover $\mathrm{acl}(A)$ is exactly the linear span of A. For two subspaces A, B, $A \mathop{\smile\!\!\!\!\!\!|}_{A\cap B} B$ holds. When they are finite dimensional, it means $\dim(A \cup B) = \dim(A) + \dim(B) - \dim(A \cap B)$.

We say T is *modular* if nonforking shares the dimensional property as mentioned.

Definition 3.1.

(1) T is called *modular* if for any A, B, we have

$$A \mathop{\smile\!\!\!\!\!\!|}_{\operatorname{acl}(A)\cap\operatorname{acl}(B)} B.$$

(2) T is said to be *1-based* if for any indiscernible sequence $I = \langle a_i |\ i < \omega\rangle$, we have $I - \{a_0\}$ is Morley over a_0.

(3) Similarly, for a type p over C, we say p is *modular* (*1-based* resp.) if after naming C, the above holds for any subsets or tuples from the solution set of p.

By so far developed techniques we can easily show the equivalence of modularity and 1-basedness.

Fact 3.2. *The following are equivalent.*[c]

(1) T is 1-based.
(2) For an amalgamation base $p = \operatorname{tp}(c_0/A)$, $\operatorname{Cb}(c_0/A) \subseteq \operatorname{acl}(c_0)$.
(3) T is modular.

Proof. (1)⇒(2) We can assume $A = \operatorname{acl}(A)$. Let $C := \operatorname{Cb}(c_0/A)(\subseteq A)$. Now there is a Morley sequence $J = \langle c_i | i \leq \omega\rangle$ of p. Then $c_\omega \mathop{\smile\!\!\!\!\!\!|}_A I$ where $I = J - \{c_\omega\}$. Since $c_\omega \mathop{\smile\!\!\!\!\!\!|}_C AI$ and $C \subseteq I$, we have $c_\omega \mathop{\smile\!\!\!\!\!\!|}_I A$. Hence $C = \operatorname{Cb}(c_\omega/A) = \operatorname{Cb}(c_\omega/I)$. By 1-basedness, $c_\omega \mathop{\smile\!\!\!\!\!\!|}_{c_0} I$. Hence $C \subseteq \operatorname{acl}(c_0)$.

(2)⇒(3) Given A, B, we have $A \mathop{\smile\!\!\!\!\!\!|}_{\operatorname{Cb}(A/B)} B$ and $\operatorname{Cb}(A/B) \subseteq \operatorname{acl}(A)$. Hence by transitivity, $A \mathop{\smile\!\!\!\!\!\!|}_{\operatorname{acl}(A)\cap\operatorname{acl}(B)} B$.

(3)⇒(1) Indiscernible $I = \langle c_i |\ i < \omega\rangle$ is given. (c) implies $c_0 \mathop{\smile\!\!\!\!\!\!|}_A c_1c_2...$ where $A = \operatorname{acl}(c_0) \cap \operatorname{acl}(c_1c_2...)$. Now, $tp(c_1c_2.../\operatorname{acl}(c_0)) = tp(c_2c_3.../\operatorname{acl}(c_0))$. Hence $A = \operatorname{acl}(c_0) \cap \operatorname{acl}(c_2c_3...)$. Again, $tp(c_0/\operatorname{acl}(c_2c_3...)) = tp(c_1/\operatorname{acl}(c_2c_3...))$, and thus $A = \operatorname{acl}(c_1) \cap \operatorname{acl}(c_2c_3..)$. Then, due to (c) again, $c_1 \mathop{\smile\!\!\!\!\!\!|}_A c_2c_3....$ From transitivity, we have $c_1 \mathop{\smile\!\!\!\!\!\!|}_{c_0A} c_2c_3....$ Then $c_1 \mathop{\smile\!\!\!\!\!\!|}_{c_0} c_2c_3...$ since $A \subseteq \operatorname{acl}(c_0)$. Thus (a) follows. □

[c] The same proof works for a type.

4. Geometries

Now we substantially get into geometric simplicity theory.

We begin this section by defining a *geometry*, a combinatorial generalization of the basis notion.

Definition 4.1. Let S be a set. If an operation $cl : \mathcal{P}(S) \to \mathcal{P}(S)$ satisfies the following properties, then we say that (S, cl) is a *geometry*[d]

(1) For $A \subseteq S$, $A \subseteq cl(A) = cl(cl(A))$.
(2) For $A \subseteq B \subseteq S$, $cl(A) \subseteq cl(B)$.
(3) For $A \subseteq S$ and $a, b \in S$, if $a \in cl(Ab) \setminus cl(A)$, then $b \in cl(Aa)$.
(4) If $a \in cl(A)$, then $a \in cl(A_0)$, for some finite $A_0 \subseteq A$.

Let (S, cl) be a geometry. It is said to be *homogeneous* if for any closure X of a finite set and $a, b \in S \setminus X$, there is an automorphism of (S, cl) which fixes X pointwise and moves a to b. We say $(S, cl) = (S_0, cl_0)$ is equivalent to (S_1, cl_1) if there is a bijection $f : S_0'/\sim_0 \to S_1'/\sim_1$ preserving closure relations, where for $a, b \in S_i' := S_i - cl_i(\emptyset)$, $a \sim_i b$ iff $cl_i(a) = cl_i(b)$.

We say $A(\subseteq S)$ is *independent* if $a \notin cl((A \setminus \{a\})$ for all $a \in A$. Given B, a subset $B_0 \subseteq B$ is said to be a *basis* for B if $B \subseteq cl(B_0)$ and B_0 is independent. It follows that any two bases for B have the same cardinality, denoted by $dim(B)$. Any $A \subseteq S$ gives a *localized* geometry on S defined by $cl_A(B) = cl(A \cup B)$, and a notion of dimension *over* A ($dim(-/A)$). For $A, B, C \subseteq S$, if $dim(A'/C) = dim(A'/B \cup C)$ for any finite $A' \subseteq A$, then we say that A is *independent from* B *over* C.

Definition 4.2. Let (S, cl) be a geometry.

(1) (S, cl) is *trivial* if $cl(A) = \bigcup\{cl(\{a\}) : a \in A\}$ for all $A \subseteq S$.
(2) (S, cl) is *modular* if X is independent from Y over $X \cap Y$ for all closed X, Y, or equivalently, if $dim(X) + dim(Y) = dim(X \cup Y) + dim(X \cap Y)$ for finite dimensional closed X and Y.
(3) (S, cl) is *locally modular* if it is modular over some point in S.
(4) (S, cl) is *locally finite* if the closure of a finite set is finite.

Each of the properties is invariant under localization. Thus modularity implies local modularity.

[d]It indeed is called a pregeometry and a *geometry* is meant to be a pregeometry where each closure is a singleton. But here we try not to distinguish.

Fact 4.3. *(S, cl) is modular iff for any subset A and $b \in S$, whenever $c \in cl(Ab)$, then $c \in cl(ab)$ for some $a \in cl(A)$.*

Now we putting the set theoretical notion of geometry to the model theoretic context. Assume T is strongly minimal (i.e. any definable subset (with parameters) of $\mathcal{M}^1$ is either finite or co-finite) so that stable. Then it is easy to see that $(\mathcal{M}^1, \mathrm{acl})$ forms a geometry, (where for a set $A \subseteq \mathcal{M}^1$, the closure is $\mathrm{acl}(A)(\cap \mathcal{M}^1)$). There, the forking independence in section 1 coincides with the dimensional independence. The following are the typical examples of such strongly minimal geometries.

Example 4.4.

(1) Infinite set (having no relation other than equality): a trivial geometry.
(2) (a) Vector space $(V, +, 0, r)_{r \in F}$ as mentioned in section 3: a modular geometry. The corresponding projective space has the equivalent geometry.
 (b) Affine space $(V, \lambda_r, G)_{r \in F}$ where $\lambda_r(u, v) = ru + (1 - r)v$, $G(u, v, w) = u - v + w$: a locally modular geometry. But this geometry is not modular as two parallel lines (each having dimension[e] 2) in a plain (having dimension 3) do not satisfy the equation in 4.2.(2).
(3) Complex field $\mathbb{C} = (\mathbb{C}, +, -, \times, 0, 1)$: a non-modular geometry.

Zilber's Principle (early 80s) roughly says that above are more or less all the examples of strongly minimal structures. Namely, any strongly minimal structure is either trivial, locally modular, or interpreting an infinite field (which must be algebraically closed field by Macintyre's result[43]).

Hrushovski (early 90s) constructed the counterexamples of Zilber's Principle using his ingenious construction method, which are not locally modular where no infinite group is interpretable(=definable in $\mathcal{M}^{eq}$).[20] But later on, he and Zilber suggested the so-called 'Zariski condition', and *proved* Zilber's Principle under the constraint.[28] It is well-known that Hrushovski (mid 90s) solved function field version Mordell-Lang conjecture in number theory, by the striking applications of the mentioned results.[21]

Zilber's Principle also makes sense in the o-minimal theory context, and Starchenko and Peterzil solved it fully and positively.[44]

Now if $\mathcal{M}^1$ is strongly minimal, then there is a unique complete 1-type p over $\emptyset$ having SU-rank 1. p indeed is a strong type. If D is the solution

[e]Here the dimension is that of the geometry.

set of the type[f], then (D, cl) where $cl(-) = \mathrm{acl}(-) \cap D$ has the equivalent geometry to $(\mathcal{M}^1, \mathrm{acl})$. Hence it suffices to pay our attention to (D, cl) to comprehend the geometry of the strongly minimal structure. Indeed for *any* SU-rank 1 strong type p (over $\emptyset$ after suppressing $\mathrm{dom}(p)$) with the solution set D, the same closure relation gives the geometry to D. Moreover inside D, $\dim(-/A) = SU(-/A)$ and $\mathrel{\raise0.2ex\hbox{$\smile$}\kern-0.75em\raise0.6ex\hbox{$\mid$}} =$ dimensional independence.

For the rest of this section, we fix such $D = (D, cl)$ in a simple theory. Understanding the geometry is the suitable generalization of the study of strong minimal structures. In the spirit of Zilber's principle for the strong minimality context, we come up with the following for the simple theory context.

Question 4.5.

(1) Relationship between D being 1-based(=modular in the sense of 3.1.1) and its geometry being (locally) modular.[g]
(2) If (D, cl) is locally modular or more generally D is 1-based, then can we recover a vector space from D?
(3) If D is not modular, then when we get a field and what kind of a field we get?

Indeed describing how much we have accomplished to solve the questions is the whole theme of this paper. Let us first see the answers of 4.5.1 for the stable case.

Definition 4.6. We call D (equivalently p) is k-*linear* ($k \geq 1$) if for any $a, b \in D^1$, and $A \subseteq D$, if $SU(ab/A) = 1$, then $SU(e) \leq k$ where $e = \mathrm{Cb}(ab/A)$. It is called *linear* if it is 1-linear.

Fact 4.7. *[45, 2.2.6] T stable. The following are equivalent.*

(1) D has the locally modular geometry (i.e. for $A, B \subseteq D$ and $d \in D^1$, $A \mathrel{\raise0.2ex\hbox{$\smile$}\kern-0.75em\raise0.6ex\hbox{$\mid$}}_{\mathrm{acl}(Ad) \cap \mathrm{acl}(Bd) \cap D} B$).
(2) D is linear.
(3) (D, cl_A) is modular for some A.
(4) D is 1-based (i.e. for $A, B \subseteq D$, $A \mathrel{\raise0.2ex\hbox{$\smile$}\kern-0.75em\raise0.6ex\hbox{$\mid$}}_{\mathrm{acl}(A) \cap \mathrm{acl}(B)} B$).

Above Fact 4.7 no longer holds in general if D is simple. Consider $V_P := (V; P)$, where V is the Affine space (4.4.2) and P is the unary generic

[f]Here $|D| = \bar{\kappa}$.
[g]For the moments, not to confuse with 3.1.1, when we refer to modularity in 4.2, we either say D has the modular geometry, or (D, cl) is modular.

predicate on V as in [10]. Note that the independence relation $\mathop{\smile\!\!\!\!\!\!\downarrow}$ computed on V and (V,P) are equal. Hence D =(the solution set of)$P(x)$ is also 1-based. But it not locally modular: Choose $a,b,c,d \in P$ such that both $a \neq b$, $c \neq d$ forms two parallel lines. Choose also $e \in P$ not in the plain generated by $\{a,b,c,d\}$. By the genericity of P, we can assume that $\{e\} = \mathrm{acl}(abe) \cap \mathrm{acl}(cde) \cap P$. Then

$$3 = \dim(abcd/e) \neq \dim(ab/e) + \dim(cd/e) - \dim(e/e) = 4,$$

and as P is the solution set of $\mathrm{tp}(e)$ in V_P, it is not locally modular.

To resolve above nuisance, we have to get missing points. For this we introduce the following notion for simple theories.

Definition 4.8. $G(D)$ denotes the collection of all SU-rank 1 elements in $D^{eq} := \mathrm{dcl}(D)$.

$D \subseteq G(D)$ and $(G(D), cl)$ also forms a geometry where again $cl(-) = \mathrm{acl}(-) \cap G(D)$.

Fact 4.9.[13,54] *D simple. The following are equivalent.*

(1) D is 1-based.
(2) D is linear.
(3) $G(D)$ has a modular geometry.
(4) $(G(D), cl_A)$ is modular, for any (some) set A.

Above fact gives more clear picture to the geometry of 1-based type. Identifying it as a locally modular geometry is rather a partial (such as only for the stable case) and mislead approach. Only after the study of general simple 1-based type, we have better and correct understanding of the type. This also answers the first question of 4.5.

4.1. *ω-categorical case*

Now we move to the second question of 4.5. For the rest of section 4, we restrict out attention mostly to the ω-categorical case, and review answers or counterexamples related to the question (2) of 4.5. Again we see the stable context first.

Theorem 4.10 (Zilber[58]). *D stable and ω-categorical.*

(1) Then D is 1-based.

(2) *If (D, cl) is non-trivial, then a vector space over a finite field having the same geometry can definably be recovered.*

Remark 4.11. Above theorem is no longer true for simple theories. In [22], Hrushovski constructs non 1-based ω-categorical simple D by using a variation of his construction method (to produce stable counterexamples of Zilber's Principle) mentioned after 4.4.

Fact 4.12.[15] *D stable.*

(1) *If D has the non-trivial modular geometry, then it is equivalent to that of some projective space over a division ring.*
(2) *If ω-categorical D has the non-modular but locally modular geometry, then it is equivalent to that of some Affine space over a finite field.*

The fact is (not) generalized for the simple context as follows.

Fact 4.13.[13] *D simple.*

(1) *If D has the non-trivial modular geometry, then again it is a projective geometry over some division ring. More generally, if D is nontrivial 1-based, then $G(D)$ has a projective geometry over some division ring.*
(2) *(de Piro) The second fact of 4.12 in ω-categorical case is not generalized to the simple context. Namely there is a simple ω-categorical D whose geometry is non-modular and locally modular, but not equivalent to that of any Affine space over a finite field.*

Even if simple D is ω-categorical, $G(D)$ need not be definable. But it is essentially so when D is 1-based. This is the topic of next subsection.

4.2. *1-based case*

Fact 4.14.[13] *D 1-based. If $u \in G(D)$, then $u \in \mathrm{acl}(u_0u_1u_2)$ for some $u_0, u_1, u_2 \in D$. More precisely if (x, y) is a fixed independent pair from D. Then $u \in acl(x'y'z)$ where $x', y', z \in D$ and $\mathrm{tp}(xy) = \mathrm{tp}(x'y')$.*

Corollary 4.15. *If D is ω-categorical and non-trivial 1-based, then D has a strongly minimal reduct, and a vector space over a finite field can definably be recovered.*

Proof of Fact 4.14. Suppose that $u \in \mathrm{dcl}(a_1...a_{n+1})$ $(a_i \in D)$ with the induction hypothesis for n. We will verify 4.14 for $n+1$. We can assume that $\{a_1, ..., a_{n+1}\}$ is independent. Now let $g = \mathrm{Cb}(ua_1/a_2...a_{n+1})$. Thus

$SU(g) = 1$ and $g \in G(D)$. Since $a_1 \notin \mathrm{acl}(g) \subseteq \mathrm{acl}(a_2...a_{n+1})$, u is in the line generated by $\{g, a_1\}$ (*). Then, since $g \in \mathrm{acl}(a_2...a_{n+1})$, by the induction hypothesis, $g \in \mathrm{acl}(bcd)$ for some $c, d \in D$ such that $tp(cd) = tp(xy)$. If g is already in $\mathrm{acl}(cd)$, then from (*), $u \in \mathrm{acl}(ga_1) \subseteq \mathrm{acl}(cda_1)$. Thus, in this case, 4.14 holds.

Therefore we only need to consider the case when $g \notin \mathrm{acl}(cd)$. We can also clearly assume that $g \notin \mathrm{acl}(b)$ (otherwise we are done). Then, since $G(D)$ is modular, we can find $v \in \mathrm{acl}(cd) \cap \mathrm{acl}(bg) \cap G(D)$ such that $g \in \mathrm{acl}(bv)$ (†), and $\dim(vg) = 2$. Also, at least one of $\{v, c\}$ or $\{v, d\}$ (say $\{v, d\}$) is independent (**). Now, as $d, a_1 \in D$, $\mathrm{Ltp}(d) = \mathrm{Ltp}(a_1)$. Hence, we can find v' such that $\mathrm{Ltp}(vd) = \mathrm{Ltp}(v'a_1)$ ($\star$). In particular, $\mathrm{Ltp}(v) = \mathrm{Ltp}(v')$. Then by (*),(**),($\star$), we can amalgamate types $\mathrm{tp}(v'/a_1)$ and $\mathrm{tp}(v/g)$, so that we obtain $v'' \models \mathrm{tp}(v'/a_1) \cup \mathrm{tp}(v/g)$. Hence, there are b', c' such that $\mathrm{tp}(vgb) = \mathrm{tp}(v''gb')$, $\mathrm{tp}(vdc) = \mathrm{tp}(v''a_1c')$ (by ($\star$)), and so by (†), $\mathrm{tp}(c'a_1) = \mathrm{tp}(xy)$, $v'' \in \mathrm{acl}(c'a_1)$ and $g \in \mathrm{acl}(b'v'')$ (‡). Then by (*),(‡),

$$u \in \mathrm{acl}(ga_1) \subseteq \mathrm{acl}(ga_1v'') \subseteq \mathrm{acl}(a_1b'v'') \subseteq \mathrm{acl}(c'a_1b').$$

Since $\mathrm{tp}(c'a_1) = \mathrm{tp}(xy)$, the $(n+1)$th induction hypothesis for Fact 4.14 is deduced. □

We can impose triviality to the whole theory.

Definition 4.16. T is *trivial* if for a, b, c, A, whenever $\{a, b, c\}$ is pairwise independent over A, then $\{a, b, c\}$ is independent over A.

Fact 4.17. *Suppose that T is 1-based. Then T is trivial if and only if all SU-rank 1 types are trivial (i.e. having the trivial geometries).*

Corollary 4.18. *Let T be non-trivial, 1-based and ω-categorical. Then an infinite dimensional vector space over a finite field, in particular the infinite additive group, is definable.*

4.3. *ω-categorical supersimple case*

In this subsection we describe related important issues of the finite rank property and CM-triviality.

Fact 4.19. *Let T be supersimple and ω-categorical. Then T is 1-based if and only if T has finite SU-rank (i.e. every complete type has finite SU-rank) and all SU-rank 1 types are 1-based.*

As pointed out, Theorem 4.10 no longer holds for simple theories. But still the following question is sensible to ask and its answer is still unknown.

Fact 4.20 (Cherlin, Harrington, Lachlan[11]). *Assume T is ω-categorical superstable. Then it has finite SU-rank. Hence due to Zilber's theorem and above Fact 4.19, T is 1-based.*

Open Problem 1. If T is ω-categorical supersimple, then does it have finite SU-rank?

Moreover all the stable and simple counterexamples mentioned in 4.11 have the following property.

Definition 4.21. T is said to be *CM-trivial* if for acl-sets A, B, C, whenever $\mathrm{acl}(A \cup C) \cap \mathrm{acl}(A \cup B) = A$ then $\mathrm{Cb}(C/A) \subseteq \mathrm{acl}(\mathrm{Cb}(C/A \cup B))$.

Any 1-based theory is CM-trivial. As said, Hrushovski's example of non 1-based ω-categorical supersimple theory is CM-trivial. Hence naturally we question below.

Open Problem 2. If T is ω-categorical supersimple, then is it CM-trivial?

The solution of Problem 1 is positive, in the case of groups or if the answer of Problem 2 is yes. Recall that a group G is said to be P-by-Q if G has a normal subgroup N having the property P such that G/N has the property Q. As usual, G' denotes the commutator subgroup of G.

Theorem 4.22 (Evans, Wagner[17]). *(1) If a group G (possibly with extra relations) over $\emptyset$ is ω-categorical supersimple, then it is (finite-by-Abelian)-by-finite (i.e. $(G^0)'$ is finite, where G^0 is the smallest $\emptyset$-definable subgroup of G having finite index, which must be normal (see an explanation after 7.1)), and it has finite SU-rank.*

(2) If T is ω-categorical supersimple and CM-trivial, then it has finite SU-rank.

Now in the following section we deal with the question 4.5.2 in the more general (not necessarily ω-categorical) context.

5. Group Configuration

We review some terminology for the type-definable groups in simple theories. Note that the notion of generic types plays central role. Roughly saying a type p in a group G (i.e. $p(x) \models G(x)$ which type-defines G) is said to be *generic* if p and G have the same ($D_{\varphi,k}$ or SU-)rank.

Definition 5.1. Let $(G, ., 1_G)$ be a group in T (so there may exist additional relations on G) type-defined say by $G(x)$, $.(x, y, z)$ over $\emptyset$ (after suppressing the domains). By $S_G(A)$, we mean the collection of complete types over A extending $G(x)$.

A type $p(x) \in S_G(A)$ is said to be *left-generic* of G over A if for any $g \models p$ and $a \in G$ with $g \mathop{\smile\!\!\!\!\!\!|}_A a$, we have $a.g \mathop{\smile\!\!\!\!\!\!|}_A a$. We say g is *left-generic* over A for G, if $\mathrm{tp}(g/A)$ is so. Right-generic types are defined similarly. A (partial) type $q(x)$ over A implying $G(x)$ is said to be *(left/right-)generic* if some generic g over A realizes $q(x)$. More generally when $(G, X, \bullet)$ is a type-definable homogeneous space over $\emptyset$, then $x \in X$ is *generic* over A, if $a \bullet x \mathop{\smile\!\!\!\!\!\!|}_A a$ holds for $a \in G$ with $a \mathop{\smile\!\!\!\!\!\!|}_A x$.

For a partial type $q(x) \models G(x)$ over A, and $g \in G$, we write $g.q$ to denote the left translation (of the solution set) of q by g, and it is easy to see that $g.q$ is a partial type over $A \cup \{g^{-1}\}$.

Fact 5.2. *We take the same notation in 5.1.*

(1) There is generic $p(x) \in S_G(A)$.
(2) $g \in G$ is left-generic over A iff it is right-generic over A.
(3) The following are equivalent.
  (a) The type p is generic.
  (b) For any $g \in G$, the type $g.p$ does not fork over $\emptyset$.
  (c) For any $g \in G$, the type $g.p$ does not fork over A.
  (d) For any $g \in G$, $g.p$ is generic.

We now again fix a solution set D of a strong type over $\emptyset$ of SU-rank 1. In the previous section we have seen that if D is ω-categorical non-trivial superstable, then Zilber showed it is 1-based and a vector space over a finite field having the same geometry can definably be recovered. Later Hrushovski extends the result as follows.

Fact 5.3. *If D is 1-based non-trivial stable, then there is a type-definable group $(G, +)$ such that $(V = G/G_0, +)$, where $G_0 = \mathrm{acl}(\mathrm{dom}(G)) \cap G$, forms a vector space over the division ring of the definable endomorphisms of V. Moreover there $\mathop{\smile\!\!\!\!\!\!|}$ is linear independence.*

In other words, Hrushovski gives positive answers to 4.5.2 for stable theories even in the non ω-categorical context. When he constructs the group, the group configuration theorem plays a pivotal role.

Definition 5.4. By *a group configuration* over e we mean a 6-tuple $C = (f_1, f_2, f_3, x_1, x_2, x_3)$ with a tuple e such that, for $\{i, j, k\} = \{1, 2, 3\}$,

(1) $f_i \in \mathrm{acl}(f_j, f_k; e)$,
(2) $x_i \in \mathrm{acl}(f_j, x_k; e)$, and
(3) all other triples from C are independent over e.

If there is $C' = (f'_1, f'_2, f'_3, x'_1, x'_2, x'_3)$ such that $\mathrm{acl}(f_i e) = \mathrm{acl}(f'_i e), \mathrm{acl}(x_i e) = \mathrm{acl}(x'_i e)$, then C' is also a group configuration over e. In this case, we say C and C' are *equivalent* over e.

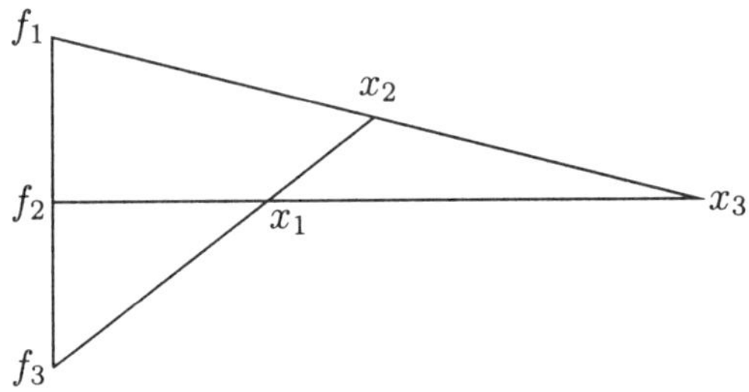

Definition 5.5. We say $C = (f_1, f_2, f_3, x_1, x_2, x_3)$ is a *group configuration from a type-definable homogeneous space* (G, X) if (G, X) and the group action of G on X are all type-definable, and $f_i \in G$, $x_i \in X$ generic elements with $f_1^{-1} f_2 = f_3$, $x_2 = f_3 x_1$, $x_3 = f_2 x_1$.

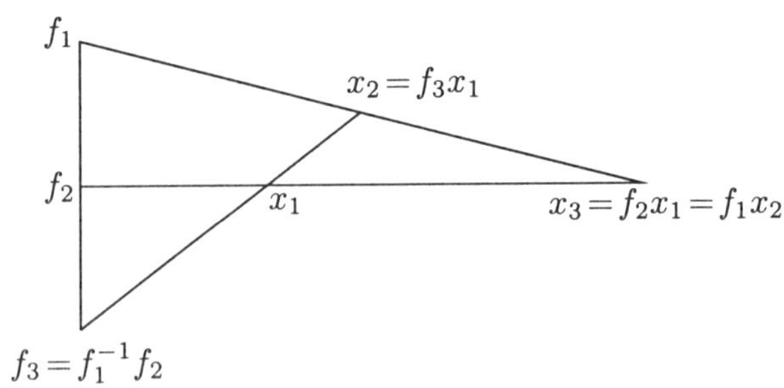

Theorem 5.6 (Hrushovski[19]). *(The group configuration theorem for stable theories) Assume T stable. A group configuration $C = (f_1, f_2, f_3, x_1, x_2, x_3)$ is given. Then a type-definable homogeneous space (G, X) over $e(\perp C)$ is constructed so that C is equivalent over e to some group configuration from (G, X).*

In the proof of above theorem, the stationarity of strong types is crucially used. As stationarity is no longer available, how to generalize the group configuration theorem to the simple theory context remains difficult problem to the researchers. Surprisingly there is a deep connection between this question and the so called notion of generalized amalgamation. We will discuss the issue in the following section.

6. Generalized Amalgamation

We introduce the concept of generalized amalgamation. This will be used to achieve the group configuration theorem for simple theories, but the notion itself is very intriguing in its own right. *In this section, tuples and the number of variables are possibly infinite.*

Definition 6.1. Let $I = \mathcal{P}(n)^-(= \mathcal{P}(n) \setminus \{n\})$, ordered by inclusion. Let $(\{A_i\}_{i\in I}, \{\pi^i_j\}_{i\le j\in I})$ be a directed family. Namely, each $\pi^i_j : A_i \to A_j$ is an elementary map; $\pi^i_i = \mathrm{id}_{A_i}$; and $\pi^j_k \circ \pi^i_j = \pi^i_k$ for $i \le j \le k \in I$. We say T has *n-amalgamation* if whenever

(1) $\{\pi^{\{i\}}_u(A_{\{i\}}) : i \in u\}$ is $\pi^{\emptyset}_u(A_\emptyset)$-independent,
(2) $A_u = \mathrm{acl}(\bigcup_{i\in u} \pi^{\{i\}}_u(A_{\{i\}}))$,

for any $u \in I$, then we can extend the direct family to the one indexed by $\mathcal{P}(n)$ (by finding A_n and π^j_n) so that (1) and (2) hold for n, too. We say T has *n-amalgamation* over A, if $A_\emptyset \coloneqq \mathrm{acl}(A) = A$.

One can restate the definition in terms of functorial terminology. Recall that by a category $\mathcal{C} = (\mathrm{Ob}(\mathcal{C}), \mathrm{Mor}(\mathcal{C}))$, we mean a class $\mathrm{Ob}(\mathcal{C})$ of members called *objects* of the category; equipped with a class $\mathrm{Mor}(\mathcal{C}) = \{\mathrm{Mor}(a,b) \mid a,b \in \mathrm{Ob}(\mathcal{C})\}$ where $\mathrm{Mor}(a,b) = \mathrm{Mor}_{\mathcal{C}}(a,b)$ is the class of *morphisms* between objects a, b (we write $f : a \to b$ to denote $f \in \mathrm{Mor}(a,b)$); and composition maps

$$\circ : \mathrm{Mor}(a,b) \times \mathrm{Mor}(b,c) \to \mathrm{Mor}(a,c)$$

for each $a, b, c \in \mathrm{Ob}(\mathcal{C})$ such that

(1) (Associativity) if $f : a \to b$, $g : b \to c$ and $h : c \to d$ then $h \circ (g \circ f) = (h \circ g) \circ f$ holds, and
(2) (Identity) for each object c, there exists a morphism $1_c : c \to c$ called the identity morphism for c, such that for $f : a \to b$, we have $1_b \circ f = f = f \circ 1_a$.

Note that any ordered set $(P, \le)$ is a category where objects are members of P, and $\mathrm{Mor}(a,b) = \{(a,b)\}$ if $a \le b$; $= \emptyset$ otherwise.

Now we recall a functor F between two categories $\mathcal{C}, \mathcal{D}$. The functor F sends an object $c \in \mathrm{Ob}(\mathcal{C})$ to $F(c) \in \mathrm{Ob}(\mathcal{D})$; and a morphism $f \in \mathrm{Mor}_{\mathcal{C}}(a,b)$ to $F(f) \in \mathrm{Mor}_{\mathcal{D}}(F(a), F(b))$ in such a way that

(1) (Associativity) $F(g \circ f) = F(g) \circ F(f)$ for $f : a \to b$, $g : b \to c$;
(2) (Identity) $F(1_c) = 1_{F(c)}$.

Definition 6.2. Let $\mathcal{C}_T$ be the category of algebraically closed substructures with elementary embeddings.

(1) A functor $a : W(\subseteq \mathcal{P}(n)) \to \mathcal{C}_T$ is given, where W is closed under subsets. For $u \subseteq w \in W$, we write $a(w) = a_w$, and a'_u (within a_w) means $a((u,w))(a_u)$. Now a is said to be *independence preserving (i.p.)* if for $w \in W$,

   (a) $\{a'_{\{i\}} | i \in w\}$ is $a'_\emptyset$-independent (within a_w);
   (b) $a_w = \operatorname{acl}(\bigcup\{a'_{\{i\}} |\ i \in w\})$.

(2) We say T has *n-complete amalgamation* (n-CA) if any i.p. functor $a : W \to \mathcal{C}_T$ can be extended to i.p. $\hat{a} : \mathcal{P}(n) \to \mathcal{C}_T$.

Equivalently we can define in terms of types.

Definition 6.3. We say T has *n-complete amalgamation* (n-CA) over a set $B = \operatorname{acl}(B)$ if the following holds: Let W be a collection of subsets of $\mathcal{P}(n)$, closed under subsets. For each $w \in W$, a complete type $r_w(x_w)$ over B is given where x_w is possibly an infinite set of variables. Suppose that

(1) for $w \subseteq w'$, we have $x_w \subseteq x_{w'}$ and $r_w \subseteq r_{w'}$.

Moreover for any $a_w \models r_w$,

(2) $\{a_{\{i\}} | i \in w\}$ is B-independent,
(3) a_w is as a set $\operatorname{acl}(\bigcup_{i \in w} a_{\{i\}} B)$ (and the map $a_w \to x_w$ is a bijection).

Then there is a complete type $r_n(x_n)$ over B such that (1),(2),(3) hold for all $w \in W \cup \{n\}$. We say T has n-CA if it has n-complete amalgamation over any algebraically closed set.

It is not hard to see that T has n-CA over B iff T has m-amalgamation over B for all $m \leq n$. Moreover the usual type-amalgamation (2.6) is indeed equivalent to 3-amalgamation. Hence any simple theory has 3-amalgamation.

One may wonder why such a bit complicated amalgamation is taken as the generalized version of type-amalgamation. Recall that the usual type-amalgamation (or the independence theorem) is as follows: For $A_0 \mathop{\smile\!\!\!\!\!\!\mid}_A A_1$ with $A \subseteq A_0, A_1$, if $c_0 \equiv^L_A c_1$ and $c_i \mathop{\smile\!\!\!\!\!\!\mid}_A A_i$ $(i = 0, 1)$, then there is $c \equiv^L_{A_i} c_i$ such that $c \mathop{\smile\!\!\!\!\!\!\mid}_A A_0 A_1$. Note that we can think of A_0, A_1 (after naming A) as two vertices of a base edge of a triangle and c a top vertex, and $\mathrm{Ltp}(c_0/A_0), \mathrm{Ltp}(c_1/A_1)$ are the 2 (edge) types to be amalgamated. One would expect generalized amalgamation to be a natural generalization of this amalgamation(=3-amalgamation), using a tetrahedron and higher dimensional simplices instead of a triangle (*). For example the candidate for 4-amalgamation would be: Let $\{A_0, A_1, A_2\}$ be independent over $A(\subseteq A_i = \mathrm{acl}(A_i))$. For $\{i,j,k\} = \{0,1,2\}$, let $c_i \equiv_{A_j} c_k$ and $c_i \downarrow_A A_j A_k$. Then there is $c \equiv^L_{A_j A_k} c_i$ such that $c \downarrow_A A_0 A_1 A_2$ (**).

However the above candidate for the definition of 4-amalgamation does not work, although it looks natural.

Example 6.4. Let M be the random graph in $\mathcal{L} = \{R\}$. Choose distinct $a_i, b_i, d_i \in M$ and imaginary elements $c_i = \{b_i, d_i\}$ $(i = 0,1,2)$. We can additionally assume that $R(b_2, a_0) \wedge R(d_2, a_1) \wedge \neg R(b_2, a_1) \wedge \neg R(d_2, a_0)$ and $\mathrm{tp}(b_2 d_2; a_0 a_1) = \mathrm{tp}(b_1 d_1; a_0 a_2) = \mathrm{tp}(b_0 d_0; a_1 a_2)$. Now it follows that $\mathrm{Ltp}(c_1/a_0) = \mathrm{Ltp}(c_2/a_0), \mathrm{Ltp}(c_0/a_1) = \mathrm{Ltp}(c_2/a_1)$ and $\mathrm{Ltp}(c_0/a_2) = \mathrm{Ltp}(c_1/a_2)$. However $\mathrm{Ltp}(c_0/a_1 a_2)$, $\mathrm{Ltp}(c_1/a_0 a_2)$, $\mathrm{Ltp}(c_2/a_0 a_1)$ have no common realization.

In above example, $\{c_0, c_1, c_2\}$ can be considered as a base triangle, and $\mathrm{Lstp}(d_0/c_0 c_1), \mathrm{Lstp}(d_1/c_1 c_2), \mathrm{Lstp}(d_2/c_2 c_0)$ form other 3 triangles attached to the base triangle. The example shows that even if the edges of the 3 triangles are compatible over the base vertices, there is no common vertex joining the 3 triangles. On the other hand, due to the nature of the random graph if we only work in the home-sort, then any desired 3 types attached on a base triangle with compatible edges will be realized. As we want the notion of generalized amalgamation to be preserved in interpreted theories, we can not take above (**) as 4-amalgamation. Moreover when we use inductive arguments for example, we often have to consider not only the algebraic closures of vertices of amalgamated types but also those of higher dimensional surfaces as well, since after naming parameters the surface dimension is increasing.

Conclusively, the accurate definition for 4-amalgamation is then the one defined in 6.1-3. But note that indeed the essential philosophy of n-amalgamation mentioned in (*) is the same. We simply need extra care not only to compatible surfaces but to the (enumerations of) algebraic closures of all the lower dimensional surfaces.

The correct 4-amalgamation can be written in this manner (***): Let $\{A_0, A_1, A_2\}$ be independent over $A(\subseteq A_i = \mathrm{acl}(A_i))$. For $\{i,j,k\} = \{0,1,2\}$, let $\overrightarrow{\mathrm{acl}(c_i A_j)} \equiv_{A_j} \overrightarrow{\mathrm{acl}(c_k A_j)}$ and $c_i \downarrow_A A_j A_k$, where $\overrightarrow{\mathrm{acl}(c_i A_j)}$ is a

subsequence of a fixed sequence $\overrightarrow{\mathrm{acl}(c_i A_j A_k)}$.
Then there are c and an enumeration $\overrightarrow{\mathrm{acl}(cA_0A_1A_2)}$ such that

$$\overrightarrow{\mathrm{acl}(cA_jA_k)} \equiv_{\overrightarrow{\mathrm{acl}(A_jA_k)}} \overrightarrow{\mathrm{acl}(c_iA_jA_k)}$$

and $c \mathop{\smile\!\!\!\!\!\!\!\;|}_A A_0A_1A_2$.

6.1. *Group configuration for simple theories*

We are now ready to speak about the group configuration theorem for simple theories. Without generalized amalgamation, I. Ben-Yaacov, I. Tomasic, and F. O. Wagner obtained the group configuration theorem as follows.

Theorem 6.5 (Ben-Yaacov, Tomasic, Wagner[2,53]).
T simple. A group configuration $C = (f_1, f_2, f_3, x_1, x_2, x_3)$ is given. Then it is equivalent to some group configuration from an invariant *homogeneous space (G, X) over $e(\mathop{\smile\!\!\!\!\!\!\!\;|} C)$.*

The proof of the theorem is very ingenious and deep. The group and the homogeneous space they produce are *not* type-definable but just automorphism invariant, hence do not completely fit into the first order context. However in the ω-categorical context, the group turns out to be definable, and using this they resolved the conjecture positively that if ω-categorical D is k-linear (4.6), then it is linear.[53] This is known to be true for stable D without the assumption of ω-categoricity, and it is still unknown whether the same holds for an arbitrary simple D.

Later the author together with T. dePiro and J. Millar obtained the type-definable group configuration theorem under the generalized amalgamation property.

Theorem 6.6 (de Piro, Kim, Millar[14]). *Assume T simple having 4-amalgamation. A group configuration $C = (f_1, f_2, f_3, x_1, x_2, x_3)$ is given. Then there is a type-definable group G over $e(\mathop{\smile\!\!\!\!\!\!\!\;|} C)$ with generics g_i such that* $\mathrm{acl}(f_ie) = \mathrm{acl}(g_ie)$.

Theorem 6.7.[33] (The group configuration theorem for simple theories) *Assume T simple having 4-amalgamation. A group configuration $C = (f_1, f_2, f_3, x_1, x_2, x_3)$ is given. Then it is equivalent to some group configuration from a type-definable homogeneous space (G, X) over $e(\mathop{\smile\!\!\!\!\!\!\!\;|} C)$.*

In section 7, as promised we will speak about the application of theorems 6.6,7 to answer the question (2) of 4.5 similarly to the stable case 5.3. Before getting into it, we speak further issues on generalized amalgamation in the next subsections.

6.2. *Generalized imaginaries*

In this subsection we restrict our attention mainly to stable theories. Unfortunately, even in stable theories, n-amalgamation need not hold over an algebraically closed set, though it always holds over a model (thus 6.6,7 can cover all stable theories if we assume 4-amalgamation over models). There is an example below.

Example 6.8. Consider $[A]^2 = \{\{a,b\} | a \neq b \in A\}$ where A infinite. Let $B = [A]^2 \times \{0,1\}$ where $\{0,1\} = \mathbb{Z}/2\mathbb{Z}$. Also let $E \subseteq A \times [A]^2$ be a membership relation, and let P be a subset of B^3 such that $((w_1,\delta_1)(w_2,\delta_2)(w_3,\delta_3)) \in P$ iff there are distinct $a_1,a_2,a_3 \in A$ such that for $\{i,j,k\} = \{1,2,3\}$, $w_i = \{a_j,a_k\}$, and $\delta_1+\delta_2+\delta_3 = 0$.

Let $M = (A, [A]^2, B;\ E, P;\ \mathrm{Pr}_1 : B \to [A]^2)$. Then since M is a reduct of $(A, \mathbb{Z}/2\mathbb{Z})^{\mathrm{eq}}$, M is stable. We work in M^{eq} and argue that the example does not have 4-amalgamation. Note first that $\mathrm{dcl}(\emptyset) = \mathrm{acl}(\emptyset)$, and for $a \in A$, $\mathrm{dcl}(a) = \mathrm{acl}(a)$. Now choose distinct $a_1,a_2,a_3,a_4 \in A$. For $\{i,j,k\} \subseteq \{1,2,3,4\}$, fix an enumeration $\overline{a_ia_j} = (b_{ij},...)$ of $\mathrm{acl}(a_ia_j)$ where $b_{ij} = (\{a_i,a_j\},\delta) \in B = [A]^2 \times \{0,1\}$. Let $r_{ij}(x_{ij}) = \mathrm{tp}(\overline{a_ia_j})$, and let x^1_{ij} be the variable for b_{ij}. Note that $b_{ij} = (\{a_i,a_j\},\delta)$ and $b'_{ij} = (\{a_i,a_j\},\delta+1)$ have the same type over a_ia_j. Hence there is $(\overline{a_ia_j})' = (b'_{ij},...)$ also realizing $r_{ij}(x_{ij})$. Therefore we have complete types $r_{ijk}(x_{ijk})$, $r'_{ijk}(x'_{ijk})$ both extending $r_{ij}(x_{ij}) \cup r_{ik}(x_{ik}) \cup r_{jk}(x_{jk})$ realized by some enumerations of $\mathrm{acl}(a_ia_ja_k)$ such that $P(x^1_{ij}x^1_{ik}x^1_{jk}) \in r_{ijk}$ whereas $\neg P(x^1_{ij}x^1_{ik}x^1_{jk}) \in r'_{ijk}$. Then it is easy to see that $r_{123} \cup r_{124} \cup r_{134} \cup r'_{234}$ is inconsistent.

In his recent preprint,[25] Hrushovski remedies the deficiency of generalized amalgamation for stable theories by introducing the so called generalized imaginaries. More precisely, if $\mathcal{M} \models T$ is stable, then he is able to construct $(\mathcal{M} \subseteq)\mathcal{M}^* \models T^*$ in some $\mathcal{L}^*(\supseteq \mathcal{L})$ such that $\mathcal{M}$ is *stably embedded* into $\mathcal{M}^*$ (i.e. every subset of $\mathcal{M}$ definable in $\mathcal{M}^*$ is already definable in $\mathcal{M}$ within $\mathcal{L}(\mathcal{M})$), and $\mathcal{M}^*$ has n-amalgamation for all n over *any* algebraically closed sets. As the new elements in $\mathcal{M}^*$ can be called *generalized imaginaries*, we may write $\mathcal{M}^*$ as $\mathcal{M}^{\mathrm{geq}}$. In short, like $\mathcal{M}^{\mathrm{eq}}$, without loss of generality, we can freely work in $\mathcal{M}^{\mathrm{geq}}$ and assume to have n-amalgamation when T is stable.

Open Problem 3. Can we construct such $\mathcal{M}^*$ for simple T? If yes, then possibly we can remove the assumption of 4-amalgamation in the group configuration theorem.

6.3. *n-simplicity*

Notice that it is very cumbersome working with terminology for the full definition of n-amalgamation. On the other hand, the usage of the natural simpler notations as pointed out in (**) before 6.4 (not bothered by the enumerations of all the lower dimensional surfaces), although not 100% correct ones, might be more transparent to conceptualize.

Hence in the following definitions of n-simplicity and $K(n)$-simplicity, for convenience, we use imprecise notations good enough however representing the essence of concepts. (The reader will be able to fully recover the correct definitions as the same manners in (***) below Example 6.4.) Also for simplicity, we describe the notions only over $\emptyset = \mathrm{acl}(\emptyset)$.

Fact 6.9. *Let $I = \langle a_i |\ i \in \omega \rangle$ be Morley, and let $b \mathop{\smile\!\!\!\!\!\!\,|}\, a_0$. Then there is $b' \equiv_{a_0} b$ such that I is Morley over b' and $b' \mathop{\smile\!\!\!\!\!\!\,|}\, I$.*

Above fact is a particular case of type amalgamation(=the independence theorem=3-amalgamation), and used crucially in showing it for simple T. Namely, showing particular case of 3-amalgamation is the way of fully obtaining 3-amalgamation. One may ask whether the same holds for n-amalgamation. In this spirit, Kolesnikov[40] introduces the notion of n-*simplicity* by generalizing above Fact 6.9.

Definition 6.10.

(1) T is $K(n)$-*simple* if for $k \le n$ and any Morley sequence $I = \langle a_i |\ i \in \omega \rangle$, whenever $I_k = \langle a_i | i < k \rangle$ is Morley over b with $b \mathop{\smile\!\!\!\!\!\!\,|}\, I_k$, there is $b' \equiv_{I_k} b$ such that I is Morley over b' and $b' \mathop{\smile\!\!\!\!\!\!\,|}\, I$.

(2) T is n-*simple* if for $k \le n$ and any Morley $I = \langle a_i |\ i \le k \rangle$, whenever $I_k = \langle a_i | i < k \rangle$ is Morley over b with $b \mathop{\smile\!\!\!\!\!\!\,|}\, I_k$, there is $b' \equiv_{I_k} b$ such that I is Morley over b' and $b' \mathop{\smile\!\!\!\!\!\!\,|}\, I$.

(3) Recall that T has n-CA if T has k-amalgamation for all $k \le n$.

As in simplicity(=1-simplicity), both n-simplicity, $K(n)$-simplicity are particular cases of $(n+2)$-CA. It then is natural to imagine that $K(n)$-simplicity, n-simplicity, and $(n+2)$-CA might be all equivalent. Type amalgamation already says that is the case when $n = 1$. Indeed Kolesnikov show in his thesis and[40] that is the case for $n = 2$ as well and leave the case $n > 2$

as questions. He also supplies examples of simple structures having n-CA but not having $(n+1)$-CA for each $n \geq 3$.

However in the author's joint paper with Kolesnikov and Tsuboi,[37] it is revealed that the harmony breaks up when $n \geq 3$. Namely, they first show that the equivalence of n-simplicity and $(n+2)$-CA for all n. But they also build, for each $n \geq 3$, an example of $K(n)$-simple but not having $(n+2)$-CA. (Indeed this is the reason why 6.10.1 is called $K(n)$-*simple* rather than n-*simple*, albeit it is more canonical generalization of 6.9): Let $\mathcal{L} = \{R\}$ where R is an n-ary relation. Let

$$\begin{aligned}\mathcal{K} := \{A \mid\ & A \text{ is a finite } R\text{-structure; } R \text{ is symmetric and irreflexive;}\\ & \text{for any } A_0 \subseteq A \text{ with } |A_0| = n+1, \text{ the no. of}\\ & n\text{-element subsets of } A_0 \text{ satisfying } R \text{ is even}\}\end{aligned}$$

Then $\mathcal{K}$ has the Fraïssé limit, and that is the simple ω-categorical counterexample.

7. 1-based Groups

As promised in this section we see the analogous result of 5.3 for the simple case as the answer for 4.5.2. This is obtained by an application of 6.6 or 6.7.

Theorem 7.1. *T 1-based nontrivial having 4-amalgamation. Then there is a type-definable group $(G,+)$ say over $\emptyset = \mathrm{dom}(G)$ such that $(V = G/G_0, +)$, where $G_0 = G \cap \mathrm{acl}(\emptyset)$, forms a vector space over a division ring of the type-definable endomorphisms of V, and $\perp$ is linear independence there.*

For the rest of this section, we sketch to explain how to obtain the vector space from the non-trivial modular structure. Note that by 4.17, there exists a non-trivial SU-rank 1 strong type p. For convenience, name $\mathrm{dom}(p)$ and assume $p \in S(\emptyset)$. Since p is non-trivial, there exists $\{a,b,c\}$ realizing p such that b,c is independent and $a \in \mathrm{acl}(b,c) \setminus (\mathrm{acl}(b) \cup \mathrm{acl}(c))$. Choose yx realizing $\mathrm{tp}(ab/c)$ with $yx \perp_c ab$. Then $\dim(ay/bx) = 1$ as $y \in \mathrm{acl}(abx)$ and $a \perp bx$. Let $z = \mathrm{Cb}(\mathrm{Ltp}(ay/bx))$, then by 1-basedness, $z \in \mathrm{acl}(ay) \cap \mathrm{acl}(bx)$. Moreover, by a straightforward rank calculation, we have $SU(z) = 1$. This gives a group configuration $C = (a,b,c,x,y,z)$. Now by Theorem 6.6, we obtain a type-definable group G over $e(\perp C)$ such that the generic types have SU-rank 1. The group G is 1-based since the underlying theory is 1-based.

To proceed, we now need to study type-definable (1-based) groups in simple theories. Hence fix the group type-definable group $G = G(x)$ over

$\emptyset$ (after suppressing e). Note that for a set A, by intersecting all the type-definable subgroups over A having bounded index, we get the smallest A-type-definable group G^0_A, called the connected component of G over A, having bounded index in G (i.e. $[G : G^0_A] = \lambda < \bar{\kappa}$ no matter how large saturated model $\mathcal{M}$ we take). It is not hard to see that G^0_A is a normal subgroup. It is well-known that in the case of stable theories, $G^0_A = G^0_\emptyset$ for any A. But for the simple case, it need not be true. For example, if G is a vector space over a finite field equipped with a bilinear map such as an inner product, then G^0_A is a subspace orthogonal to all the vectors in A.

Now reset G by $G^0 = G^0_\emptyset$. Hence *G has no proper $\emptyset$-type-definable subgroup of bounded index* ($\star$). *For the rest of this section, G is 1-based.* We recall some more notation.

- For $g, h \in G$, we write $h^g := g^{-1}.h.g$.
- The *commutator* of $g, h \in G$ is $[g,h] = g^{-1}.h^{-1}.g.h$. For $A, B \subseteq G$, $[A,B] := \{[g,h] |\ g \in A,\ h \in B\}$. Recall that the commutator subgroup G' is the subgroup generated by $[G,G]$.
- The *center* of G is $Z(G) = \{g \in G |\ g.h = h.g \text{ for all } h \in G\}$.
- The *centralizer* of $g \in G$ is $C_G(g) = \{h \in G | g.h = h.g\}$.
- We say two type-definable groups H_0, H_1 are *commensurate* if the index $[H_i : H_0 \cap H_1]$ is bounded for each $i = 0, 1$. That two groups are commensurate is an equivalence relation.

Results in this subsection are due to F. O. Wagner.[55]

Fact 7.2. *(G 1-based) Any type-definable subgroup of G^n is commensurate with one over* $\operatorname{acl}(\emptyset)$.

Theorem 7.3.

(1) G' is bounded. Indeed, $G' \leq Z(G)$.
(2) Hence G is bounded-by-Abelian, and the group $G_0 = G \cap \operatorname{acl}(\emptyset) (\geq G')$ is normal.

Proof. We sketch the proof. (2) is automatically deduced from (1). So we will show (1). Note that $Z(G)$ need not be type-definable where as its approximation $\tilde{Z} = \tilde{Z}(G) := \{g \in G |\ [G : C_G(g)] \text{ is bounded}\}$ is. It follows $\tilde{Z}$ is a normal subgroup of G.

Claim 1. $\tilde{Z}' \subseteq \operatorname{acl}(\emptyset)$, and $\tilde{Z}' \leq Z(G)$: For $g \in \tilde{Z}$, $|[g,G]| = |[G,g]| = [G : C_G(g)]$ is bounded, hence $[g,G] \subseteq \operatorname{acl}(g)$. Then for $h, g \in \tilde{Z}$ with $h \mathop{\smile\!\!\!\!\!\!\,|}\, g$, we have $[h,g] \in \operatorname{acl}(g) \cap \operatorname{acl}(h) = \operatorname{acl}(\emptyset)$. For arbitrary $h, g \in \tilde{Z}$, there are $h_i(\mathop{\smile\!\!\!\!\!\!\,|}\, g)$ such that $h = h_1.h_2$, and the repeated applications yield $[h, g] \in \mathrm{acl}(\emptyset)$ as well. Hence $\tilde{Z}' \subseteq \mathrm{acl}(\emptyset)$. Moreover since $\tilde{Z}$ is normal in G, so is $\tilde{Z}'$. Hence for $h \in \tilde{Z}'$, we have $[G, h] \subseteq \tilde{Z}'$. Therefore $[G : C_G(h)] = |[G, h]|$ is bounded and due to the assumption $(\star)$ above, $C_G(h) = G$. Thus $\tilde{Z}' \subseteq Z(G)$ and Claim 1 is proved.

It suffice to show the following claim.

Claim 2. $\tilde{Z}$ has bounded index in G. Hence $\tilde{Z} = G$: Note that for $h, g, g' \in G$, we have

$$h^g = h^{g'} \quad \text{iff } hg'g^{-1} = g'g^{-1}h$$
$$\text{iff } h \in C_G(g'g^{-1}) \text{ iff } (h, h^g) = (h, h^{g'}) \in H_g \cap H_{g'};$$

where $H_g = \{(h, h^g) | \ h \in G\}$ is a type-definable subgroup of $G \times G$. It follows H_g and $H_{g'}$ are commensurate iff $\{h \in G | \ h^g = h^{g'}\} = C_G(g'.g^{-1})$ has bounded index in G iff $g'.g^{-1} \in \tilde{Z}(G)$ iff g', g are in the same right coset of $Z(G)$.

Now due to 7.2, there are only boundedly many commensurate equivalence classes of subgroups of G^2. Hence $Z(G)$ must have bounded index. Therefore Claim 2 follows. □

We have shown that G/G_0 is Abelian. *We now use the fact that $SU(g) = 1$ for generic $g \in G$ $(\star\star)$.*

Definition 7.4. By an *endogeny* of G/G_0, we mean an endomorphism f of G/G_0 such that the graph of f is *type-definably induced* over $\mathrm{acl}(\emptyset)$, i.e. there is a type-definable subset S_f over $\mathrm{bdd}(\emptyset)$ of $G \times G$ such that

$$\{(a + G_0, b + G_0) | \ (a, b) \in S_f\}$$

forms the graph of f.

Let $\mathrm{End}(G) := \{f | \ f \text{ is an endogeny of } G/G_0\}$. Using $(\star\star)$, it is not hard to check $\mathrm{End}(G)$ with addition and composition forms a division ring.

Theorem 7.5. *For $a_0, ..., a_n \in G \setminus G_0$, the following are equivalent.*

(1) $\{a_0, ..., a_n\}$ is (forking) dependent.
(2) $a_i \in \mathrm{acl}(\{a_0, ..., a_n\} \setminus \{a_i\})$ for some $i \leq n$.
(3) There are $f_0, ..., f_n \in \mathrm{End}(G)$ not all zero such that $f_0(a_0 + G_0) + ... + f_n(a_n + G_0) = 0$ in G/G_0.

Hence $(V, +)$ with $V := G/G_0$ forms a vector space over the division ring $\mathrm{End}(G)$, in which $\mathop{\smile\!\!\!\!\!\!\,|}$ = linear independence.

To prove this we need the important fact below.

Fact 7.6.

- *Any strong type in G^n is generic for some coset of a type-definable subgroup of G^n.*

The analogous result for the stable case is:

- *(Hrushovski, Pillay[26]) (T stable.) Any type-definable subset of G^n is a finite Boolean combination of cosets of type-definable subgroups of G^n.*

Proof sketch of 7.5. (3)⇒(2)⇒(1) Clear.

(1)⇒(3). By induction we can suppose that any proper subtuple of $\bar{a} = (a_0, ..., a_n)$ is independent. Due to Fact 7.6, there is $H(\leq G^{n+1})$ a type-definable subgroup over $\mathrm{acl}(\emptyset)$ such that $\mathrm{tp}(a_0, ..., a_n/\mathrm{acl}(\emptyset))$ is generic for some $H + \bar{h}$. For $i < n$, let

$$F_i := \{(x_i, x_n) \in G \times G |\ (0, ..., 0, x_i, 0, ..., 0, x_n) \in H\}.$$

Then one can show

$$f_i := \{(a + G_0, b + G_0) |\ (a, b) \in F_i\}$$

is the graph of a nonzero endogeny of G/G_0, and

$$f_0(a_0 + G_0) + ... + f_{n-1}(a_{n-1} + G_0) = -a_n + G_0. \qquad \square$$

8. Fields Having Simple Theories

As said in section 4 regarding to the question 4.5.3 for stable theories, although there are counterexamples to Zilber's Principle in its face value, Hrushovski and Zilber show that under the so called 'Zariski constraint' if a strongly minimal structure is not modular, then an algebraically closed field is interpretable. Strongly minimal sets in many important algebraic structures such as DCF, ACFA (1.8) are Zariski, so there Zilber's Principle hold. Lots of spectacular applications have been made to algebraic geometry and number theory[21,23,48–50] using this trichotomy principle. Arguably the solution to Zilber's Principle in the Zariski context and its applications to core mathematics is the most culminant events in the history of model theory.

Almost nothing has been done in respect to the generalization of Hrushovski and Zilber trichotomy to the simple theory context, i.e. the study for the question (3) of 4.5 in unstable simple theories is not cultivated yet. It is even not known yet what kinds of fields should be supersimple. We are only a little soothed by the achievement of the group configuration theorem for simple theories which, in the stable case, plays a pivotal role. In this last section, hence instead we summarize so far known results on fields having simple theories. In particular when a field is PAC, the suitable classification is made in the context of subclasses of simple theories.

Definition 8.1. Let F be a field.

(1) F is said to be *pseudo-algebraically closed* (PAC) if any absolutely irreducible algebraic set V defined over F has F-rational point.
(2) F is *pseudo-finite* if it is an infinite field of the theory of all finite fields.
(3) An extension field E of F is *separable* if it is algebraic over F, and for each $e \in E$, the irreducible polynomial of e over F has no multiple roots.
(4) F *perfect* if every algebraic extension is separable; equivalently either the characteristic of F is 0, or $x \mapsto x^p$ is onto where p is the characteristic of F.
(5) F is *bounded* for each $n > 1$, in its algebraic closure, there are only finitely many separably algebraic extensions of degree n.
(6) F is *separably closed* if it has no proper separable extension.

Fact 8.2. *Assume F is PAC.*

(1) (Chatzidakis[7]) F is simple iff F is bounded.
(2) (Hrshovski;[24] Pillay, Poizat[46]) F is supersimple iff F is perfect and bounded.
(3) (Duret,[16] Wood[57]) F is stable iff F is separably closed.

Historically the first simple unstable fields studied seriously are pseudo-finite fields by J. Ax[1] in late 60s.

Fact 8.3. *A field F is pseudo-finite if and only if it is perfect, PAC, and for each n it has exactly one algebraic extension of degree n.*

Hence pseuo-finite fields are unstable supersimple (of SU-rank 1).

Fact 8.4.

(1) (Macintyre;[43] Cherlin, Shelah[12]) Superstable division ring is an algebraically closed field.
(2) (Pillay, Scanlon, Wagner[47]) Supersimple division ring is a field.

Open Problem 4.

(1) Is any stable field separably closed?
(2) Is any supersimple field PAC?

References

1. J. Ax, 'The elementary theory of finite fields', *Annals of Math.* 88 (1968) 239-271.
2. I. Ben-Yaacov, I. Tomasic and F. Wagner, 'Constructing an almost hyperdefinable group', *Journal of Math. Logic* 4 (2004) 181-212.
3. S. Buechler, 'Lascar strong types in some simple theories', *J. of Symbolic Logic* 64 (1999) 817-824.
4. S. Buechler, A. Pillay and F. O. Wagner 'Supersimple theories', *J of AMS* 14 (2000) 109-124.
5. E. Casanovas and B. Kim, 'A supersimple nonlow theory', *Notre Dame Journal of Formal Logic* 39 (1998) 507-518.
6. E. Casanovas and F. O. Wagner, 'Local supersimplicity and related concepts', *Journal of Symbolic Logic* 67 (2002) 744-758.
7. Z. Chatzidakis, 'Simplicity and Independence for Pseudo-algebraically closed fields', in *Models and Computability*, London Math. Soc. Lecture Notes Series 259, Cambridge University Press, Cambridge (1999) 41-61.
8. Z. Chatzidakis and E. Hrushovski, 'Model theory of difference fields', *Transactions of AMS* 351 (1999) 2997-3071.
9. Z. Chatzidakis, A. Macintyre and L. van den Dries, 'Definable sets in finite fields', *J. reine angew. Math.* 427 (1992) 107-135.
10. Z. Chatzidakis and A. Pillay, 'Generic structures and simple theories', *Ann. Pure and Applied Logic* 95 (1998) 71-92.
11. G. Cherlin, L. Harrington, and A. H. Lachlan, '$\aleph_0$-categorical, $\aleph_0$-stable structures', *Ann. Pure and Applied Logic* 28 (1985) 103-135.
12. G. Cherlin and S. Shelah, 'Superstable fields and groups', *Ann. of Math. Logic* 18 (1980) 227-270.
13. T. de Piro and B. Kim, 'The geometry of 1-based minimal types', *Transactions of American Math. Soc.* 355 (2003) 4241-4263.
14. T. de Piro, B. Kim and J. Millar 'Constructiong the hyperdefinable group from the group configuration', *Journal of Math. Logic* 6 (2006) 121-139.
15. J. Doyen and X. Hubaut, 'Finite regular locally projective spaces', *Math. Z.* 119 (1971) 83-88 .
16. J. L. Duret, 'Les corps faiblement algébriquement clos non séparablement clos ont la propriété d'indépendance', in *Modle theory of algebra and arithmetic*, Springer Lecture Note 834 (1980) 135-157.
17. D. Evans and F. O. Wagner, 'Supersimple ω-caterorical groups and theories', *J. of Symbolic Logic* 65 (2000) 767-776.
18. B. Hart, B. Kim, and A. Pillay, 'Coordinatization and canonical bases in simple theories', *J. of Symbolic Logic* 65 (2000) 293-309.
19. E. Hrushovski, 'Unimodular minimal theories', *Journal of London Math. Soc.* 46 (1992) 385-396.
20. E. Hrushovski, 'A new strongly minimal set', *Ann. Pure and Applied Logic*

62 (1993) 147-166.

21. E. Hrushovski, 'The Mordell-Lang conjecture for function fields', *J. of AMS* 9 (1996) 667-690.
22. E. Hrushovski, 'Simplicity and the Lascar group', preprint (1998).
23. E. Hrushovski, 'Difference fields and Manin-Mumford conjecture', *Ann. of Pure and Applied Logic* 112 (2001) 43 -115.
24. E. Hrushovski, 'Pseudo-finite fields and related structures', *Quaderni di Matematica* 11 (2002) 151-212.
25. E. Hrushovski, 'Groupoids, imaginaries and internal covers', preprint (2006).
26. E. Hrushovski and A. Pillay, 'Weakly normal groups', in *Logic Colloquium 85*, North Holland, Amsterdam (1987) 233-244.
27. E. Hrushovski and A. Pillay, 'Groups definable in local fields and pseudo-finite fields', *Israel J. Math.* 85 (1994) 203-262.
28. E. Hrushovski and B. Zilber, 'Zariski geometries', *Journal of AMS* 9 (1996) 1-56.
29. W. M. Kantor, M. W. Liebeck, and H. D. Macpherson, '$\aleph_0$-categorical structures smoothly approximated by finite structures', *Proc. London Math. Soc.* (3) 59 (1989) 439-463.
30. B. Kim, *Simple first order theories*, Ph. D. Thesis, Univ. of Notre Dame, 1996.
31. B. Kim, 'Forking in simple unstable theories', *J. of London Math. Soc.* 57 (1998) 257-267.
32. B. Kim, 'A note on Lascar strong types in simple theories', *J. of Symbolic Logic* 63 (1998) 926-936.
33. B. Kim, 'On the number of countable models of a countable supersimple theory', *J. of London Math. Soc.* (2) 60 (1999) 641-645.
34. B. Kim, 'Simplicity, and stability in there', *Journal of Symbolic Logic* 66 (2001) 822-836.
35. B. Kim 'Recovering the hyperdefinable group action in the group configuration theorem', will appear in *J. of Symbolic Logic.*
36. B. Kim, *Simplicity theory*, will be published by Oxford Univ. Press.
37. B. Kim, A. Kolesnikov and A. Tsuboi 'Generalized amalgamation and n-simplicity,' *Annals of Pure and Applied Logic* 155 (2008) 97-114.
38. B. Kim and A. Pillay, 'Simple theories', *Ann. Pure and Applied Logic* 88 (1997) 149-164.
39. B. Kim and A. Pillay, 'From stability to simplicity', *Bulletin of Symbolic Logic* 4 (1998) 17-36.
40. A. Kolesnikov, 'n-simple theories', *Ann. Pure and Applied Logic* 131 (2005) 227-261.
41. A. H. Lachlan, 'On the number of countable models of a countable superstable theory', in *Logic, Methodology, Philosophy of Science IV*, North Holland, Amsterdam (1973) 45-56.
42. D. Lascar, 'On the category of models of a complete theory', *J. of Symbolic Logic* 47 (1982) 249-266.
43. A. Macintyre, 'On ω_1-categorical theories of fields' *Fundamenta Mathematicae* 71 (1971) 1-25.

44. Y. Peterzil and S. Starchenko, 'A trichotomy theorem for o-minimal structures', *Proceedings of London Math. Soc.* (3) 77 (1998) 481-523.
45. A. Pillay, *Geometric stability theory*, Oxford University Press, Oxford (1996).
46. A. Pillay and B. Poizat, 'Corps et chirurgie' *J of Symbolic Logic* 60 (1995) 528-533.
47. A. Pillay, T. Scanlon and F. O. Wagner, 'Supersimple fields and division rings', *Mathematical Research Letters* 5 (1998) 473-483.
48. T. Scanlon, 'The *abc*-theorem for commutative algebraic groups in characteristic *p*', *International Math. Research Notices* 18 (1997) 881-898.
49. T. Scanlon, '*p*-adic distance from torsion points of semi-abelian varieties', *J. reine angew. Math.* 499 (1998) 225-236.
50. T. Scanlon, 'Local André-Oort conjecture for the universal abelian variety', *Inventiones mathematicae* 163 (2006) 191-211.
51. S. Shelah, 'Simple unstable theories', *Ann. Math. Logic* 19 (1980) 177-203.
52. S. Shelah, *Classification theory*, 2nd edn, North-Holland, Amsterdam (1990).
53. I. Tomasic and F. O. Wagner, 'Applications of the group configuration theorem in simple theories', *Journal of Math. Logic* 3 (2003) 239-255.
54. E. Vassiliev, 'Generic pairs of SU-rank 1 structures', *Ann. Pure and Applied Logic* 120 (2003) 103-149.
55. F. O. Wagner, 'Groups in simple theories', in *Logic Colloquium 2001*, Lecture Notes in Logic 20, A K Peters, Wellesley, USA (2005) 440-467.
56. F. O. Wagner, *Simple theories*, Kluwer Academic Publishers, Dordrecht (2000).
57. C. Wood, 'Notes on the stability of separably closed fields', *J. of Symbolic Logic* 44 (1979) 412-416.
58. B. I. Zilber, *Uncountably categorical theories*, AMS Translations of Mathematical Monographs, Vol. 117.

Weak weak König's lemma in constructive reverse mathematics

Takako Nemoto
E-mail: sa4m20@math.tohoku.ac.jp
Mathematical Institute, Tohoku University,
Sendai, 980-8578, Japan

Mathematical Institute, Tohoku University,
Sendai, 980-8578, Japan
E-mail: sa4m20@math.tohoku.ac.jp
www.math.tohoku.ac.jp/~sa4m20/

In this paper we show in a constructive setting that weak weak König's lemma, the assertion that every binary tree $T \subseteq \{-1,1\}^{<\mathbb{N}}$ without infinite paths satisfies $\lim_{n\to\infty} \sharp\{t \in T : \mathit{length}(t) = n\}/2^n = 0$, is equivalent to the following assertion: Every positive uniformly continuous function $f : [0,1] \to \mathbb{R}$ satisfies $\lim_{\delta\to 0} \mu(\{x \in [0,1] : f(x) < \delta\}) = 0$.

Keywords: constructive reverse mathematics; Brouwer's fan theorem; weak weak König's lemma; uniform continuous function.

1. Introduction

In classical reverse mathematics, weak König's lemma plays a quite important role. It asserts that every binary *tree* or *fan* $T \subseteq \{-1,1\}^{<\mathbb{N}}$, (i.e., a set of finite sequences closed under restriction) with branches of any finite length has an infinite path. It is known that it implies many basic properties of continuous function, e.g., the uniform continuity of continuous function on $[0,1]$.

Some important consequences of weak König's lemma are known to be equivalent to a weaker version of weak König's lemma (cf. 5);

Weak weak König's lemma: Every tree $T \subseteq \{-1,1\}^{<\mathbb{N}}$ without infinite paths satisfies

$$\lim_{n\to\infty} \frac{\sharp\{t \in T : \mathit{length}(t) = n\}}{2^n} = 0.$$

On the other hand, some roles of weak König's lemma in classical reverse mathematics are played by the following assertion in constructive reverse mathematics. A binary tree T is *detachable* if $t \in T$ or $t \notin T$ holds for every $t \in 2^{<\mathbb{N}}$:

> **Binary version of Brouwer's fan theorem:** Every detachable binary tree $T \subseteq \{-1,1\}^{<\mathbb{N}}$ without infinite paths is bounded.

Note that Brouwer's fan theorem does not hold in Bishop's constructive mathematics.

We can check that, even in the constructive setting, the binary version of Brouwer's fan theorem implies weak weak König's lemma (in the same formulation given above in the classical setting), as Brouwer's fan theorem asserts that, for every tree $T \subseteq \{-1,1\}$ without infinite paths, there exists n such that $\sharp\{t \in T : \mathit{length}(t) = n\} = 0$.

Therefore, the following is a natural question: What consequences of (the binary version of) Brouwer's fan theorem can be implied by or are equivalent to weak weak König's lemma in Bishop's constructive mathematics?

Some basic properties of uniformly continuous function on $[0,1]$ are known to be consequences of (the binary version of) Brouwer's fan theorem: 3 proved the following equivalence over Bishop's constructive mathematics:

(1) Binary version of Brouwer's fan theorem
(2) Every positive uniformly continuous function on $[0,1]$ has a positive infimum.

As the main result of this paper, we have the following equivalence to a weaker property of uniformly continuous function:

Theorem 1.1. *The following assertions are equivalent in Bishop's constructive mathematics.*

(1) Weak weak König's lemma: Every tree $T \subseteq \{-1,1\}^{<\mathbb{N}}$ without infinite paths satisfies

$$\lim_{n\to\infty} \frac{\sharp\{t \in T : \mathit{length}(t) = n\}}{2^n} = 0.$$

(2) Every positive, uniformly continuous function $f : [0,1] \to \mathbb{R}$ satisfies the following property: For any $\varepsilon > 0$, there exists $\delta > 0$ such that $\mu(\{x : f(x) < \delta\})$ is defined and $\mu(\{x : f(x) < \delta\}) < \varepsilon$.

2. A proof of (2) → (1)

In this section, we prove that (2) implies (1) in Theorem 1.1.

Definition 2.1. The *complete binary fan* C is the set $\{-1,1\}^{<\mathbb{N}}$. We call $a \in C$ a *node*. $|a|$ denotes the *length* of a and $a(n)$ denotes the $(n+1)$-th element of a for $n < |a|$. If $|a| < m$, then the restriction of a to m is $a[m] = \langle a(0), a(1), ..., a(m-1) \rangle$. For nodes a and b, $a \subseteq b$ means that a is an initial segment of b, i.e., $|a| \leq |b|$ and there is $m < |b|$ with $a = b[m]$. For $i \in \{-1,1\}$, $\langle i \rangle^n$ is the node a of length n with $a(m) = i$ for all $m < n$. The *concatenation* of a and b, denoted by $a * b$, is $\langle a(0), a(1), ..., a(|a|-1), b(0), b(1), ..., b(|b|-1) \rangle$.

A *branch* α of C is an element $\{-1,1\}$. $\alpha(m)$ denotes the $(m+1)$-th element of α. The restriction of α to m is defined in the same way as the restriction of a node and denoted $\alpha[m]$.

Definition 2.2. For each node a and $A \subseteq C$, we define x_a, y_a and $P(A)$ as

$$x_a = \frac{1}{2} + \sum_{i=0}^{|a|-1} \frac{a(i)}{2^{i+2}}, \quad y_a = \frac{1}{2} - \sum_{i=0}^{|a|-1} \frac{1}{2^{i+2}} = \frac{1}{2^{|a|+1}} \text{ and}$$

$$P(A) = \{(x_a, y_a) : a \in A\}.$$

Definition 2.3 (Binary fans). *A binary fan A is a subset of C which is closed under restriction, i.e., $a \in A$ and $b \subseteq a$ implies $b \in A$. A binary fan A is branch-bounded if each branch α of C, there exists m with $\alpha[m] \notin A$. A binary fan is bounded if there is $m \in \mathbb{N}$ such that, for all $a \in A$, $|a| < m$. A set $B \subseteq C$ is a bar if, for each branch, there exists n with $\alpha[n] \in B$.*

Brouwer's fan theorem, which is a non-constructive statement, and so it is not included in base system, implies that every branch-bounded fan is bounded.

A set $X \subseteq \mathbb{R}^k$ is *totally bounded* if, for any $r \in \mathbb{R}$, there exists $n \in \mathbb{N}$ and $x_0, ..., x_n \in \mathbb{R}^k$ such that, for any $x \in X$, there exists $m \leq n$ with $|x - x_n| < r$.

Lemma 2.1.

(1) *A binary fan A is detachable from C if and only if $P(A)$ is totally bounded.*

(2) *Let x be a real number and α a branch of C. Suppose, for some n, we have $\alpha(n) = 1$ and $x < x_{\alpha[n]}$, or $\alpha(n) = -1$ and $x_{\alpha[n]} < x$. Then $|x - x_{\alpha[m]}| > 1/2^{n+2}$ for all $m > n$.*

Proof. 1 is proved essentially in Theorem 2.1 in 3. 2 is easy. □

Lemma 2.2. *Let a be a node and let α be a branch. If $x_{\alpha[|a|]} > x_a$, then $x_{\alpha[n]} - x_a > 1/2^{|a|+1}$ for all $n \geq |a|$. If $x_a > x_{\alpha[|a|]}$ then $x_a - x_{\alpha[n]} > 1/2^{|a|+1}$ for all $n \geq |a|$.*

Proof. Trivial. □

In the rest of this section, we fix a detachable binary branch-bounded fan F.

Definition 2.4. For a node a and $i \in \{-1, 1\}$,

$$a^i = \begin{cases} \subseteq\text{-minimal } b \in C \text{ such that } b = a * \langle i \rangle * \langle -i \rangle^n \text{ and } b \notin F & \text{if } a * \langle i \rangle \in \\ a * \langle i \rangle & \text{otherwise} \end{cases}$$

Set $\delta_a = 1/2^{\max\{|a^1|, |a^{-1}|\}+2} = \min\{|x_a - x_{a^1}|, |x_a - x_{a^{-1}}|\}$. Note that $x_{a^{-1}} < x_a < x_{a^1}$ and that there is no $b \in F$ with $x_{a^{-1}} < x_b < x_{a^1}$.

For any (x, y) and (x', y') in $\mathbb{R}^2$, $\rho((x, y), (x', y'))$ denotes the ordinary distance between (x, y) and (x', y'), i.e., $\sqrt{(x - x')^2 + (y - y')^2}$. For $X \subseteq \mathbb{R}$, $\rho((x, y), X)$ is the distance $\inf\{\rho((x, y), (x', y')) : (x', y') \in X\}$, if it is defined.

Lemma 2.3. *For a node $a \in F$, if $|x - x_a| < \delta_a$, then $(x, 0)$ is bounded away from $P(F)$ by $\delta_a/2$.*

Proof. Since $|x - x_a| < \delta_a$, $x_{a^{-1}} < x < x_{a^1}$. Assume $b \in F$. If $|b| < \max\{|a^1|, |a^{-1}|\}$ then $y_b = 1/2^{|a|+1} > \delta_a/2$, and so $\rho((x, 0), (x_a, y_a)) > \delta_a/2$. Suppose $|b| \geq \max\{|a^1|, |a^{-1}|\}$. Since neither $a^1 \in F$ nor $a^{-1} \in F$, there exists $n < |b|$ such that either $b(n) < a^{-1}(n)$ or $a^1(n) < b(n)$. Take such the least n. If $b(n) < a^{-1}(n)$, then then $x_b < x_{a^{-1}} < x$, and so $x - x_b > x_{a^{-1}} - x_b > 1/2^{|a^{-1}|+1}$ by Lemma 2.2. Similarly, if $a^1(n) < b(n)$, then $x < x_{a^1} < x_b$, and so $x_b - x > x_b - x_{a^1} > 1/2^{|a^1|+1}$ by Lemma 2.2. □

Lemma 2.4. *For a real number $x \in [0, 1]$, there is k such that $(x, 0)$ is bounded away from $P(F)$ by $1/2^k$.*

Proof. For given $x \in [0, 1]$, define a branch α so that $x_{\alpha[n]} - (\delta_{\alpha[n]}/2) < x$ if $\alpha(n) = 1$, $x < x_{\alpha[n]} + (\delta_{\alpha[n]}/2)$ if $\alpha(n) = -1$. Assume $\alpha[m + 1] \notin F$. We may assume $a[m] \in F$. Since, for any $n < m + 1$, either $|x - x_{\alpha[n]}| < \delta_{\alpha[n]}$ or $|x - x_{\alpha[n]}| > \delta_{\alpha[n]}/2$ holds, the following two cases are possible.

(i) $|x - x_{\alpha[n]}| < \delta_{\alpha[n]}$ for some $n < m + 1$.
Then $(x, 0)$ is bounded away from $P(F)$ by $\delta_{\alpha[n]}/2$ by Lemma 2.3.

(ii) $|x - x_{\alpha[n]}| > \delta_{\alpha[n]}/2$ for all $n < m + 1$.
In this case, for any $n \leq m$, $x_{\alpha[n]} < x$ holds if $\alpha(n) = 1$ and $x < x_{\alpha[n]}$ holds if $\alpha(n) = -1$. Assume $a \in F$. If $|a| \leq m$, $y_a = 1/2^{|a|+1} \geq 1/2^{m+1}$. If $|a| > m + 1$, there is $n \leq m$ such that $\alpha(n) \neq a(n)$. Take such the least n. If $\alpha(n) = 1$, then $x - x_{\alpha[n]} > \delta_{\alpha[n]}/2$ and $a(n) = -1$ hold. Since $x_{a[n+1]} < x_{\alpha[n+1]}$, $1/2^{n+2} < x_{\alpha[n]} - x_a$ holds by 2 of Lemma 2.1, and so $|x - x_a| > 1/2^{n+2}$. Also in the case $\alpha(n) = 1$, it can be proved $|x - x_a| > 1/2^{n+2}$ in a similar way. Therefore $\rho((x, 0), (x_a, y_a)) > 1/2^{m+2}$.

Therefore x is bounded away from $P(F)$ by $1/2^k$, where $k = \max(\{m + 2\} \cup \{|(\alpha[n])^1| + 1 : n \leq m\} \cup \{|(\alpha[n])^{-1}| + 1 : n \leq m\}) + 1$. □

Lemma 2.5. *For any $a \in F$ and for any $x \in [x_a - 1/2^{|a|+1}, x_a + 1/2^{|a|+1}]$, $\rho((x, 0), P(F)) \leq \sqrt{2}/2^{|a|+1}$ holds.*

Proof. Trivial. □

Lemma 2.6. *We can define $f : [0, 1] \to \mathbb{R}$ by $f(x) = \rho((x, 0), P(F))$ and f is a positive, uniformly continuous function.*

Proof. Since $P(F)$ is totally bounded, it is located (cf. [2, Chapter 4, (4.4) Proposition]), i.e., $f(x)$ is defined for each $x \in [0, 1]$. f is uniformly continuous, since we have

$$\rho(x, P(F)) \leq \rho(x, y) + \rho(y, P(F)),$$

which implies $|\rho(x, P(F)) - \rho(y, P(F))| < |x - y|$ (cf. [2, p88]). By Lemma 2.4, f is positive. □

Now we use constructive Lebesgue measure μ on $[0, 1]$ defined in 1. The properties of μ that we need here are as follows.

Fact 2.1. *μ satisfies the following properties:*

(1) If $X \subseteq Y \subseteq [0, 1]$ and if both $\mu(X)$ and $\mu(Y)$ are defined, then $\mu(X) \leq \mu(Y)$.

(2) Let $X_i \subset [0, 1]$ $(i < n)$. If $\mu(X_i)$ is defined for all $i < n$, then $\mu(\bigcup_{i<n} X_i)$ is defined.

(3) There exists a sequence $\langle x_i : in \in \mathbb{N}\rangle$ of real numbers in $[0, 1]$ such that, if $x < y$ and $x \neq x_i$ and $y \neq x_i$ for all $i \in \mathbb{N}$, then $\mu([x, y])$ is defined and equals $(y - x)$.

(4) For any uniformly continuous function $f : [0,1] \to \mathbb{R}$, there exists a sequence $\langle x_i : in \in \mathbb{N}\rangle$ of real numbers in $[0,1]$ such that if $t \neq x_i$ for every $i \in \mathbb{N}$, then all $\mu(\{x : f(x) \geq t\})$, $\mu(\{x : f(x) > t\})$, $\mu(\{x : f(x) \leq t\})$ and $\mu(\{x : f(x) < t\})$ are definable.

For detailed properties of μ, see [1] and [2, Chapter 6].

Now, we have one direction of the main result.

Theorem 2.1. *Assume that every positive, uniformly continuous function satisfies that there, for any $\varepsilon > 0$, there exists δ with $\mu(\{x \in [0,1] : f(x) < \delta\})$ is defined and that $\mu(\{x \in [0,1] : f(x) < \delta\}) < \varepsilon$. Then F satisfies*

$$\lim_{n\to\infty} \frac{\sharp\{a \in F : |a| = n\}}{2^n} = 0,$$

where $\sharp\{a \in F : |a| = n\}$ is the number of elements of $\{a \in F : |a| = n\}$. (Note that $\{a \in F : |a| = n\}$ is finite for any n, since F is detachable.)

Proof. Since F is a detachable, branch-bounded binary fan, as shown in Lemma 2.6, $f(x) = \rho((x,0), P(F))$ is a positive, uniformly continuous function. Assume that for any $\varepsilon > 0$, there exists δ with $\mu(\{x \in [0,1] : f(x) < \delta\})$ is defined and $\mu(\{x \in [0,1] : f(x) < \delta\}) < \varepsilon$. For given $\varepsilon > 0$, take δ such that $\mu(\{x \in [0,1] : f(x) < \delta\})$ is defined and that $\mu(\{x \in [0,1] : f(x) < \delta\}) < \varepsilon$. Take m so that $\sqrt{2}/2^{m+1} < \delta$. Then, if a node a of length m in in F, $[x_a - 1/2^{m+1}, x_a + 1/2^{m+1}] \subseteq \{x \in [0,1] : f(x) < \delta\}$ by Lemma 2.5, and so $\sharp\{a \in F : |a| = m\} < \lfloor 2^m \cdot \varepsilon \rfloor$. Then

$$\frac{\sharp\{a \in F : |a| = m\}}{2^m} < \frac{\lfloor 2^m \cdot \varepsilon \rfloor}{2^m} < \varepsilon. \qquad \square$$

3. A proof of (1) → (2)

In this section we prove that (1) implies (2) in Theorem 1.1.

We need another interpretation of $\{-1,1\}^{\mathbb{N}}$ to $[0,1]$.

Definition 3.1. Let α be a branch and let a be a node. Define x'_α and x'_a by

$$x'_\alpha = \frac{1}{2} - \sum_{i\in\mathbb{N}}^{\infty} \frac{\alpha(i)}{2^{i+2}}, \quad x'_a = \frac{1}{2} - \sum_{i=0}^{|b|-1} \frac{b(i)}{2^{i+2}}.$$

Fact 3.1 ([3, Lemma 1.2]). *Let B be a countable set of nodes. Then there exists a detachable set B' of nodes such that each element of B' has a restriction in B, and for each node b of B, there exists $n > |b|$ such that* *every node a of length n extending b has a restriction in B'. In particular, if B is a bar, then so is B'.*

Theorem 3.1. *If every detachable, branch-bounded fan F satisfies*

$$\lim_{n\to\infty} \frac{\sharp\{b \in F : |b| = n\}}{2^n} = 0 \tag{1}$$

then every positive, uniformly continuous function $f : [0,1] \to \mathbb{R}$ satisfies the following property:

For any $\varepsilon > 0$, there exists $\delta > 0$ such that $\mu(\{x : f(x) < \delta\})$ is defined and $\mu(\{x : f(x) < \delta\}) < \varepsilon$.

Proof. Let f be a positive uniformly continuous function. Then there exists a function ω from positive rational numbers to positive rational numbers such that $|x - y| < \delta$ implies $|f(x) - f(y)| < \omega(\delta)$ and $\lim_{\delta\to 0} \omega(\delta) = 0$. Set $B = \{b \in C : \omega(1/2^{|b|}) < f(x'_b)\}$. Since the relation $r_1 < r_2$ for real numbers is of the form $\exists n P(n)$ for a decidable predicate P, the set B is countable. Let B' be a detachable set associated with B by Fact 3.1 and let $F = \{b \in C : \forall n < |b|(b[n] \notin B')\}$. Then we can check that F is branch-bounded as follows: Since f is positive and $\lim_{\delta\to 0} \omega(\delta) = 0$, for any branch α, there exists n with $2 \cdot \omega(1/2^n) < f(x'_\alpha)$. Together with the fact $|x'_\alpha - x'_{\alpha[n]}| < 1/2^{n+1}$, we have $\omega(1/2^n) < f(x'_\alpha) - \omega(1/2^n) \leq f(x'_\alpha) - \omega(1/2^{n+1}) < f(x'_{\alpha[n]})$, and so $\alpha[n] \in B$. By Fact 3.1, there exists $m > n$ such that every node a of length n extending $\alpha[n]$ has a restriction in B', in particular, α has a restriction in B'.

By the hypothesis of our theorem, F satisfies Eq. (1). Fix ε and let n be such that

$$\frac{\sharp\{b \in F : |b| = n\}}{2^n} < \varepsilon.$$

We prove that there exists δ' such that $\mu(\{x \in [0,1] : f(x) \geq \delta'\}) \geq 1-\varepsilon$, which implies that, for every $\delta < \delta'$, $\mu(\{x \in [0,1] : f(x) < \delta\}) < \varepsilon$, if it is defined. Note that, by Fact 2.1.4, we can take $\delta < \delta'$ with $\mu(\{x \in [0,1] : f(x) < \delta\})$ is defined.

Since $b \notin F$, there exists a natural number $m' < |b|$ with $b[m'] \in B'$. By the definition of B', there is $m \leq m'$ with $b[m] \in B$. Then, $\omega(1/2^{|b|}) < f(x'_{b[m]})$. Take a positive rational δ^b with $f(x'_{b[m]}) - \omega(1/2^m) = \delta^b > 0$. Then $f(x) > \delta^b$ for all $x \in [x'_{b[m]} - 1/2^{m+1}, x'_{b[m]} + 1/2^{m+1}]$, since

$$\begin{aligned} 0 < \delta^b &< f(x'_{b[m]}) - \omega(1/2^m) \\ &\leq f(x'_{b[m]}) - \omega(1/2^{m+1}) \\ &< f(x). \end{aligned}$$

Note that $[x'_b - 1/2^{|b|+1}, x'_b + 1/2^{|b|+1}] \subseteq [x'_{b[m]} - 1/2^{m+1}, x'_{b[m]} + 1/2^{m+1}]$, which implies $f(x) > \delta^b$ for all $x \in [x'_b - 1/2^{|b|+1}, x'_b + 1/2^{|b|+1}]$. Since F is detachable, $\{b \notin F : |b| = n\}$ is finite, and so we can take a positive real number δ' such that $\delta' \leq \min\{\delta^b : b \notin F \wedge |b| = n\}$ and $\mu(\{x \in [0,1] : f(x) > \delta'\})$ is definable. Therefore $f(x) > \delta'$ for all $b \notin F$ of length n and $x \in [x'_b - 1/2^{|b|}, x'_b + 1/2^{|b|}]$. Therefore

$$\begin{aligned}\mu(\{x \in [0,1] : f(x) > \delta'\}) &\geq \sharp\{b \notin F : |b| = n\} \cdot \frac{1}{2^n} \\ &= 1 - \sharp\{b \in F : |b| = n\} \cdot \frac{1}{2^n} \\ &> 1 - \varepsilon \end{aligned}$$

□

Acknowledgements

The author would to express her gratitude to Prof. Hajime Ishihara for introducing the work of Julian and Richman to her and the anonymous referee for careful reading and picking up a number of mistakes. She was partly supported by the Grant-in-Aid for JSPS Fellows, The Ministry of Education, Culture, Sports, Science and Technology, Japan.

References

1. Erret Bishop and Henry Cheng, *Constructive Measure Theory*, Memoirs of the American Mathematical Society 116 1972.
2. Erret Bishop and Douglas Bridges, *Constructive analysis*, Springer, 1985.
3. William Julian and Fred Richman, A uniformly continuous function on $[0,1]$ that is everywhere different from its infimum, *Pacific Journal of Mathematics*, vol. 111 (1984), no. 2, pp. 333–340.
4. Stephen G. Simpson, *Subsystems of Second Order Arithmetic*, Springer, 1991.
5. Xiaokang Yu and Stephen G. Simpson, Measure theory and weak König's lemma, *Archive for Mathematical Logic*, vol. 30 (1990), pp. 171–180.

On preservation and reflection of stationary subsets of $\mathcal{P}_\kappa\lambda$ when $\mathrm{cf}(\lambda) < \kappa$

Hiroshi Sakai
Graduate School of Engineering, Kobe University,
Kobe, Hyogo, Japan
E-mail: hsakai@people.kobe-u.ac.jp

We study the preservation under $< \kappa$-closed forcing extensions and the reflection of stationary subsets of $\mathcal{P}_\kappa\lambda$ in the case when λ is a singular cardinal with $\mathrm{cf}(\lambda) < \kappa$. In particular we discuss those under GCH.

Keywords: Stationary Reflection Principle; Singular Cardinal.

1. Introduction

The following stationary reflection principle in $\mathcal{P}_\kappa\lambda$, which is often called the weak reflection principle in $\mathcal{P}_\kappa\lambda$, has been studied by many set theorists so far:

Definition 1.1. For a regular uncountable cardinal κ, a cardinal $\lambda > \kappa$ and a stationary $T \subseteq \mathcal{P}_\kappa\lambda$, let $\mathsf{SR}_{\kappa\lambda}(T)$ be the following assertion:

$\mathsf{SR}_{\kappa\lambda}(T) \equiv$ For every stationary $S \subseteq T$ there exists $W \subseteq \lambda$ such that $|W| = \kappa \subseteq W$ and such that $S \cap \mathcal{P}_\kappa W$ is stationary in $\mathcal{P}_\kappa W$.

We let $\mathsf{SR}_{\kappa\lambda}$ denote $\mathsf{SR}_{\kappa\lambda}(\mathcal{P}_\kappa\lambda)$.

(Nowadays there are two notions of stationary subsets of $\mathcal{P}_\kappa\lambda$. We adopt the one introduced by Jech.[8] More precisely, see Section 2.)

The importance of the above stationary reflection principle was revealed by Foreman–Magidor–Shelah.[5] They showed that $\mathsf{SR}_{\omega_1\lambda}$ is implied by Martin's maximum for every $\lambda > \omega_1$ and holds if a λ-supercompact cardinal is Lévy collapsed to ω_2. Moreover they showed that $\mathsf{SR}_{\omega_1\lambda}$ has interesting consequences such as Chang's conjecture and the presaturation of the non-stationary ideal over ω_1. It is also known that $\mathsf{SR}_{\omega_1\lambda}$ has the influence on the cardinal arithmetic. Todorčević showed that $\mathsf{SR}_{\omega_1\omega_2}$ implies $2^\omega \le \omega_2$,[19] and Shelah showed that $\mathsf{SR}_{\omega_1\lambda}$ implies SCH below λ.[15] Besides these, the stationary reflection principle in $\mathcal{P}_{\omega_1}\lambda$ and its strengthening have been studied extensively.[1,2,6,20,etc.]

The stationary reflection principle in $\mathcal{P}_\kappa\lambda$ for $\kappa \ge \omega_2$ also have been studied so far.[3,4,6,16,17,etc.] Now, due to Feng–Magidor, Foreman–Magidor and Shelah–Shioya, it is known that $\mathsf{SR}_{\kappa\lambda}$ is not consistent if $\omega_2 \le \kappa < \lambda$.[3,4,16] On the other hand it was already shown by Foreman–Magidor–Shelah that the above mentioned facts on $\mathsf{SR}_{\omega_1\lambda}$ can be partially generalized to the case when $\kappa \ge \omega_2$.[5] More precisely, the stationary reflection principle in $\mathcal{P}_\kappa\mathcal{H}_\theta$ below internally approachable sets is consistent for regular cardinals $\theta > \kappa$, and this partial stationary reflection principle implies the precipitousness of the nonstationary ideal over κ. Here we review this more closely in the context of $\mathcal{P}_\kappa\lambda$. (Subsection 3.2 of this paper contains the proof of facts below.)

Let κ be a regular uncountable cardinal, and λ be a cardinal $> \kappa$ with $\lambda^{<\kappa} = \lambda$. Note that $|{}^{<\kappa}(\mathcal{P}_\kappa\lambda)| = \lambda$. Hence we can take a bijection $\psi : {}^{<\kappa}(\mathcal{P}_\kappa\lambda) \to \lambda$. Then $x \in \mathcal{P}_\kappa\lambda$ is said to be *ψ-internally approachable* (ψ-i.a.) if there is a $\subseteq$-increasing sequence $\langle x_\xi \mid \xi < \zeta\rangle$ (ζ: limit ordinal $< \kappa$) such that $\bigcup_{\xi<\zeta} x_\xi = x$ and such that $\psi(\langle x_\xi \mid \xi < \zeta'\rangle) \in x$ for every $\zeta' < \zeta$. (We call such sequence $\langle x_\xi \mid \xi < \zeta\rangle$ a *ψ-internally approaching sequence to x*.) Moreover let $S^\psi_{\kappa\lambda}$ be the set of all ψ-i.a. $x \in \mathcal{P}_\kappa\lambda$.

Note that $S^\psi_{\kappa\lambda}$ is independent of the choice of a bijection ψ modulo the nonstationary ideal over $\mathcal{P}_\kappa\lambda$. In fact, if ψ and φ are both bijections from ${}^{<\kappa}(\mathcal{P}_\kappa\lambda)$ to λ, then being ψ-i.a. and being φ-i.a. are equivalent for all $x \in \mathcal{P}_\kappa\lambda$ which is closed under $\varphi\circ\psi^{-1}$ and $\psi\circ\varphi^{-1}$. We omit the superscript ψ and write $S_{\kappa\lambda}$ for $S^\psi_{\kappa\lambda}$ because the difference of a nonstationary set does not matter in our context. What is important on $S_{\kappa\lambda}$ is the following:

Fact 1.2 (Foreman–Magidor–Shelah[5]). *Let κ be a regular uncountable cardinal, and λ be a cardinal $> \kappa$ with $\lambda^{<\kappa} = \lambda$. Then all stationary subsets of $S_{\kappa\lambda}$ remain stationary in all $<\kappa$-closed forcing extensions.*

It was also shown by Foreman–Magidor that the complement of $S_{\kappa\lambda}$ becomes nonstationary in the forcing extension by $\mathrm{Col}(\kappa, \lambda)$.[4] Here $\mathrm{Col}(\kappa, \lambda)$ denotes the forcing notion ${}^{<\kappa}\lambda$ ordered by reverse inclusions. Hence $\mathrm{Col}(\kappa, \lambda)$ is a $<\kappa$-closed forcing notion which forces that $|\lambda| = \kappa$.

We obtain the consistency of $\mathsf{SR}_{\kappa\lambda}(S_{\kappa\lambda})$ from Fact 1.2 and Fact 1.3 below. In Fact 1.3 $\mathrm{Col}(\kappa, <\nu)$ denotes the Lévy collapse forcing ν to be κ^+.

Fact 1.3 (Foreman–Magidor–Shelah[5]). *Let κ be a regular uncountable cardinal and λ be a cardinal $> \kappa$. Suppose that ν is a λ-supercompact cardinal with $\kappa < \nu < \lambda$. In $V^{\mathrm{Col}(\kappa,<\nu)}$ assume that T is a stationary subset of $\mathcal{P}_\kappa\lambda$ such that all stationary subsets of T remain stationary in all $<\kappa$-closed forcing extensions. Then $\mathsf{SR}_{\kappa\lambda}(T)$ holds in $V^{\mathrm{Col}(\kappa,<\nu)}$.*

Corollary 1.4 (Foreman–Magidor–Shelah[5]). *Let κ be a regular uncountable cardinal and λ be a cardinal $> \kappa$ with $\lambda^{<\kappa} = \lambda$. Suppose that ν is a λ-supercompact cardinal with $\kappa < \nu < \lambda$. Then $\mathsf{SR}_{\kappa\lambda}(S_{\kappa\lambda})$ holds in $V^{\mathrm{Col}(\kappa,<\nu)}$.*

Furthermore it was shown by Foreman–Magidor–Shelah that if λ is sufficiently large then $\mathsf{SR}_{\kappa\lambda}(S_{\kappa\lambda})$ implies the precipitousness of the nonstationary ideal over κ.[5]

Above we have briefly reviewed the stationary reflection principle in $\mathcal{P}_\kappa\lambda$ in the case when $\lambda^{<\kappa} = \lambda$. Then the question naturally arises. What about the stationary reflection principle in $\mathcal{P}_\kappa\lambda$ in the case when $\lambda^{<\kappa} > \lambda$? Moreover recall that Fact 1.2 was the key of Cor.1.4. What about the preservation of stationary subsets of $\mathcal{P}_\kappa\lambda$ under $<\kappa$-closed forcing extensions when $\lambda^{<\kappa} > \lambda$? Note that if λ is a singular cardinal with $\mathrm{cf}(\lambda) < \kappa < \lambda$, then $\lambda^{<\kappa} > \lambda$. Shioya asked how is the preservation and the reflection of stationary subsets of $\mathcal{P}_\kappa\lambda$ when λ is a singular cardinal with $\mathrm{cf}(\lambda) < \kappa < \lambda$.[18]

In this paper we discuss the preservation and the reflection of stationary subsets of $\mathcal{P}_\kappa\lambda$ in the case when λ is a singular cardinal with $\mathrm{cf}(\lambda) < \kappa < \lambda$. In particular we study those under GCH.

For this we introduce the notion of the semi-internally approachability:

Notation . Let κ be a regular uncountable cardinal. A cardinal λ is said to be $< \kappa$*-strong limit* if $\nu^{<\kappa} < \lambda$ for every $\nu < \lambda$. Note that if λ is a $<\kappa$-strong limit cardinal, then $|\bigcup_{\lambda'<\lambda} {}^{<\kappa}(\mathcal{P}_\kappa\lambda')| = \lambda$.

Definition 1.5. Let κ be a regular uncountable cardinal, and λ be a $< \kappa$-strong limit cardinal with $\mathrm{cf}(\lambda) < \kappa < \lambda$. Suppose that $\varphi : \bigcup_{\lambda'<\lambda} {}^{<\kappa}(\mathcal{P}_\kappa\lambda') \to \lambda$ is a bijection. Let x be an element of $\mathcal{P}_\kappa\lambda$.

For a limit ordinal $\zeta < \kappa$, x is said to be *φ-semi-internally approachable* (φ-s.i.a.) of length ζ if there exists a $\subseteq$-increasing sequence $\langle x_\xi \mid \xi < \zeta\rangle$ in $\mathcal{P}_\kappa\lambda$ such that

(i) $\bigcup_{\xi<\zeta} x_\xi = x$,
(ii) $\varphi(\langle x_\xi \cap \lambda' \mid \xi < \zeta'\rangle) \in x$ for every $\zeta' < \zeta$ and every $\lambda' < \lambda$.

A sequence $\langle x_\xi \mid \xi < \zeta \rangle$ as above is called a *φ-semi-internally approaching sequence to x*. x is said to be *φ-semi-internally approachable* if x is φ-s.i.a. of length ζ for some limit $\zeta < \kappa$.

It is easy to see that if θ is a regular cardinal $> \lambda$, $M \in \mathcal{P}_\kappa \mathcal{H}_\theta$ is internally approachable of length ζ, and $M \prec \langle \mathcal{H}_\theta, \in \kappa, \lambda, \varphi \rangle$, then $M \cap \lambda$ is φ-s.i.a. of length ζ. (See Lem.3.6.)

As in the case when $\lambda^{<\kappa} = \lambda$, we consider the set of all φ-s.i.a. sets. But we need divide this set according to the length of approaching sequence:

Definition 1.6. Let κ, λ and φ be as in Def.1.5. Then let

$$T^{\varphi,0}_{\kappa\lambda} := \{x \in \mathcal{P}_\kappa\lambda \mid x \text{ is } \varphi\text{-s.i.a. of length } \mathrm{cf}(\lambda)\}$$
$$T^{\varphi,1}_{\kappa\lambda} := \{x \in \mathcal{P}_\kappa\lambda \mid x \text{ is } \varphi\text{-s.i.a. but is not } \varphi\text{-s.i.a. of length } \mathrm{cf}(\lambda)\}$$

As is the case with $S^{\psi}_{\kappa\lambda}$, both $T^{\varphi,0}_{\kappa\lambda}$ and $T^{\varphi,1}_{\kappa\lambda}$ do not depend on the choice of φ modulo the nonstationary ideal over $\mathcal{P}_\kappa\lambda$. Hence we often write $T^0_{\kappa\lambda}$ and $T^1_{\kappa\lambda}$ for $T^{\varphi,0}_{\kappa\lambda}$ and $T^{\varphi,1}_{\kappa\lambda}$, respectively.

As we see in Lem.3.7, if $\kappa = \omega_1$, then $T^0_{\omega_1\lambda}$ contains a club in $\mathcal{P}_{\omega_1}\lambda$. Moreover if $\kappa \geq \omega_2$, then $T^0_{\kappa\lambda}$, $T^1_{\kappa\lambda}$ and $\mathcal{P}_\kappa\lambda \setminus (T^0_{\kappa\lambda} \cup T^1_{\kappa\lambda})$ are all stationary in $\mathcal{P}_\kappa\lambda$.

It will be proved in Cor.3.14 that $\mathcal{P}_\kappa\lambda \setminus (T^0_{\kappa\lambda} \cup T^1_{\kappa\lambda})^V$ becomes nonstationary in $V^{\mathrm{Col}(\kappa,\lambda)}$. As for the preservation and the reflection below $T^0_{\kappa\lambda}$ we will obtain the following. In particular $\mathsf{SR}_{\kappa\lambda}(T^0_{\kappa\lambda}) + \mathsf{GCH}$ is consistent:

Theorem 1.7. *Let κ be a regular uncountable cardinal and λ be a $<\kappa$-strong limit cardinal with $\mathrm{cf}(\lambda) < \kappa < \lambda$.*

(1) All stationary subsets of $T^0_{\kappa\lambda}$ remain stationary in all $<\kappa$-closed forcing extensions.

(2) Suppose that ν is a λ-supercompact cardinal with $\kappa < \nu < \lambda$. Then $\mathsf{SR}_{\kappa\lambda}(T^0_{\kappa\lambda})$ holds in $V^{\mathrm{Col}(\kappa,<\nu)}$.

On the other hand we will obtain the following as for the preservation and the reflection below $T^1_{\kappa\lambda}$:

Theorem 1.8. *Assume* GCH. *Let κ be a regular cardinal $\geq \omega_2$ and λ be a singular cardinal with $\mathrm{cf}(\lambda) < \kappa < \lambda$.*

(1) $T^1_{\kappa\lambda}$ (defined in V) remains stationary in all $<\kappa$-closed forcing extensions. But for every stationary $T \subseteq T^1_{\kappa\lambda}$ there exists stationary $S \subseteq T$ which becomes nonstationary in $V^{\mathrm{Col}(\kappa,\lambda)}$.

(2) $\mathsf{SR}_{\kappa\lambda}(T)$ fails for every stationary $T \subseteq T^1_{\kappa\lambda}$.

Thm.1.7 and 1.8 will be proved in Section 4. In Section 3 we study the basic relationship among the notion of s.i.a. and i.a. and the preservation and the reflection of stationary subsets of $\mathcal{P}_\kappa\lambda$. As we mentioned before, Subsection 3.2 contains the proof of Fact 1.2 and 1.3.

2. Preliminaries

Here we present our notation and basic facts used in this paper. For those which are not presented below, consult Jech's book and Kanamori's book.[9,10]

First we give our notation and basic facts on subsets of $\mathcal{P}_\kappa W$. For a regular cardinal κ and a set $W \supseteq \kappa$, $\mathcal{P}_\kappa W$ denotes the set $\{x \subseteq W \mid |x| < \kappa\}$. $C \subseteq \mathcal{P}_\kappa W$ is said to be *club* if C is $\subseteq$-cofinal in $\mathcal{P}_\kappa W$ and closed under unions of $\subseteq$-increasing sequence of length $< \kappa$. $S \subseteq \mathcal{P}_\kappa W$ is said to be *stationary* if $S \cap C \neq \emptyset$ for every club $C \subseteq \mathcal{P}_\kappa W$.

We use the following facts without any mention. The proof can be found in Jech's book.[9]

Fact 2.1 (Kueker[11]). *Let κ be a regular uncountable cardinal, and W be a set $\supseteq \kappa$. The following are equivalent for $S \subseteq \mathcal{P}_\kappa W$:*

(I) S is stationary in $\mathcal{P}_\kappa W$.

(II) For every function $F : [W]^{<\omega} \to \mathcal{P}_\kappa W$ there exists $x \in S$ such that $F(a) \subseteq x$ for all $a \in [x]^{<\omega}$.

(III) For every function $F : [W]^{<\omega} \to W$ there exists $x \in S$ such that $F(a) \in x$ for all $a \in [x]^{<\omega}$ and such that $x \cap \kappa \in \kappa$.

(IV) For every structure $\mathcal{M}$ of a countable language with its universe W there exists $x \in S$ such that $x \prec \mathcal{M}$ and $x \cap \kappa \in \kappa$.

Fact 2.2 (Menas[12]). *Let κ be a regular uncountable cardinal, and suppose that $\kappa \subseteq W \subseteq \bar{W}$. Then $S \subseteq \mathcal{P}_\kappa W$ is stationary if and only if the set $\{\bar{x} \in \mathcal{P}_\kappa \bar{W} \mid \bar{x} \cap W \in S\}$ is stationary in $\mathcal{P}_\kappa \bar{W}$.*

Next we present notation and a fact on structures. Suppose that $\mathcal{M}$ is a structure for a countable language and that there is a well-ordering of the universe of $\mathcal{M}$ which can be defined over $\mathcal{M}$. For $x \subseteq \mathcal{M}$ let $\mathrm{skull}^{\mathcal{M}}(x)$ denotes the Skolem hull of x in $\mathcal{M}$, i.e. the smallest M with $x \subseteq M \prec \mathcal{M}$.

We often use the following lemma:

Lemma 2.3. *Let κ and θ be regular uncountable cardinals with $\kappa \leq \theta$. Suppose that $M \prec \langle \mathcal{H}_\theta, \in \rangle$ and that $M \cap \kappa \in \kappa$. Then $y \subseteq M$ for every $y \in M \cap \mathcal{P}_\kappa \mathcal{H}_\theta$.*

Proof. There is a bijection $\pi \in M$ from $|y|$ to y by the elementarity of M. Note that $|y| \subseteq M$ because $|y| \in M \cap \kappa \in \kappa$. Thus $y = \pi[|y|] \subseteq M$. $\square$

Next we give our notation on forcing notions for collapsing cardinalities. Let κ be a regular uncountable cardinal. For an ordinal λ let $\mathrm{Col}(\kappa, \lambda)$ denotes the forcing notion ${}^{<\kappa}\lambda$ ordered by reverse inclusions. Note that $\mathrm{Col}(\kappa, \lambda)$ is a $<\kappa$-closed forcing notion and collapses the cardinality of λ to be $\leq \kappa$. For an ordinal ν let $\mathrm{Col}(\kappa, <\nu)$ denotes the Lévy collapse which collapses the cardinality of all ordinals $< \nu$ to be $\leq \kappa$. That is, $\mathrm{Col}(\kappa, <\nu)$ is the $<\kappa$-support product of $\langle \mathrm{Col}(\kappa, \upsilon) \mid \upsilon < \nu \rangle$.

Next we present our notation on scales, which are the central notion of Shelah's PCF theory. Let λ be a singular cardinal, and let $\vec{\lambda} = \langle \lambda_\eta \mid \eta < \mathrm{cf}(\lambda) \rangle$ be an increasing sequence of regular cardinals converging to λ. We let $\Pi\vec{\lambda}$ denote $\Pi_{\eta<\mathrm{cf}(\lambda)}\lambda_\eta$, that is, the set of all functions f on $\mathrm{cf}(\lambda)$ such that $f(\eta) \in \lambda_\eta$ for each $\eta < \mathrm{cf}(\lambda)$. Moreover let J be an ideal over $\mathrm{cf}(\lambda)$. For $g, h \in \Pi\vec{\lambda}$ let

$$g <_J h \Leftrightarrow \{\eta < \mathrm{cf}(\lambda) \mid g(\eta) \geq h(\eta)\} \in J$$
$$g \leq_J h \Leftrightarrow \{\eta < \mathrm{cf}(\lambda) \mid g(\eta) > h(\eta)\} \in J$$
$$g < h \Leftrightarrow g(\eta) < h(\eta) \text{ for all } \eta < \mathrm{cf}(\lambda)$$
$$g \leq h \Leftrightarrow g(\eta) \leq h(\eta) \text{ for all } \eta < \mathrm{cf}(\lambda)$$

If υ is a regular cardinal and $\vec{h} = \langle h_\alpha \mid \alpha < \upsilon \rangle$ is a $<_J$-increasing cofinal sequence in $\Pi\vec{\lambda}$, then we call $\vec{h}$ a $(\vec{\lambda}, J)$-*scale of length* υ. For a regular cardinal υ let

$$\mathrm{tcf}(\Pi\vec{\lambda}/J) = \upsilon \overset{\text{def}}{\Leftrightarrow} \text{there exists a } (\vec{\lambda}, J)\text{-scale of length } \upsilon.$$

At the end of this section we recall the ideal $I[\kappa]$ over a regular uncountable cardinal κ, introduced by Shelah.[13] Let κ be a regular uncountable cardinal. For a sequence $\vec{b} = \langle b_\xi \mid \xi < \kappa \rangle$ of bounded subsets of κ and a limit ordinal $\zeta < \kappa$, we say that ζ is *approachable with respect to* $\vec{b}$ if there exists an unbounded $b \subseteq \zeta$ of order-type $\mathrm{cf}(\zeta)$ such that $b \cap \rho \in \{b_\xi \mid \xi < \zeta\}$ for every $\rho < \zeta$. For each $E \subseteq \kappa$ let

$E \in I[\kappa] \overset{\text{def}}{\Leftrightarrow}$ there exist a sequence $\vec{b} = \langle b_\xi \mid \xi < \kappa \rangle$ of bounded subsets of κ and a club $C \subseteq \kappa$ such that every $\zeta \in E \cap C$ is approachable with respect to $\vec{b}$.

It was shown by Shelah that $I[\kappa]$ is a normal ideal over κ. $I[\kappa]$ is not proper, i.e. $\kappa \in I[\kappa]$, in many cases.[13] For example if μ is a regular cardinal with $\mu^{<\mu} = \mu$, then it is easy to see that $I[\mu^+]$ is not proper. On the other hand there may be a regular cardinal κ with $I[\kappa]$ proper. For example it is shown by Shelah that if μ is a singular cardinal such that there exists a μ-supercompact cardinal between $\mathrm{cf}(\mu)$ and μ, then $I[\mu^+]$ is proper.[13]

Here note that the set of all regular cardinals $< \kappa$ belongs to $I[\kappa]$. Hence if $I[\kappa]$ is proper, then there exists a regular cardinal $\mu < \kappa$ such that $\{\zeta < \kappa \mid \mathrm{cf}(\zeta) = \mu\} \notin I[\kappa]$ by the normality of $I[\kappa]$.

3. s.i.a, i.a, preservation and reflection

Here we overview the basic relationship among the notion of s.i.a. and i.a. and the preservation and the reflection of stationary subsets of $\mathcal{P}_\kappa\lambda$. (Lem.3.6, 3.11 and Prop.3.15.)

In Subsection 3.1 we present basic facts on s.i.a. and i.a. sets and the relationship between them. In Subsection 3.2 we review the basic relationship among the notion of i.a and the preservation and the reflection of stationary subsets of $\mathcal{P}_\kappa\lambda$.

3.1. *s.i.a. and i.a. sets*

First we recall the notion of internally approachability:

Definition 3.1. Let x be a set. For a limit ordinal ζ, x is said to be *internally approachable* (i.a.) of length ζ if there exists an $\subseteq$-increasing sequence $\langle x_\xi \mid \xi < \zeta \rangle$ such that $\bigcup_{\xi<\zeta} x_\xi = x$ and such that $\langle x_\xi \mid \xi < \zeta' \rangle \in x$ for every $\zeta' < \zeta$.

A sequence $\langle x_\xi \mid \xi < \zeta \rangle$ as above is called an *internally approaching sequence to x*. x is said to be *internally approachable* if x is internally approachable of length ζ for some limit ordinal ζ.

Recall that there are stationary many i.a. sets:

Lemma 3.2. *Let μ, κ and θ be regular cardinals with $\mu < \kappa \leq \theta$. Then the set $\{M \in \mathcal{P}_\kappa\mathcal{H}_\theta \mid M$ is i.a. of length $\mu\}$ is stationary in $\mathcal{P}_\kappa\mathcal{H}_\theta$.*

Proof. Take an arbitrary function $F : [\mathcal{H}_\theta]^{<\omega} \to \mathcal{H}_\theta$. It suffices to find an i.a. $M \in \mathcal{P}_\kappa\mathcal{H}_\theta$ of length μ such that $M \cap \kappa \in \kappa$ and M is closed under F.

By induction on $\xi < \mu$ we can easily construct a $\subseteq$-increasing sequence $\langle M_\xi \mid \xi < \mu \rangle$ in $\mathcal{P}_\kappa\mathcal{H}_\theta$ such that $\langle M_\eta \mid \eta < \xi \rangle \in M_\xi$, $M_\xi \cap \kappa \in \kappa$ and M_ξ is closed under F for each $\xi < \mu$.

Let $M := \bigcup_{\xi<\mu} M_\xi \in \mathcal{P}_\kappa\mathcal{H}_\theta$. Clearly M is closed under F, and $M \cap \kappa \in \kappa$. Moreover $\langle M_\xi \mid \xi < \mu \rangle$ witnesses that M is i.a. of length μ. □

Next we present easy lemmata on i.a. and s.i.a. sets, which we will use later without any mention:

Lemma 3.3. *Let κ and θ be regular cardinals with $\kappa \leq \theta$. Suppose that $M \in \mathcal{P}_\kappa \mathcal{H}_\theta$ is such that $M \cap \kappa \in \kappa$ and $M \prec \langle \mathcal{H}_\theta, \in, \kappa \rangle$. Suppose also that M is i.a. of length ζ, where ζ is a limit ordinal $< \kappa$. Then* $\mathrm{cf}(\sup(M \cap \nu)) = \mathrm{cf}(\zeta)$ *for every regular cardinal ν with $\kappa \leq \nu \leq \theta$.*

Proof. Fix a regular cardinal ν with $\kappa \leq \nu \leq \theta$. Let $\langle M_\xi \mid \xi < \zeta \rangle$ be an i.a. sequence to M. First note that $M_\xi \in M$ for each $\xi < \zeta$ because M_ξ is the last element of $\langle M_\eta \mid \eta \leq \xi \rangle \in M$. From this, it follows that $\sup(M_\xi \cap \nu) \in M \cap \nu$ for each $\xi < \zeta$: First note that $\sup(M_\xi \cap \nu) < \nu$ because ν is a regular cardinal $\geq \kappa$, and $M_\xi \in \mathcal{P}_\kappa \mathcal{H}_\theta$. We show that $\sup(M_\xi \cap \nu) \in M$. Let $\upsilon := \min(M \setminus \nu)$ if $\nu < \sup(M \cap \theta)$, and $\upsilon := \theta$ otherwise. Note that $M_\xi \cap \nu = M_\xi \cap \upsilon$ because $M_\xi \subseteq M$. Moreover $\sup(M_\xi \cap \upsilon) \in M$ by the elementarity of M, the choice of υ and the fact that $M_\xi \in M$. Thus $\sup(M_\xi \cap \nu) = \sup(M_\xi \cap \upsilon) \in M$.

Hence $\sup(M_\xi \cap \nu) < \sup(M \cap \nu)$ for each $\xi < \zeta$. Moreover $\langle \sup(M_\xi \cap \nu) \mid \xi < \zeta \rangle$ converges to $\sup(M \cap \nu)$ because $\bigcup_{\xi<\zeta} M_\xi = M$. Thus $\mathrm{cf}(\sup(M \cap \nu)) = \mathrm{cf}(\zeta)$. □

Lemma 3.4. *Let κ be a regular uncountable cardinal, let λ be a $<\kappa$-strong limit cardinal with $\mathrm{cf}(\lambda) < \kappa < \lambda$, and let θ be a regular cardinal $> \lambda$. Moreover let $\varphi : \bigcup_{\lambda'<\lambda} {}^{<\kappa}(\mathcal{P}_\kappa \lambda') \to \lambda$ be a bijection. Suppose that $M \in \mathcal{P}_\kappa \mathcal{H}_\theta$ is such that $M \cap \kappa \in \kappa$ and $M \prec \langle \mathcal{H}_\theta, \in, \kappa, \lambda, \varphi \rangle$. Let ζ be a limit ordinal $< \kappa$, and suppose that $\vec{x} = \langle x_\xi \mid \xi < \zeta \rangle$ is a $\subseteq$-increasing sequence with $\bigcup_{\xi<\zeta} x_\xi = M \cap \lambda$. Then the following are equivalent:*

- *(I) $\vec{x}$ is a φ-s.i.a. sequence to $M \cap \lambda$.*
- *(II) $\langle x_\xi \cap \lambda' \mid \xi < \zeta' \rangle \in M$ for every $\zeta' < \zeta$ and $\lambda' < \lambda$.*
- *(III) For every $\lambda' < \lambda$ there exists $\lambda'' \geq \lambda'$ such that $\langle x_\xi \cap \lambda'' \mid \xi < \zeta' \rangle \in M$ for every $\zeta' < \zeta$.*

Proof. (I)⇒(II) follows from the fact that M is closed under φ^{-1}. (II)⇒(III) is clear. We prove (III)⇒(I).

Assume (III). It suffices to show that $\varphi(\langle x_\xi \cap \lambda' \mid \xi < \zeta' \rangle) \in M \cap \lambda$ for every $\zeta' < \zeta$ and every $\lambda' < \lambda$. Fix $\zeta' < \zeta$ and $\lambda' < \lambda$.

By (III) take $\lambda'' \geq \lambda'$ such that $\langle x_\xi \cap \lambda'' \mid \xi < \zeta' \rangle \in M$. Moreover let $\nu := \min(M \cap \mathrm{On} \setminus \lambda')$. Note that $x_\xi \cap \lambda' = (x_\xi \cap \lambda'') \cap \nu$ for each $\xi < \zeta'$. Hence $\langle x_\xi \cap \lambda' \mid \xi < \zeta' \rangle \in M$ because $\nu, \langle x_\xi \cap \lambda'' \mid \xi < \zeta' \rangle \in M$. Then $\varphi(\langle x_\xi \cap \lambda' \mid \xi < \zeta' \rangle) \in M \cap \lambda$ because M is closed under φ. □

Lemma 3.5. *Let κ, λ, θ, φ and M be as in Lem.3.4. Suppose also that $M \cap \lambda$ is φ-s.i.a. of length ζ, where ζ is a limit ordinal $< \kappa$. Then* $\mathrm{cf}(\sup(M \cap \nu)) = \mathrm{cf}(M \cap \kappa)$ *for every regular cardinal ν with $\kappa \leq \nu < \lambda$.*

Proof. Similar as the proof of Lem.3.3 using Lem.3.4. □

Here we make a remark on the length of i.a. and s.i.a. sequences:

Let κ and θ be regular cardinals with $\kappa < \theta$, and suppose that $M \in \mathcal{P}_\kappa \mathcal{H}_\theta$, $M \prec \langle \mathcal{H}_\theta, \in, \kappa \rangle$ and $M \cap \kappa \in \kappa$. Then the cofinality of the length of i.a. sequences to M is unique. It must be $\mathrm{cf}(M \cap \kappa)$ by Lem.3.3. But the length of i.a. sequences to M may not be unique.

For example, suppose that $M \cap \kappa$ is not regular and that M is i.a. of length $\mathrm{cf}(M \cap \kappa)$. Then there are many $\zeta < M \cap \kappa$ with $\mathrm{cf}(\zeta) = \mathrm{cf}(M \cap \kappa)$. Moreover M is i.a. of length ζ for all such ζ: Fix such ζ. Let $\langle M_\eta \mid \eta < \mathrm{cf}(M \cap \kappa) \rangle$ be an i.a. sequence to M, and take an increasing sequence $\langle \zeta_\eta \mid \eta < \mathrm{cf}(M \cap \kappa) \rangle \in M$ converging to ζ. Moreover for each $\xi < \zeta$, taking the least $\eta < \mathrm{cf}(M \cap \kappa)$ with $\xi \leq \zeta_\eta$, let $M'_\xi := M_\eta$. Then it is easy to see that $\langle M'_\xi \mid \xi < \zeta \rangle$ is an i.a. sequence to M.

The similar is the case for the length of s.i.a. sequences.

Now we present the relationship between the notion of s.i.a. and i.a:

Lemma 3.6. *Let κ, λ, θ, φ and M be as in Lem.3.4. Let ζ and ρ be limit ordinals $< \kappa$.*

(1) If M is i.a. of length ζ, then $M \cap \lambda$ is φ-s.i.a. of length ζ.
(2) If M is i.a. but is not i.a. of length ρ, then $M \cap \lambda$ is φ-s.i.a. but is not φ-s.i.a. of length ρ.

Proof. (1) Let $\langle M_\xi \mid \xi < \zeta \rangle$ be an i.a. sequence to M. Then $\langle M_\xi \cap \lambda \mid \xi < \zeta \rangle$ is a φ-s.i.a. sequence to $M \cap \lambda$ by Lem.3.4.

(2) By (1) it suffices to show that if M is i.a. but is not i.a. of length ρ, then $M \cap \lambda$ is not φ-s.i.a. of length ρ. We prove that if M is i.a. and $M \cap \lambda$ is φ-s.i.a. of length ρ, then M is i.a. of length ρ. Suppose that M is i.a. of length ζ, and let $\langle M_\xi \mid \xi < \zeta \rangle$ be an i.a. sequence to M. Moreover suppose that $M \cap \lambda$ is φ-s.i.a. of length ρ, and let $\langle x_\eta \mid \eta < \rho \rangle$ be a φ-s.i.a. sequence to $M \cap \lambda$.

Note that if $\eta < \rho$, then $\sup(x_\eta \cap \kappa) \in M \cap \kappa$, and so there exists $\xi < \zeta$ with $\sup(x_\eta \cap \kappa) \leq \sup(M_\xi \cap \kappa)$. For each $\eta < \rho$ let ξ_η be the least $\xi < \zeta$ with $\sup(x_\eta \cap \kappa) \leq \sup(M_\xi \cap \kappa)$. Then $\langle \xi_\eta \mid \eta < \rho \rangle$ is increasing clearly. Moreover it is cofinal in ζ because $\bigcup_{\eta<\rho} x_\eta \cap \kappa = M \cap \kappa$. Here note that if $\rho' < \rho$, then $\langle \xi_\eta \mid \eta < \rho' \rangle$ can be recovered from $\langle x_\eta \cap \kappa \mid \eta < \rho' \rangle$ and $\langle M_\xi \mid \xi \leq \xi_{\rho'} \rangle$, both of which belong to M. Hence $\langle \xi_\eta \mid \eta < \rho' \rangle \in M$ for every $\rho' < \rho$.

Then it is easy to see that $\langle M_{\xi_\eta} \mid \eta < \rho \rangle$ is an i.a. sequence to M. Therefore M is i.a. of length ρ. □

Combining above lemmata we obtain the following:

Lemma 3.7. *Let κ be a regular uncountable cardinal and λ be a $<\kappa$-strong limit cardinal with $\mathrm{cf}(\lambda) < \kappa < \lambda$.*

(1) If $\kappa = \omega_1$, then $T^0_{\kappa\lambda}$ contains a club in $\mathcal{P}_\kappa\lambda$.
(2) If $\kappa \geq \omega_2$, then $T^0_{\kappa\lambda}$, $T^1_{\kappa\lambda}$ and $\mathcal{P}_\kappa\lambda \setminus (T^0_{\kappa\lambda} \cup T^1_{\kappa\lambda})$ are all stationary in $\mathcal{P}_\kappa\lambda$.

Proof. Let $\varphi : \bigcup_{\lambda' < \lambda} {}^{<\kappa}(\mathcal{P}_\kappa\lambda') \to \lambda$ be a bijection, and let $T^0_{\kappa\lambda}$ and $T^1_{\kappa\lambda}$ denote $T^{\varphi,0}_{\kappa\lambda}$ and $T^{\varphi,1}_{\kappa\lambda}$, respectively. Moreover let θ be a regular cardinal $> \lambda$, and let $\mathcal{M}$ be the structure $\langle \mathcal{H}_\theta, \in, \kappa, \lambda, \varphi \rangle$.

(1) Suppose $\kappa = \omega_1$. Note that $\mathrm{cf}(\lambda) = \omega$. Let $\bar{T}$ be the set of all $M \in \mathcal{P}_{\omega_1}\mathcal{H}_\theta$ with $M \prec \mathcal{M}$. Note that if $M \in \bar{T}$, then $M \cap \omega_1 \in \omega_1$ and M is i.a. of length ω. Hence $T^0_{\omega_1\lambda} \supseteq \{M \cap \lambda \mid M \in \bar{T}\}$ by Lem.3.6 (1). Then $T^0_{\omega_1\lambda}$ contains a club in $\mathcal{P}_{\omega_1}\lambda$ because $\bar{T}$ is a club in $\mathcal{P}_{\omega_1}\mathcal{H}_\theta$.

(2) Suppose $\kappa \geq \omega_2$. Then there is a regular cardinal $\mu < \kappa$ with $\mu \neq \mathrm{cf}(\lambda)$. Let

$$\bar{T}^0 := \{M \in \mathcal{P}_\kappa\mathcal{H}_\theta \mid M \cap \kappa \in \kappa \ \wedge \ M \prec \mathcal{M} \ \wedge \ M \text{ is i.a. of length } \mathrm{cf}(\lambda)\}\,,$$
$$\bar{T}^1 := \{M \in \mathcal{P}_\kappa\mathcal{H}_\theta \mid M \cap \kappa \in \kappa \ \wedge \ M \prec \mathcal{M} \ \wedge \ M \text{ is i.a. of length } \mu\}\,.$$

Then both $\bar{T}^0$ and $\bar{T}^1$ are stationary by Lem.3.2. First note that $T^0_{\kappa\lambda} \supseteq \{M \cap \lambda \mid M \in \bar{T}^0\}$ by Lem.3.6 (1). Hence $T^0_{\kappa\lambda}$ is stationary. Next note that every $M \in \bar{T}^1$ is not i.a. of length $\mathrm{cf}(\lambda)$ by Lem.3.3. Thus $T^1_{\kappa\lambda} \supseteq \{M \cap \lambda \mid M \in \bar{T}^1\}$ by Lem.3.6. Therefore $T^1_{\kappa\lambda}$ is stationary.

It is easy to see that the set $S := \{x \in \mathcal{P}_\kappa\lambda \mid \mathrm{cf}(\sup(x \cap \kappa^+)) \neq \mathrm{cf}(\sup(x \cap \kappa))\}$ is stationary in $\mathcal{P}_\kappa\lambda$. But both $S \cap T^0_{\kappa\lambda}$ and $S \cap T^1_{\kappa\lambda}$ are nonstationary by Lem.3.5. Hence $\mathcal{P}_\kappa\lambda \setminus (T^0_{\kappa\lambda} \cup T^1_{\kappa\lambda})$ is stationary. □

At the end of this section we present lemmata on the length of i.a. and s.i.a. sequences:

Lemma 3.8. *Let κ and θ be regular uncountable cardinals with $\kappa \leq \theta$. Suppose that $M \in \mathcal{P}_\kappa\mathcal{H}_\theta$ is such that $M \cap \kappa \in \kappa$ and $M \prec \langle \mathcal{H}_\theta, \in, \kappa \rangle$.*

Suppose also that M is i.a. Then M is either i.a. of length $M \cap \kappa$ or i.a. of length $\mathrm{cf}(M \cap \kappa)$.

Proof. We show that if M is not i.a. of length $M \cap \kappa$, then M is i.a. of length $\mathrm{cf}(M \cap \kappa)$. Assume that M is not i.a. of length $M \cap \kappa$.

Suppose that M is i.a. of length ζ and that $\langle M_\xi \mid \xi < \zeta \rangle$ is an i.a. sequence to M. Note that if $\zeta' < \zeta$, then ζ' is the length of the sequence $\langle M_\xi \mid \xi < \zeta' \rangle \in M$, and thus $\zeta' \in M \cap \kappa$. Hence $\zeta \leq M \cap \kappa$. But $\zeta \neq M \cap \kappa$ by our assumption. Therefore $\zeta \in M \cap \kappa$.

Note that $\mathrm{cf}(\zeta) = \mathrm{cf}(M \cap \kappa)$ by Lem.3.3. Because $\zeta \in M$ we can take an increasing sequence $\langle \xi_\eta \mid \eta < \mathrm{cf}(M \cap \kappa) \rangle \in M$ converging to ζ. Then it is easy to see that $\langle M_{\xi_\eta} \mid \eta < \mathrm{cf}(M \cap \kappa) \rangle$ is an i.a. sequence to M. Therefore M is i.a. of length $\mathrm{cf}(M \cap \kappa)$. □

Lemma 3.9. *Let κ, λ, θ, φ and M be as in Lem.3.4. Suppose also that $M \cap \lambda$ is φ-s.i.a. Then $M \cap \lambda$ is either φ-s.i.a. of length $M \cap \kappa$ or φ-s.i.a. of length* $\mathrm{cf}(M \cap \kappa)$.

Proof. Similar as the proof of Lem.3.8 using Lem.3.4. □

Note that there are stationary many $M \in \mathcal{P}_\kappa \mathcal{H}_\theta$ which is i.a. of length $\mathrm{cf}(M \cap \kappa)$ by Lem.3.2 and 3.3. Then there are also stationary many $x \in \mathcal{P}_\kappa \lambda$ which is φ-s.i.a. of length $\mathrm{cf}(x \cap \kappa)$ by Lem.3.6 (1). Below we prove that if $I[\kappa]$ is proper, then there are stationary many $M \in \mathcal{P}_\kappa \mathcal{H}_\theta$ which is i.a. of length $M \cap \kappa$ but is not i.a. of length $\mathrm{cf}(M \cap \kappa)$. Then the similar holds for φ-s.i.a. by Lem.3.6:

Lemma 3.10. *Let μ and κ are regular cardinals with $\mu < \kappa$, and suppose that $\{\zeta < \kappa \mid \mathrm{cf}(\zeta) = \mu\} \notin I[\kappa]$.*

(1) Suppose that θ is a regular cardinal $\geq \kappa$. Then there are stationary many $M \in \mathcal{P}_\kappa \mathcal{H}_\theta$ such that $M \cap \kappa \in \kappa$, $\mathrm{cf}(M \cap \kappa) = \mu$ and M is i.a. of length $M \cap \kappa$ but is not i.a. of length μ.

(2) Suppose that λ is a $<\kappa$-strong limit cardinal with $\mathrm{cf}(\lambda) < \kappa < \lambda$, and let $\varphi : \bigcup_{\lambda' < \lambda} {}^{<\kappa}\mathcal{P}_\kappa \lambda' \to \lambda$ be a bijection. Then there are stationary many $x \in \mathcal{P}_\kappa \lambda$ such that $x \cap \kappa \in \kappa$, $\mathrm{cf}(x \cap \kappa) = \mu$ and x is φ-s.i.a. of length $x \cap \kappa$ but is not φ-s.i.a. of length μ.

Proof. Note that (2) follows from (1) and Lem.3.6. We prove (1).

Take an arbitrary function $F : [\mathcal{H}_\theta]^{<\omega} \to \mathcal{H}_\theta$. It suffices to find $M \in \mathcal{P}_\kappa \mathcal{H}_\theta$ such that

(i) M is closed under F,
(ii) $M \cap \kappa \in \kappa$ and $\mathrm{cf}(M \cap \kappa) = \mu$,
(iii) M is i.a. of length $M \cap \kappa$ but is not i.a. of length μ.

Take a well-ordering Δ of $\mathcal{H}_\theta$, and let $\mathcal{M}$ be the structure $\langle \mathcal{H}_\theta, \in, \Delta, F \rangle$. By induction on $\xi < \kappa$ define $M_\xi \in \mathcal{P}_\kappa \mathcal{H}_\theta$ as follows: If ξ is a limit ordinal, then let $M_\xi := \bigcup_{\eta<\xi} M_\eta$. If ξ is a successor ordinal, then let

$$M_\xi \;:=\; \mathrm{skull}^{\mathcal{M}}(M_{\xi-1} \cup \xi \cup \{\langle M_\eta \mid \eta < \xi \rangle\}) \,.$$

Note that $M_\xi \prec \mathcal{M}$ for every $\xi < \kappa$ and thus that M_ξ is closed under F for every $\xi < \kappa$. Note also that if ζ is a limit ordinal $< \kappa$, then $\langle M_\xi \mid \xi < \zeta \rangle$ is an i.a. sequence to M_ζ.

Let $M_\kappa := \bigcup_{\xi<\kappa} M_\xi$, and take an enumeration $\langle b_\xi \mid \xi < \kappa \rangle$ of all bounded subsets of κ which belong to M_κ. Moreover let C be the set of all limit $\zeta < \kappa$ such that $M_\zeta \cap \kappa = \zeta$ and such that $\{b_\xi \mid \xi < \zeta\}$ is equal to the set of all bounded subsets of κ which are in M_ζ. It is easy to see that C is a club in κ.

By our assumption on $I[\kappa]$ we can take $\zeta \in C$ of cofinality μ which is not approachable with respect to $\langle b_\xi \mid \xi < \kappa \rangle$. Then $M := M_\zeta$ satisfies the properties (i) and (ii) above and is i.a. of length $\zeta = M \cap \kappa$. Hence it suffices to show that M is not i.a. of length μ.

Assume that M is i.a. of length μ, and let $\langle N_\eta \mid \eta < \mu \rangle$ is an i.a. sequence to M. Let $b := \{\sup(N_\eta \cap \kappa) \mid \eta < \mu\}$. Then b is an unbounded subset of $M \cap \kappa = \zeta$ of order-type μ. Moreover all proper initial segments of b belong to M because all proper initial segments of $\langle N_\eta \mid \eta < \mu \rangle$ belong to M. Then all proper initial segments of b belong to $\{b_\xi \mid \xi < \zeta\}$ because $\zeta \in C$. This contradicts that ζ is not approachable with respect to $\langle b_\xi \mid \xi < \kappa \rangle$. □

3.2. *i.a, preservation and reflection*

First we review the relationship between the notion of i.a. and the preservation of stationary subsets of $\mathcal{P}_\kappa\lambda$ under $<\kappa$-closed forcing extensions, due to Foreman–Magidor–Shelah and Foreman–Magidor:[4,5]

Lemma 3.11 (Foreman–Magidor–Shelah, Foreman–Magidor). *Let κ be a regular uncountable cardinal and λ be a cardinal $\geq \kappa$. The following are equivalent for a stationary $S \subseteq \mathcal{P}_\kappa\lambda$:*

(I) S remains stationary for all $<\kappa$-closed forcing extensions.
(II) S remains stationary in the forcing extension by $\mathrm{Col}(\kappa, \lambda)$.

(III) For every regular cardinal $\theta \geq \lambda$ the set of all $M \in \mathcal{P}_\kappa\mathcal{H}_\theta$ with the following properties is stationary in $\mathcal{P}_\kappa\mathcal{H}_\theta$:

(i) *there exist a limit ordinal $\zeta < \kappa$ and a $\subseteq$-increasing sequence $\langle x_\xi \mid \xi < \zeta\rangle$ in $\mathcal{P}_\kappa\lambda$ such that $\bigcup_{\xi<\zeta} x_\xi = M \cap \lambda$ and such that $\langle x_\xi \mid \xi < \zeta'\rangle \in M$ for every $\zeta' < \zeta$.*

(ii) *$M \cap \lambda \in S$.*

(IV) For every regular cardinal $\theta \geq \lambda$ the set

$$\{M \in \mathcal{P}_\kappa\mathcal{H}_\theta \mid M \text{ is i.a.} \wedge M \cap \lambda \in S\}$$

is stationary in $\mathcal{P}_\kappa\mathcal{H}_\theta$.

To prove the lemma above we need the following lemma:

Lemma 3.12. *Let κ be a regular uncountable cardinal and W be a set with $|W| = \kappa \subseteq W$. Then all stationary subset of $\mathcal{P}_\kappa W$ remain stationary in all $<\kappa$-closed forcing extensions.*

Proof. Roughly speaking, the lemma follows from the facts that stationary subsets of $\mathcal{P}_\kappa W$ can be identified with stationary subsets of κ and that all stationary subsets of κ remain stationary in all $<\kappa$-closed forcing extensions. But here we give a direct proof of the lemma.

Suppose that $S \subseteq \mathcal{P}_\kappa W$ is stationary and that $\mathbb{P}$ is a $<\kappa$-closed forcing notion. We show that S remains stationary in $V^{\mathbb{P}}$. Take an arbitrary $\mathbb{P}$-name $\dot{F}$ for a function from $[W]^{<\omega}$ to W and an arbitrary $p \in \mathbb{P}$. It suffices to find $p^* \leq p$ and $x \in S$ such that $x \cap \kappa \in \kappa$ and $p^* \Vdash$ "x is closed under $\dot{F}$".

Using the facts that $|W| = \kappa$ and that $\mathbb{P}$ is $<\kappa$-closed, we can easily construct a descending sequence $\langle p_\xi \mid \xi < \kappa\rangle$ in $\mathbb{P}$ below p such that for every $a \in [W]^{<\omega}$ some p_ξ decides $\dot{F}(a)$. For each $a \in [W]^{<\omega}$ let ξ_a be the least $\xi < \kappa$ such that p_ξ decides $\dot{F}(a)$. Moreover let $F : [W]^{<\omega} \to W$ be such that $F(a)$ is the value of $\dot{F}(a)$ decided by p_{ξ_a}.

Then there exists $x \in S$ such that $x \cap \kappa \in \kappa$ and x is closed under F because S is stationary. Take $\xi^* < \kappa$ such that $\xi_a \leq \xi^*$ for every $a \in [x]^{<\omega}$. Then $p_{\xi^*} \leq p$. Moreover $p_{\xi^*} \Vdash$ "x is closed under $\dot{F}$" because x is closed under F and $p_{\xi^*} \Vdash$ "$\dot{F} \restriction [x]^{<\omega} = F \restriction [x]^{<\omega}$". Thus p_{ξ^*} and x are what we seek. □

Proof of Lem.3.11. We prove (IV)⇒(III)⇒(II)⇒(I)⇒(IV). Fix a stationary $S \subseteq \mathcal{P}_\kappa\lambda$.

(IV)⇒(III):

Assume (IV). Let θ be a regular cardinal $\geq \lambda$. Let $\bar{S}$ be the set of all $M \in \mathcal{P}_\kappa \mathcal{H}_\theta$ such that $M \prec \langle \mathcal{H}_\theta, \in, \lambda \rangle$, M is i.a. and $M \cap \lambda \in S$. Then $\bar{S}$ is stationary by (IV). Moreover if $M \in \bar{S}$ and $\langle M_\xi \mid \xi < \zeta \rangle$ is an i.a. sequence to M, then it is easy to see that $\langle M_\xi \cap \lambda \mid \xi < \zeta \rangle$ witnesses that M satisfies (i) in (III). Hence the set in (III) includes $\bar{S}$, and therefore it is also stationary.

(III)$\Rightarrow$(II):

Assume (III). To prove (II) take an arbitrary $\mathrm{Col}(\kappa, \lambda)$-name $\dot{F}$ of a function from $[\lambda]^{<\omega}$ to λ and an arbitrary $p \in \mathrm{Col}(\kappa, \lambda)$. It suffices to find $x \in S$ and $p^* \leq p$ such that $x \cap \kappa \in \kappa$ and $p^* \Vdash_{\mathrm{Col}(\kappa,\lambda)}$ "x is closed under $\dot{F}$".

Take a sufficiently large regular cardinal θ and a well-ordering Δ of $\mathcal{H}_\theta$. Let $\mathcal{M}$ be the structure $\langle \mathcal{H}_\theta, \in, \Delta, \kappa, \lambda, p, \dot{F} \rangle$. Then, by (III), we can take $M \in \mathcal{P}_\kappa \mathcal{H}_\theta$ such that $M \prec \mathcal{M}$, $M \cap \kappa \in \kappa$ and M satisfies (i) and (ii) in (III). It suffices to find $p^* \leq p$ which forces that $x := M \cap \lambda$ is closed under $\dot{F}$.

Suppose that a sequence $\langle x_\xi \mid \xi < \zeta \rangle$ witnesses (i) for M. By induction on $\xi < \zeta$ let $p_\xi \in \mathrm{Col}(\kappa, \lambda)$ be the Δ-least lower bound of $\{p\} \cup \{p_\eta \mid \eta < \xi\}$ such that p_ξ decides $\dot{F}(a)$ for all $a \in [x_\xi]^{<\omega}$. We can take such p_ξ by the $<\kappa$-closure of $\mathrm{Col}(\kappa, \lambda)$.

Here note that $p_\xi \in M$ for each $\xi < \zeta$ because p_ξ can be definable in $\mathcal{M}$ from a parameter $\langle x_\eta \mid \eta \leq \xi \rangle \in M$. Hence p_ξ decides $\dot{F}(a)$ to be in $M \cap \lambda = x$ for each $a \in [x_\xi]^{<\omega}$.

Now take a lower bound p^* of $\{p_\xi \mid \xi < \zeta\}$. Then $p^* \leq p$, and p^* forces that x is closed under $\dot{F}$.

(II)$\Rightarrow$(I):

Assume (II). Take an arbitrary $<\kappa$-closed forcing notion $\mathbb{P}$. We must show that S remains stationary in $V^{\mathbb{P}}$.

Let υ be a regular cardinal with $\lambda, 2^{|\mathbb{P}|} < \upsilon$. In $V^{\mathrm{Col}(\kappa,\upsilon)}$, for every $p \in \mathbb{P}$, there exists a $\mathbb{P}$-generic filter over V containing p. Hence it suffices to show that S remains stationary in $V^{\mathrm{Col}(\kappa,\upsilon)}$. But note that $\mathrm{Col}(\kappa, \upsilon) \cong \mathrm{Col}(\kappa, \lambda) \times \mathrm{Col}(\kappa, \upsilon)$. So it suffices to show that S remains stationary in $V^{\mathrm{Col}(\kappa,\lambda) \times \mathrm{Col}(\kappa,\upsilon)}$.

First, S remains stationary in $V^{\mathrm{Col}(\kappa,\lambda)}$ by (II). Here note that $|\lambda| = \kappa$ in $V^{\mathrm{Col}(\kappa,\lambda)}$. Moreover $\mathrm{Col}(\kappa, \upsilon)$ is absolute between V and $V^{\mathrm{Col}(\kappa,\lambda)}$, and so $V^{\mathrm{Col}(\kappa,\lambda) \times \mathrm{Col}(\kappa,\upsilon)}$ is a $<\kappa$-closed forcing extension of $V^{\mathrm{Col}(\kappa,\lambda)}$. Therefore S remains stationary in $V^{\mathrm{Col}(\kappa,\lambda) \times \mathrm{Col}(\kappa,\upsilon)}$ by Lem.3.12.

(I)$\Rightarrow$(IV):

Assume (I). Let θ be a regular cardinal $\geq \lambda$ in V, let $H := \mathcal{H}_\theta{}^V$, and let $|H|^V = \upsilon$. Here note that $\mathcal{P}_\kappa H$ and the internally approachability of sets are absolute between V and $V^{\mathrm{Col}(\kappa,\upsilon)}$. Moreover $V \subseteq V^{\mathrm{Col}(\kappa,\upsilon)}$. Hence it suffices to show that the set $\{M \in \mathcal{P}_\kappa H \mid M \text{ is i.a.} \wedge M \cap \lambda \in S\}$ is stationary in $V^{\mathrm{Col}(\kappa,\upsilon)}$. But S remains stationary in $V^{\mathrm{Col}(\kappa,\upsilon)}$ by (1). So it suffices to prove that the set $\{M \in \mathcal{P}_\kappa H \mid M \text{ is i.a.}\}$ contains a club in $V^{\mathrm{Col}(\kappa,\upsilon)}$. We work in $V^{\mathrm{Col}(\kappa,\upsilon)}$.

Let $F : \kappa \to H$ be a surjection. Note that $\langle F[\xi] \mid \xi < \zeta\rangle \in H$ for each $\zeta < \kappa$ by the $<\kappa$-closure of $\mathrm{Col}(\kappa, \upsilon)$. Let C be the set of all limit $\zeta < \kappa$ such that for every $\zeta' < \zeta$ there exists ζ'' with $F(\zeta'') = \langle F[\xi] \mid \xi < \zeta'\rangle$. Then C is a club in κ. Hence the set $\{F[\zeta] \mid \zeta \in C\}$ contains a club in $\mathcal{P}_\kappa H$. Moreover $\langle F[\xi] \mid \xi < \zeta\rangle$ is an i.a. sequence to $F[\zeta]$ for every $\zeta \in C$. Thus $F[\zeta]$ is i.a. for every $\zeta \in C$. □

Fact 1.2 and what we mentioned below Fact 1.2 easily follow from Lem.3.11:

Corollary 3.13. *Let κ be a regular uncountable cardinal and λ be a cardinal with $\lambda^{<\kappa} = \lambda$.*

(1) All stationary subsets of $S_{\kappa\lambda}$ remain stationary in all $<\kappa$-closed forcing extensions.

(2) $\mathcal{P}_\kappa\lambda \setminus (S_{\kappa\lambda})^V$ becomes nonstationary in $V^{\mathrm{Col}(\kappa,\lambda)}$.

Proof. Let $\psi : {}^{<\kappa}(\mathcal{P}_\kappa\lambda) \to \lambda$ be a bijection, and let $S_{\kappa\lambda}$ denote $S^\psi_{\kappa\lambda}$. Take an arbitrary regular cardinal $\theta \geq \lambda$. Let $\bar{C}$ be the set of all $M \in \mathcal{P}_\kappa\mathcal{H}_\theta$ with $M \prec \langle \mathcal{H}_\theta, \in, \kappa, \lambda, \psi\rangle$. Moreover for each $S \subseteq \mathcal{P}_\kappa\lambda$ let $\bar{S}$ be the set of all $M \in \mathcal{P}_\kappa\mathcal{H}_\theta$ satisfying the properties (i) and (ii) in (III) of Lem.3.11.

Note that for each $M \in \bar{C}$ and each sequence $\langle x_\xi \mid \xi < \zeta\rangle$ in $\mathcal{P}_\kappa\lambda$, $\langle x_\xi \mid \xi < \zeta\rangle$ is a ψ-i.a. sequence to $M \cap \lambda$ if and only if it witnesses that M satisfies (i) in (III) of Lem.3.11. Hence

$$\bar{C} \cap \bar{S} \;=\; \bar{C} \cap \{M \in \mathcal{P}_\kappa\mathcal{H}_\theta \mid M \cap \lambda \in S \cap S_{\kappa\lambda}\}\,.$$

Here note that $\bar{C}$ is a club. Thus $\bar{S}$ is stationary in $\mathcal{P}_\kappa\mathcal{H}_\theta$ if and only if the set $\{M \in \mathcal{P}_\kappa\mathcal{H}_\theta \mid M \cap \lambda \in S \cap S_{\kappa\lambda}\}$ is stationary in $\mathcal{P}_\kappa\mathcal{H}_\theta$. But the latter is equivalent to that $S \cap S_{\kappa\lambda}$ is stationary in $\mathcal{P}_\kappa\lambda$. Therefore $\bar{S}$ is stationary in $\mathcal{P}_\kappa\mathcal{H}_\theta$ if and only if $S \cap S_{\kappa\lambda}$ is stationary in $\mathcal{P}_\kappa\lambda$. Both (1) and (2) easily follow from this fact and Lem.3.11. □

The corollary below immediately follows from Lem.3.6 (1) and 3.11:

Corollary 3.14. *Let κ be a regular uncountable cardinal, and λ be a $<\kappa$-strong limit cardinal with* $\mathrm{cf}(\lambda) < \kappa < \lambda$. *Then* $\mathcal{P}_\kappa\lambda \setminus (T^0_{\kappa\lambda} \cup T^1_{\kappa\lambda})^V$ *becomes nonstationary in* $V^{\mathrm{Col}(\kappa,\lambda)}$.

Finally we review the relationship between the preservation and the reflection of stationary subsets of $\mathcal{P}_\kappa\lambda$. We give the proof of Fact 1.3:

Proposition 3.15 (Foreman–Magidor–Shelah[5]). *Let κ be a regular uncountable cardinal and λ be a cardinal $> \kappa$. Suppose that ν is a λ-supercompact cardinal with $\kappa < \nu < \lambda$. In $V^{\mathrm{Col}(\kappa,<\nu)}$ assume that T is a stationary subset of $\mathcal{P}_\kappa\lambda$ such that all stationary subsets of T remain stationary in all $<\kappa$-closed forcing extensions. Then* $\mathsf{SR}_{\kappa\lambda}(T)$ *holds in* $V^{\mathrm{Col}(\kappa,<\nu)}$.

Proof. We make a remark before starting the proof. Several models of ZFC will appear in the proof below. But ${}^{<\kappa}\mathrm{On}$ is absolute among all models, and thus notions such as $\mathrm{Col}(\kappa,*)$ and $\mathcal{P}_\kappa *$ are absolute among them.

Now we start the proof. In V let $j : V \to M$ be a λ-supercompact embedding with $\mathrm{crit}(j) = \nu$, where crit denotes the critical point. Let G be a $\mathrm{Col}(\kappa, <\nu)$-generic filter over V, and in $V[G]$ let T be a stationary subset of $\mathcal{P}_\kappa\lambda$ such that all stationary subsets of T remain stationary in all $<\kappa$-closed forcing extensions of $V[G]$. We show that $\mathsf{SR}_{\kappa\lambda}(T)$ holds in $V[G]$. For this take an arbitrary stationary $S \subseteq T$ in $V[G]$. We must show that the following holds in $V[G]$:

(I) There exists $W \subseteq \lambda$ such that $|W| = \kappa \subseteq W$ and such that $S \cap \mathcal{P}_\kappa W$ is stationary.

First note that

$$j(\mathrm{Col}(\kappa,<\nu)) \;=\; \mathrm{Col}(\kappa,<j(\nu)) \;\cong\; \mathrm{Col}(\kappa,<\nu) \times \mathrm{Col}(\kappa,[\nu,j(\nu)))\,,$$

where $\mathrm{Col}(\kappa,[\nu,j(\nu)))$ is the $<\kappa$-support product of $\langle \mathrm{Col}(\kappa,\nu') \mid \nu \le \nu' < j(\nu)\rangle$. Let H be a $\mathrm{Col}(\kappa,[\nu,j(\nu)))$-generic filter over $V[G]$. We work in $V[G*H]$ below.

Note that $G*H$ is $j(\mathrm{Col}(\kappa,<\nu))$-generic over M. Then by the standard argument, $j : V \to M$ can be extended to the elementary embedding $j^* : V[G] \to M[G*H]$. For proving (I) in $V[G]$, it suffices to show that the following holds in $M[G*H]$ by the elementarity of j^*:

(II) There exists $W \subseteq j^*(\lambda)$ such that $|W| = \kappa \subseteq W$ and $j^*(S) \cap \mathcal{P}_\kappa W$ is stationary.

(Here note that $\kappa < \nu = \mathrm{crit}(j^*)$ and thus that $j^*(\kappa) = \kappa$.)

We show that $W := j^*[\lambda] \in M[G * H]$ witnesses (II). Clearly $|j^*[\lambda]| = |\lambda| = \kappa \subseteq j^*[\lambda]$ in $M[G * H]$. We prove that $j^*(S) \cap \mathcal{P}_\kappa j^*[\lambda]$ is stationary in $M[G * H]$.

First note that S remains stationary in $\mathcal{P}_\kappa\lambda$ in $V[G * H]$ by the assumption on T. Hence the set $\{j^*[x] \mid x \in S\}$ is stationary in $\mathcal{P}_\kappa j^*[\lambda]$ in $V[G * H]$. But $j^*(x) = j^*[x]$ for each $x \in S$ because $|x| < \kappa < \nu = \mathrm{crit}(j^*)$. Thus $\{j^*[x] \mid x \in S\} \subseteq j^*(S) \cap \mathcal{P}_\kappa j^*[\lambda]$. So $j^*(S) \cap \mathcal{P}_\kappa j^*[\lambda]$ is stationary in $V[G * H]$. Then $j^*(S) \cap \mathcal{P}_\kappa j^*[\lambda]$ is stationary also in $M[G * H]$ because $M[G * H] \subseteq V[G * H]$.

This completes the proof. □

4. Preservation and reflection below $T^0_{\kappa\lambda}$ and $T^1_{\kappa\lambda}$

In this section we prove Thm.1.7 and 1.8. We prove Thm.1.7 in Subsection 4.1 and prove Thm.1.8 in Subsection 4.2.

4.1. $T^0_{\kappa\lambda}$

Here we prove Thm.1.7:

Theorem 1.7. *Let κ be a regular uncountable cardinal and λ be a $<\kappa$-strong limit cardinal with $\mathrm{cf}(\lambda) < \kappa < \lambda$.*

(1) All stationary subsets of $T^0_{\kappa\lambda}$ remain stationary in all $<\kappa$-closed forcing extensions.

(2) Suppose that ν is a λ-supercompact cardinal with $\kappa < \nu < \lambda$. Then $\mathsf{SR}_{\kappa\lambda}(T^0_{\kappa\lambda})$ holds in $V^{\mathrm{Col}(\kappa,<\nu)}$.

Note that Thm.1.7 (2) follows from Thm.1.7 (1) and Prop.3.15. We prove Thm.1.7 (1). The point is that elements of $T^0_{\kappa\lambda}$ is essentially internally approachable:

Proof of Thm.1.7 (1). Let κ be a regular uncountable cardinal and λ be a $<\kappa$-strong limit cardinal with $\mathrm{cf}(\lambda) < \kappa < \lambda$. Fix a bijection $\varphi : \bigcup_{\lambda'<\lambda} {}^{<\kappa}(\mathcal{P}_\kappa\lambda') \to \lambda$, and let $T^0_{\kappa\lambda}$ be $T^{\varphi,0}_{\kappa\lambda}$. Let S be an arbitrary stationary subset of $T^0_{\kappa\lambda}$. It suffices to show that S satisfies (III) in Lem.3.11.

Let θ be a regular cardinal $> \lambda$, and let $\bar{S}$ be the set of all $M \in \mathcal{P}_\kappa\mathcal{H}_\theta$ such that $M \prec \langle \mathcal{H}_\theta, \in, \kappa, \lambda, \varphi\rangle$, $M \cap \lambda \in S$ and $M \cap \kappa \in \kappa$. Then $\bar{S}$ is stationary because S is stationary. We prove that if $M \in \bar{S}$, then M satisfies the properties (i) and (ii) in (III) of Lem.3.11. Fix $M \in \bar{S}$. All we have to show is that M satisfies (i).

Because $M \cap \lambda \in S \subseteq T^0_{\kappa\lambda}$ there exists a φ-s.i.a. sequence $\langle y_\xi \mid \xi < \mathrm{cf}(\lambda)\rangle$ to $M \cap \lambda$. Moreover take an increasing sequence $\langle \lambda_\xi \mid \xi < \mathrm{cf}(\lambda)\rangle \in M$ converging to λ. Note that $\mathrm{cf}(\lambda) \subseteq M$ because $\mathrm{cf}(\lambda) \in M \cap \kappa \in \kappa$. Hence all initial segments of $\langle \lambda_\xi \mid \xi < \mathrm{cf}(\lambda)\rangle$ belong to M.

Let $x_\xi := y_\xi \cap \lambda_\xi$ for each $\xi < \mathrm{cf}(\lambda)$. Then $\langle x_\xi \mid \xi < \mathrm{cf}(\lambda)\rangle$ is a $\subseteq$-increasing sequence in $\mathcal{P}_\kappa\lambda$ with $\bigcup_{\xi<\mathrm{cf}(\lambda)} x_\xi = M \cap \lambda$. Moreover $\langle x_\xi \mid \xi < \zeta\rangle \in M$ for every $\zeta < \mathrm{cf}(\lambda)$ because $\langle x_\xi \mid \xi < \zeta\rangle$ can be recovered from $\langle y_\xi \cap \lambda_\zeta \mid \xi < \zeta\rangle$ and $\langle \lambda_\xi \mid \xi < \zeta\rangle$, both of which belong to M. Therefore $\langle x_\xi \mid \xi < \mathrm{cf}(\lambda)\rangle$ witnesses that M satisfies (i) in (III) of Lem.3.11. □

4.2. $T^1_{\kappa\lambda}$

Here we prove Thm.1.8:

Theorem 1.8. *Assume* GCH. *Let κ be a regular cardinal $\geq \omega_2$ and λ be a singular cardinal with* $\mathrm{cf}(\lambda) < \kappa < \lambda$.

(1) $T^1_{\kappa\lambda}$ (defined in V) remains stationary in all $<\kappa$-closed forcing extensions. But for every stationary $T \subseteq T^1_{\kappa\lambda}$ there exists stationary $S \subseteq T$ which becomes nonstationary in $V^{\mathrm{Col}(\kappa,\lambda)}$.

(2) $\mathsf{SR}_{\kappa\lambda}(T)$ fails for every stationary $T \subseteq T^1_{\kappa\lambda}$.

Note that the first statement of Thm.1.8 (1) follows from Lem.3.2, 3.6 and 3.11. In fact we do not need the assumption of GCH for the first statement of Thm.1.8 (1):

Proposition 4.1. *Let κ be a regular cardinal $\geq \omega_2$ and λ be a $<\kappa$-strong limit cardinal with $\mathrm{cf}(\lambda) < \kappa < \lambda$. Then $(T^1_{\kappa\lambda})^V$ remains stationary in $\mathcal{P}_\kappa\lambda$ in all $<\kappa$-closed forcing extensions.*

Below we prove the remaining part of Thm.1.8. In fact we prove it from a weaker assumption than GCH. We use the following assumption:

Definition 4.2. For a singular cardinal λ let Ψ_λ be the following assertion:

$\Psi_\lambda \equiv 2^\lambda$ is a regular cardinal, and there exist an increasing sequence $\vec{\lambda} = \langle \lambda_\eta \mid \eta < \mathrm{cf}(\lambda)\rangle$ of regular cardinals converging to λ and a $\mathrm{cf}(\lambda)$-complete proper ideal J over $\mathrm{cf}(\lambda)$ such that $\mathrm{tcf}(\Pi\vec{\lambda}/J) = 2^\lambda$.

Note that if $2^\lambda = \lambda^+$, then $\mathrm{tcf}(\Pi\vec{\lambda}/\mathrm{BDD}_{\mathrm{cf}(\lambda)}) = 2^\lambda$ for every increasing sequence $\vec{\lambda}$ of regular cardinals converging to λ. Here $\mathrm{BDD}_{\mathrm{cf}(\lambda)}$ denotes the bounded ideal over $\mathrm{cf}(\lambda)$, i.e. the ideal consisting of all bounded subsets of $\mathrm{cf}(\lambda)$. Hence if GCH holds, then Ψ_λ holds for every singular cardinal λ.

Ψ_λ holds in some other situations. Suppose that $\lambda = \omega_\omega$ is strong limit. Then

$$2^{\omega_\omega} = \mathrm{cf}(\mathcal{P}_{\omega_2}\omega_\omega, \subseteq) = \max \mathrm{pcf}(\{\omega_n \mid n \in \omega\}) .$$

The former equality follows from the fact that ω_ω is strong limit. The latter is due to Shelah,[14] and its proof can be also found in §8.4 of Holz–Steffens–Weitz's book.[7] Hence there exists a maximal ideal J over ω such that $\mathrm{tcf}(\Pi\langle\omega_n \mid n \in \omega\rangle / J) = 2^{\omega_\omega}$. Here note that every ideal is ω-complete. Therefore Ψ_{ω_ω} holds if ω_ω is strong limit. See Shelah's book for other situation in which Ψ_λ holds.[14]

We prove the following proposition which implies Thm.1.8:

Proposition 4.3. *Let κ be a regular uncountable cardinal, and let λ be a $<\kappa$-strong limit cardinal such that $\mathrm{cf}(\lambda) < \kappa < \lambda$ and such that Ψ_λ holds. Then the following hold:*

(1) For every stationary $T \subseteq T^1_{\kappa\lambda}$ there exists a stationary $S \subseteq T$ such that S becomes nonstationary in $V^{\mathrm{Col}(\kappa,\lambda)}$.

(2) Assume also that $\kappa^{<\kappa} = \kappa$. Then $\mathsf{SR}_{\kappa\lambda}(T)$ fails for every stationary $T \subseteq T^1_{\kappa\lambda}$.

The rest of this subsection is devoted to the proof of Prop.4.3. We need some preliminaries. First we fix our notation in the rest of this subsection. Let

κ: regular uncountable cardinal,

λ: $<\kappa$-strong limit cardinal with $\mathrm{cf}(\lambda) < \kappa < \lambda$ and Ψ_λ.

Note that if $\vec{\lambda} = \langle\lambda_\eta \mid \eta < \mathrm{cf}(\lambda)\rangle$ and J witnesses Ψ_λ, then $J \supseteq \mathrm{BDD}_{\mathrm{cf}(\lambda)}$, i.e. $\rho \in J$ for every $\rho < \mathrm{cf}(\lambda)$. This is because if $J \not\supseteq \mathrm{BDD}_{\mathrm{cf}(\lambda)}$, then there exists $\eta^* < \mathrm{cf}(\lambda)$ with $\{\eta^*\} \notin J$ by the $\mathrm{cf}(\lambda)$-completeness of J, and then there are no $<_J$-increasing sequence of length $> \lambda_{\eta^*}$. Hence we can take a pair of witness $\vec{\lambda} = \langle\lambda_\eta \mid \eta < \mathrm{cf}(\lambda)\rangle$ and J of Ψ_λ so that $\lambda_0 > \kappa$. Let

$\vec{\lambda} = \langle\lambda_\eta \mid \eta < \mathrm{cf}(\lambda)\rangle$: increasing sequence of regular cardinals converging to λ with $\lambda_0 > \kappa$,

J: $\mathrm{cf}(\lambda)$-complete proper ideal over $\mathrm{cf}(\lambda)$,

$\vec{h} = \langle h_\alpha \mid \alpha < 2^\lambda\rangle$: $(\vec{\lambda}, J)$-scale.

For each $x \in \mathcal{P}_\kappa\lambda$ let

ch_x := the characteristic function for x with respect to $\vec{\lambda}$, i.e. the function on $\mathrm{cf}(\lambda)$ such that $\mathrm{ch}_x(\eta) = \sup(x \cap \lambda_\eta)$ for each $\eta < \mathrm{cf}(\lambda)$,

β_x := the least $\beta < 2^\lambda$ with $\mathrm{ch}_x \leq_J h_\beta$.

Finally let

$\varphi : \bigcup_{\lambda' < \lambda} {}^{<\kappa}(\mathcal{P}_\kappa \lambda') \to \lambda$: bijection, $\quad T^1_{\kappa\lambda} := T^{\varphi,1}_{\kappa\lambda}$,

θ: regular cardinal $> 2^\lambda$, $\quad \Delta$: well-ordering of $\mathcal{H}_\theta$,

$\mathcal{M} := \langle \mathcal{H}_\theta, \in, \Delta, \kappa, \lambda, \vec{\lambda}, J, \vec{h}, \varphi \rangle$.

$\mathcal{T} := \{x \in T^1_{\kappa\lambda} \mid x \cap \kappa \in \kappa \wedge \mathrm{skull}^{\mathcal{M}}(x) \cap \lambda = x\}$

Note that $T^1_{\kappa\lambda} \setminus \mathcal{T}$ is nonstationary. Note also that $\mathrm{cf}(\lambda), \{\lambda_\eta \mid \eta < \mathrm{cf}(\lambda)\} \subseteq x$ for every $x \in \mathcal{T}$ by Lem.2.3.

Next we present lemmata on elements of $\mathcal{T}$:

Lemma 4.4. *Let M be an element of $\mathcal{P}_\kappa \mathcal{H}_\theta$ such that $M \cap \lambda \in \mathcal{T}$ and $M \prec \mathcal{M}$. Suppose that $\langle x_\xi \mid \xi < \zeta \rangle$ is a φ-s.i.a. sequence to $M \cap \lambda$, and let $\gamma^* := \sup_{\xi<\zeta} \beta_{x_\xi} < 2^\lambda$. Moreover let g be an element of $\Pi\vec{\lambda}$ such that $g \restriction \rho \in M$ for every $\rho < \mathrm{cf}(\lambda)$. Then $g \leq_J h_{\gamma^*}$.*

Proof. First of all note that $g(\eta) \in M \cap \lambda$ for every $\eta < \mathrm{cf}(\lambda)$ by the assumption on g. Thus for every $\eta < \mathrm{cf}(\lambda)$ there exists $\xi < \zeta$ with $g(\eta) \in x_\xi$.

The proof of the lemma splits into two cases by the cofinality of ζ:

Case 1: $\mathrm{cf}(\zeta) < \mathrm{cf}(\lambda)$.

Take an increasing sequence $\langle \xi_k \mid k < \mathrm{cf}(\zeta) \rangle$ cofinal in ζ. For each $k < \mathrm{cf}(\zeta)$ let $A_k := \{\eta < \mathrm{cf}(\lambda) \mid g(\eta) \in x_{\xi_k}\}$. Note that $\mathrm{cf}(\lambda) = \bigcup_{k<\mathrm{cf}(\zeta)} A_k$ by the remark at the beginning of the proof.

Moreover let $B_k := \{\eta \in A_k \mid g(\eta) > h_{\gamma^*}(\eta)\}$ for each $k < \mathrm{cf}(\zeta)$. Note that $B_k \in J$ for every $k < \mathrm{cf}(\zeta)$. This follows from the facts that $g(\eta) \leq \mathrm{ch}_{x_{\xi_k}}(\eta)$ for every $\eta \in A_k$ and that $\mathrm{ch}_{x_{\xi_k}} \leq_J h_{\gamma^*}$. Thus $B := \bigcup_{k<\mathrm{cf}(\zeta)} B_k \in J$ by the $\mathrm{cf}(\lambda)$-completeness of J.

Here note that $B = \{\eta < \mathrm{cf}(\eta) \mid g(\eta) > h_{\gamma^*}(\eta)\}$ because $\bigcup_{k<\mathrm{cf}(\zeta)} A_k = \mathrm{cf}(\lambda)$. Therefore $B \in J$ means that $g \leq_J h_{\gamma^*}$.

Case 2: $\mathrm{cf}(\zeta) \geq \mathrm{cf}(\lambda)$.

In this case we claim that there exists $\xi^* < \zeta$ with $g[\mathrm{cf}(\lambda)] \subseteq x_{\xi^*}$. Note that if such ξ^* exists, then $g \leq \mathrm{ch}_{x_{\xi^*}} \leq_J h_{\gamma^*}$, and so $g \leq_J h_{\gamma^*}$. Thus it suffices to show that such ξ^* exists.

First note that for each $\eta < \mathrm{cf}(\lambda)$ there exists $\xi < \zeta$ with $g[\eta] \subseteq x_\xi$. This follows from the remark at the beginning of this proof and the fact that $\mathrm{cf}(\zeta) \geq \mathrm{cf}(\lambda)$. For each $\eta < \mathrm{cf}(\lambda)$ let ξ_η be the least $\xi < \zeta$ with $g[\eta] \subseteq x_\xi$.

Clearly $\langle \xi_\eta \mid \eta < \mathrm{cf}(\lambda)\rangle$ is an increasing sequence below ζ. Moreover note that $\langle \xi_\eta \mid \eta < \rho\rangle \in M$ for every $\rho < \mathrm{cf}(\lambda)$ because $\langle \xi_\eta \mid \eta < \rho\rangle$ can be recovered from $g \restriction \rho$ and $\langle x_\xi \cap \lambda_\rho \mid \xi \leq \xi_\rho\rangle$ both of which belong to M. Hence $\langle x_{\xi_\eta} \cap \lambda' \mid \eta < \rho\rangle \in M$ for every $\rho < \mathrm{cf}(\lambda)$ and every $\lambda' < \lambda$.

Thus if $\sup_{\eta<\mathrm{cf}(\lambda)} \xi_\eta = \zeta$, then it is easy to see that $\langle x_{\xi_\eta} \mid \eta < \mathrm{cf}(\lambda)\rangle$ becomes a φ-s.i.a. sequence to x, which contradicts that $x \in T^1_{\kappa\lambda}$. Therefore $\xi^* := \sup_{\eta<\mathrm{cf}(\lambda)} \xi_\eta < \zeta$. Then $g[\mathrm{cf}(\lambda)] \subseteq x_{\xi^*}$. □

Lemma 4.5. *Suppose that $x \in \mathcal{T}$ and that both $\langle x_\xi \mid \xi < \zeta\rangle$ and $\langle y_\iota \mid \iota < \epsilon\rangle$ are φ-s.i.a. sequence to x. Then $\sup_{\xi<\zeta} \beta_{x_\xi} = \sup_{\iota<\epsilon} \beta_{y_\iota}$.*

Proof. It suffices to show that $\sup_{\xi<\zeta} \beta_{x_\xi} \geq \sup_{\iota<\epsilon} \beta_{y_\iota}$. Let $\gamma^* := \sup_{\xi<\zeta} \beta_{x_\xi}$, and let $M := \mathrm{skull}^{\mathcal{M}}(x)$. Note that $M \cap \lambda = x$ and thus that $\langle x_\xi \mid \xi < \zeta\rangle$ is a φ-s.i.a. sequence to $M \cap \lambda$. Note also that if $\iota < \epsilon$, then $\mathrm{ch}_{y_\iota} \restriction \rho \in M$ for every $\rho < \mathrm{cf}(\lambda)$ because $\langle \lambda_\eta \mid \eta < \rho\rangle$, $y_\iota \cap \lambda_\rho \in M$. Hence $\mathrm{ch}_{y_\iota} \leq_J h_{\gamma^*}$, i.e. $\beta_{y_\iota} \leq \gamma^*$, for every $\iota < \epsilon$ by Lem.4.4. Therefore $\sup_{\iota<\epsilon} \beta_{y_\iota} \leq \gamma^*$. □

For each $x \in \mathcal{T}$, taking a φ-s.i.a. sequence $\langle x_\xi \mid \xi < \zeta\rangle$ to x, let

$$\gamma_x := \sup_{\xi<\zeta} \beta_{x_\xi} < 2^\lambda .$$

γ_x is independent of the choice of a φ-s.i.a. sequence to x by the above lemma. Note that if $x \in \mathcal{T}$ and $\langle x_\xi \mid \xi < \zeta\rangle$ is a φ-s.i.a. sequence to x, then $\beta_{x_\xi} \leq \beta_x$ for each $\xi < \zeta$. Hence

$$\gamma_x \leq \beta_x$$

for every $x \in \mathcal{T}$.

Lemma 4.6. *Let M be an element of $\mathcal{P}_\kappa \mathcal{H}_\theta$ such that $M \cap \lambda \in \mathcal{T}$, $M \prec \mathcal{M}$ and M is i.a. Then $\sup(M \cap 2^\lambda) = \gamma_{M\cap\lambda}$.*

Proof. Suppose that $\langle M_\xi \mid \xi < \zeta\rangle$ is an i.a. sequence to M. Note that $\langle M_\xi \cap \lambda \mid \xi < \zeta\rangle$ is a φ-s.i.a. sequence to $M \cap \lambda$. Hence $\gamma_{M\cap\lambda} = \sup_{\xi<\zeta} \beta_{M_\xi\cap\lambda}$. Moreover $\beta_{M_\xi\cap\lambda} \in M \cap 2^\lambda$ for every $\xi < \zeta$ because $M_\xi \cap \lambda \in M \prec \mathcal{M}$. Therefore $\gamma_{M\cap\lambda} \leq \sup(M \cap 2^\lambda)$.

To see the converse inequality, take an arbitrary $\beta \in M \cap 2^\lambda$. It suffices to show that $\beta \leq \gamma_{M\cap\lambda}$.

First note that $h_\beta \in M$ because $\beta \in M \prec \mathcal{M}$. Hence $h_\beta \restriction \rho \in M$ for every $\rho < \mathrm{cf}(\lambda)$. Then $h_\beta \leq_J h_{\gamma_{M\cap\lambda}}$ by Lem.4.4. Therefore $\beta \leq \gamma_{M\cap\lambda}$. □

Lemma 4.7. *$\{\gamma_x \mid x \in T\}$ is unbounded in 2^λ for every stationary $T \subseteq \mathcal{T}$.*

Proof. Let T be a stationary subset of $\mathcal{T}$. Take an arbitrary $\alpha < 2^\lambda$. We must find $x \in T$ with $\gamma_x \geq \alpha$.

Because T is stationary, we can take $M \in \mathcal{P}_\kappa \mathcal{H}_\theta$ such that $M \cap \lambda \in T$, $M \prec \mathcal{M}$ and $\alpha \in M \cap 2^\lambda$. Then $h_\alpha \restriction \rho \in M$ for every $\rho < \mathrm{cf}(\lambda)$. Then $h_\alpha \leq_J h_{\gamma_{M\cap\lambda}}$ by Lem.4.4. Therefore $\alpha \leq \gamma_{M\cap\lambda}$. This completes the proof because $M \cap \lambda \in T$. □

To prove Prop.4.3 (1), for each stationary $T \subseteq \mathcal{T}$ we must construct a stationary $S \subseteq T$ which becomes nonstationary in $V^{\mathrm{Col}(\kappa,\lambda)}$. To prove (2) we must construct a nonreflecting stationary $S \subseteq T$ for each stationary $T \subseteq \mathcal{T}$. The constructions of such S are the same. Here we present the construction. For each stationary $T \subseteq \mathcal{T}$ define $S_T \subseteq T$ as follows:

First let $\langle F_\alpha \mid \alpha \in 2^\lambda \rangle$ be the Δ-least enumeration of all functions from $[\lambda]^{<\omega}$ to $\mathcal{P}_\kappa \lambda$. By induction on $\alpha < 2^\lambda$ let z^T_α be the Δ-least $z \in T$ such that

- z is closed under F_α, i.e. $F_\alpha(a) \subseteq z$ for every $a \in [z]^{<\omega}$,
- $\gamma_z > \sup_{\alpha' < \alpha} \beta_{z_{\alpha'}}$.

Note that the set $\{z \in T \mid z \text{ is closed under } F_\alpha\}$ is stationary. Hence we can take such $z^T_\alpha \in T$ by Lem.4.7. Then let

$$S_T := \{z^T_\alpha \mid \alpha < 2^\lambda\} .$$

Here we present basic properties of S_T:

Lemma 4.8. *Let T be a stationary subset of $\mathcal{T}$.*

(1) S_T is a stationary subset of T.
(2) The set $\{\gamma_z \mid z \in S_T\}$ is nonstationary in 2^λ.
(3) Suppose that ζ is a limit ordinal $< \kappa$ and that $\langle y_\xi \mid \xi < \zeta \rangle$ is a strictly $\subseteq$-increasing sequence of elements of S_T. Then there are no $z \in S_T$ with $\gamma_z = \sup_{\xi<\zeta} \beta_{y_\xi}$.

Proof. Throughout this proof we let z_α denote z^T_α for each $\alpha < 2^\lambda$.

(1) For every function $F : [\lambda]^{<\omega} \to \mathcal{P}_\kappa \lambda$ there exists $z \in S_T$ closed under F by the first property of the choice of each z_α. This means that S_T is stationary. Clearly $S_T \subseteq T$.

(2) First of all recall that $\gamma_z \leq \beta_z$ for every $z \in \mathcal{T}$. Hence $\gamma_{z_\alpha} > \sup_{\alpha'<\alpha} \gamma_{z_{\alpha'}}$ for every $\alpha < 2^\lambda$ by the second property of the choice of z_α.

Thus the set $\{\gamma_{z_\alpha} \mid \alpha < 2^\lambda\} = \{\gamma_z \mid z \in S_T\}$ is nonstationary in 2^λ.

(3) First of all note that if $\alpha' < \alpha < 2^\lambda$, then $\gamma_{z_{\alpha'}} \leq \beta_{z_{\alpha'}} < \gamma_{z_\alpha} \leq \beta_{z_\alpha}$ by the choice of z_α and the fact that $\gamma_z \leq \beta_z$ for every $z \in \mathcal{T}$.

For each $\xi < \zeta$ let $\alpha_\xi < 2^\lambda$ be such that $y_\xi = z_{\alpha_\xi}$. Then $\langle \alpha_\xi \mid \xi < \zeta \rangle$ is strictly increasing because both $\langle y_\xi \mid \xi < \zeta \rangle$ and $\langle \beta_{z_\alpha} \mid \alpha < 2^\lambda \rangle$ are strictly increasing. Let $\alpha^* := \sup_{\xi<\zeta} \alpha_\xi$. If $\alpha \geq \alpha^*$, then $\gamma_{z_\alpha} > \sup_{\xi<\zeta} \beta_{y_\xi}$ by the second property of the choice of z_α. On the other hand if $\alpha < \alpha^*$, then $\gamma_{z_\alpha} < \sup_{\xi<\zeta} \beta_{y_\xi}$ by the remark above. Therefore there are no $z \in S_T$ with $\gamma_z = \sup_{\xi<\zeta} \beta_{y_\xi}$. □

Below we prove that S_T becomes nonstationary in $V^{\mathrm{Col}(\kappa,\lambda)}$ and is nonreflecting for each stationary $T \subseteq \mathcal{T}$. First we prove that S_T becomes nonstationary in $V^{\mathrm{Col}(\kappa,\lambda)}$. This implies Prop.4.3 (1):

Lemma 4.9. *S_T becomes nonstationary in $V^{\mathrm{Col}(\kappa,\lambda)}$ for all stationary $T \subseteq \mathcal{T}$.*

Proof. Fix a stationary $T \subseteq \mathcal{T}$. Let $\bar{S}$ be the set of all $M \in \mathcal{P}_\kappa \mathcal{H}_\theta$ such that $M \cap \lambda \in S_T$, $M \prec \mathcal{M}$ and M is i.a. By Lem.3.11 it suffices to show that $\bar{S}$ is nonstationary in $\mathcal{P}_\kappa \mathcal{H}_\theta$.

Note that $\{\sup(M \cap 2^\lambda) \mid M \in \bar{S}\} \subseteq \{\gamma_z \mid z \in S_T\}$ by Lem.4.6 and that $\{\gamma_z \mid z \in S_T\}$ is nonstationary in 2^λ by Lem.4.8 (2). Hence $\{\sup(M \cap 2^\lambda) \mid M \in \bar{S}\}$ is nonstationary. This implies that $\bar{S}$ is nonstationary in $\mathcal{P}_\kappa \mathcal{H}_\theta$. □

Now we have proved Prop.4.3 (1). For (2) we must prove that S_T is nonreflecting. We prove this by splitting $\mathcal{T}$ into the following two sets:

$\mathcal{T}_0$:= the set of all $x \in \mathcal{T}$ which is φ-s.i.a. of length ζ for some $\zeta < \kappa$ with $\mathrm{cf}(\zeta) = \mathrm{cf}(\lambda)$.

$\mathcal{T}_1$:= the set of all $x \in \mathcal{T}$ which is φ-s.i.a. of length ζ for some $\zeta < \kappa$ with $\mathrm{cf}(\zeta) \neq \mathrm{cf}(\lambda)$.

Note that $\mathcal{T}_1$ is stationary because $\kappa \geq \omega_2$. In fact we proved that $\mathcal{T}_1$ is stationary in the proof of Lem.3.7 (2). On the other hand, note that elements of $\mathcal{T}_0$ are φ-s.i.a. of length ζ for some ζ with $\mathrm{cf}(\zeta) = \mathrm{cf}(\lambda)$ but are not i.a. of length $\mathrm{cf}(\lambda)$. By Lem.3.10, $\mathcal{T}_0$ is stationary if the set $\{\zeta < \kappa \mid \mathrm{cf}(\zeta) = \mathrm{cf}(\lambda)\}$ does not belong to $I[\kappa]$. Note also that if $x \in \mathcal{T}_0$, then x is φ-s.i.a. of length $x \cap \kappa$ by Lem.3.9.

Clearly the following two lemmata imply Prop.4.3 (2):

Lemma 4.10. *Assume that $\kappa^{<\kappa} = \kappa$, and suppose that $T \subseteq \mathcal{T}_0$ is stationary. Then $S_T \cap \mathcal{P}_\kappa W$ is nonstationary for every $W \subseteq \lambda$ with $|W| = \kappa \subseteq W$.*

Lemma 4.11. *Suppose that $T \subseteq \mathcal{T}_1$ is stationary. Then $S_T \cap \mathcal{P}_\kappa W$ is nonstationary for every $W \subseteq \lambda$ with $|W| = \kappa \subseteq W$.*

First we prove Lem.4.11. We use the following lemma, due to Shelah:

Lemma 4.12 (Shelah). *Suppose that ζ is a limit ordinal $< \kappa$ with $\mathrm{cf}(\zeta) \neq \mathrm{cf}(\lambda)$ and that $\langle x_\xi \mid \xi < \zeta \rangle$ is a $\subseteq$-increasing sequence. Let $x := \bigcup_{\xi<\zeta} x_\xi$. Then $\beta_x = \sup_{\xi<\zeta} \beta_{x_\xi}$.*

Proof. Clearly $\beta_x \geq \sup_{\xi<\zeta} \beta_{x_\xi}$. We prove that $\beta_x \leq \sup_{\xi<\zeta} \beta_{x_\xi}$. Let $\gamma^* := \sup_{\xi<\zeta} \beta_{x_\xi}$. All we have to show is that $\mathrm{ch}_x \leq_J h_{\gamma^*}$. The proof splits into two cases:

Case 1: $\mathrm{cf}(\zeta) < \mathrm{cf}(\lambda)$.

Take an increasing cofinal sequence $\langle \xi_k \mid k < \mathrm{cf}(\zeta) \rangle$ in ζ. Note that $B_k := \{\eta < \mathrm{cf}(\lambda) \mid \mathrm{ch}_{x_{\xi_k}}(\eta) > h_{\gamma^*}(\eta)\} \in J$ because $\beta_{x_{\xi_k}} \leq \gamma^*$. Then $B := \bigcup_{k<\mathrm{cf}(\zeta)} B_k \in J$ by the $\mathrm{cf}(\lambda)$-completeness of J.

Here note that $\mathrm{ch}_x(\eta) = \sup_{k<\mathrm{cf}(\zeta)} \mathrm{ch}_{x_{\xi_k}}(\eta)$ for every $\eta < \mathrm{cf}(\lambda)$. Thus $\{\eta < \mathrm{cf}(\lambda) \mid \mathrm{ch}_x(\eta) > h_{\gamma^*}(\eta)\} = B \in J$. Therefore $\mathrm{ch}_x \leq_J h_{\gamma^*}$.

Case 2: $\mathrm{cf}(\zeta) > \mathrm{cf}(\lambda)$.

Assume that $\mathrm{ch}_x \not\leq_J h_{\gamma^*}$. Then $A := \{\eta < \mathrm{cf}(\lambda) \mid \mathrm{ch}_x(\eta) > h_{\gamma^*}(\eta)\} \notin J$. For each $\eta \in A$ we can take $\xi_\eta < \zeta$ such that $\mathrm{ch}_{x_{\xi_\eta}}(\eta) > h_{\gamma^*}(\eta)$ because $\mathrm{ch}_x(\eta) = \sup_{\xi<\zeta} \mathrm{ch}_{x_\xi}(\eta)$.

Then $\xi^* := \sup_{\eta\in A} \xi_\eta < \zeta$ because $\mathrm{cf}(\zeta) > \mathrm{cf}(\lambda)$. Moreover $\{\eta < \mathrm{cf}(\lambda) \mid \mathrm{ch}_{x_{\xi^*}}(\eta) > h_{\gamma^*}(\eta)\} = A \notin J$. Hence $\mathrm{ch}_{x_{\xi^*}} \not\leq_J h_{\gamma^*}$. This contradicts that $\beta_{x_{\xi^*}} \leq \sup_{\xi<\zeta} \beta_{x_\xi} = \gamma^*$. □

The following is immediate from the above lemma:

Corollary 4.13. *$\beta_x = \gamma_x$ for every $x \in \mathcal{T}_1$.*

Now we prove Lem.4.11:

Proof of Lem.4.11. Suppose that $T \subseteq \mathcal{T}_1$ is stationary. We claim the following:

Claim . *Suppose that ζ is a limit ordinal $< \kappa$ and that $\langle y_\xi \mid \xi < \zeta \rangle$ is a $\subseteq$-increasing sequence in S_T with $\langle y_\xi \cap \kappa \mid \xi < \zeta \rangle$ strictly increasing. Then $\bigcup_{\xi<\zeta} y_\xi \notin S_T$.*

$\vdash$ Let $y := \bigcup_{\xi<\zeta} y_\xi$. The proof splits into two cases by the cofinality of ζ.

First suppose that $\mathrm{cf}(\zeta) = \mathrm{cf}(\lambda)$. Then $\mathrm{cf}(y \cap \kappa) = \mathrm{cf}(\lambda)$, and thus $y \notin \mathcal{T}_1$ by Lem.3.5. Therefore $y \notin S_T$ because $S_T \subseteq T \subseteq \mathcal{T}_1$.

Next suppose that $\mathrm{cf}(\zeta) \neq \mathrm{cf}(\lambda)$. Assume that $y \in S_T$. Then

$$\sup_{\xi<\zeta} \beta_{y_\xi} = \beta_y = \gamma_y$$

by Lem.4.12 and Cor.4.13. This contradicts Lem.4.8 (3) because $y \in S_T$ and $\langle y_\xi \mid \xi < \zeta \rangle$ is a strictly $\subseteq$-increasing sequence in S_T. Therefore $y \notin S_T$. $\dashv$

Take an arbitrary $W \subseteq \lambda$ with $|W| = \kappa \subseteq W$. We show that $S_T \cap \mathcal{P}_\kappa W$ is nonstationary.

We can take a $\subseteq$-increasing continuous cofinal sequence $\langle x_\xi \mid \xi < \kappa \rangle$ in $\mathcal{P}_\kappa W$ with $\langle x_\xi \cap \kappa \mid \xi < \kappa \rangle$ strictly increasing. Let $E := \{\xi < \kappa \mid x_\xi \in S_T\}$. By the claim above, if ζ is a limit ordinal $< \kappa$ with $\sup(E \cap \zeta) = \zeta$, then $\zeta \notin E$. This implies that E is nonstationary in κ. Therefore $S_T \cap \mathcal{P}_\kappa W$ is nonstationary. □

We turn our attention to Lem.4.10. First we extend the notion of the φ-semi-internally approachability to subsets of λ of cardinality κ:

Definition 4.14. Suppose that W is a subset of λ with $|W| = \kappa$. Then W is said to be *φ-semi-internally approachable of length* κ if there exists a $\subseteq$-increasing sequence $\langle w_\xi \mid \xi < \kappa \rangle$ in $\mathcal{P}_\kappa \lambda$ such that $\bigcup_{\xi<\kappa} w_\xi = W$ and such that $\varphi(\langle w_\xi \cap \lambda' \mid \xi < \zeta \rangle) \in W$ for every $\zeta < \kappa$ and every $\lambda' < \lambda$. We call a sequence $\langle w_\xi \mid \xi < \kappa \rangle$ as above a *φ-semi-internally approaching sequence to W*.

Clearly Lem.4.10 follows from the two lemmata below:

Lemma 4.15. *Suppose that $T \subseteq \mathcal{T}_0$ is stationary. If $|W| = \kappa \subseteq W \subseteq \lambda$ and W is φ-s.i.a. of length κ, then $S_T \cap \mathcal{P}_\kappa W$ is nonstationary.*

Lemma 4.16. *Assume that $\kappa^{<\kappa} = \kappa$. If $|W| = \kappa \subseteq W \subseteq \lambda$ and $\mathcal{T}_0 \cap \mathcal{P}_\kappa W$ is stationary, then W is φ-s.i.a. of length κ.*

First we prove Lem.4.15:

Proof of Lem.4.15. Suppose that $|W| = \kappa \subseteq W \subseteq \lambda$ and that W is φ-s.i.a. of length κ. Assume that $S_T \cap \mathcal{P}_\kappa W$ is stationary. We work for a contradiction.

First note that $\mathrm{skull}^{\mathcal{M}}(W) \cap \lambda = W$ because $\mathrm{skull}^{\mathcal{M}}(x) \cap \lambda = x$ for each $x \in S_T$ and $S_T \cap \mathcal{P}_\kappa W$ is stationary. Then we can take a φ-s.i.a. sequence $\langle w_\xi \mid \xi < \kappa \rangle$ to W which is strictly $\subseteq$-increasing continuous. (First take a φ-s.i.a. sequence $\langle w'_\xi \mid \xi < \kappa \rangle$ to W. Let B be the set of all $\xi < \kappa$ such that

$w_\xi \cap \kappa \supsetneq w_\eta \cap \kappa$ for every $\eta < \xi$. Note that B is a club in κ. Let $\langle \rho_\xi \mid \xi < \kappa \rangle$ be the increasing enumeration of B, and let $w_\xi := \bigcup_{\eta < \rho_\xi} w'_\eta$ for each $\xi < \kappa$. Note that if $\zeta < \kappa$ and $\kappa \leq \lambda' < \lambda$, then $\langle w_\xi \cap \lambda' \mid \xi < \zeta \rangle \in \mathrm{skull}^{\mathcal{M}}(W)$ because this sequence can be recovered from $\langle w'_\xi \cap \lambda' \mid \xi \leq \rho_\zeta \rangle \in \mathrm{skull}^{\mathcal{M}}(W)$. Then it is easy to see that $\langle w_\xi \mid \xi < \kappa \rangle$ is a φ-s.i.a. sequence to W which is strictly $\subseteq$-increasing continuous.)

Next let $E := \{\xi < \kappa \mid w_\xi \in S_T\}$. Then E is stationary in κ. Moreover define a club $C \subseteq \kappa$ as follows:

First, for each $\eta < \mathrm{cf}(\lambda)$, let C_η be the set of all limit $\zeta < \kappa$ such that $\varphi(\langle w_\xi \cap \lambda_\eta \mid \xi < \zeta' \rangle) \in w_\zeta$ for every $\zeta' < \zeta$. Note that C_η is a club in κ for each $\eta < \mathrm{cf}(\lambda)$. Next let C' be the set of all $\zeta < \kappa$ such that $\mathrm{skull}^{\mathcal{M}}(w_\xi) \cap \lambda = w_\xi$. C' is also a club because $\mathrm{skull}^{\mathcal{M}}(W) \cap \lambda = W$. Finally let $C := C' \cap \bigcap_{\eta < \mathrm{cf}(\lambda)} C_\eta$. Then C is a club in κ. Here note that $\langle w_\xi \mid \xi < \zeta \rangle$ is a φ-s.i.a. sequence to w_ζ for every $\zeta \in C$ by Lem.3.4.

Because E is stationary we can take $\zeta \in E \cap C$ such that $\sup(E \cap \zeta) = \zeta$. Then $\langle w_\xi \mid \xi < \zeta \rangle$ is a φ-s.i.a. sequence to w_ζ, and so

$$\gamma_{w_\zeta} = \sup_{\xi < \zeta} \beta_{w_\xi} = \sup_{\xi \in E \cap \zeta} \beta_{w_\xi} .$$

But $w_\zeta \in S_T$, and $\langle w_\xi \mid \xi \in E \cap \zeta \rangle$ is a strictly $\subseteq$-increasing sequence in S_T. Hence this contradicts Lem.4.8 (3). □

For Lem.4.16 we introduce the notion of φ-semi-internally unboundedness:

Definition 4.17. Let W be a subset of λ with $|W| = \kappa \subseteq W$. We say that W is *$\mathcal{M}$-semi-internally unbounded* ($\mathcal{M}$-s.i.u.) if it satisfies the following:

(i) $\mathrm{skull}^{\mathcal{M}}(W) \cap \lambda = W$.
(ii) For any $y \in \mathcal{P}_\kappa W$ there exists $x \in \mathcal{P}_\kappa W$ such that $y \subseteq x$ and such that $x \cap \lambda' \in \mathrm{skull}^{\mathcal{M}}(W)$ for every $\lambda' < \lambda$.

Lem.4.16 splits into the following two lemmata:

Lemma 4.18. *If $|W| = \kappa \subseteq W \subseteq \lambda$ and $\mathcal{T}_0 \cap \mathcal{P}_\kappa W$ is stationary, then W is $\mathcal{M}$-s.i.u.*

Lemma 4.19. *Assume that $\kappa^{<\kappa} = \kappa$. If $|W| = \kappa \subseteq W \subseteq \lambda$ and W is $\mathcal{M}$-s.i.u, then W is φ-s.i.a. of length κ.*

First we prove Lem.4.18. Lem.4.18 easily follows from the lemma below. Recall that if $x \in \mathcal{T}_0$, then x is φ-s.i.a. of length $x \cap \kappa$ by Lem.3.9:

Lemma 4.20. *Suppose that $x, y \in \mathcal{T}_0$, $y \subseteq x$ and $y \cap \kappa < x \cap \kappa$. Let $\langle x_\xi \mid \xi \in x \cap \kappa \rangle$ be a φ-s.i.a. sequence to x. Then $y \subseteq x_\xi$ for some $\xi < x \cap \kappa$.*

Proof. Let $\langle y_\iota \mid \iota < y \cap \kappa \rangle$ be a φ-s.i.a. sequence to y. Moreover let $M :=$ skull$^{\mathcal{M}}(x)$. Note that $\langle y_\iota \cap \lambda' \mid \iota < \zeta \rangle \in M$ for every $\zeta < y \cap \kappa$ and $\lambda' < \lambda$ because $y \subseteq x$.

We use the following claim:

Claim . *For every $\lambda' < \lambda$ and every $\iota < y \cap \kappa$ there exists $\xi < x \cap \kappa$ with $y_\iota \cap \lambda' \subseteq x_\xi$.*

$\vdash$ Fix $\lambda' < \lambda$ and $\iota < y \cap \kappa$. Assume that $y_\iota \cap \lambda' \not\subseteq x_\xi$ for every $\xi < \kappa$. We work for a contradiction.

Note that $y_\iota \cap \lambda' \in M$. In M take a bijection $\pi : |y_\iota \cap \lambda'| \to y_\iota \cap \lambda'$. Moreover let ζ be the least ordinal $\leq |y_\iota \cap \lambda'|$ such that $\pi[\zeta] \not\subseteq x_\xi$ for every $\xi < x \cap \kappa$. Note that $\mathrm{cf}(\zeta) = \mathrm{cf}(x \cap \kappa) = \mathrm{cf}(\lambda)$. Moreover $\zeta \in M \cap \kappa$ because $\zeta \leq |y_\iota \cap \lambda'| \in M \cap \kappa \in \kappa$. Hence we can take an increasing sequence $\langle \zeta_\eta \mid \eta < \mathrm{cf}(\lambda) \rangle \in M$ converging to ζ. For each $\eta < \mathrm{cf}(\lambda)$ let ξ_η be the least $\xi < x \cap \kappa$ such that $\pi[\zeta_\eta] \subseteq x_\xi$.

Then $\langle \xi_\eta \mid \eta < \mathrm{cf}(\lambda) \rangle$ is an increasing cofinal sequence in $x \cap \kappa$. Moreover $\langle \xi_\eta \mid \eta < \rho \rangle \in M$ for every $\rho < \mathrm{cf}(\lambda)$ because $\langle \xi_\eta \mid \eta < \rho \rangle$ can be recovered from $\langle x_\xi \cap \lambda' \mid \xi \leq \xi_\rho \rangle$ and $\langle \pi[\zeta_\eta] \mid \eta < \rho \rangle$ both of which belong to M. From these it is easy to see that $\langle x_{\xi_\eta} \mid \eta < \mathrm{cf}(\lambda) \rangle$ is a φ-s.i.a. sequence to x. Hence x is φ-s.i.a. of length $\mathrm{cf}(\lambda)$. This contradicts that $x \in \mathcal{T}_0$. $\dashv$

We proceed to the proof of the lemma. Assume that there are no $\xi < x \cap \kappa$ with $y \subseteq x_\xi$. We work for a contradiction.

First note that $y \cap \kappa \in M \cap \kappa$ and that $\mathrm{cf}(y \cap \kappa) = \mathrm{cf}(\lambda)$. Take an increasing sequence $\langle \iota_\eta \mid \eta < \mathrm{cf}(\lambda) \rangle \in M$ converging to $y \cap \kappa$. By the claim above, for each $\eta < \mathrm{cf}(\lambda)$, let ξ_η be the least $\xi < x \cap \kappa$ such that $y_{\iota_\eta} \cap \lambda_\eta \subseteq x_\xi$.

Here note that $\langle y_{\iota_\eta} \cap \lambda_\eta \mid \eta < \mathrm{cf}(\lambda) \rangle$ is a $\subseteq$-increasing sequence with its union y. Note also that $\langle y_{\iota_\eta} \cap \lambda_\eta \mid \eta < \rho \rangle \in M$ for each $\rho < \mathrm{cf}(\lambda)$ because $\langle y_{\iota_\eta} \cap \lambda_\eta \mid \eta < \rho \rangle$ can be recovered from $\langle y_\iota \cap \lambda_\rho \mid \iota < \iota_\rho \rangle$, $\langle \iota_\eta \mid \eta < \rho \rangle$ and $\langle \lambda_\eta \mid \eta < \rho \rangle$, all of which belong to M.

Then $\langle \xi_\eta \mid \eta < \mathrm{cf}(\lambda) \rangle$ is an increasing cofinal sequence in $x \cap \kappa$ by the remark above and the fact that there are no $\xi < x \cap \kappa$ with $y \subseteq x_\xi$. Note also that $\langle \xi_\eta \mid \eta < \rho \rangle \in M$ for each $\rho < \mathrm{cf}(\lambda)$ because $\langle \xi_\eta \mid \eta < \rho \rangle$ can be recovered from $\langle x_\xi \cap \lambda_\rho \mid \xi \leq \xi_\rho \rangle$ and $\langle y_{\iota_\eta} \cap \lambda_\eta \mid \eta < \rho \rangle$ both of which belong to M.

Then it is easy to see that $\langle x_{\xi_\eta} \mid \eta < \mathrm{cf}(\lambda)\rangle$ is a φ-s.i.a. sequence to x. Thus x is φ-s.i.a. of length $\mathrm{cf}(\lambda)$. This contradicts that $x \in \mathcal{T}_0$. $\square$

Now we prove Lem.4.18:

Proof of Lem.4.18. Suppose that $|W| = \kappa \subseteq W \subseteq \lambda$ and that $\mathcal{T}_0 \cap \mathcal{P}_\kappa W$ is stationary. First note that $\mathrm{skull}^{\mathcal{M}}(W) \cap \lambda = W$ because $\mathrm{skull}^{\mathcal{M}}(x) \cap \lambda = x$ for each $x \in \mathcal{T}_0$.

To prove (ii) in Def.4.17 take an arbitrary $y \in \mathcal{P}_\kappa W$. We will find $x^* \in \mathcal{P}_\kappa W$ such that $y \subseteq x^*$ and such that $x^* \cap \lambda' \in \mathrm{skull}^{\mathcal{M}}(W)$ for every $\lambda' < \lambda$. Because $\mathcal{T}_0 \cap \mathcal{P}_\kappa W$ is stationary we may assume that $y \in \mathcal{T}_0$.

We can take $x \in \mathcal{T}_0 \cap \mathcal{P}_\kappa W$ with $y \subseteq x$ and $y \cap \kappa < x \cap \kappa$. Let $\langle x_\xi \mid \xi < x \cap \kappa\rangle$ be a φ-s.i.a. sequence to x. By Lem.4.20 there exists $\xi < x \cap \kappa$ with $y \subseteq x_\xi$. Note that $x_\xi \cap \lambda' \in \mathrm{skull}^{\mathcal{M}}(x) \subseteq \mathrm{skull}^{\mathcal{M}}(W)$ for every $\lambda' < \lambda$. Therefore $x^* := x_\xi$ is what we seek. $\square$

Finally we prove Lem.4.19:

Proof of Lem.4.19. Assume that $\kappa^{<\kappa} = \kappa$. Suppose that $|W| = \kappa \subseteq W \subseteq \lambda$ and that W is $\mathcal{M}$-s.i.u. Take an arbitrary $\subseteq$-increasing cofinal sequence $\langle w_\xi \mid \xi < \kappa\rangle$ in $\mathcal{P}_\kappa W$. We show that $\langle w_\xi \mid \xi < \kappa\rangle$ is a φ-s.i.a. sequence to W. For this it suffices to show that $\langle w_\xi \cap \lambda' \mid \xi < \zeta\rangle \in \mathrm{skull}^{\mathcal{M}}(W)$ for every $\zeta < \kappa$ and every $\lambda' < \lambda$. Let $M := \mathrm{skull}^{\mathcal{M}}(W)$.

Fix $\zeta < \kappa$ and $\lambda' < \lambda$. Then there exists $x \in \mathcal{P}_\kappa \lambda'$ such that $\bigcup_{\xi<\zeta} w_\xi \cap \lambda' \subseteq x \in M$ by the φ-s.i.u. of W. Here note that $\langle w_\xi \cap \lambda' \mid \xi < \zeta\rangle \in {}^{<\kappa}\mathcal{P}(x)$. Note also that $|{}^{<\kappa}\mathcal{P}(x)| = \kappa$ because $\kappa^{<\kappa} = \kappa$. Moreover ${}^{<\kappa}\mathcal{P}(x) \in M$ and $\kappa \subseteq M$. Therefore ${}^{<\kappa}\mathcal{P}(x) \subseteq M$. Thus $\langle w_\xi \cap \lambda' \mid \xi < \zeta\rangle \in M$. $\square$

This completes the proof of Prop.4.3.

References

1. Bekkali, *Topics in Set Theory*, Springer-Verlag, Berlin, 1991.
2. Q. Feng and T. Jech, *Projectively stationary sets and a strong reflection principle*, J. London Math. Soc. (2) **58** (1998), no.2, 271-283.
3. Q. Feng and M. Magidor, *On reflection of stationary sets*, Fund. Math. **140** (1992), no.2, 175-181.
4. M. Foreman and M. Magidor, *Large cardinals and definable counterexamples to the continuum hypothesis*, Ann. of Pure and Appl. Logic **76** (1995), no.1, 47-97.
5. M. Foreman, M. Magidor and S. Shelah, *Martin's maximum, saturated ideals and nonregular ultrafilters I*, Ann. of Math. (2) **127** (1988), no.1, 1-47.

6. M. Foreman and S. Todorčević, *A new Löwenheim-Skolem theorem*, Trans. Amer. Math. Soc. **357** (2005), no.5, 1693-1715.
7. M. Holz, K. Steffens and E. Weitz, *Introduction to Cardinal Arithmetic*, Birkhäuser-Verlag, Basel, 1999.
8. T. Jech, *Some combinatorial problems concerning uncountable cardinals*, Ann. Math. Logic **5** (1972/1973), no.3, 165-198.
9. T. Jech, *Set Theory*, Third edition, Springer-Verlag, Berlin, 2002.
10. A. Kanamori, *The Higher Infinite*, Second edition, Springer-Verlag, Berlin, 2003.
11. D. W. Kueker, *Countable approximations and Löwenheim-Skolem theorems*, Ann. Math. Logic **11** (1977), no.1, 57-103.
12. T.K. Menas, *On strong compactness and supercompactness*, Ann. Math. Logic **7** (1974/75), 327-359.
13. S. Shelah, *On successors of singular cardinals*, Logic Colloquium '78 (Mons, 1978), 357-380, North-Holland, Amsterdam-New York, 1979.
14. S. Shelah, *Cardinal Arithmetic*, Oxford University Press, Oxford, 1994.
15. S. Shelah, *Reflection implies the SCH*, Fund. Math. **198** (2008), no.2, 95-111.
16. S. Shelah and M. Shioya, *Nonreflecting stationary sets in* $\mathcal{P}_\kappa\lambda$, Adv. Math. **199** (2006), no.1, 185-191.
17. M. Shioya, *Stationary reflection and the club filter*, J. Math. Soc. Japan **59** (2007), no.4, 1045-1065.
18. M. Shioya, *Some application of stationary reflection in* $\mathcal{P}_\kappa\lambda$, RIMS Kôkyûroku **1471** (2006), 54-73.
19. T. Todorčević, *Reflecting stationary sets*, Handwritten note.
20. V. Veličković, *Forcing axioms and stationary sets*, Adv. Math. **94** (1992), no.2, 256-284.

TURING DEGREES AND THE ERSHOV HIERARCHY*

FRANK STEPHAN[†,§], YUE YANG[†,¶] and LIANG YU[‡]

[†]*Department of Mathematics, Faculty of Science, National University of Singapore, 2 Science Drive 2, Singapore 117543*

[‡]*Institute of Mathematical Science, Nanjing University, Nanjing, Jiangsu province 210093, P.R. of China*

[§]*E-mail: fstephan@comp.nus.edu.sg*

[¶]*E-mail: matyangy@nus.edu.sg*

[‡]*E-mail: yuliang.nju@gmail.com*

An n-r.e. set can be defined as the symmetric difference of n recursively enumerable sets. The classes of these sets form a natural hierarchy which became a well-studied topic in recursion theory. In a series of ground-breaking papers, Ershov generalized this hierarchy to transfinite levels based on Kleene's notations of ordinals and this work lead to a fruitful study of these sets and their many-one and Turing degrees. The Ershov hierarchy is a natural measure of complexity of the sets below the halting problem. In this paper, we survey the early work by Ershov and others on this hierarchy and present the most fundamental results. We also provide some pointers to concurrent work in the field.

1. Introduction

This paper aims to achieve two goals. One is to give a survey on the Turing degree structure of the Ershov hierarchy, including some recent results on elementary equivalence and elementary substructure problems. The other is presenting short proofs to some selected results of transfinite levels of the Ershov hierarchy. In his three classic papers Refs. 10, 11 and 12, Ershov studied a hierarchy of sets generated by the recursively enumerable sets. At the finite levels, the hierarchy is also referred as difference hierarchy, which forms a proper subclass of the Δ^0_2-sets. The transfinite extension of the difference hierarchy exhausts all Δ^0_2-sets. Although the finite levels of

*This work was supported by NUS Academic Research Grants R146-000-114-112 (F. Stephan and Y. Yang) and R252-000-308-112 (F. Stephan); National Natural Science Fund of China, No. 10701041 (L. Yu) and Research Fund for the Doctoral Program of Higher Education, No. 20070284043 (L. Yu)

the Ershov hierarchy have been studied by recursion theorists extensively, the same cannot be said on transfinite levels. Things started to change only recently. The transfinite levels of the Ershov hierarchy appeared naturally in many areas, for example, in recursion theory and recursive model theory, see Downey and Gale[14] and Khoussainov, Stephan and Yang,[18] and in inductive inference, see Ambainis, Freivalds and Smith,[2] Carlucci, Case and Jain[6] and Freivalds and Smith.[13] Yet we still feel that Ershov's theorems did not achieve the popularity which they deserved. It is a pity that such a treasure remains obscure due to partially the limited availability of Ershov's original papers and the notations used almost 40 years ago. We hope our selection of results and our presentation would make the transfinite levels of the Ershov hierarchy more accessible for beginners. We refer the interested reader to the excellent and detailed article of Arslanov[3] about the same topic which goes in depth much beyond our short survey.

The paper is organized as follows. In Section 2, we discuss the elementary equivalent problems among finite levels of the Ershov hierarchy; in Section 3, we remind the readers some basic facts about ordinal notations which is necessary for the later part; in Section 4, we present some basic results on transfinite levels of Ershov hierarchy; in the Section 5 we discuss some applications to the area of inductive inference. We conclude with two interesting open problems.

2. The Finite Levels of the Ershov Hierarchy

The finite levels of the Ershov hierarchy cover those sets which are the symmetric difference of finitely many r.e. sets. This class coincides with the sets which are bounded truth-table reducible to the halting problem $K = \{e : \varphi_e(e)\downarrow\}$.

2.1. *The basic definitions*

Let us begin with the well-known Limit Lemma which characterizes the class of sets Turing reducible to K as the class of the limit-recursive sets.

Theorem 2.1 (Shoenfield). *Let $A \subseteq \omega$. The following are equivalent:*

- *(a) $A \leq_T K$.*
- *(b) A is Δ^0_2.*
- *(c) There is a recursive function $f : \omega \times \omega \to \{0, 1\}$ such that*
  - *(1) for all $x \in \omega$, $f(x, 0) = 0$;*
  - *(2) for all $x \in \omega$, $\lim_{s\to\infty} f(x, s) = A(x)$, here we identified A with its characteristic function.*

Definition 2.2. Let n be a fixed natural number. If a set $A \subseteq \omega$ satisfies condition (c) in Limit Lemma together with

(3) for all x, $|\{s : f(x, s+1) \neq f(x, s)\}| \leq n$

then A is called an *n-r.e.* set.

The classes of n-r.e. sets form the finite levels of Ershov hierarchy. Clearly a 1-r.e. set is just a r.e. set in the usual sense. A 2-r.e. set can be expressed as the difference of two r.e. sets, hence they are often called *d-r.e.* sets. In fact, any n-r.e. set can be expressed as the symmetric difference of n recursively enumerable sets. This is the reason why finite levels of the Ershov hierarchy is also called the *difference hierarchy.* An easy diagonalization argument shows that if $n < m$ then there is an m-r.e. set which is not n-r.e. Consequently the hierarchy is proper.

Traditionally when n-r.e. sets were studied, the so-called ω-r.e. sets were normally included also, even if strictly speaking they do not belong to the finite levels of the Ershov hierarchy. In section 4, we will look at the same class of sets again but with a different definition. Below is the definition of ω-r.e. sets used in this section.

Definition 2.3. Let $g : \omega \to \omega$ be a fixed recursive function. If a set A satisfies condition (c) in Limit Lemma together with

(3) for all x, $|\{s : f(x, s+1) \neq f(x, s)\}| \leq g(x)$

then A is called an *ω-r.e.* set.

Historically the notion of n-r.e. sets was first introduced by Putnam[23] and Gold[15] in the middle of 1960's. It should be noted that we do not state the definitions as in the original literature.

The notions of n-r.e. and ω-r.e. sets can be naturally extended to degrees. Ershov had studied the m-degrees intensively in his three papers. We will study their Turing degrees instead. Let D_n and D_ω denote the class of n-r.e. and ω-r.e. sets respectively. What we are interested in is the partially ordered structures $\mathcal{D}_n = (D_n, \leq)$ and $\mathcal{D}_\omega = (D_\omega, \leq)$ where $\leq$ is the Turing reducibility. For historical reasons, the notation $\mathcal{R}$ is often used to denote $\mathcal{D}_1$. Clearly the class of n-r.e. and ω r.e. degrees are contained inside $\mathcal{D}(\leq \mathbf{0}')$ which are the Δ^0_2-degrees.

The study of n-r.e. Turing degrees can be traced back to Cooper[7] who showed that the degrees of difference hierarchy do not collapse.

Theorem 2.4 (Cooper). *There is a d-r.e. set D such that for all r.e. set W, $D \not\equiv_T W$.*

The proof of Cooper's theorem can be generalized to show the proper containment of the following nested degree structures.

$$\mathcal{R} \subsetneq \mathcal{D}_2 \subsetneq \cdots \subsetneq \mathcal{D}_\omega \subsetneq \mathcal{D}(\leq \mathbf{0}'). \tag{1}$$

2.2. *Elementary equivalence problems*

Thus when the number of changes allowed in the approximation increases, we get more and more degrees. One natural question to ask is whether the corresponding partial order structures also become more and more complicated. More precisely:

(a) Are degrees appeared in (1) elementarily equivalent?
(b) If not, at which level does the difference show up?

Recall that two structures $\mathfrak{A}$ and $\mathfrak{B}$ over the same language are called *elementarily equivalent* (written $\mathfrak{A} \equiv \mathfrak{B}$) if for any sentence σ in that language,

$$\mathfrak{A} \models \sigma \quad \text{iff} \quad \mathfrak{B} \models \sigma.$$

If the above holds only for Σ_k-sentences, then we say that $\mathfrak{A}$ is Σ_k-elementarily equivalent to $\mathfrak{B}$, written $\mathfrak{A} \equiv_{\Sigma_k} \mathfrak{B}$ or simply $\mathfrak{A} \equiv_k \mathfrak{B}$.

The elementary differences among structures in (1) would not occur at the Σ_1-level, which is a fact first observed by Sacks.

Theorem 2.5. *For any natural numbers $m, n \geq 1$, $\mathcal{D}_m \equiv_1 \mathcal{D}_n$, $\mathcal{D}_m \equiv_1 \mathcal{D}_\omega$ and $\mathcal{D}_m \equiv_1 \mathcal{D}(\leq \mathbf{0}')$.*

The outline of the proof is as follows: By modifying the proof of Friedberg-Muchnik Theorem, we can embed all countable partial orders into any of the above degrees structures, say $\mathcal{D}_*$. Therefore for any Σ_1-sentence σ, $\mathcal{D}_*$ satisfies σ if and only if σ is consistent with the theory of partial orderings. In other words, σ is true in one of the structures if and only if it is true in any other structures.

The same cannot be said at Σ_2-level. Indeed we are going to quote contrasting pairs of results about degree structures and conclude that most of the degree structures in (1) are not elementarily equivalent and the differences show up at level two.

We begin with separating $\mathcal{D}(\leq \mathbf{0}')$ with the rest.

Theorem 2.6 (Sacks,[26] 1961). *There is a minimal degree less than* $\mathbf{0}'$.

In fact the minimal degree built by Sacks is in $\mathcal{D}_\omega$.

Theorem 2.7 (Sacks,[27] 1964). $\mathcal{R}$ *is dense.*

Corollary 2.8. $\mathcal{R} \not\equiv_2 \mathcal{D}_\omega$ *and* $\mathcal{R} \not\equiv_2 \mathcal{D}(\leq \mathbf{0}')$.

One can also use the existence of minimal degrees to separate $\mathcal{D}_\omega$ from $\mathcal{D}_n$ by the following theorem.

Theorem 2.9 (Lachlan). *Let $n \geq 2$ be a natural number. For any $\mathbf{d} \in \mathcal{D}_n$, if $\mathbf{d} \neq \mathbf{0}$ then there is $\mathbf{c} \in \mathcal{D}_{n-1}$ such that $\mathbf{0} < \mathbf{c} \leq \mathbf{d}$. Moreover $\mathbf{d}$ is r.e. in $\mathbf{c}$. Consequently for all natural number n, $\mathcal{D}_n$ is downward dense.*

Proof. We prove the case when $n = 2$ to illustrate the idea. Let D be any nonrecursive d-r.e. set. Fix a recursive approximation D_s of D, such that $D_0 = \emptyset$, $D_s(x)$ can at most change twice and at any stage only one element can enter or leave D. Let

$$A = \{s : (\exists x)[x \text{ enters } D \text{ at stage } s \text{ and } \exists t > s(x \notin D_t)]\}.$$

It's easy to verify that A is r.e., $A \leq_T D$ and D is r.e. in A. If D is properly d-r.e., then the set A cannot be recursive, since otherwise D is r.e. in a recursive set, contradicting D being properly d-r.e. To finish the proof of downward density, if D is not properly d-r.e., then it follows from Sacks Density Theorem. □

Corollary 2.10. *For any natural number $n \geq 1$, $\mathcal{D}_n \not\equiv_2 \mathcal{D}_\omega$ and $\mathcal{D}_n \not\equiv_2 \mathcal{D}(\leq 0')$.*

Next we separate $\mathcal{R}$ from $\mathcal{D}_n$ $(n \geq 2)$. The clearest distinction is the non-density theorem of n-r.e. degrees:

Theorem 2.11 (Cooper, Harrington, Lachlan, Lempp and Soare,[8] 1991). *For every natural number $n \geq 2$ there is a maximal element in $\mathcal{D}_n$.*

The technical theorem they showed is actually stronger:

Theorem 2.12. *There is a d-r.e. degree $\mathbf{d}$ such that no ω-r.e. degree $\mathbf{a}$ can have $\mathbf{d} < \mathbf{a} < \mathbf{0}'$.*

Corollary 2.13. *For every natural number* $n \geq 2$,

$$\mathcal{R} \not\equiv_2 \mathcal{D}_n.$$

It also gives another instance of

$$\mathcal{R} \not\equiv_2 \mathcal{D}_\omega.$$

Furthermore, observe that for any Δ_2-set $D <_T K$, K is r.e. in D. Thus by relativizing Sacks Density Theorem, we see that $\mathcal{D}(\leq 0')$ is upward dense, thus we can separate $\mathcal{D}_n$ and $\mathcal{D}_\omega$ from $\mathcal{D}(\leq \mathbf{0}')$.

Corollary 2.14. *For each natural number* $n \geq 2$

$$\mathcal{D}_n \not\equiv_2 \mathcal{D}(\leq \mathbf{0}') \text{ and } \mathcal{D}_\omega \not\equiv_2 \mathcal{D}(\leq \mathbf{0}').$$

Historically, it was Downey[9] who first demonstrated a Σ_2-difference between $\mathcal{R}$ and $\mathcal{D}_n$ by showing that diamond exists in $\mathcal{D}_n$, in contrast with Lachlan's Nondiamond Theorem,[20] which says there is no diamond in $\mathcal{R}$. The result motivated Downey to propose the following Conjecture 2.15:

Conjecture 2.15 (Downey[9]). *For any natural number* $n, m \geq 2$, *the structures* $\mathcal{D}_n$ *and* $\mathcal{D}_m$ *are elementarily equivalent.*

It is said that Downey did not believe that $\mathcal{D}_n$ and $\mathcal{D}_m$ $(n \neq m)$ are elementarily equivalent. He proposed his conjecture in this way to challenge people to explore the structural differences among $\mathcal{D}_n$ and $\mathcal{D}_m$ for $n, m > 1$. It was one of the major problems in the area of difference hierarchy.

Recently Arslanov, Kalimullin and Lempp[4] announced a negative solution to Conjecture 2.15.

Theorem 2.16 (Arslanov, Kalimullin and Lempp[4]). $\mathcal{D}_2 \not\equiv_2 \mathcal{D}_3$.

The technical statement of the difference goes as follows.

Theorem 2.17 (Arslanov, Kalimullin and Lempp[4]).

(a) *There are 3-r.e. degrees* $\mathbf{f} > \mathbf{e} > \mathbf{d} > \mathbf{0}$ *such that any 3-r.e. degree* $\mathbf{u} \leq \mathbf{e}$ *is comparable with* $\mathbf{d}$, *and any 3-r.e. degree* $\mathbf{u}$ *with* $\mathbf{d} \leq \mathbf{u} \leq \mathbf{f}$ *is comparable with* $\mathbf{e}$.

(b) *There are no d-r.e. degrees* $\mathbf{f} > \mathbf{e} > \mathbf{d} > \mathbf{0}$ *such that any d-r.e. degree* $\mathbf{u} \leq \mathbf{e}$ *is comparable with* $\mathbf{d}$, *and any d-r.e. degree* $\mathbf{u}$ *with* $\mathbf{d} \leq \mathbf{u} \leq \mathbf{f}$ *is comparable with* $\mathbf{e}$.

The question whether $\mathcal{D}_n \equiv_2 \mathcal{D}_m$ for $2 < n < m$ remains open, though Arslanov, Kalimullin and Lempp suggested that their techniques might be generalized to settle it negatively.

2.3. *Elementary substructure problems*

What remains is whether one structure can be a Σ_1-elementary substructure of the other.

Recall: When $\mathfrak{A}$ is a substructure of $\mathfrak{B}$, there is a finer notion to gauge the structural differences by allowing parameters from the universe of $\mathfrak{A}$. More precisely, let $\mathcal{L}_A$ be the extended language $\mathcal{L} \cup \{\mathbf{a} : a \in A\}$ obtained by adding a constant symbol $\mathbf{a}$ for each element a in A (here A denotes the universe of $\mathfrak{A}$).

Definition 2.18. Let n be a natural number. We say that $\mathfrak{A}$ is a Σ_n-*substructure* of $\mathfrak{B}$, written $\mathfrak{A} \preccurlyeq_n \mathfrak{B}$, if for all Σ_n-formulas $\varphi(x_1, x_2, ..., x_m)$ and all $a_1, a_2, ..., a_m \in A$,

$$\mathfrak{A} \models \varphi(\mathbf{a}_1, \mathbf{a}_2, ..., \mathbf{a}_m) \text{ if and only if } \mathfrak{B} \models \varphi(\mathbf{a}_1, \mathbf{a}_2, ..., \mathbf{a}_m).$$

In Ershov hierarchy the natural question is: Can one structure in (1) be a Σ_1-elementary substructure of another?

The first breakthrough related to the Σ_1-elementary substructure problem was obtained by Slaman[30] in 1983.

Theorem 2.19 (Slaman).

(i) *There are r.e. sets A, B and C and a Δ^0_2-set X such that*

  (a) $\emptyset <_T X \leq_T A$;
  (b) $C \not\leq_T B \oplus X$;
  (c) *for all r.e. set W, $\emptyset <_T W \leq_T A \Rightarrow C \leq_T B \oplus W$.*

(ii) *For each natural number $n \geq 1$, $\mathcal{D}_n \not\preccurlyeq_1 \mathcal{D}(\leq \mathbf{0}')$.*

The r.e. sets A, B and C in Theorem 2.19 form the so-called *Slaman triple.* If one tries to use Slaman triples to separate $\mathcal{R}$ and $\mathcal{D}_2$, one immediately runs into the following difficulty. By Theorem 2.9 of Lachlan, every nonrecursive n-r.e. degree bounds a nonrecursive r.e. degree, one cannot hope that any n-r.e. degree D plays the role of X as in Theorem 2.19. In 2006, Yang and Yu[31] solved the difficulty by introducing another parameter to control the r.e. degrees below the degree of D. The technical statement is as follows:

Theorem 2.20 (Yang and Yu).

(i) There are r.e. sets A, B, C and E and a d.r.e. set D such that

- *(a) $D \leq_T A$ and $D \not\leq_T E$;*
- *(b) $C \not\leq_T B \oplus D$;*
- *(c) for all r.e. set W, if $W \leq_T A$ then either $C \leq_T B \oplus W$ or $W \leq_T E$.*

(ii) For each natural number $n \geq 2$, $\mathcal{R} \not\preccurlyeq_1 \mathcal{D}_n$.

Shore and Slaman (2007) have announced that $\mathcal{D}_m \not\preccurlyeq_1 \mathcal{D}_n$ for natural numbers $m < n$, thus completely settled the Σ_1-elementary equivalence problem for finite levels of the Ershov hierarchy. However, many questions on transfinite levels of the Ershov hierarchy are still open.

3. Ordinal Notations

In this section, we remind the readers some basic definitions and facts about ordinal notations, which are necessary for the development of the Ershov hierarchy to transfinite levels. We will only prove Markwald Theorem (Theorem 3.5) in detail, because its technique will be used in Section 4. Other materials can be found in standard reference books such as Rogers,[24] Sacks[25] and Ash and Knight.[5]

3.1. *Kleene's $\mathcal{O}$*

We define a set of notations $\mathcal{O} \subset \omega$, a function $|\text{--}|_{\mathcal{O}}$ taking each $a \in \mathcal{O}$ to an ordinal $\alpha = |a|_{\mathcal{O}}$ and a strict partial ordering $<_{\mathcal{O}}$ on $\mathcal{O}$ simultaneously. The elements of $\mathcal{O}$ are the notations and $|a|_{\mathcal{O}}$ is the ordinal represented by the notation a.

- We let 1 be the notation for 0; that is, $|1|_{\mathcal{O}} = 0$.
- If a is a notation for α, then 2^a is a notation for $\alpha + 1$, i.e., $|2^a|_{\mathcal{O}} = \alpha + 1$. Let $b <_{\mathcal{O}} 2^a$ if $b <_{\mathcal{O}} a$ or $b = a$.
- If φ_e is a total recursive function such that for each $n \in \omega$, we have defined $|\varphi_e(n)|_{\mathcal{O}} = \alpha_n$ and $\varphi_e(n) <_{\mathcal{O}} \varphi_e(n+1)$, then $3 \cdot 5^e$ is a notation for $\alpha = \lim_n \alpha_n$, i.e., $|3 \cdot 5^e|_{\mathcal{O}} = \alpha$. Let $b <_{\mathcal{O}} 3 \cdot 5^e$ if there exists some n such that $b <_{\mathcal{O}} \varphi_e(n)$.

That finishes the definition of our system of notations.

The initial part of $\mathcal{O}$ looks like the figure below, where the vertical line to the left indicates the ordinal line:

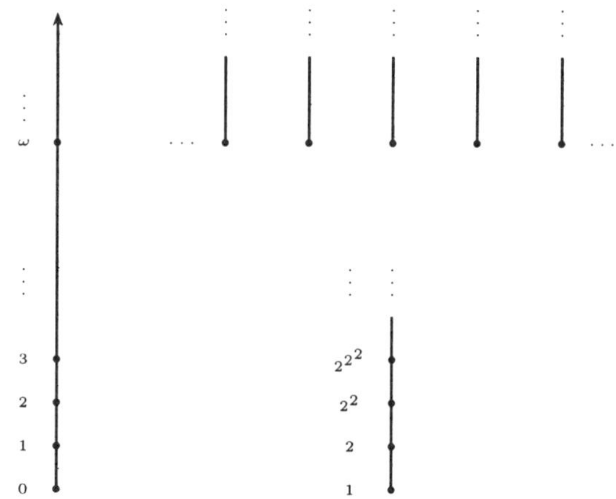

We will also need the following results on effective addition on $\mathcal{O}$.

Lemma 3.1. *There is a 2-place total recursive function $+_{\mathcal{O}}$ such that for all $a, b \in \mathcal{O}$,*

$$|a +_{\mathcal{O}} b|_{\mathcal{O}} = |a|_{\mathcal{O}} + |b|_{\mathcal{O}}.$$

3.2. *Constructive and recursive ordinals*

Definition 3.2. The ordinals having notations in $\mathcal{O}$ are called *constructive ordinals.*

Since $\mathcal{O}$ is countable, there are countably many constructive ordinals. The first nonconstructive ordinal is denoted by ω_1^{CK} for Church-Kleene.

Definition 3.3. An ordinal α is *recursive* if it is finite or it is isomorphic to some recursive well-ordering of ω.

Using Recursion Theorem and transfinite induction up to ω_1^{CK}, one can show that

Theorem 3.4. *Fix an enumeration of all r.e. sets $\{W_e : e \in \omega\}$.*

1. *(1) There exists a total recursive function p such that for all b in $\mathcal{O}$,*

$$W_{p(b)} = \{a : a <_{\mathcal{O}} b\}.$$

2. *(2) There exists a total recursive function q such that for all b in $\mathcal{O}$,*

$$W_{q(b)} = \{\langle u, v\rangle : u <_{\mathcal{O}} v <_{\mathcal{O}} b\}.$$

Therefore all constructive ordinals are recursive because r.e. linear orderings are recursive. The converse is also true:

Theorem 3.5 (Markwald). *Every recursive ordinal is constructive.*

The proof follows from the two Lemmas 3.6 and 3.7, which we give detailed proof in a moment. Assuming the two Lemmas, the argument goes as follows. For any recursive ordinal α, first embed it into a recursive well-ordering $\mathfrak{B}$ as specified in Lemma 3.6; then into $\mathcal{O}$ as in Lemma 3.7. Since $|\mathfrak{B}| \geq \alpha$ as ordinals, α is embedded too.

Lemma 3.6. *Given a recursive index for a linear ordering $\mathfrak{A}$, we can find a recursive index for a linear order $\mathfrak{B}$ of type $\omega \cdot (1 + \mathfrak{A}) + 1$ in which we can apply uniform effective procedure to*

- *(a) determine whether a given element b is first, last, a successor, or a limit point;*
- *(b) for any successor element b, find all the immediate predecessor, and for any limit b, determine an increasing sequence $l(b, i)$ with limit b.*

Furthermore, we can effectively determine a recursive index for the embedding that takes a in $\mathfrak{A}$ to the first element of the corresponding copy of ω in $\mathfrak{B}$.

Note if $\mathfrak{A}$ is a linear ordering then $\mathfrak{B}$ is a well-ordering if and only if $\mathfrak{A}$ is a well-ordering. Moreover if $\mathfrak{A}$ is well-ordered of order type α then the order type of $\mathfrak{B}$ is $\beta > \alpha$.

Proof. We may suppose that the universe of $\mathfrak{A}$ is $\omega \setminus \{0\}$ or a finite initial segment. Let $\mathfrak{A}^*$ be the result of adding 0 at the front of $\mathfrak{A}$. Let $\mathfrak{B}$ be the linear order which is the product of $\mathfrak{A}^*$ and ω with lexicographic order, with an extra last point ∞.

Now $\mathfrak{B}$ is recursive and its limit points are exactly $(x, 0)$ for $x > 1$ and ∞. Also its successors are exactly the newly add elements, (x, y) for $y \neq 0$. Hence the predecessor of (x, y) is $(x, y - 1)$.

It remains to show the second part of (b). If b is a limit point in $\mathfrak{B}$, define $b_{i+1} = \mu\langle x, y\rangle[b_i <_{\mathfrak{B}} \langle x, y\rangle <_{\mathfrak{B}} b]$, where the μ-operator searches through the normal order of ω. It's easy to see that for any $b' <_{\mathfrak{B}} b$ there is i such that $b' <_{\mathfrak{B}} b_i <_{\mathfrak{B}} b$. □

Lemma 3.7. *Given an index for a recursive linear ordering $\mathfrak{B}$ with the special properties in Lemma 3.6 (there are first, last element, the sets of* *successor and limit elements are recursive with recursive witnessing functions and the index for $\mathfrak{B}$ tells us how to compute them), we can find a recursive index for a function g defined on all of $\mathfrak{B}$ such that*

(a) *if* $\text{pred}(x)$*, which is the set* $\{y : y <_{\mathfrak{B}} x\}$*, is well-ordered, then* $g(x)$ *is a notation in* $\mathcal{O}$ *for the height* x*;*
(b) *if* $\text{pred}(x)$ *is not well-ordered then* $g(x) \notin \mathcal{O}$*.*

Proof. We begin by defining a partial recursive function f such that if e is a recursive index for a function with the behavior we want for g on $\text{pred}(x)$ (thinking of $\mathfrak{B}$ as well-ordered), then $f(e,x)$ has the value we want for $g(x)$. We consider three cases.

Case 1. If x is the first element of $\mathfrak{B}$, then $f(e,x) = 1$.

Case 2. If x is the successor of y in $\mathfrak{B}$, then $f(e,x) = 2^{\varphi_e(y)}$.

Case 3. If x is a limit element of $\mathfrak{B}$ and $l(x,i)$ is the recursive increasing sequence with limit b defined in the proof of Lemma 3.6, let $h(e,b)$ be the recursive function given by s-m-n Theorem such that

$$\varphi_{h(e,x)}(i) = \varphi_e(l(e,i)),$$

then $f(e,x) = 3 \cdot 5^{h(e,x)}$.

Note that $f(e,x)$ is defined for all e and all $x \in \mathfrak{B}$, even when $\mathfrak{B}$ is not well-ordered or the function with index e is badly behaved. Now $f(e,x) = \varphi_{k(e)}(x)$ for some k, by Recursion Theorem, there is an n such that for all $x \in \mathfrak{B}$,

$$\varphi_n(x) = \varphi_{k(n)}(x) = f(n,x).$$

Let $g(x) = \varphi_n(x)$. We show that g is what we wanted. Note that $g(x)$ is defined for all $x \in \mathfrak{B}$ since f is.

Claim 1. If $\text{pred}(x)$ is well-ordered and x has height α, then $|g(x)|_{\mathcal{O}} = \alpha$.

Claim 1 can be proven by transfinite induction on α.

Claim 2. If $g(x) \in \mathcal{O}$, then $\text{pred}(x)$ is well-ordered in $\mathfrak{B}$.

Claim 2 can be proven by showing that whenever $|a|_{\mathcal{O}} = \alpha$ and $g(x) = a$ then x has height α in $\mathfrak{B}$. This is done by transfinite induction on α, here we used the definition of f and the choice of n. □

3.3. *$\mathcal{O}$ is Π^1_1-complete*

Recall that a set $X \subseteq \omega$ is called Σ^1_1 if $X = \{z : \exists f \forall y R(f,y,z)\}$ where R is recursive and the quantifiers $\exists f$ and $\forall y$ are ranging over functions and numbers respectively. X is Π^1_1 if the complement of X is Σ^1_1. X is Δ^1_1 if X is both Σ^1_1 and Π^1_1. The connection between Π^1_1-sets and well-orderings can be seen below. Kleene[19] showed that $\mathcal{O}$ is Π^1_1-complete and hence $\mathcal{O}$ is not Σ^1_1.

Theorem 3.8 (Kleene[19]).

(1) Every Π^1_1-set is many-one reducible to $\mathcal{O}$.
(2) $\mathcal{O}$ is Π^1_1.

Theorem 3.9 (Kleene Boundedness Theorem). *Suppose $X \subseteq \mathcal{O}$ and X is Σ^1_1. Then $X \subseteq \mathcal{O}_b$ for some $b \in \mathcal{O}$, where $\mathcal{O}_b = \{a : a \in \mathcal{O} \wedge |a|_\mathcal{O} < |b|_\mathcal{O}\}$.*

Note that in Kleene's $\mathcal{O}$, each finite ordinal has a unique notation, while for an infinite ordinal if there is one notation, then there are infinitely many. In other words, the notations of many ordinals are not unique. This is a consequence of the fact that $\mathcal{O}$ is some type of universal structure in which every well-ordering with possibilities to identify successors and limit-ordinals can be embedded. Harrison[17] showed that one can have all recursive ordinals without losing uniqueness.

Theorem 3.10 (Harrison[17]). *There is a recursive linear ordering $\sqsubset$ on the natural numbers such that for every recursive ordinal there is exactly one representative n such that $(\{m : m \sqsubset n\}, \sqsubset)$ is order isomorphic to this ordinal.*

It follows that an initial segment of the natural numbers with the ordering $\sqsubset$ is order isomorphic to ω_1^{CK} but this initial segment does not have a supremum with respect to $\sqsubset$. Note that one can improve this structure in order to get more operations on ordinals to be recursive.

Remark 3.11. Given Harrison's recursive linear ordering $\sqsubset$, let

$$L = \{(a_1, a_2, \ldots, a_n) : n \in \omega \wedge a_1 \sqsupset a_2 \sqsupset \ldots \sqsupset a_n\}$$

and

$$\begin{aligned}
&(a_1, a_2, \ldots, a_n) \sqsubset (b_1, b_2, \ldots, b_m) \\
&\quad \Leftrightarrow \exists k \leq 1 + \min\{m, n\}\ [\forall \ell \in \{1, 2, \ldots, k-1\} \\
&\quad [a_\ell = b_\ell \wedge (n < k \leq m \vee (k \leq n, m \wedge a_k < b_k))]]; \\
&(a_1, a_2, \ldots, a_n) + (b_1, b_2, \ldots, b_m) \\
&\quad = (a_1, a_2, \ldots, a_k, b_1, b_2, \ldots, b_m) \\
&\quad \text{for the maximal } k \leq n \text{ with } a_k \leq b_1.
\end{aligned}$$

The empty sequence λ is also an element of L and satisfies $\lambda \sqsubset z$ for all $z \in L - \{\lambda\}$ and $\lambda + z = z + \lambda = z$ for all $z \in L$.

One can show that for any $(a_1, a_2, \ldots, a_n) \in L$ where all a_k represent a recursive ordinal α_k in Harrison's given ordering, the set $\{z \in L : z \sqsubset (a_1, a_2, \ldots, a_n)\}$ is order isomorphic to the ordinal $\omega^{\alpha_1} + \omega^{\alpha_2} + \ldots + \omega^{\alpha_n}$. Hence $(L, \sqsubset)$ is a recursive linear ordering where the initial part represents all recursive ordinals and where one can test for the ordinals in the initial segment how they are composed of ω-powers and what the coefficients of the lowest ordinals $\omega^n, \omega^{n-1}, \ldots, \omega^1, \omega^0$ for a fixed constant n are.

This remark shows that Kleene's $\mathcal{O}$ is not the only system where one can represent all recursive ordinals adequately and $(L, \sqsubset)$ has the advantage that it is a linear ordering with unique representatives for the recursive ordinals where various natural operations on the ordinals are recursive.

Somehow, such a system of representatives would have the short-coming that not every notation in $\mathcal{O}$ could be translated into a notation in it effectively and so one could argue that it picks the representatives of the ordinals a bit arbitrarily out of $\mathcal{O}$. Furthermore, as already said, the set of the ordinals in a linear ordering can never by Π^1_1-complete; indeed, such a set is even not many-one hard for the halting problem.

4. Transfinite Levels of the Ershov Hierarchy

We are now ready to extend the Ershov hierarchy to transfinite levels. We begin with extending the notions of n-r.e. sets and degrees to the transfinite levels and then prove some important theorems of the Ershov.

4.1. *a-r.e. Sets and Degrees*

In the following, we will define D_a for each $a \in \mathcal{O}$. As we shall see the definition is heavily notation dependent. For example, for ordinals $\alpha \geq \omega^2$ we may have a and b both are notations for α, yet D_a and D_b are different. This is the reason why we do not define α-r.e. sets for recursive ordinals α, instead we can only define a-r.e. sets for ordinal notations $a \in \mathcal{O}$. It should also be noted that our phrasing is different from the original definition by Ershov. In fact, there are several different versions of the same notion in the literature.

Definition 4.1. For each $a \in \mathcal{O}$, a subset A of ω is a-r.e. if and only if there are recursive functions $f : \omega \times \omega \to \{0, 1\}$ and $o : \omega \times \omega \to \mathcal{O}$ such that

(1) For all x, $f(x,0)=0$ and $o(x,0) <_{\mathcal{O}} a$.
(2) For all x and s, $o(x,s+1) \leq_{\mathcal{O}} o(x,s)$.
(3) For all x and s, if $f(x,s+1) \neq f(x,s)$ then $o(x,s+1) \neq o(x,s)$.
(4) For all x, $\lim_s f(x,s) = A(x)$.

We use D_a to denote the class of all a-r.e. sets.

It is easy to see that a set A is ω-r.e. as in Definition 2.3 if and only if A is a-r.e. for some a such that $|a|_{\mathcal{O}} = \omega$.

There is slight inconsistency of notations between the finite and transfinite levels. In this section, we only talk about a-r.e. sets where a is a notation of an infinite ordinal α.

It is easy to see that for all $a \in \mathcal{O}$ there is a a-r.e. set which is not b-r.e. for any $b <_{\mathcal{O}} a$. Selivanov[28] transferred this result to Turing degrees.

Theorem 4.2 (Selivanov[28]). *For all $a \in \mathcal{O}$, there is a set $X \in D_a$ such that for all $Y \in \bigcup\{D_b : b <_{\mathcal{O}} a\}$, $X \not\equiv_T Y$.*

The major steps of the proof can be outlined as follows: For each fixed notation a, by Theorem 3.4 we can effectively enumerate all $\{b : b <_{\mathcal{O}} a\}$. For each fixed b we can effectively enumerate all sets in D_b. Thus we can effectively enumerate all sets in $\bigcup\{D_b : b <_{\mathcal{O}} a\}$. Once we have the enumeration, we can apply Cooper's technique used in the proof of Theorem 2.4 to find an $X \in D_a$ such that $Y \not\equiv_T Y$.

Let $\mathcal{D}_a$ to denote class of a-r.e. Turing degrees. Selivanov's Theorem tells us that for any fixed notation $a \in \mathcal{O}$, the set $\bigcup\{\mathcal{D}_b : b <_{\mathcal{O}} a\}$ will not exhaust all Δ^0_2-degrees.

4.2. *Ershov's Theorems*

We now present some interesting theorems by Ershov. We begin with the one which roughly says all Δ^0_2-sets have appeared at level ω^2.

Theorem 4.3 (Ershov). *Any Δ^0_2-set A is a-r.e. for some $a \in \mathcal{O}$ with $|a|_{\mathcal{O}} = \omega^2$.*

Proof. Let A be a Δ^0_2-set with recursive approximation $f(x,s)$ as stated in the Limit Lemma. We need to find a notation $a \in \mathcal{O}$ and the corresponding $o(x,s) : \omega \times \omega \to a$ for $f(x,s)$. We do it in two steps.

Step 1: Obtain a recursive well-ordering of order type ω which "records" the changes of the approximation of f. The main idea informally is: When $f(x,s+1) = f(x,s)$ we put the pair $(x,s+1)$ after (x,s); when

$f(x, s+1) \neq f(x, s)$ we put $(x, s+1)$ before all other pairs (x, y) where $y \leq s$. Fix an recursive ω-ordering $<$ on $\omega \times \omega$ as in the pairing function:

$$(0,0) < (0,1) < (1,0) < (0,2) < (1,1) < (2,0) < \dots$$

Define $\prec$ on $\omega \times \omega$ by recursion with respect to $<$: Suppose we have done all pairs before (x, y) with respect to the $<$-order. Assume the numbers between $(x-1, 0)$ and $(x, 0)$ are

$$(x-1,0) \prec (u_1, v_1) \prec \cdots \prec (u_k, v_k) \prec (x, y_i) \prec (x, y_{i-1}) \prec \cdots \prec (x, 0)$$

where $u_i < x$ for all $i \in \{1, 2, \dots, k\}$. Consider the pair (x, y). If $y = 0$ or ($y \neq 0$ and $f(x, y) = f(x, y-1)$), then put (x, y) in the list following the same $<$-order. If $y \neq 0$ and $f(x, y) \neq f(x, y-1)$ then insert (x, y) between (u_k, v_k) and (x, y_i).

Clearly $\prec$ is recursive. Since A is Δ^0_2, $\prec$ is a well-order of order type ω.

Step 2. We obtain a notation for $\prec$ by following the proof of Lemmas 3.6 and 3.7, the process of getting the $\mathfrak{B}$ as in Lemma 3.6 gives us the extra factor ω.

Let g be the embedding of $(\omega \times \omega, \prec)$ into $\mathcal{O}$. Then $g : \omega \times \omega \to a$ for some a with $|a|_{\mathcal{O}} = \omega^2$. Define $o : \omega \times \omega \to a$ by $o(x, 0) = g(x, 0)$ and $o(x, y+1) = o(x, y)$ if $f(x, y+1) = f(x, y)$; and $o(x, y+1) = g(x, y+1)$ if $f(x, y+1) \neq f(x, y)$. Then o is what we wanted. □

The next theorem shows that ω^2 is the first level where one can exhaust all Δ^0_2-sets.

Theorem 4.4 (Ershov). *There is a Δ_2-set A such that $A \not\equiv_T B$ for any $B \in \bigcup\{D_b : b \in \mathcal{O} \wedge |b|_{\mathcal{O}} < \omega^2\}$.*

Proof. We first show two claims which tell us that D_α makes sense for $\alpha < \omega^2$. The point is that there are only finitely many limit points below α, thus we can code the information (nonuniformly).

Claim 1. If $a, b \in \mathcal{O}$ and $|a|_{\mathcal{O}} = |b|_{\mathcal{O}} = \alpha < \omega^2$, then there is a partial recursive function ϕ such that $\phi \restriction \{k : k <_{\mathcal{O}} a\}$ is an isomorphism from $\{k : k <_{\mathcal{O}} a\} \to \{k : k <_{\mathcal{O}} b\}$.

Claim 1 can be proven as follows: Suppose $\alpha = \omega \cdot i + j$ for some $i, j \in \omega$. Let $n_1, \dots, n_i$ and $m_1, \dots, m_i$ be the notations of $\omega, \dots, \omega \cdot i$ below a and b respectively. Define ϕ such that $\phi(n_l) = m_l$ and naturally extend to the notations of successor ordinals.

Claim 2. If $a, b \in \mathcal{O}$ and $|a|_{\mathcal{O}} = |b|_{\mathcal{O}} = \alpha < \omega^2$, then $D_a = D_b$.

For a proof of Claim 2, assume that f and o_1 are the witnessing functions for a set A being in D_a. Then f and $o_2 = \phi \circ o_1$ witness A being in D_b, where ϕ is the isomorphism in Claim 1.

Finally fix any notation a_0 such that $|a_0|_{\mathcal{O}} = \omega^2$. By Selivanov Theorem, there is $A \in D_{a_0}$ such that for any $B \in \bigcup\{D_a : a <_{\mathcal{O}} a_0\}$, $A \not\equiv_T B$. By Claim 2, diagonalizing against all $\{D_a : a <_{\mathcal{O}} a_0\}$ is as good as diagonalizing against all $\{D_a : |a|_{\mathcal{O}} < \omega^2\}$. Thus this A is what we wanted. □

Since there are uncountably many paths of height $\geq \omega^2$ on $\mathcal{O}$, and $\mathcal{O}$ is countable, there are many paths of height ω^2. We now study the classes D_a when a belongs to some path T on $\mathcal{O}$.

Theorem 4.5 (Ershov).

(a) There is a path $T \subset \mathcal{O}$ with $|T| = \omega^3$ such that $A \in \Delta^0_2$ if and only if $A \in D_a$ for some $a \in T$.

(b) For each path $T \subset \mathcal{O}$, if $|T| < \omega^3$, then there is some Δ^0_2-set A such that A is not in $\bigcup\{D_a : a \in T\}$.

Proof. We just give a proof sketch. For (a), list all Δ_2-sets $\{X_i : i \in \omega\}$ and choose a notation a_i such that $|a_i|_{\mathcal{O}} = \omega^2$ and $X_i \in \mathcal{D}_i$. a_i exists by Theorem 4.3. Note that the listing and choice are done noneffectively. Then $T = \{a_0 +_{\mathcal{O}} \cdots +_{\mathcal{O}} a_n : n \in \omega\}$ is the path that we wanted.

For (b), first observe that for any given notation a_0, we can mimic the proof of Theorem 4.4 to get a notation a_1 with $|a_1|_{\mathcal{O}} = \omega^2$ and a set $A \in D_{a_0 +_{\mathcal{O}} a_1}$ which is not in

$$\bigcup\{D_a : a_0 <_{\mathcal{O}} a \text{ and } |a_0 +_{\mathcal{O}} a|_{\mathcal{O}} < |a_0|_{\mathcal{O}} + \omega^2\}.$$

Suppose that $|T| = \omega^2 \cdot i + \alpha$ for some $i \in \omega$, $\alpha < \omega^2$. If $\alpha \neq 0$, let $a_0 \in T$ be such that $|a_0|_{\mathcal{O}} = \omega^2 \cdot i$; if $\alpha = 0$, then $i \geq 1$ by Theorem 4.4, let $a_0 \in T$ be a notation such that $|a_0|_{\mathcal{O}} = \omega^2 \cdot (i - 1)$. Apply the above observation to a_0, the result follows. □

One can also get the following result about Δ^0_2-sets and long paths through $\mathcal{O}$.

Theorem 4.6 (Ershov).

(a) There is an unbounded path T through $\mathcal{O}$ such that for any Δ^0_2-set A, there is an $a \in T$ such that $A \in D_a$. Furthermore, this T can be chosen recursive in $\mathcal{O}$.

(b) Such a path cannot be Π^1_1.

Proof. Fix an enumeration of Δ_2^0-sets A_n, we build a path T recursively in $\mathcal{O}$ in stages as follows: At stage s, suppose we have had $T_s \in \mathcal{O}$. Let $a_s \in \mathcal{O}$ be such that $|a_s| = \omega^2$ and $A_s \in D_{a_s}$. Let

$$T_{s+1} = \begin{cases} T_s +_{\mathcal{O}} a_s +_{\mathcal{O}} s, & \text{if } s \in \mathcal{O}; \\ T_s +_{\mathcal{O}} a_s, & \text{otherwise.} \end{cases}$$

Clearly $T = \bigcup_s \{a : a <_{\mathcal{O}} T_s\}$ is what we wanted.

For (b), suppose that T is Π_1^1. Define a function $h : \omega \to T$ by $h(n) =$ the $<_{\mathcal{O}}$-least $a \in T$ such that A_n appears in D_a. Then h is a total Π_1^1-function, therefore Δ_1^1. By Theorem 3.9 of Kleene, the range of h is bounded, contradiction. □

5. Applications to Inductive Inference

In inductive inference, transfinite ordinals were initially used to generalize the notion of counters. The general setting is that a learner M — which in the following discussion is always assumed to be recursive — reads from an infinite tape more and more data and eventually outputs a finite sequence of indices of r.e. sets such that for the last index e in this sequence it holds that W_e enumerates exactly all the elements the learner finds on the tape.[16] A class of r.e. sets is explanatorily learnable iff there is a single learner M which learns every set in the class from every possible way in which its elements can be written down on the tape; such tape contents are called texts in learning theory. Not every class is learnable.

Theorem 5.1 (Gold[16]). *The class of the set of all natural numbers plus all of its finite subsets is not explanatorily learnable.*

In his paper, Gold introduced the initial definitions and results of inductive inference which were then the basis for work which spans now already more than four decades. Freivalds and Smith[13] investigated restrictions on the number of mind changes or hypotheses by adding to the learner a counter.[13] This counter takes initially a fixed value $a \in \mathcal{O}$ before starting the learning process and outputting any hypothesis; then, whenever the learner outputs a hypothesis, the current value in the counter, say b, is replaced by a new value c with $c <_{\mathcal{O}} b$; if such a c does not exist, the learner does not output a new conjecture. Note that $b, c \leq_{\mathcal{O}} a$ by induction. Such a learner can be called an a-shot learner, as a bounds the number of attempts the learner can make to output a hypothesis. Note that it is enforced that an a-shot learner outputs only finitely many conjectures. There are learnable classes which cannot be learnt by any a-shot learner for any $a \in \mathcal{O}$. An example of such a class is the class of all finite sets. A characterization for such classes is based on the following notion: Osherson, Stob and Weinstein[22] called a learner M confident iff M converges on every text, even on those texts which are for languages which do not belong to the class to be learnt or which are even not recursively enumerable.

Theorem 5.2 (Ambainis, Freivalds and Smith[2]). *A class $\mathcal{L}$ of r.e. sets is confidently learnable iff there is an $a \in \mathcal{O}$ such that $\mathcal{L}$ is learnable by an a-shot learner.*

Sharma, Stephan and Ventsov[29] considered generalizations of this result and also characterized when a class is learnable by a learner converging on all recursive texts and so on. Freivalds and Smith[13] showed that the hierarchy of learning from ordinal notations does not collapse.

Theorem 5.3 (Freivalds and Smith[13]). *For every notation $a \in \mathcal{O}$ there is a class which can be learnt by an a-shot learner but not by a b-shot learner whenever $b \in \mathcal{O}$ and $|b|_{\mathcal{O}} < |a|_{\mathcal{O}}$.*

Proof. This can be proven by using the class of all sets $L_c = \{x : c \leq_{\mathcal{O}} x <_{\mathcal{O}} a\}$ with $c <_{\mathcal{O}} a$. The learner for this class is quite straight-forward: whenever it has seen a finite set D of data, it computes the minimum c of D with respect to $<_{\mathcal{O}}$ and it conjectures L_c. The ordinal counter is set to c. As any new hypothesis d after c would satisfy that $L_d \supset L_c$ and $d = \min(L_d) <_{\mathcal{O}} c = \min(L_c)$, the ordinal counter is counted down whenever the hypothesis is revised. On the other hand, if the ordinal counter would start with $c = b$ instead of a, one could feed the learner always with data for the L_c until it outputs a conjecture for L_c and counts down the ordinal from c to d; then one starts to feed more data into the learner which are then the data for the set L_d. By this method, one always feeds data for a set the learner is not conjecturing and when it revises the conjecture, the set of the data fed into the learner gets increased. This process can go on only finitely often and will get stuck in the learner conjecturing a different set than from which the data are; hence the learner would not succeed. It follows that the class of all L_c with $c <_{\mathcal{O}} a$ is learnable by an a-shot learner but not by a b-shot learner. □

Note that the class $\{L_c : c <_{\mathcal{O}} a\}$ is even uniformly recursively enumerable. Ordinal counters serve as a measure of complexity in various areas of inductive inference.

There is also a connection to a-r.e. sets. Given an a-r.e. set A with $|a|_\mathcal{O} \geq \omega$ and an approximation f as defined in Definition 4.1, let

$$\begin{aligned} L_{x,y} &= \{\langle x,0\rangle, \langle x,1\rangle, \ldots, \langle x,y\rangle\} \text{ and} \\ \mathcal{L} &= \{L_{x,y} : x, y \in \omega \wedge (y = 0 \vee f(x,y-1) \neq f(x,y))\}. \end{aligned}$$

Now $\mathcal{L}$ is learnable by an a-shot learner. Furthermore, whenever $\mathcal{L}$ is learnable by a b-shot learner then A is a b-r.e. set. The converse does not hold, as one could make the approximation of $A = \emptyset$ having so many changes that it does not even witness that $\emptyset$ is ω-r.e.; the corresponding class has then no ω-shot learner although A is an ω-r.e. set.

Carlucci, Case and Jain[6] called approximations as the f in Definition 4.1 an a-correction-grammar. They considered the learning of classes of r.e. languages where the learner is permitted to output conjectures which are a-correction-grammars for the set to be learnt. In their work, they considered more general learners than before, namely explanatory learners (as already defined above) and behaviourally correct learners. Behaviourally correct learners are more general than explanatory learners: given a text for a set L to be learnt, the behaviourally correct learner outputs an infinite sequence $e_0, e_1, e_2, \ldots$ of a-correction-grammars such that almost all e_n are an a-correction grammar for L; but, in contrast to explanatory learning, no syntactic convergence is required and it is permtitted that there are infinitely many n with $e_{n+1} \neq e_n$. A behaviourally correct learner learns a whole class $\mathcal{L}$ iff it learns every $L \in \mathcal{L}$ this way. Note that behaviourally correct learners are more powerful than explanatory learners. Given an undecidable r.e. set K, for example the halting problem, the class $\{K \cup D : D \text{ is finite}\}$ is behaviourally correctly but not explanatorily learnable. Now Carlucci, Case and Jain[6] investigate how the usage of a-correction-grammars relates to explanatory and behaviourally correct learning.

For explanatory learning, they showed that there is no optimal hypothesis space; more precisely, given ordinal notations $a, b \in \mathcal{O}$ with $a <_\mathcal{O} b$ there is a class which is explanatorily learnable using b-correction grammars but not using a-correction grammars.

In the case of behaviourally correct learning, the situation is different and all hypothesis spaces which permit at least "ω-many" corrections are equivalent and optimal: They showed that for all notations $a, b \in \mathcal{O}$ with $|b|_\mathcal{O} \geq \omega$ and every class $\mathcal{L}$ of r.e. sets, if the class $\mathcal{L}$ is learnable using a-correction-grammars then $\mathcal{L}$ is also learnable using b-correction grammars.

Other applications of ordinals in learning theory refer to measuring how far the hypothesis can go off the minimal index of the target language.[1]

Some papers also abstract from the requirement that learners have to be recursive and can then use ordinals directly in place of their notations.[21]

6. Conclusion

In this paper we reviewed the usage of recursive ordinals in recursion and learning theory. We first reviewed the finite levels of the Ershov hierarchy and the degree structures of these sets. Then we turned to the concept of recursive ordinals and ways to represent them, with a particular emphasis on Kleene's $\mathcal{O}$ and notations of ordinals. The notation a used to represent a given recursive ordinal has a strong impact on which limit-recursive sets are a-r.e. and hence the world of a-r.e. sets is not only a study of recursive ordinals but also of notations of recursive ordinals. We summarize the ground-breaking results of Ershov and show some selected results on transfinite levels of the Ershov hierarchy. At the end we survey the usage of ordinals in inductive inference, where the notion of a recursive ordinal turned out to be a natural method to measure convergence speed and learning complexity.

We end our survey with two open problems. The first one can be viewed as a generalized version of Downey's Conjecture:

Conjecture 6.1. *For all* $a \neq b \in \mathcal{O}$, *if* $\omega \leq |a|_{\mathcal{O}}, |b|_{\mathcal{O}} < \omega^2$ *then* $\mathcal{D}_a \equiv \mathcal{D}_b$.

We know that $\bigcup\{\mathcal{D}_a : a \in \mathcal{O} \wedge |a|_{\mathcal{O}} < \omega^2\}$ is a proper subset of all Δ^0_2-degrees by Theorem 4.4, however, is there any degree theoretic difference?

Conjecture 6.2. $\bigcup\{\mathcal{D}_a : a \in \mathcal{O} \wedge |a|_{\mathcal{O}} < \omega^2\} \equiv \mathcal{D}(\leq \mathbf{0}')$.

Acknowledgments. We would like to thank John Case, Sanjay Jain and Theodore A. Slaman for fruitful discussions and information about the topics of this paper.

References

1. Andris Ambainis, John Case, Sanjay Jain and Mandayam Suraj. Parsimony hierarchies for inductive inference. *The Journal of Symbolic Logic*, 69(1):287–327, 2004.
2. Andris Ambainis, Rūsiņš Freivalds and Carl H. Smith. Inductive inference with procrastination: back to definitions. *Fundamenta Informaticae*, 40(1): 1–16, 1999.
3. Marat M. Arslanov. The Ershov hierarchy. *Computation and Logic in the Real World*, S.B. Cooper and A. Sorbi, eds., to appear.

4. Marat M. Arslanov, Iskander Sh. Kalimullin and Steffen Lempp. On Downey's conjecture. Preprint, 2005.
5. Christopher J. Ash and Julia Knight. *Computable structures and the hyperarithmetical hierarchy*, volume 144 of *Studies in Logic and the Foundations of Mathematics*. North-Holland Publishing Co., Amsterdam, 2000.
6. Lorenzo Carlucci, John Case and Sanjay Jain. Learning Correction Grammars. *Proceedings of the 20th Annual Conference on Computational Learning Theory*, COLT 2007, *Springer Lecture Notes in Artificial Intelligence*, 4539:203–217, 2007.
7. S. Barry Cooper. *Degrees of Unsolvability*. PhD thesis, Leicester University, 1971.
8. S. Barry Cooper, Leo Harrington, Alistair H. Lachlan, Steffen Lempp and Robert I. Soare. The d.r.e. degrees are not dense. *Annals of Pure and Applied Logic*, 55(2):125–151, 1991.
9. Rodney G. Downey. *D*-recursively enumerable degrees and the nondiamond theorem. *Bulletin of the London Mathematical Society*, 21:43–50, 1989.
10. Yuri L. Ershov. A certain hierarchy of sets. I. *Algebra i Logika*, 7(1):47–74, 1968.
11. Yuri L. Ershov. A certain hierarchy of sets. II. *Algebra i Logika*, 7(4):15–47, 1968.
12. Yuri L. Ershov. A certain hierarchy of sets. III. *Algebra i Logika*, 9:34–51, 1970.
13. Rūsiņš Freivalds and Carl H. Smith. On the role of procrastination in machine learning. *Information and Computation*, 107(2):237–271, 1993.
14. Amy Gale and Rod Downey. On genericity and Ershov's hierarchy. *Mathematical Logic Quarterly*, 47(2):161–182, 2001.
15. E. Mark Gold. Limiting recursion. *The Journal of Symbolic Logic*, 30(1):28–48, March 1965.
16. E. Mark Gold. Language identification in the limit. *Information and Control*, 10:447–474, 1967.
17.
17. Joseph Harrison. Recursive pseudo well-orderings. *Transactions of the American Mathematical Society*, 131:526–543, 1968.
18. Bakhadyr Khoussainov, Frank Stephan and Yue Yang. Computable categoricity and the Ershov hierarchy. to appear in *Annals of Pure and Applied Logic*.
19. Stephen C. Kleene. On the forms of the predicates in the theory of constructive ordinals. II. *The American Journal of Mathematics*, 77:405–428, 1955.
20. Alistair H. Lachlan. Lower bounds for pairs of recursively enumerable degrees. *Proceedings of the London Mathematical Society (3)*, 16:537–569, 1966.
21. Eric Martin, Arun Sharma and Frank Stephan. On ordinal VC-dimension and some notions of complexity. *Theoretical Computer Science*, 364:62-76, 2006.
22. Daniel Osherson, Michael Stob and Scott Weinstein. *Systems That Learn, An Introduction to Learning Theory for Cognitive and Computer Scientists*. Bradford — The MIT Press, Cambridge, Massachusetts, 1986.

23. Hillary Putnam. Trial and error predicates and the solution to a problem of Mostowski. *The Journal of Symbolic Logic*, 30(1):49–57, March 1965.
24. Hartley Rogers, Jr. *Theory of Recursive Functions and Effective Computability.* McGraw-Hill, New York, 1967.
25. Gerald E. Sacks. *Higher Recursion Theory.* Perspectives in Mathematical Logic. Springer–Verlag, Heidelberg, 1990.
26. Gerald E. Sacks. A minimal degree below $\mathbf{0}'$. *Bulletin of the American Mathematical Society*, 67:416–419, 1961.
27. Gerald E. Sacks. The recursively enumerable degrees are dense. *Annals of Mathematics*, 80:300–312, 1964.
28. Victor L. Selivanov. Ershov's hierarchy and T-jumps. *Algebra i Logika*, 27(4):464–478, 498–499, 1988.
29. Arun Sharma, Frank Stephan and Yuri Ventsov. Generalized notions of mind change complexity. *Information and Computation*, 189:235-262, 2004.
30. Theodore A. Slaman. The recursively enumerable degrees as a substructure of the Δ_2^0-degrees. Hand written notes, 1983.
31. Yue Yang and Liang Yu. On Σ_1-structural differences among finite levels of the Ershov hierarchy. *The Journal of Symbolic Logic*, 71(4):1223–1236, 2006.

Weak randomness, genericity and Boolean decision trees

Toshio Suzuki

Department of Mathematics and Information Sciences,
Tokyo Metropolitan University,
Minami-Ohsawa, Hachioji, Tokyo 192-0397, Japan
Corresponding author. E-mail: toshio-suzuki@tmu.ac.jp

Masahiro Kumabe

Faculty of Liberal Arts, The Open University of Japan,
31-1, Ōoka 2, Minami-ku, Yokohama 232-0061, Japan
E-mail: kumabe@u-air.ac.jp

We regard a sequence of copies of a Boolean decision tree as a mapping of a given oracle X (of leaves) to the other oracle Y (of roots), and we ask which weak-randomness property (of the leaf oracle) is conserved through the mapping. As is easily seen, each of the following properties (1), (2) of X is not necessarily shared by Y. (1) "Being Martin-Löf random", (2) "Being partially Martin-Löf random". We show that the following property (3) of X is conserved through the mapping of X to Y. (3) "Being r-generic in the sense of [Dowd 1992] (*Inform. and comput.*) for every positive integer r" (that is, for each r, there exists a polynomial that bounds the minimum sizes of Cohen-type forcing conditions associated with the r-query tautologies with respect to X; this property is different from r-genericity of arithmetical forcing). It is known that the class $\mathcal{D}$ of all oracles having property (3) has Lebesgue measure one [Dowd 1992]. We show that every Martin-Löf random oracle (more precisely, every Schnorr random oracle) has property (3). And, we show that $\mathcal{D}$ is closed under PTIME bounded-truth-table reductions having both of the following properties: (i) Different inputs correspond to disjoint sets of queries, and (ii) Every truth table is a Boolean function onto $\{0, 1\}$.

Mathematics Subject Classification: 68Q15, 03D15.

Keywords: Martin-Löf randomness; Dowd-type generic oracle; Boolean decision tree.

1. Introduction

The concept of Boolean decision tree is important in game theory and artificial intelligence. Since Yao[1] applies von Neuman's mini-max theorem to complexity of tree algorithms, many researchers have studied computational complexity issues of randomized Boolean decision trees, where a randomized Boolean decision tree denotes a Boolean decision tree having a probability distribution on the truth values of the leaves.

In a randomized Boolean decision tree, each leaf has value 1 (throughout the paper, 1 denotes true and 0 denotes false) with certain probability. According to the truth values of leaves, the truth value of the root is probabilistically determined.

Among literatures on complexity of such trees, the paper of Liu and Tanaka[2] discusses a special distribution, called eigen-distribution, on random assignments for a class of binary AND-OR game trees. They study the issue in two separate cases: (1) the case where the assignments to leaves are independently distributed (ID); and (2) the case where the assignments to leaves are correlatively distributed (CD).

In the present paper, we are interested in an "algorithmically random" version of the above case (1). Instead of a probability distribution on the truth values of the leaves, we consider a fixed oracle that is algorithmically random, namely an oracle that is Martin-Löf random, and an oracle satisfying a certain necessary condition for Martin-Löf randomness.

In order to explain our setting, we review the concept of k-round tree T_2^k discussed in Ref. 2, where the subscript "2" denotes "binary". Throughout the present paper, each AND gate ($\wedge$) of a tree has two child nodes, and so does an OR gate ($\vee$). The tree T_2^1, the binary AND-OR tree of round 1, is a tree whose (unique) root is an AND-gate, and each child of the root is OR-gate, and each child of an OR gate is a leaf. The tree T_2^{k+1}, the binary AND-OR tree of round $k+1$, is defined through replacing every leaf of T_2^1 by T_2^k.

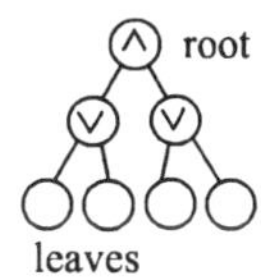

Fig. 1. T_2^1

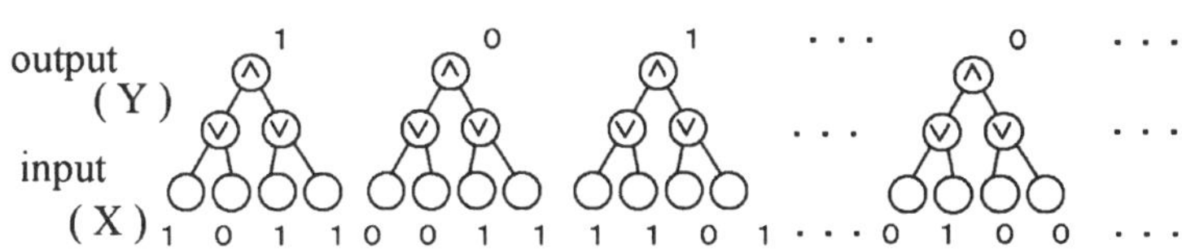

Fig. 2. a system consisting of infinitely many copies of T_2^1

We consider a system consisting of infinite copies of T_2^k. Fig. 2 illustrates the case where $k = 1$. A subset of $\omega = \{0, 1, 2, \cdots\}$ is called an *oracle*. An oracle is identified with its characteristic function. Thus, we may regard an oracle X as an infinite binary sequence $X(0)X(1)X(2)\cdots$. Given an oracle

X, we assign X to the leaves of the system in such a way that, for each $i \in \omega$, the leaves of the $(i+1)$st T_2^k from the left are assigned $X(4^k i)$, $X(4^k i+1)$, $\cdots$, and $X(4^k(i+1)-1)$, respectively, in this order. Then, by calculating the Boolean function, we get an infinite sequence $Y(0)Y(1)Y(2)\cdots$ of the root values. More precisely, for each $i \in \omega$, $Y(i)$ denotes the root value of the $(i+1)$st T_2^k from the left. In such a way, we regard a sequence of copies of a Boolean decision tree as a mapping of a given oracle X (of leaves) to the other oracle Y (of roots).

Now, which weak-randomness property (of the leaf oracle) is conserved through such a mapping?

Each of the following properties (1), (2) of X is not necessarily shared by Y. (1) "Being Martin-Löf 1-random", (2) "Being Martin-Löf s-random for some s such that $0 < s \leq 1$". Informally speaking, the reason why Martin-Löf randomness is not conserved is because probability of having value 1 at the root of T_2^k is not $1/2$ when each leaf has probability $1/2$. We observe the reason more precisely in section 2.

In the present paper, we prove that the following property (3) of X is conserved through such a mapping of X to Y. (3) "Being r-generic in the sense of Dowd[3] for every positive integer r". More precisely, (3) means: "For each r, there exists a polynomial that bounds the minimum sizes of Cohen-type forcing conditions associated with r-query tautologies with respect to X." This property is different from r-genericity of arithmetical forcing. We review definition of Dowd-type genericity in section 2.

It is known that the class $\mathcal{D}$ of all oracles having property (3) has Lebesgue measure one.[3] In section 3, we show that every Martin-Löf random oracle (more precisely, every Schnorr random oracle) has property (3).

In section 4, we show that $\mathcal{D}$ is closed under PTIME bounded-truth-table reductions having both of the following properties: (i) Different inputs correspond to disjoint sets of queries, and (ii) Every truth table is an onto Boolean function $\{0,1\}^n \to \{0,1\}$, where n is the norm. As a special case of this result, it is proved that property (3) of X is conserved through the mapping defined by means of copies of T_2^k.

In section 4, we show elementary examples of such reductions. An advanced example is shown in section 5.

Strictly speaking, property (3) is not "randomness", because the notion (3) is "not stochastic" in the following sense. There exists an oracle X such that X has property (3) and that the following holds.

$$\lim_{n\to\infty} \frac{X(0)+X(1)+\cdots+X(n-1)}{n} \neq \frac{1}{2}.$$

More generally, for every bit string $u = u_1 \cdots u_m$, there exists an oracle X such that X has property (3) and that the following holds, which is shown in section 5.

$$\lim_{n\to\infty} \frac{X(0)+X(1)+\cdots+X(n-1)}{n} = \sum_{i=1}^{m} \frac{u_i}{2^i}.$$

The present paper is a refinement of the following talks: [Refs. 4–6].

2. Notation and basic concepts

2.1. *Strings and sets*

$\mathbb{N}_+$ denotes the set of all positive integers, that is, $\mathbb{N}_+ = \omega - \{0\}$. We let $\{0,1\}^*$ denote the set of all finite binary strings. For each natural number n, we let $\{0,1\}^n$ denote the set of all binary strings of length n. We represent the empty string by λ. For each natural number n, we let $z(n)$ denote the $(n+1)$st string in length-lexicographic order. For example, $z(0) = \lambda, z(1) = 0, z(2) = 1, z(3) = 00, z(4) = 01$. For each string u, we represent its length by $|u|$. For a set A, we let $|A|$ denote the cardinality of A. We believe that this abuse of the symbol "| |" does not confuse the reader. The set of all oracles is called the *Cantor space*, and denoted by $\{0,1\}^\omega$.

For a function f, we denote its domain by dom$f$. For a subset A of dom$f$, we let $f \restriction A$ denote the restriction of f to A, that is, its domain is A and for every $x \in A$, $(f \restriction A)(x) = f(x)$. When we apply this notation to the case where f is an oracle X, we regard X as its characteristic function.

2.2. *Martin-Löf randomness*

We review concepts of algorithmically random sets, in particular, Martin-Löf randomness[7] and Schnorr randomness.[8]

Definition 2.1. [9]

(1) [9, Def.5.9 (p.109)] A computably enumerable set $V \subseteq \{0,1\}^* \times \mathbb{N}_+$ is called *Martin-Löf test* if the following two properties hold true:

(1a) $V_{m+1} \subseteq V_m$, for all $m \geq 1$, where V_m denotes $\{x \in \{0,1\}^* : (x, m) \in V\}$.

(1b) $|\{0,1\}^n \cap V_m| < 2^{n-m}$, for all $n \geq m \geq 1$.

(2) [9, Def.6.12 (p.158)] A Martin-Löf test $V \subseteq \{0,1\}^* \times \mathbb{N}_+$ is called *sequential* if it satisfies the additional sequential property:

For every positive integer m, if $x \in V_m$ and x is a prefix (an initial segment) of y, then $y \in V_m$.

(3) [9, Def.6.25 (p.169)] An oracle X is called *random* if for every sequential Martin-Löf test V, there exists m such that no initial segment of X belongs to V_m.

Definition 2.2. [10, Def.3.1] A collection of sets $\mathcal{A} \subseteq \{0,1\}^\omega$ is *Martin-Löf null* if there is a uniformly c.e. sequence $\{U_i\}_{i\in\omega}$ of Σ^0_1-classes (called a *Martin-Löf test*) such that $\forall i \in \omega(\mu(U_i) \leq 2^{-i})$ and $\mathcal{A} \subseteq \bigcap_i U_i$. An oracle A is *Martin-Löf random*, or *1-random*, if $\{A\}$ is not Martin-Löf null.

In the above definition, "uniformly c.e. sequence $\{U_i\}_{i\in\omega}$ of Σ^0_1-classes" means that there exists an oracle Turing machine M, which may not halt for some input strings, such that for every oracle X and for every natural number i, the following holds.

$$X \in U_i \text{ if and only if } M^X(i) = 1. \tag{1}$$

It is known that an oracle A is Martin-Löf random if and only if it is random in the sense of Definition 2.1.

Tadaki[11] investigates partial Martin-Löf randomness. This concept is defined similarly to the original Martin-Löf randomness, but is defined by means of a certain special kind of measure, instead of the usual Lebesgue measure.

Definition 2.3. [11] (see also [10, Def.15.8]) Suppose that s is a real number such that $0 < s \leq 1$.

(1) Suppose that V is an open set of the Cantor space. We think of V as a prefix-free collection of strings. Then, we define its s-measure $\mu_s(V)$ as follows.

$$\mu_s(V) = \sum_{u\in V} 2^{-s|u|}.$$

Thus, 1-measure is the usual Lebesgue measure.

(2) A uniformly c.e. sequence $\{U_i\}_{i\in\omega}$ of Σ^0_1-classes is called a *weak Martin-Löf s-test* if $\forall i \in \omega(\mu_s(U_i) \leq 2^{-i})$.

(3) An oracle A is *weak Martin-Löf s-random*, if $A \notin \bigcap_i U_i$ for every weak Martin-Löf s-test $\{U_i\}_{i\in\omega}$. Thus, weak Martin-Löf 1-randomness is equivalent to the usual Martin-Löf randomness.

Example 2.1. [10, p.483] Suppose that an oracle X is Martin-Löf random. Let Y be the oracle defined as the following infinite sequence.

$$X(0)\ 0\ X(1)\ 0\ X(2)\ 0\ X(3) \cdots X(n)\ 0\ X(n+1)\ 0 \cdots$$

Then, Y is weak Martin-Löf (1/2)-random.

Definition 2.4. [12, p.396] An oracle A is called *Schnorr random* if for every Martin-Löf test $\{U_i\}_{i \in \omega}$ such that $\forall i \in \omega(\mu(U_i) = 2^{-i})$, there exists i such that $A \notin U_i$.

It is immediate that every Martin-Löf random oracle is Schnorr random.

2.3. *Probability and density*

For an oracle X, we discuss its "density"

$$\lim_{n \to \infty} \frac{X(0) + X(1) + \cdots + X(n-1)}{n}.$$

It is not hard to see that there exists an oracle X for which the limit diverges, and that if an oracle X is randomly chosen then with probability 1, the limit converges to $1/2$.

Proposition 2.1. *[13, Lemma 11.9] (The Chernoff Bound) Suppose that* $x_1, \cdots$ *and* x_n *are independent random variables taking the values 1 and 0 with probabilities* p *and* $1-p$*, respectively, and consider their sum* $X = \sum_{i=1}^{n} x_i$*. Then for all* $0 \leq \theta \leq 1$, $\text{Prob}[X \geq (1+\theta)pn] \leq \exp(-\theta^2 pn/3)$.

Example 2.2. [9] By using Proposition 2.1, it is shown that the following class $\mathcal{S}_1$ of oracles is Martin-Löf null (in the sense of Def. 2.2).

$$\mathcal{S}_1 = \{X : \lim_{n \to \infty} \frac{X(0) + X(1) + \cdots + X(n-1)}{n} \neq \frac{1}{2}\}.$$

More formally, $\mathcal{S}_1$ is the class of all oracles X such that either the limit diverges or the limit converges to a value not equal to $1/2$. Now, since $\mathcal{S}_1$ is Martin-Löf null, it holds that for every Martin-Löf oracle X, we have:

$$\lim_{n \to \infty} \frac{X(0) + X(1) + \cdots + X(n-1)}{n} = \frac{1}{2}.$$

Example 2.3. By using an argument similar to Example 2.2, it is shown that the following class $\mathcal{S}_2$ of oracles is Martin-Löf null.

$$\mathcal{S}_2 = \{X : \lim_{n \to \infty} \frac{X(0)X(1) + X(1)X(2) + \cdots + X(2n-2)X(2n-1)}{n} \neq \frac{1}{4}\}.$$

Therefore, for every Martin-Löf oracle X, it holds that

$$\lim_{n \to \infty} \frac{X(0)X(1) + X(1)X(2) + \cdots + X(2n-2)X(2n-1)}{n} = \frac{1}{4}.$$

Martin-Löf randomness is not conserved through the mapping defined in section 1 by means of T_2^k. The property of "being weak Martin-Löf s-random for some s $(0 < s \leq 1)$" is not conserved through the mapping, either.

Proposition 2.2. *Let f be the mapping of X (the leaf oracle) to Y (the root oracle) introduced in section* 1 *by means of infinite many copies of T_2^k, the k-round binary AND-OR tree. Let $k = 1$.*

(1) Suppose that an oracle X is Martin-Löf random. Let $Y = f(X)$. Then, Y is not Martin-Löf random.
(2) There exists an oracle X such that X is weak Martin-Löf $(1/2)$-random and such that for every s' $(0 < s' \leq 1)$, $Y = f(X)$ is not weak Martin-Löf s'-random.

Proof. (1) Since X is Martin-Löf random, by using an argument similar to Example 2.3, it holds that

$$\lim_{n\to\infty} \frac{Y(0) + X(1) + \cdots + Y(n-1)}{n} = \frac{9}{16}.$$

Hence, by Example 2.2, Y is not Martin-Löf random.

(2) Let A be a Martin-Löf random oracle. Let X be the oracle defined as the following infinite sequence.

$$A(0)\ A(1)\ 0\ 0\ A(2)\ A(3)\ 0\ 0 \cdots A(n)\ A(n+1)\ 0\ 0 \cdots$$

Similarly to Example 2.1, X is weak Martin-Löf (1/2)-random. Let $Y = f(X)$. Then, for all natural number n, we have $Y(n) = 0$. Hence, for every s' $(0 < s' \leq 1)$, Y is not weak Martin-Löf s'-random. □

2.4. *Dowd-type generic oracle*

We have seen that partial Martin-Löf randomness is not conserved through the mapping of X to Y introduced section 1. In this section, we review the concept of Dowd-type genericity, which is a necessary condition for Martin-Löf randomness (see section 3), but is different from partial Martin-Löf randomness.

Dowd-type generic oracle is introduced by Dowd[3] in his study of NP =?coNP question (A certain conjecture on Dowd-type generic oracle implies the unrelativized conjecture NP $\neq$ coNP). In the definition of this genericity concept, the concept of forcing condition is the same as arithmetical forcing, that is, Cohen-type forcing condition. However, the concept of forcing relation is much simpler than that in arithmetical forcing. In the study of arithmetical forcing, a certain property of an oracle is determined by its finite portion. On the other hand, in the current setting, a certain property of a forcing condition (of exponential size) is determined by its polynomial-sized portion. Dowd introduced a system of Boolean formulas with query symbols, called the relativized propositional calculus, by which he represents a certain property of an oracle.

A function is called *a forcing condition* if its domain is a finite subset of $\{0,1\}^*$ and its range is a subset of $\{0,1\}$. By adding a set of query symbols $\{\xi^1, \xi^2, \xi^3, \cdots\}$ to the usual propositional calculus, we get *the relativized propositional calculus.*[3] For each n, the symbol ξ^n is an n-ary connective. If $q_1, q_2, \cdots, q_n$ are propositional variables, the interpretation of $\xi^n(q_1, q_2, \cdots, q_n)$ is, roughly speaking, given by "the string $q_1 \cdots q_n$ belongs to a given oracle". For technical reason, for an oracle A and each positive integer n, we define an n-ary Boolean function A^n so that the relation

$$A^{n+1}(0, x_1, \cdots, x_n) = A^n(x_1, \cdots, x_n)$$

holds for every $x_1, \cdots, x_n \in \{0,1\}$. Then, we interpret ξ^n as A^n. More precisely, we define A^n as follows. For each $j < 2^n$,

$$A^n(\text{ the } (j+1)\text{st } n\text{-bit string }) = A(\text{ the } (j+1)\text{st string }).$$

Thus, for example,

$$\begin{array}{llll} A^3(000) = A(\lambda), & A^3(001) = A(0), & A^3(010) = A(1), & A^3(011) = A(00), \\ A^3(100) = A(01), & A^3(101) = A(10), & A^3(110) = A(11), & A^3(111) = A(000). \end{array}$$

When a formula F of the relativized propositional calculus and an oracle X are given, the concept of "F is a tautology with respect to X" is naturally defined. If S is a forcing condition and F is a tautology with respect to every oracle X that is an extension of S, then we say that S *forces* F. Suppose that r is a positive integer. An r-query formula is, roughly speaking, a formula that has just r-many occurrences of query symbols. More precisely, a formula of the relativized propositional calculus is called *an r-query formula* if it is of the following form.

$$[(q_0^{(1)} \Leftrightarrow \xi^n(q_1^{(1)}, \cdots, q_n^{(1)}) \wedge \cdots \wedge (q_0^{(r)} \Leftrightarrow \xi^n(q_1^{(r)}, \cdots, q_n^{(r)}))] \Rightarrow H,$$

where each of $q_j^{(i)}$ is a propositional variable and H is a query free formula. H may have occurrences of each of $q_j^{(i)}$, and may have occurrences of other propositional variables.

For a relativized formula F, we let $\ell(F)$ denote the length of F. More formally, $\ell(F)$ denotes the total number of occurrences of propositional variables, constants (0 and 1), logical connectives, query symbols (ξ^n), punctuation marks and parentheses. For example, if F is

$$\big(q_0 \Leftrightarrow \xi^3(q_1, q_2, q_3)\big) \Rightarrow \big((q_1 \wedge q_2) \vee (q_0 \wedge q_3)\big)$$

then $\ell(F) = 26$.

An oracle D is called *an r-Dowd oracle* if there exists a polynomial p of the following property: For every r-query formula F that is a tautology with respect to D, there exists a forcing condition S such that S is a sub-function of D, the cardinality of the domain of S is at most $p(\ell(F))$ and such that S forces F. It is known that for every positive integer r, the class of all r-Dowd oracles has Lebesgue measure one (sketch of a proof is in Ref. 3, and more rigorous proofs are in Refs. 14,15). Though Dowd used the terminology "an r-generic oracle" to denote an r-Dowd oracle, the concept of r-Dowd oracles is different from the concept of r-generic oracles in arithmetical forcing.

For more formal treatment of Dowd-type genericity, see Refs. 15,16.

3. ML randomness and Dowd-type genericity

In this section, we show the following theorem.

Theorem 3.1. *Suppose that X is Martin-Löf random, and that r is a positive integer. Then, X is r-Dowd.*

In the paper [17, p.38–39], for each $r \geq 2$, we defined a class $r\mathrm{GEN}_1$ of oracles which are r-Dowd in a weak sense. The class is the collection of all oracles X of the following property: there exists a polynomial p such that for every "nice" r-query formula F, there exists a sub-function S of X such that the size of S (the cardinality of its domain) is at most $p(\ell(F))$ ($\ell(F)$ is the length of F, see sub-section 2.4), and S "partially forces" F.

We use the class in our proof of the theorem. The class is easier to analyze than the class of all r-Dowd oracles. The class $r\mathrm{GEN}_1$ is defined by means of countable many sets, $r\mathrm{GEN}_1(k, c, n)$, of forcing conditions, where k, c, n are natural numbers. For each k, c, n, every element of $r\mathrm{GEN}_1(k, c, n)$ is a forcing condition whose domain is $\{z(0), z(1), \cdots, z(2^n - 1)\}$, that is, the first 2^n strings in length-lexicographic order. If a forcing condition, whose domain is as above, is randomly chosen, then the probability of not belonging to $r\mathrm{GEN}_1(k, c, n)$ is small enough; an upper bound for the probability is given in Lemma 3.1 of this section. In [17, Theorem 4.5], it is shown that $r\mathrm{GEN}_1$ is equal to the class of all r-Dowd oracles.

Now, we define the above terminologies formally. We begin with defining "nice r-query formula".

Definition 3.1. [14,17]

(1) Suppose that n is a positive integer and that each of $\alpha_1, \cdots, \alpha_n$ is either a propositional variable or a constant (0 or 1). Then, the formula

$$\xi^n(\alpha_1, \cdots, \alpha_n)$$

is called a *query.*

(2) Two queries $\xi^n(\alpha_1, \cdots, \alpha_n)$ and $\xi^n(\beta_1, \cdots, \beta_n)$ are *disjoint* if there exists $i \in \{1, \cdots, n\}$ such that both α_i and β_i are constants and that $\alpha_i \neq \beta_i$. Note that if they are disjoint then the following sets of strings are disjoint.

$$\{x_1 \cdots x_n \in \{0,1\}^n : \forall i\,(\alpha_i \text{ is a constant} \to x_i = \alpha_i)\}, \text{ and}$$
$$\{x_1 \cdots x_n \in \{0,1\}^n : \forall i\,(\beta_i \text{ is a constant} \to x_i = \beta_i)\}.$$

(3) Suppose that $r \geq 2$. A relativized formula F is called *a nice r-query formula* if it is of the form

$$[(q_0^{(1)} \Leftrightarrow \xi^n(\cdots)) \wedge \cdots \wedge (q_0^{(r)} \Leftrightarrow \xi^n(\cdots))] \Rightarrow H,$$

where each of $q_0^{(1)}, \cdots, q_0^{(r)}$ is a propositional variable, H is a query-free formula, each $\xi^n(\cdots)$ is a query, and the queries are pairwise disjoint.

Example 3.1. Let F_1 be the following formula.

$$[(q_0^{(1)} \Leftrightarrow \xi^3(0,0,q_3^{(1)})) \wedge (q_0^{(2)} \Leftrightarrow \xi^3(0,1,q_3^{(2)})) \wedge (q_0^{(3)} \Leftrightarrow \xi^3(1,0,q_3^{(3)}))]$$
$$\Rightarrow [(q_0^{(1)} \vee q_0^{(2)}) \wedge q_0^{(3)}].$$

In addition, let F_2 be the following formula.

$$[(q_0^{(1)} \Leftrightarrow \xi^3(0,0,q_3^{(1)})) \wedge (q_0^{(2)} \Leftrightarrow \xi^3(0,1,q_3^{(2)})) \wedge (q_0^{(3)} \Leftrightarrow \xi^3(q_3^{(1)},1,q_3^{(3)}))]$$
$$\Rightarrow [(q_0^{(1)} \vee q_0^{(2)}) \wedge q_0^{(3)}].$$

Then, F_1 is a nice 3-query formula. On the other hand, F_2 is not a nice 3-query formula because the second query and the third query are not disjoint.

Next, we define the concept of "to force partially".

Definition 3.2. Suppose that n is a positive integer and that X is a forcing condition.

(1) In sub-section 2.4, for each oracle A, we define the n-ary function A^n. In the same way, we define an n-ary Boolean (partial) function X^n. That is, For each $j < 2^n$ such that $z(j) \in \mathrm{dom}X$,

$$X^n(\text{ the } (j+1)\text{st } n\text{-bit string }) = X(\text{ the } (j+1)\text{st string }).$$

(2) For each string u of length n, letting $u = u_1 \cdots u_n$, we define meta-symbols $*_{u,1}, \cdots, *_{u,n}$ as follows. For each $j = 1, \cdots, n$,

$$*_{u,j} \text{ denotes } \begin{cases} \text{(empty expression), if } u_j = 1, \\ \neg \text{ (negation symbol), otherwise.} \end{cases}$$

In addition, if $u \in \mathrm{dom}X^n$, we define a meta-symbol and $*_u$ as follows.

$$*_u \text{ denotes } \begin{cases} \text{(empty expression), if } X^n(u) = 1, \\ \neg \text{ (negation symbol), otherwise.} \end{cases}$$

(3) Suppose that q is a propositional variable, $\xi^n(\alpha_1, \cdots, \alpha_n)$ is a query in the sense of Definition 3.1 (1). Let A be the set of all strings u such that $|u| = n$ and such that the formula $*_{u,1}\ \alpha_1 \wedge \cdots \wedge *_{u,n}\ \alpha_n$ is satisfiable. Assume that $A \subseteq \mathrm{dom}X^n$. Then, we let meta-symbol

$$[\![q \Leftrightarrow \xi^n(\alpha_1, \cdots, \alpha_n)]\!]^X$$

denote the conjunction of all the formulas of the form

$$(*_{u,1}\ \alpha_1 \wedge \cdots \wedge *_{u,n}\ \alpha_n) \Rightarrow *_u\ q,$$

where $u \in A$.

(4) Suppose that $r \geq 2$ is a natural number and that F is a nice r-query formula of the form

$$[(q_0^{(1)} \Leftrightarrow \xi^n(\alpha_1^{(1)}, \cdots, \alpha_n^{(1)})) \wedge (q_0^{(2)} \Leftrightarrow \xi^n(\alpha_1^{(2)}, \cdots, \alpha_n^{(2)})) \wedge \cdots$$
$$\wedge\ (q_0^{(r)} \Leftrightarrow \xi^n(\alpha_1^{(r)}, \cdots, \alpha_n^{(r)}))] \Rightarrow H.$$

In addition, suppose that S is a forcing condition and that the domain of X^n contains all strings relevant to the queries of F from the second query onward. Then, the sentence "*S partially forces F with respect to X*" denotes that S forces the following 1-query formula.

$$[(q_0^{(1)} \Leftrightarrow \xi^n(\alpha_1^{(1)}, \cdots, \alpha_n^{(1)})) \wedge [\![q_0^{(2)} \Leftrightarrow \xi^n(\alpha_1^{(2)}, \cdots, \alpha_n^{(2)})]\!]^X \wedge \cdots$$
$$\wedge\ [\![q_0^{(r)} \Leftrightarrow \xi^n(\alpha_1^{(r)}, \cdots, \alpha_n^{(r)})]\!]^X] \Rightarrow H.$$

Note that each of the following predicates $\Psi_1(F)$ and $\Psi_2(S, F, X)$ is primitive recursive.

$\Psi_1(F) \equiv$ "F is a nice r-query formula",

$\Psi_2(S, F, X) \equiv$ "S partially forces F with respect to X".

Remark: For more formal treatments of Definition 3.1 and Definition 3.2, see [17, §4.2], in which "a nice r-query formula" is called "a disentangled r-query formula", and the technical term "partially forces" is not explicitly used. In Ref. 15, a symbol $r\mathrm{DO}_1$ is used instead of $r\mathrm{GEN}_1$. The motive for these names are "r-**D**owd **o**racle" and "r-**gen**eric oracle in the sense of Dowd", respectively.

Definition 3.3. [17, p.38–39] Suppose that $r \geq 2$.

(1) Suppose that F is a relativized formula having at least one occurrence of a query symbol. Then, the dimension of F denotes the maximum natural number n such that the query symbol ξ^n has at least one occurrence in F. The dimension is denoted by $\dim F$.
(2) For each positive integer n, we let $\mathrm{Func}(n)$ denote the set of all functions from $\{z(0), z(1), \cdots, z(2^{n-1})\}$ to $\{0, 1\}$.
(3) For each positive integers r, k, c, n, we let $r\mathrm{GEN}_1(k, c, n)$ denote the set of all functions $S \in \mathrm{Func}(n)$ with the following property: For every r-query formula F such that

$$\begin{cases} \ell(F) \geq c, \quad \dim F = n, \quad F \text{ is nice, and} \\ F \text{ is a tautology with respect to S,} \end{cases}$$

there exists a function S' such that

$$\begin{cases} S' \text{ is a sub-function of } S, \quad \left|\mathrm{dom} S'\right| \leq \ell(F)^k, \text{ and} \\ S' \text{ partially forces } F \text{ with respect to } S, \end{cases}$$

where $\ell(F)$ is the length of F (see sub-section 2.4).
(4) For each positive integers r, k, c, we let $r\mathrm{GEN}_1(k, c)$ denote the set of all oracles X such that

$$\forall n \geq 1, X \upharpoonright \{z(0), z(1), \cdots, z(2^n - 1)\} \in r\mathrm{GEN}_1(k, c, n),$$

where the symbol $\upharpoonright$ denotes restriction of a function to a given subset of its domain (see sub-section 2.1).
(5) We define $r\mathrm{GEN}_1$ as follows, where the union extends over all positive integers k and c.

$$r\mathrm{GEN}_1 = \bigcup_{k,c \geq 1} r\mathrm{GEN}_1(k, c).$$

Lemma 3.1. *[17, p.41] Suppose that r, k, c and n are positive integers. Suppose that $r \geq 2$, and that k, c are large enough. Let N denote 2^n. Then, the following holds.*

$$\left|\text{Func}(n) - r\text{GEN}_1(k,c,n)\right| \leq 2^{N-(n+2)} \tag{2}$$

Lemma 3.2. *[17, Theorem 4.5] For each $r \geq 2$, the set of all r-Dowd oracles is equal to $r\text{GEN}_1$.*

In the rest of this section, we fix a natural number $r \geq 2$, and fix natural numbers k and c which are large enough.

Definition 3.4. For each positive integer n, let U_n be the class of all oracles X with the following property (see sub-section 2.1 for the symbol $\upharpoonright$).

$$\exists m \geq n \ \big(X \upharpoonright \{z(0), z(1), \cdots, z(2^m - 1)\} \notin r\text{GEN}_1(k,c,m)\ \big).$$

Lemma 3.3. *For each positive integer n, the following hold.*

(1) $U_{n+1} \subseteq U_n$.
(2) $\mu(U_n) \leq 2^{-(n+1)}$, where μ denotes Lebesgue measure.

Thus, $\{U_n\}_{n\in\omega}$ is a Martin-Löf test in the sense of Definition 2.2.

Proof. (1) is obvious.

(2) By (2), the following holds.

$$\mu(U_n) \leq \sum_{m=n}^{\infty} 2^{-(m+2)} = 2^{-(n+2)} \frac{1}{1-\frac{1}{2}} = 2^{-(n+1)}. \qquad \square$$

Proof of Theorem 3.1. For an oracle X, the following are equivalent.

(i) $\forall n \geq 1\ X \in U_n$.
(ii) $\forall n \geq 1\, \exists m \geq n \ \big(X \upharpoonright \{z(0), z(1), \cdots, z(2^m - 1)\} \notin r\text{GEN}_1(k,c,m)\ \big)$.

Now, suppose that X is Martin-Löf random. Then, by Lemma 3.3, (i) does not hold. Therefore, (ii) does not hold. Hence, there exists a positive integer n with the following property.

$$\forall m \geq n \ \big(X \upharpoonright \{z(0), z(1), \cdots, z(2^m - 1)\} \in r\text{GEN}_1(k,c,m)\ \big).$$

We fix such n. Then, for every r-query formula F such that

$$\begin{cases} \ell(F) \geq c, \quad \dim F \geq n, \quad F \text{ is nice, and} \\ F \text{ is a tautology with respect to } X \upharpoonright \{z(0), z(1), \cdots, z(2^{\dim F} - 1)\}, \end{cases}$$

there exists a function S such that

$$\begin{cases} S \text{ is a sub-function of } X \restriction \{z(0), z(1), \cdots, z(2^{\dim F} - 1)\}, \\ |\mathrm{dom} S| \leq \ell(F)^k, \text{ and } S \text{ partially forces } F \text{ with respect to } X. \end{cases} \tag{3}$$

On the other hand, for every r-query formula F such that

$$\begin{cases} \ell(F) \geq 2^{(n-1)/k} \text{ (See the following remark.)}, \\ \dim F \leq n-1, \quad F \text{ is nice, and} \\ F \text{ is a tautology with respect to } X \restriction \{z(0), z(1), \cdots, z(2^{\dim F} - 1)\}, \end{cases} \tag{4}$$

equation (3) holds for (a certain sub-function of) the following S.

$$S := X \restriction \{z(0), z(1), \cdots, z(2^{\dim F} - 1)\}.$$

(Remark: Assume equation (4). Then "$\ell(F)^k \geq |\{z(0), \cdots, z(2^{n-1} - 1)\}| \geq |\{z(0), \cdots, z(2^{\dim F} - 1)\}| = |\mathrm{dom} S|$". And, S forces F.)

Thus, let

$$c' := \max\{c, 2^{(n-1)/k}\}.$$

Then, the following holds.

$$\forall m \geq 1 \ \big(X \restriction \{z(0), z(1), \cdots, z(2^m - 1)\} \in r\mathrm{GEN}_1(k, c', m) \big).$$

Hence,

$$X \in r\mathrm{GEN}_1(k, c') \subseteq r\mathrm{GEN}_1.$$

Therefore, by Lemma 3.2, X is r-Dowd. □

The following corollary is proved through an argument similar to the proof of Theorem 3.1 (Add elements to each U_n in order to make its measure just 2^{-n}).

Corollary 3.1. *Suppose that X is Schnorr random, and that r is a positive integer. Then, X is r-Dowd.*

4. Conservation of Dowd-type genericity

Suppose that X is an oracle. We let $\Phi_D(X)$ denote the assertion of "For every positive integer r, X is r-Dowd." In this section, we show that the class $\{X : \Phi_D(X) \text{ holds.}\}$ is closed under a certain type of bounded truth-table reduction.

Now, we review the concept of truth table reduction, which is a special type of Turing reduction. Suppose that X and Y are oracles. A function f :

$\{0,1\}^* \to \{0,1\}^*$ is called *a truth-table reduction of* Y *to* X (tt-reduction, for short) if the following (1) and (2) hold.

(1) For every $u \in \{0,1\}^*$, the $f(u)$ is of the form $\langle \varphi_u, s_{u,1}, \cdots, s_{u,a} \rangle$, where a is a positive integer (called *norm*), φ_u is an a-ary Boolean function, and $s_{u,1}, \cdots, s_{u,a}$ are strings.
(2) For every $u \in \{0,1\}^*$, it holds that $u \in Y$ if and only if $\varphi_u(X(s_{u,1}), \cdots, X(s_{u,a})) = 1$.

The concept of *bounded truth-table reduction* (btt-reduction, for short) is defined as the special case of tt-reduction where the norm a is the same for all inputs. For a formal treatment of polynomial-time btt-reduction, see Ref. 18.

Definition 4.1. Suppose that X and Y are oracles and that $f : \{0,1\}^* \to \{0,1\}^*$ is a PTIME bounded-truth-table reduction of Y to X. Let a be the norm of f. For each string u, we represent $f(u)$ as $\langle \varphi_u, s_{u,1}, \cdots, s_{u,a} \rangle$, where φ_u is an a-ary Boolean function and $s_{u,1}, \cdots, s_{u,a}$ are strings. We say "*f is independent*" if the following (1) and (2) hold.

(1) For each strings u, v such that $u \neq v$, it holds that

$$\{s_{u,1}, \cdots, s_{u,a}\} \cap \{s_{v,1}, \cdots, s_{v,a}\} = \emptyset.$$

(2) For every $u \in \{0,1\}^*$, it holds that $\varphi_u : \{0,1\}^a \to \{0,1\}$ is onto.

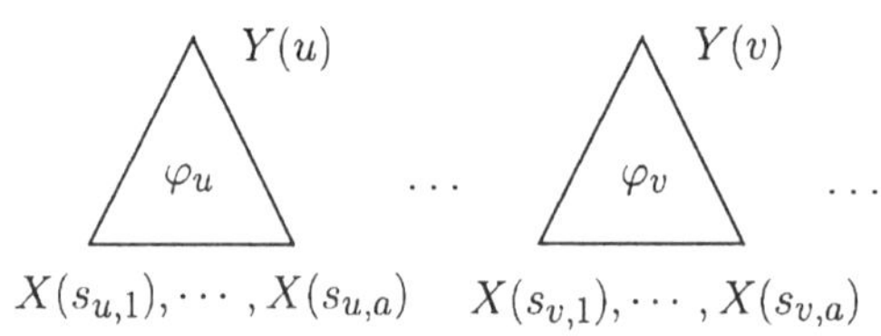

Fig. 3. Independent PTIME btt-reduction of Y to X

Example 4.1. Let $f^2_\wedge$, $f^2_\vee$ and f_{XOR}, respectively, be the following functions.

$$\begin{aligned} f^2_\wedge(z(m)) &= \langle \mathrm{AND}^2, z(2m), z(2m+1) \rangle, \\ f^2_\vee(z(m)) &= \langle \mathrm{OR}^2, z(2m), z(2m+1) \rangle, \\ f_{\mathrm{XOR}}(z(m)) &= \langle \mathrm{XOR}, z(2m), z(2m+1) \rangle \quad \text{for } m = 0, 1, 2, \cdots, \end{aligned}$$

where Boolean functions AND^2, OR^2 and XOR are 2-ary conjunction, 2-ary disjunction and exclusive or, respectively. Then, each of $f^2_\wedge$, $f^2_\vee$ and f_{XOR} is a PTIME independent btt-reduction.

Example 4.2. Suppose that c is a positive integer and that T is a Boolean decision tree that has c leaves. In addition, suppose that $T : \{0,1\}^c \to \{0,1\}$ is an onto mapping, where we regard T as a mapping of truth values of leaves to truth values of the root.

Let f be the function such that, for each natural number m, the following holds.

$$f(z(m)) = \langle T, z(cm), z(cm+1), \cdots, z(c(m+1)-1)\rangle.$$

Then, f is a PTIME independent btt-reduction.

Theorem 4.1. *The class $\{X : \Phi_D(X) \text{ holds.}\}$ is closed under PTIME independent btt-reductions.*

In other words, if A, B are oracles such that $\Phi_D(A)$ holds and such that B is PTIME btt-reducible to A via an independent reduction, then $\Phi_D(B)$ holds as well.

Proof. Suppose that A and B are oracles such that $\Phi_D(A)$ holds and suppose that f is a PTIME independent btt-reduction from B to A. We discuss the special case where the norm of f is 2, and for each natural number m, $f(z(m))$ is of the form $\langle \varphi_m, z(2m), z(2m+1)\rangle$. (We write as "$\varphi_m$" instead of "$\varphi_{z(m)}$" for simplicity of notation.) We can show general case in the same way. Now, for each m, it holds that

$$B(z(m)) = \varphi_m(A(z(2m)), A(z(2m+1))).$$

Let r be a positive integer. Then, there exists a polynomial p of the following property. For every r-query formula F, there exists a $(2r)$-query formula F' such that the following (i) and (ii) hold ($\ell(F)$ is the length of F, see sub-section 2.4).

(i) It holds that $\ell(F') \leq p(\ell(F))$.
(ii) For every pair (X, Y) of oracles such that

$$\forall m \in \mathbb{N}\ Y(z(m)) = \varphi_m(X(z(2m)), X(z(2m+1))),$$

the following two assertions (iia) and (iib) are equivalent:

(iia) F is a tautology with respect to Y,
(iib) F' is a tautology with respect to X.

For example, if F is

$$(q_0 \Leftrightarrow \xi^3(q_1, q_2, q_3)) \Rightarrow H,$$

where each of q_j is a propositional variable and H is a query free formula, then let F' denote the following formula:

$$\Big((a_0^{(1)} \Leftrightarrow \xi^4(a_1^{(1)}, a_2^{(1)}, a_3^{(1)}, a_4^{(1)})) \wedge (a_0^{(2)} \Leftrightarrow \xi^4(a_1^{(2)}, a_2^{(2)}, a_3^{(2)}, a_4^{(2)}))\Big)$$
$$\Rightarrow \Big[(G_1 \wedge G_2) \Rightarrow H\Big],$$

where each of $a_j^{(i)}$ is a propositional variable that does not have an occurrence in F; G_1 is a formula asserting "Letting m be the natural number such that $q_1q_2q_3$ is the $(m+1)$st string of length 3, $a_1^{(1)}a_2^{(1)}a_3^{(1)}a_4^{(1)}$ and $a_1^{(2)}a_2^{(2)}a_3^{(2)}a_4^{(2)}$ are the $(2m+1)$st and $(2m+2)$nd string of length 4, respectively"; and G_2 is a formula asserting "$q_0 \Leftrightarrow \varphi_m(a_0^{(1)}, a_0^{(2)})$".

Now, suppose that F is an r-query tautology with respect to B. Then, the $(2r)$-query formula F' is a tautology with respect to A. Since A is $(2r)$-Dowd, there exists a forcing condition S_1 such that S_1 is a sub-function of A, the cardinality of the domain of S_1 is small enough, and S_1 forces F'. Without loss of generality, we may assume that for every natural number m, either $\{z(2m), z(2m+1)\} \subseteq \mathrm{dom}S_1$ or $\{z(2m), z(2m+1)\} \cap \mathrm{dom}S_1 = \emptyset$ holds.

We define a forcing condition S_2 as follows.

- $\mathrm{dom}S_2 = \{z(m) : z(2m) \in \mathrm{dom}S_1\}$,
- $S_2(z(m)) = \varphi_m(S_1(z(2m)), S_1(z(2m+1)))$, for each $z(m) \in \mathrm{dom}S_2$.

Then, S_2 is a sub-function of B, and the cardinality of the domain of S_2 is small enough. Now, suppose that an oracle Y is an extension of S_2. Since the reduction f is independent in the sense of Definition 4.1 (in particular, note that for each natural number m, it holds that $\varphi_m : \{0,1\}^2 \to \{0,1\}$ is an onto-mapping), there exists an oracle X such that X is an extension of S_1 and such that for every natural number m, it holds that $Y(z(m)) = \varphi_m(X(z(2m)), X(z(2m+1)))$. Since S_1 forces F', it holds that F' is a tautology with respect to X. Hence, F is a tautology with respect to Y. Since Y is chosen as an arbitrary extension of S_2, it holds that S_2 forces F. Therefore, B is r-Dowd.

Since the positive integer r is arbitrary, $\Phi_D(B)$ holds. □

Corollary 4.1. *Suppose that a Boolean decision tree T and a function f are as described in Example 4.2 and that A, B are oracles. If $\Phi_D(A)$ holds and B is PTIME btt-reducible to A via f, then $\Phi_D(B)$ also holds.* □

Theorem 4.1 is generalized to reduction of the following type.

Definition 4.2. Suppose that X and Y are oracles and that $f : \{0,1\}^* \to \{0,1\}^*$ is a PTIME bounded-truth-table reduction of Y to X. Let a be the norm of f. For each string u, we represent $f(u)$ as $\langle \varphi_u, s_{u,1}, \cdots, s_{u,a} \rangle$, where φ_u is an a-ary Boolean function and $s_{u,1}, \cdots, s_{u,a}$ are strings. We say "*f is weakly independent*" if there exists a PTIME computable function g_1 on $\{0,1\}^*$ such that the following (1)–(4) hold.

(1) For each string u, $g_1(u)$ is a finite set of strings such that $u \in g_1(u)$ and that $|g_1(u)| \leq a$.
In the following, for each string u, we let $g_2(u)$ denote $\{s_{u,1}, \cdots, s_{u,a}\}$.
(2) If $g_1(u) \neq g_1(v)$ then it holds that $g_1(u) \cap g_1(v) = g_2(u) \cap g_2(v) = \emptyset$.
(3) If $g_1(u) = g_1(v)$ then it holds that $g_2(u) = g_2(v)$.
(4) For each $u \in \{0,1\}^*$ and for each mapping $h_1 : g_1(u) \to \{0,1\}$, there exists a mapping $h_2 : g_2(u) \to \{0,1\}$ such that

$$\forall v \in g_1(u), \ \ h_1(v) = \varphi_v(h_2(s_{v,1}), \cdots, h_2(s_{v,a})).$$

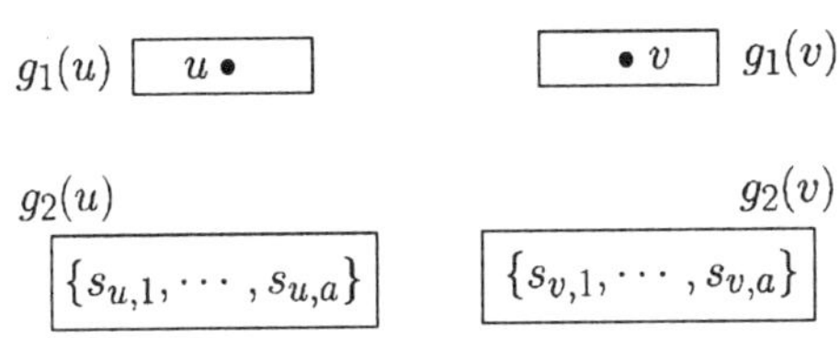

Fig. 4. Weakly independent PTIME btt-reduction of Y to X

Theorem 4.2. *The class $\{X : \Phi_D(X)$ holds.$\}$ is closed under PTIME weakly independent btt-reductions.*

Proof. The proof is similar to that of Theorem 4.1. □

Proposition 4.1.

(1) Polynomial-time independent btt-reduction satisfies the transitive law. In other words, if A_i is PTIME independent btt-reducible to A_{i+1} for $i = 1, 2$, then A_1 is PTIME independent btt-reducible to A_3.

(2) Polynomial-time weakly independent btt-reduction satisfies the following weak version of the transitive law. Suppose that A_1, A_2 and A_3 are oracles and that the following (a) and (b) hold.

(a) For each $i = 1, 2$, $f^{(i)}$ is a PTIME weakly independent btt-reduction of A_i to A_{i+1}, and that $g_1^{(i)}$ and $g_2^{(i)}$ are as described in Definition 4.2.

(b) For every string u and for every $x \in g_2^{(1)}(u)$, it holds that $g_1^{(2)}(x) \subseteq g_2^{(1)}(u)$.

Then A_1 is PTIME btt-reducible to A_3 via a weakly independent reduction.

Proof. Straightforward. □

5. Density of an oracle

We have shown that a PTIME weakly independent btt-reduction conserves Dowd-type genericity. However, note that such a reduction may change density of an oracle. In this section, we present two remarks on density of an oracle. One is on density of a Dowd-type generic oracle, and the other is on an algorithm that changes density of an oracle.

5.1. *Density of a Dowd-type generic oracle*

Strictly speaking, Dowd-type genericity is not "randomness", because a Dowd-type generic oracle may have density not equal to 1/2. Recall that $\Phi_D(X)$ denotes the assertion of "For every positive integer r, X is r-Dowd".

Proposition 5.1. *Suppose that m is a positive integer and u is a bit string of length m. Then, there exists an oracle A such that $\Phi_D(A)$ holds and that the following holds.*

$$\lim_{n\to\infty} \frac{A(0) + A(1) + \cdots + A(n-1)}{n} = \sum_{i=1}^{m} \frac{u_i}{2^i}. \tag{5}$$

Proof. We discuss the special case where $u = 11$. In this case, the right-hand-side of (5) is 3/4. We can show general case in the same way.

Let $\varphi(x_1, x_2, x_3)$ denote the propositional formula $x_1 \vee (x_2 \wedge x_3)$, and we identify φ with the Boolean function defined by φ. We define a class $\mathcal{S}$ of oracles as follows.

$$\mathcal{S} = \{X : \lim_{n\to\infty} \frac{1}{n} \sum_{i=0}^{n-1} \varphi(X(3i), X(3i+1), X(3i+2)) \neq \frac{3}{4}\}.$$

Now, let B be a Martin-Löf random oracle. We define an oracle A as follows. For each natural number n, $A(n) = \varphi(B(3n), B(3n+1), B(3n+2))$.

By Theorem 3.1, $\Phi_D(B)$ holds. By Example 4.2, A is PTIME independent btt-reducible to B. Therefore, by Theorem 4.1, $\Phi_D(A)$ holds.

Similarly to Example 2.2 and Example 2.3, by means of Chernoff bound (Proposition 2.1), we can show that $\mathcal{S}$ is Martin-Löf null. Therefore, B does not belong to $\mathcal{S}$. Hence, (5) holds. □

5.2. *An algorithm that changes density of an oracle*

We observe a case where changing density of an oracle is a useful property of a btt-reduction.

Kawanishi[19] extracts (pseudo) random strings from bitmap files of hand-drawing curves such as Fig. 5 and Fig. 6. In such a bitmap, density (the rate of the number of black pixels per total number of pixels) is not 1/2. On the other hand, in his experiment, extracted strings have density approximately 1/2. His algorithm is a variation of run-length compression, where a *run* in a string is a maximum substring of consecutive characters taking the same value. For example, runs in a string 000111100 are 000, 1111 and 00.

Fig. 5.

Fig. 6.

Suzuki[6,19] observes that a block-wise run-length compression, a modification of Kawanishi's algorithm, produces results very similar to Kawanishi's. Example 5.1 is (a slightly modified version of) the core of Suzuki's algorithm. This algorithm is a PTIME weakly independent btt-reduction. Therefore, if a bitmap is irregular enough, we may expect that the extracted string has irregularity.

Some experimental results may be found in Refs. 6,19.

Example 5.1. Block-wise run-length compression. We define functions

$$\psi_1(x_1, \cdots, x_{60}) : \{0,1\}^{60} \to \{0,1\}^5 \text{ and}$$
$$\psi_2(x_1, \cdots, x_{120}) : \{0,1\}^{120} \to \{0,1\}^5$$

as follows, and by means of these functions, we define a PTIME weakly independent btt-reduction.

We define $\psi_1(x_1, \cdots, x_{60})$ as the output of Algorithm 5.1.

Algorithm 5.1.

input (a bit string of length 60***)*** $x = x_1 \cdots x_{60}$;

STEP1: *Let* $r_0, r_1, \cdots, r_j, \cdots$ *be the lengths of runs in* x.

STEP2: *For each* j, *let* r'_j *be the binary expression of* $r_j + 2$.

STEP3: *For each* j, *remove the leftmost 1 from* r'_j, *and let* s_j *be the resulting string.*

STEP4: *Let* s *be the concatenation of* $s_0, s_1, \cdots, s_j, \cdots$, *and let* t *be its leftmost* 5 *bits. Output* t.

/* *Note that* $5 \leq \log_2(60+2)$, *where '60' is the length of input, and '2' appears at STEP2. Hence, for any* x, *we can get leftmost* 5 *bits of* s. */

/* *Example. Suppose* $x = 0001111001\cdots$. *Then* $r_0 = 3$, $r_1 = 4$, $r_3 = 2$, $r'_0 = 101$, $r'_1 = 110$ *and* $r'_3 = 100$. *We have* $s_0 = 01$, $s_1 = 10$ *and* $s_3 = 00$. *Finally, we get* $t = 01100$. */

End, Algorithm 5.1.

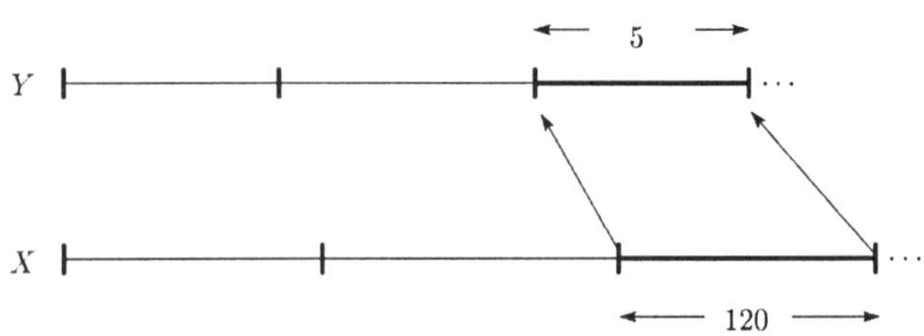

Fig. 7. Block-wise operation

For a given string $x' = x_1 \cdots x_{120}$, let $a_1 \cdots a_5 = \psi_1(x_1 \cdots x_{60})$ and let $b_1 \cdots b_5 = \psi_1(x_{61} \cdots x_{120})$. We define $\psi_2(x')$ as to be $y_1 \cdots y_5$, where $y_i = a_i$ XOR b_i for $i = 1, \cdots, 5$.

Now, for a given oracle X, we define an oracle Y as follows. For each natural number i,

$$Y(5i)Y(5i+1)\cdots Y(5i+4) = \psi_2(X(120i), X(120i+1), \cdots, X(120i+119)).$$

It is not hard to see that Y is PTIME weakly independent btt-reducible to X.

Acknowledgments

The authors were partially supported by Grant-in-Aid for Scientific Research (No. 17540131), Japan Society for the Promotion of Science. They thank Akio Kawanishi* for useful discussions during his period as a graduate student at Osaka Prefecture University (* current affiliation: Pixela corporation), and they thank to the anonymous referee for useful comments.

References

1. Yao, A.C.-C.: Probabilistic computations: towards a unified measure of complexity, in: *Proc. 18th annual IEEE symposium on foundations of computer science (FOCS)*, pp.222–227 (1977).
2. Liu, C.-G. and Tanaka, K.: Eigen-distribution on random assignments for game trees. *Inform. Process. Lett.*, **104** pp.73–77 (2007)
3. Dowd, M.: Generic oracles, uniform machines, and codes. *Inf. Comput.*, **96**, pp. 65–76 (1992).
4. Kumabe, M. and Suzuki, T.: Weak randomness, genericity and Boolean decision trees. A talk at *the 10th Asian logic conference (ALC10)*, Kobe University, Japan, September 1–6, 2008.
5. Suzuki, T. and Kawanishi, A.: Random extraction from freehand drawings and its semantics based on forcing complexity. A talk at *Franco-Canadian Workshop on Combinatorial Algorithm (COMAL 2005)*, McMaster University, Hamilton, Canada, August 18–20, 2005.
6. Suzuki, T. and Kawanishi, A.: Random extraction from freehand drawings and its semantics based on forcing complexity. A talk at *the 6th symposium on algebra and computation (AC2005)*, Tokyo Metropolitan University, Japan, November 15–18, 2005. An extended abstract may be available at the following site.
   `ftp://tnt.math.metro-u.ac.jp/pub/ac05/SuzukiKawanishi/SuzukiKawanishi.pdf`
7. Martin-Löf, P.: The definition of random sequences. *Information and Control*, **9**, pp.602–619 (1966).
8. Schnorr, C. P.: Zufälligkeit und Warscheinlichkeit. In: *Lecture notes in mathematics 218*, Springer, New York, 1971.
9. Calude, C.S.: *Information and randomness: an algorithmic perspective, 2nd ed.*, Springer, Berlin, 2002.
10. Downey, R., Hirschfeldt, D. R., Nies, A. and Terwijn, S. A.: Calibrating randomness. *Bull. Symb. Log.*, **12**, pp.411–491 (2006).
11. Tadaki, K.: A generalization of Chaitin's halting probability Ω and halting self-similar sets. *Hokkaido Mathematical Journal*, **31**, pp.219–253 (2002).
12. Miller, J. S. and Nies, A.: Randomness and computability: open questions. *Bull. Symb. Log.*, **12**, pp.390–410 (2006).
13. Papadimitriou, C. H.: *Computational complexity.* Addison-Wesley, Massachusetts (1994).
14. Suzuki, T.: Forcing complexity: minimum sizes of forcing conditions. *Notre Dame J. Formal Logic*, **42**, pp. 117–120 (2001).
15. Suzuki, T.: Degrees of Dowd-type generic oracles. *Information and Computation*, **176**, pp.66–87 (2002).
16. Kumabe, M., Suzuki, T. and Yamazaki, T.: Does truth-table of linear norm

reduce the one-query tautologies to a random oracle? *Arch. Math. Logic*, **47**, pp.159–180 (2008).

17. Suzuki, T.: *Computational complexity of Boolean formulas with query symbols.* Doctoral dissertation, Institute of Mathematics, University of Tsukuba, Tsukuba-City, Japan (1999). `http://www.ac.auone-net.jp/~bellp/toshio_suzuki_phd.pdf`
18. Ladner, R. E., Lynch, N. A. and Selman, A. L.: A comparison of polynomial time reducibilities. *Theor. Comput. Sci.*, **1**, pp.103–123 (1975).
19. Kawanishi, A.: *Random extraction from freehand drawings.* (in Japanese, with English abstract) Master thesis, Department of Mathematics and Information Sciences, Osaka Prefecture University, Sakai-City, Japan (2006). `http://www.comp.tmu.ac.jp/mathsuzuki/akio_kawanishi_master_thesis0322.pdf`

A note on multiple inductive definitions

Kazuyuki Tanaka

Mathematical Institute, Tohoku University,
Aoba, Sendai, 980-8578, JAPAN,
E-mail: tanaka@math.tohoku.ac.jp

In Ref. 8, we introduced a new axiom scheme for multiple inductive definitions to capture the exact strength of Δ^0_3-determinacy in second order arithmetic. In this paper, we show this complicated axiom turns out to be equivalent to an axiom of single inductive definitions. This observation is useful to characterize Π^1_2-CA_0 and related systems.

Keywords: inductive definitions; second order arithmetic; determinacy.

1. Introduction

In second order arithmetic, the strength of determinacy of infinite game has been investigated up to Δ^0_3 (Ref. 4, Ref. 5, Ref. 6, Ref. 7, Ref. 8). In the last paper Ref. 8, we have indeed established (over $\mathsf{ACA}_0 + \Pi^1_3$-TI) the equivalence between the determinacy of the games whose complexity lies in the α-level of difference hierarchy over Π^0_2, and the axiom which asserts the existence of the sets inductively defined by an iterative combination of α-many Σ^1_1-operators for each countable ordinal α.

In this note, we show a finite combination of inductive definitions in a class $\mathcal{C}$ can be emulated by a single inductive definition in the arithmetical closure of $\mathcal{C}$, and vice versa. We work in the language $\mathcal{L}_2$ of second order arithmetic, and mainly over the base system ACA_0, which consists of the ordered semi-ring axioms for $(\mathbb{N}, +, \cdot, 0, 1, <)$, Σ^0_1-CA and Σ^0_1-induction. See Ref. 3 for further information on second order arithmetic.

2. Multiple Inductive Definitions

We say that an operator $\Gamma : P(\mathbb{N}) \to P(\mathbb{N})$ is in the class $\mathcal{C}$ (e.g., Σ^1_1) if its graph $\{(x, X) : x \in \Gamma(X)\}$ is described by a formula in $\mathcal{C}$. For a binary relation R, let R_y and $R_{<y}$ denote $\{x : (x, y) \in R\}$ and $\{x : (x, y) \in R \land (y, x) \notin R\}$, respectively.

An axiom of *inductive definition* asserts the existence of a pre-wellordering constructed by iterative application of a given operator.

Definition 2.1. (Ref. 5) Let $\mathcal{C}$ be a set of $\mathcal{L}_2$ formulas. $\mathcal{C}$-ID asserts that for any operator $\Gamma \in \mathcal{C}$, there exists $W \subset \mathbb{N} \times \mathbb{N}$ such that

(1) W is a pre-wellordering on its field F,
(2) $\forall x \in F \quad W_x = \Gamma(W_{<x}) \cup W_{<x}$,
(3) $\Gamma(F) \subset F$.

In Ref. 5 and Ref. 8, we have shown that Σ^1_1-ID is equivalent to Σ^0_2-Det, which stands for the determinacy of Σ^0_2 games. Then to characterize the determinacy of Boolean combinations of Σ^0_2-formulas, we introduced the axiom of multiple inductive operators.

Let us consider how to combine two operators Γ_0 and Γ_1. For a better understanding, we first recall a simpler combination due to Richter and Aczel in Ref. 2. In their definition, to get a common fixed point of the two operators, a (pre-)wellordered structure is constructed as follows:

(1) Start with the empty set,
(2) Repeatedly apply Γ_0 until a fixed point is reached,
(3) Apply Γ_1 once. Go to 2 unless closure under Γ_1 is reached.

It is easy to see that this construction by two operators can be emulated by a single operator Γ defined by

$$n \in \Gamma(X) \Leftrightarrow n \in \Gamma_0(X) \vee (\Gamma_0(X) \subset X \wedge n \in \Gamma_1(X)).$$

Thus, a finite combination (à la Richter-Aczel) of Σ^1_1 operators can be viewed as a single operator in arithmetical combination of Σ^1_1 formulas.

However, to characterize the determinacy of Boolean combinations of Σ^0_2-formulas, the Richter-Aczel combination does not seem to be powerful enough. So, in Ref. 8, we devised the following sophisticated version. Suppose Γ_0 has a distinct parameter X_0, and thus also denoted $\Gamma_0^{X_0}$. We start with $X_0 = \emptyset$. Then, repeatedly apply $\Gamma_0^{X_0}$ from the empty set until we get a fixed point, say F_0. Then apply Γ_1 once to get a new $X_0 := \Gamma_1(F_0) \cup X_0$. Note that we need not have $F_0 \subset X_0$. For the second stage, re-start iterating $\Gamma_0^{X_0}$, and get a fixed point F_1. Then apply Γ_1 once, and so on. The procedure stops when we get a fixed point F such that if F' is the least fixed point of Γ_0^F then $\Gamma_1(F') \subset F$. This can be depicted as follows:

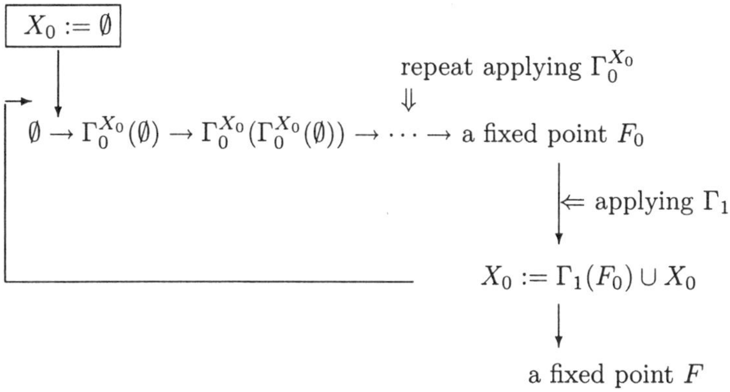

We can easily see that the Richter-Aczel combination is a special case of our multiple inductive definitions. In fact, to emulate their combination of Γ_0 and Γ_1 by this method, we may simply use a pair of $\tilde{\Gamma}_0^X(Y) = \Gamma_0(X \cup Y)$ and $\tilde{\Gamma}_1(X) = \Gamma_1(X) \cup X$.

By $[\Gamma](X)$ or simply $[\Gamma]$, we denote the fixed point of inductive operator Γ^X. Then, our multiple inductive definition of two operators Γ_0 and Γ_1 is indeed an inductive definition by the composition of $[\Gamma_0]$ and Γ_1. In Ref. 8, we write $[\Gamma_0, \Gamma_1]$ for the fixed point of such a multiple inductive definition, and in general $[\Gamma_0, \cdots, \Gamma_{k-1}, \Gamma_k]$ for $[[\Gamma_0, \cdots, \Gamma_{k-1}], \Gamma_k]$. Furthermore, for k collections $\mathcal{C}_0, \cdots, \mathcal{C}_{k-1}$ of operators, let $[\mathcal{C}_0, \mathcal{C}_1, \cdots, \mathcal{C}_{k-1}]$-ID denote the axiom which asserts the existence of the set $[\Gamma_0, \cdots, \Gamma_{k-1}]$ for any $\Gamma_0 \in \mathcal{C}_0, \cdots, \Gamma_k \in \mathcal{C}_{k-1}$. Finally, $[\mathcal{C}]^{<\omega}$-ID denotes $\cup_k [\mathcal{C}_0, \mathcal{C}_1, \cdots, \mathcal{C}_k]$-ID where $\mathcal{C}_0 = \mathcal{C}_1 = \cdots = \mathcal{C}_k = \mathcal{C}$.

For an alternative form of multiple inductive definition of Γ_0 and Γ_1, we can take the composition of $[\Gamma_0]$ and $[\Gamma_1]$ instead of $[\Gamma_0]$ and Γ_1. The fixed point of such a composition, $[\Gamma_0, [\Gamma_1]]$, is also denoted as $[[\Gamma_0, \Gamma_1]]$. Moreover, $[[\mathcal{C}]]^{<\omega}$-ID is defined to be the axiom which asserts the existence of the fixed point for any finite [ ]-composition of operators in $\mathcal{C}$, such as $[\Gamma_0, [\Gamma_1, \Gamma_2]]$. However, as shown in the next section, for any reasonable choice of $\mathcal{C}$, $[[\mathcal{C}]]^{<\omega}$-ID turns out to be equivalent to $[\mathcal{C}]^{<\omega}$-ID.

3. Main Results

A main purpose of this note is to show that if $\mathcal{C}$ is a class of formulas with certain closure conditions (e.g., Σ_1^1), $[\mathcal{C}]^{<\omega}$-ID (or $[[\mathcal{C}]]^{<\omega}$-ID) is essentially equivalent to $\mathcal{C}_{\mathsf{A}}$-ID, where $\mathcal{C}_{\mathsf{A}}$ denotes the set of arithmetical combinations of formulas in $\mathcal{C}$. For simplicity, without stating the closure conditions on $\mathcal{C}$, we would rather assume that $\mathcal{C}$ is one of Σ^1_n's, since our major interests are the cases $\mathcal{C} = \Sigma^1_1$ and Σ^1_0 (i.e., the set of arithmetical formulas). We here note that $(\Sigma^1_0)_{\mathsf{A}} = \Sigma^1_0$. Finally, we adopt the following notation: $X \uplus Y = \{2x : x \in X\} \cup \{2y+1 : y \in Y\}$.

Lemma 3.1. (ACA_0) *For any class* $\mathcal{C} = \Sigma^1_n$ $(n \geq 0)$, $[[\mathcal{C}]]^{<\omega}$*-ID is equivalent to* $[\mathcal{C}]^{<\omega}$*-ID.*

Proof. Since $[\mathcal{C}]^{<\omega}$-ID obviously follows from $[[\mathcal{C}]]^{<\omega}$-ID, it suffices to show the converse. We here only treat a typical instance of $[[\mathcal{C}]]^{<\omega}$-ID under the assumption of $[\mathcal{C}]^{<\omega}$-ID, and leave the general formulation to the reader. We indeed show by $[\mathcal{C}]^{<\omega}$-ID that $[\Gamma_0, [\Gamma_1, \Gamma_2]]$ (with each Γ_i in $\mathcal{C}$) exists.

We first prove that the composition of $[\Gamma_0]$ and $[\Gamma_1]$ (in $\mathcal{C}$) can be obtained by $[\mathcal{C}]^{<\omega}$-ID. We modify the operators Γ_0, Γ_1 as follows:

$$\tilde{\Gamma}_0^{X \uplus Y}(Z) = \Gamma_0^X(Z) \uplus Y,$$

$$\tilde{\Gamma}_1(X \uplus Y) = \emptyset \uplus \Gamma_1^X(Y).$$

Given any X, Y $(= \emptyset$, mostly at the start), we have the fixed point $[\tilde{\Gamma}_0^{X \uplus Y}] = [\Gamma_0^X] \uplus Y$, and then letting $F = [\Gamma_0^X]$, $\tilde{\Gamma}_1(F \uplus Y) = \emptyset \uplus \Gamma_1^F(Y)$, which is the end of the first round. For the second round, noticing $X \uplus (Y \cup \Gamma_1^F(Y)) = (X \uplus Y) \cup (\emptyset \uplus \Gamma_1^F(Y))$, we have $[\tilde{\Gamma}_0^{X \uplus (Y \cup \Gamma_1^F(Y))}] = [\Gamma_0^X] \uplus (Y \cup \Gamma_1^F(Y))$, and then $\tilde{\Gamma}_1(F \uplus (Y \cup \Gamma_1^F(Y))) = \emptyset \uplus \Gamma_1^F((Y \cup \Gamma_1^F(Y)))$. Iterating the combination of $[\tilde{\Gamma}_0]$ and $\tilde{\Gamma}_1$, we eventually reach $X \uplus F'$ with $F' \subset \Gamma_1^F(F')$, from which we obtain the composition of $[\Gamma_0]$ and $[\Gamma_1]$. Thus, the composition of $[\Gamma_0]$ and $[\Gamma_1]$ can be viewed as $[\tilde{\Gamma}_0, \tilde{\Gamma}_1]$, whose existence is guaranteed by $[\mathcal{C}]^{<\omega}$-ID.

Now, we treat $[\Gamma_0, [\Gamma_1, \Gamma_2]]$. Similarly to the above, we modify the three operators as follows:

$$\tilde{\Gamma}_0^{(W \uplus X) \uplus Y}(Z) = (W \uplus \Gamma_0^X(Z)) \uplus Y,$$

$$\tilde{\Gamma}_1((W \uplus X) \uplus Y) = (W \uplus \emptyset) \uplus \Gamma_1^{W \cup X}(Y),$$

$$\tilde{\Gamma}_2((W \uplus X) \uplus Y) = (\Gamma_2(Y) \uplus \emptyset) \uplus \emptyset.$$

Then, $[\tilde{\Gamma}_0, \tilde{\Gamma}_1]$ gives us $(W \uplus X) \uplus F'$ with $F' \subset \Gamma_1^{W \cup F}(F')$. By applying $\tilde{\Gamma}_2$, we have $((W \cup \Gamma_2(F')) \uplus X) \uplus Y$ for the starting values of the second round. Similarly, we have $((W \cup \Gamma_2(F') \cup \Gamma_2(F'')) \uplus X) \uplus Y$ for the third round, where $F'' \subset \Gamma_1^{W \cup \Gamma_2(F')}(F'')$. Thus, the composition of $[\Gamma_0]$ and $[\Gamma_1, \Gamma_2]$ is given by $[\tilde{\Gamma}_0, \tilde{\Gamma}_1, \tilde{\Gamma}_2]$, and its fixed point also exists by $[\mathcal{C}]^{<\omega}$-ID. More complicated combinations can be treated analogously. □

Lemma 3.2. (ACA$_0$) *For any class $\mathcal{C}$ of formulas, $[\mathcal{C}]^{<\omega}$-ID and $[[\mathcal{C}]]^{<\omega}$-ID can be deduced from $\mathcal{C}_{\mathsf{A}}$-ID, where $\mathcal{C}_{\mathsf{A}}$ denotes the set of arithmetical combinations of formulas in $\mathcal{C}$.*

Proof. By induction, it suffices to show that $[\mathcal{C}_{\mathsf{A}}, \mathcal{C}_{\mathsf{A}}]$-ID can be deduced from $\mathcal{C}_{\mathsf{A}}$-ID. To begin with, we adopt the following notation. Let $Y_x = \{y : (x,y) \in Y\}$ and $Y_X = \cap\{Y_x : x \in X\}$. In particular, if $X = \emptyset$, then we set $Y_X = \emptyset$. In what follows, Y_X is actually equal to Y_x for some $x \in X$, but it suffices to see it arithmetical in X, Y.

Given two operators Γ_0 and Γ_1 in $\mathcal{C}$, we define an operator Γ (obviously in $\mathcal{C}_{\mathsf{A}}$) as follows.

$$\Gamma(X \uplus Y) = \begin{cases} X \uplus (X \times \Gamma_0^X(Y_X)) \text{ if } & \Gamma_0^X(Y_X) \not\subset Y_X \\ \Gamma_1(Y_X) \uplus Y & \text{otherwise} \end{cases} \tag{1}$$

For simplicity, we may assume $X \neq \emptyset$ at the beginning. Then we can easily see that the inductive construction by $[\Gamma_0, \Gamma_1]$ from X can be emulated by Γ from $X \uplus \emptyset$. In order to start with $X = \emptyset$, we need slightly modify the definition of Γ, which we leave to the reader. Now, let us see how Γ works.

By iterating Γ, we first obtain $X \uplus \emptyset, X \uplus (X \times \Gamma_0^X(\emptyset)), X \uplus (X \times (\Gamma_0^X(\emptyset) \cup \Gamma_0^X(\Gamma_0^X(\emptyset))), \ldots$ until $X \uplus (X \times F)$ with $F \subset \Gamma_0^X(F)$. By focusing on the change of Y_X in this process, we have $\emptyset, \Gamma_0^X(\emptyset), \Gamma_0^X(\emptyset) \cup \Gamma_0^X(\Gamma_0^X(\emptyset)), \ldots$, which can be viewed as a process made by iterating Γ_0^X from the empty set. Now, we may assume that there exists an $x \in \Gamma_1(F) - X$. Otherwise, we have reached a fixed point of Γ. Then, we go into the second round of the construction with X replaced by a strictly larger set $X \cup \Gamma_1(F)$ and $Y = (\text{old } X) \times F$. Noticing that $Y_X = Y_x = \emptyset$, by iterating Γ, Y_X changes as $\emptyset, \Gamma_0^X(\emptyset), \Gamma_0^X(\emptyset) \cup \Gamma_0^X(\Gamma_0^X(\emptyset)), \ldots$ until $X \uplus (X \times F)$ with $F \subset \Gamma_0^X(F)$. Thus, the inductive construction by Γ is essentially the same as the construction by $[\Gamma_0, \Gamma_1]$. Therefore, $[\mathcal{C}_{\mathsf{A}}, \mathcal{C}_{\mathsf{A}}]$-ID is deducible from $\mathcal{C}_{\mathsf{A}}$-ID. □

Next, consider the converse.

Lemma 3.3. (ACA$_0$) *For any class $\mathcal{C} = \Sigma_n^1$ $(n \geq 0)$, $\mathcal{C}_{\mathsf{A}}$-ID can be deduced from $[[\mathcal{C}]]^{<\omega}$-ID and hence also from $[\mathcal{C}]^{<\omega}$-ID.*

Proof. We use induction on the complexity of arithmetical combinations of operators in $\mathcal{C}_{\mathsf{A}}$. A formula in $\mathcal{C}_{\mathsf{A}}$ is built from formulas in $\mathcal{C}$ by means of $\wedge, \neg, \forall x$. And it is said to have *degree* n if it is built from at most n formulas in $\mathcal{C}$. If Γ is defined by a formula of degree 0, $[\Gamma]$ exists by $[\mathcal{C}]^{<\omega}$-ID, since such a formula belongs to $\mathcal{C}$.

Let Γ be an inductive operator defined by $\varphi_0(n,X)\wedge\varphi_1(n,X)$ in $\mathcal{C}_{\mathsf{A}}$. Let Γ_0 and Γ_1 be the operators defined by $\varphi_0(n,X)$ and $\varphi_1(n,X)$, respectively. Then we want to emulate Γ by a multiple inductive definition of these two operators. This multiple inductive definition may be taken as in Richter-Aczel's formulation, and hence also fits for us.

First, we modify two operators Γ_0, Γ_1 as follows:

$$\tilde{\Gamma}_0(X \uplus Y) = X \uplus (X \times \Gamma_0(X)),$$

$$\tilde{\Gamma}_1(X \uplus Y) = (Y_X \cap \Gamma_1(X)) \uplus Y.$$

It is easy to see that the degrees of those operations have not been changed after the modification.

We first note that $\tilde{\Gamma}_0(X \uplus Y)$ reaches a fixed point by one step. So, in the combination, $\tilde{\Gamma}_0$ and $\tilde{\Gamma}_1$ are applied alternatively. We again assume $X \neq \emptyset$ at the beginning. Then by one round of application of $\tilde{\Gamma}_0$ and $\tilde{\Gamma}_1$, we obtain $X \uplus \emptyset, X \uplus (X \times \Gamma_0(X)), (\Gamma_0(X) \cap \Gamma_1(X)) \uplus (X \times \Gamma_0(X))$. For the second round, X is replaced by $(\Gamma_0(X) \cap \Gamma_1(X)) \cup X$ and $Y =$ (old X)$\times$(some set). We may assume that there exists an $x \in$ new X $-$ old X. Otherwise, we have reached a fixed point. By applying $\tilde{\Gamma}_0$ to the new (X, Y), we obtain $X \uplus ((X \times \Gamma_0(X)) \cup$ (old X)$\times$(some set)). Then by applying $\tilde{\Gamma}_1$, we obtain $(\Gamma_0(X) \cap \Gamma_1(X)) \uplus (X \times \Gamma_0(X))$, since $((X \times \Gamma_0(X)) \cup$ (old X) $\times$ (some set)$)_X = ((X \times \Gamma_0(X)) \cup$ (old X) $\times$ (some set)$)_x = \Gamma_0(X)$. Now, we go into the third round of the construction, and so on. In this way, Γ is emulated by a combination of $\tilde{\Gamma}_0$ and $\tilde{\Gamma}_1$.

Next, we take care of negation. Let Γ be an inductive operator defined by $\neg\varphi(n,X) \in \mathcal{C}_{\mathsf{A}}$. Then we want to emulate Γ by a multiple inductive definition (not in Richter-Aczel's) of two operators Γ_0 and Γ_1 with less complexity. In fact, we define

$$n \in \Gamma_0^X(Y) \Leftrightarrow \varphi(n,X),$$

$$n \in \Gamma_1(X) \Leftrightarrow n \notin X.$$

Obviously, $\Gamma_0^X(Y)$ reaches a fixed point by one step. So, in this combination, Γ_0 and Γ_1 are applied alternatively. Since $\Gamma_1(\Gamma_0^X(\emptyset)) = \Gamma(X)$, the inductive construction by Γ is essentially the same as the construction by $[\Gamma_0, \Gamma_1]$.

Finally, we take care of arithmetical generalization. Let Γ be an inductive operator defined by $\forall x \varphi(n,x,X) \in \mathcal{C}_{\mathsf{A}}$. We can emulate Γ by a combination of two operators Γ_0 and Γ_1, defined by

$$(n,x) \in \Gamma_0^X(Y) \Leftrightarrow \varphi(n,x,X),$$

$$n \in \Gamma_1(X) \Leftrightarrow \forall x(n,x) \in X.$$

Then as in the case of negation, the inductive construction by Γ is essentially the same as the construction by $[\Gamma_0, \Gamma_1]$. Therefore, $\mathcal{C}_{\mathsf{A}}$-ID can be deduced from $[[\mathcal{C}]]^{<\omega}$-$\mathsf{ID}$ and hence also from $[\mathcal{C}]^{<\omega}$-ID. □

By the above lemmas, we have

Theorem 3.1. (ACA_0) *For any class* $\mathcal{C} = \Sigma^1_n$ $(n \geq 0)$, $\mathcal{C}_{\mathsf{A}}$*-ID,* $\mathcal{C}_{\mathsf{A}}$*-ID and* $[[\mathcal{C}]]^{<\omega}$*-ID are pairwise equivalent.*

We conclude by the following corollary.

Corollary 3.1. *The following axioms are proof theoretically equivalent:*

- $(\Sigma^0_2)_{<\omega}$-Det_0.
- $(\Sigma^1_1)_{\mathsf{A}}$-$\mathsf{ID}_0$.
- Δ^1_2-ID_0.
- Π^1_2-CA_0.

The proof is straightforward from the result of Heinatsch and Möllerfeld in Ref. 1 together with those in Ref. 7, Ref. 8 and the above results of this note.

Acknowledgments

The author would like to thank Dr. Takako Nemoto for useful discussions and comments.

References

1. C. Heinatsch and M. Möllerfeld, *The determinacy strength of* Π^1_2*-comprehension,* preprint.
2. W. Richter and P. Aczel, *Inductive definitions and reflecting properties of admissible ordinals,* J.E. Fenstad and P.G. Hinman, eds., General Recursion Theory (1974), pp. 301-381.
3. S.G. Simpson, *Subsystems of Second Order Arithmetic,* Springer (1999).
4. K. Tanaka, *Weak axioms of determinacy and subsystems of analysis I:* Δ^0_2*-games,* Zeitschrift für mathematische Logik und Grundlagen der Mathematik 36 (1990), pp. 481-491.
5. K. Tanaka, *Weak axioms of determinacy and subsystems of analysis II:* Σ^0_2*-games,* Annals of Pure and Applied Logic 52 (1991), pp. 181-193.
6. T. Nemoto, M.O. MedSalem and K. Tanaka, *Infinite games in the Cantor space and subsystems of second order arithmetic,* Math. Logic Quarterly 53 (2007), pp. 226-236.

7. M.O. MedSalem and K. Tanaka, Δ_3^0-*determinacy, comprehension and induction,* Journal of Symbolic Logic 72 (2007) pp. 452-462.
8. M.O. MedSalem and K. Tanaka, *Weak determinacy and iterations of inductive definitions,* in the Proc. of Computational Prospects of Infinity, World Scientific (2008), pp. 333-353.

ISOLATION IN THE D.C.E. DEGREES

GUOHUA WU*

Division of Mathematical Sciences
School of Physical and Mathematical Sciences
Nanyang Technological University
Singapore 637371, Republic of Singapore
guohua@ntu.edu.sg

Keywords: Isolation, Lattice embeddings, Ershov hierarchy, CEA hierarchy

1. Introduction

The isolation phenomenon, a structural interaction property among different levels in the Ershov Hierarchy, was first observed and studied by Cooper and Yi in their seminal paper [14], where a d.c.e. degree $\mathbf{d}$ is defined to be isolated by a c.e. degree $\mathbf{a}$ if $\mathbf{a} < \mathbf{d}$ is the greatest c.e. degree below $\mathbf{d}$. Following this, research has been done mainly on exploring the distribution of the isolated degrees, the nonisolated degrees, and the nonisolating degrees, etc. Recent work shows that such a phenomenon can be applied to investigate other local properties of degrees, such as lattice embeddings, branching, boundings/nonboundings, and so on. In this paper, we briefly review various aspects of the isolation phenomenon and illustrate how to apply this to the study of local properties of the degree structure.

The paper is organized as follows. In section 2, we introduce Cooper and Yi's definition of the isolated/nonisolated degrees, and provide basic ideas of the constructions. Some nonbounding phenomena are introduced. In section 3, we consider the density problem of the nonisolating degrees, where Salts' construction of intervals of isolating degrees and the density of such intervals are introduced. We also include in this section the construction of nonisolating degrees, which can be implemented above any incomplete

*The author is partially supported by a research grant No. RG58/06 from NTU.

c.e. degree. In section 4, we show how the isolation phenomenon can be applied to embed lattices into d.c.e. degrees preserving 0 and 1. An alternative proof of Downey's lattice embeddings is provided. In section 5, we show how to obtain the isolating/isolated degrees from Kaddah's branching theorem. The dual of this approach is proposed. In section 6, we consider the usage of the Lachlan set of n-CEA sets in the study of local properties of the degree structures.

Some of the results contained in this paper are new, such as the density of isolating intervals (section 2), embedding the nondistributive lattices $1-n-1$ into the d.c.e. degrees preserving 0 and 1 (section 4), the density of isolated n-c.e. degrees (section 6), and so on. Many open problems are listed.

Our notation and terminology are standard and generally follow Soare [32]. The readers can refer Cooper's paper [10] and Arslanov's paper [2] for the general idea on local degree theory.

2. Isolated vs. nonisolated degrees

Recall that a Turing degree is properly d.c.e. if it contains a d.c.e. set, but no c.e. set. Cooper proved in [8] the existence of properly d.c.e. degrees, and later in [12], Cooper, Lempp and Watson proved that properly d.c.e. degrees exist between any two c.e. degrees. In the early stage, Lachlan observed that any nonzero d.c.e. degree bounds a nonzero c.e. degree, the first result on the structural relation between the d.c.e. degrees and the c.e. degrees. That is, given a d.c.e. set D with an effective approximation $\{D_s : s \in \omega\}$, the associated set

$$E = \{\langle x, s\rangle : x \in D_s - D\}$$

is obviously c.e., and is Turing reducible to D, while D is c.e. in E. Furthermore, if D has a proper d.c.e. degree, then E is not computable. *E is called the Lachlan set of D, and sometimes, we denote it as $\tilde{D}$.* Lachlan's observation has several applications in the study of the structure of the d.c.e. degrees, especially when combined with the isolation phenomenon. We will describe such applications in section 6.

Arslanov's cupping theorem [1] and Downey's diamond theorem ([16]) show that the structures of the d.c.e. degrees and the c.e. degrees are not elementarily equivalent. The most striking difference between these two structures is the following nondensity result:

Theorem 1. (Nondensity Theorem for $\mathcal{D}_2$) (Cooper, Harrington, Lachlan, Lempp and Soare [11]) *There exists a maximal d.c.e. degree* $\mathbf{d} < \mathbf{0}'$, *and hence the d.c.e. degrees are not dense.*

The proof of Theorem 1 actually shows that between the constructed $\mathbf{d}$ and $\mathbf{0}'$, there is no ω-c.e. degree. It involves an extremely complicated $0'''$ argument, which makes it not easy to modify for other purposes, such as resolving whether below any high c.e. degree $\mathbf{h}$, we can find a d.c.e. degree $\mathbf{d}$ such that no d.c.e. degree is between $\mathbf{d}$ and $\mathbf{h}$.

In contrast to this nondensity theorem, Ishmukhametov [21], and independently, Cooper and Yi [14], proved that any interval $(\mathbf{a}, \mathbf{d})$, with $\mathbf{a}$ c.e., contains a d.c.e. degree, a weak density theorem for d.c.e. degrees.

Theorem 2. (Ishmukhametov [21]; Cooper and Yi [14]) If $\mathbf{d}$ is a d.c.e. degree and $\mathbf{a} < \mathbf{d}$ is a c.e. degree, then there is a d.c.e. degree $\mathbf{c}$ strictly between $\mathbf{a}$ and $\mathbf{d}$.

Cooper and Yi pointed out in [14] that we cannot require the degree $\mathbf{c}$ above be c.e., as they noticed the existence of d.c.e. degrees having all the c.e. predecessors bounded by c.e. degrees (the greatest ones). This led Cooper and Yi to study the isolation phenomenon in the d.c.e. degrees.

Definition 3 (Cooper and Yi [14]) A d.c.e. degree $\mathbf{d}$ is isolated by a c.e. degree $\mathbf{a}$ if $\mathbf{a} < \mathbf{d}$ is the greatest c.e. degree below $\mathbf{d}$. A d.c.e. degree $\mathbf{d}$ is isolated, if it is isolated by some c.e. degree $\mathbf{a}$.

The construction of isolated degrees is a "wait and catch an error" process: we are constructing a d.c.e. set D and a c.e. set A such that $D \not\leq_T A$ and such that each c.e. set W reducible to $A \oplus D$ (via a given p.c. functional Φ) is also reducible to A (via a p.c. functional Γ constructed by us), and for the latter, we extend Γ on more and more arguments at the expansionary stages. If at some expansionary stage s_1, $\Gamma^A(x)[s_1] \neq W_{s_1}(x)$ appears, then we know that x enters W recently, and we will recover D to its previous status, at the last expansionary stage, s_0 say, to make $\Phi^{A,D}(x) = 0$. This creates a disagreement between $\Phi^{A,D}$ and W because

$$\Phi^{A,D}(x) = 0 \neq 1 = W(x).$$

For the consistency of the strategies, we also enumerate s_0 into A, which keeps this disagreement strategy described above, and also all the isolation strategies with higher priority, unaffected.

After showing the existence of the isolated degrees, Cooper and Yi continued to show the existence of the nonisolated degrees, where a d.c.e. degree **d** is nonisolated if no c.e. degree below **d** can isolate **d**. Obviously, it becomes nontrivial only when **d** is properly d.c.e. Cooper and Yi actually proved the existence of a properly d.c.e. degree as a minimal upper bound of a uniformly c.e. sequence of degrees, an even stronger result. Nonisolated degrees can be constructed directly as follows: construct a d.c.e. set D with proper d.c.e. degree (Cooper's construction developed in his Ph.D. thesis [8]) with the following property that for any c.e. set W reducible to D via a p.c. functional Φ, we construct a c.e. set C_W reducible to D but not reducible to W. C_W is constructed to satisfy the following requirements:

$$\mathcal{S}_{\Phi,W,\Psi} : C_W \neq \Psi^W,$$

where W is a computably enumerable set, and Ψ is a partial computable functional. To satisfy $\mathcal{S}_{W,\Phi,\Psi}$, we select a number (big) x, and wait for $\Psi^W(x)$ to converge to 0, and also Φ^D and W agree on the numbers no larger than the associated use $\psi(x)$. We then put x into both D and C_W. Here the enumeration of x into D is for the sake of reducing C_W to D, and this enumeration may change some computation $\Phi^D(y)$ at some $y \leq \psi(x)$, which can allow W to change at y (we will see it at the next expansionary stage). If it happens, then we remove x from D — this removal makes C_W not reducible to D, but we do not care about this now, as we have a global win on this requirement: $\Phi^D(y) = 0 \neq 1 = W(y)$.

In both constructions, the requirements can be satisfied via diagonalization arguments. This feature admits the possibility of showing that both the isolated degrees and the nonisolated degrees are dense in the structure of c.e. degrees.

Theorem 4. (Ding and Qian [15]; LaForte [27]; Arslanov, Lempp and Shore [5]) Both the isolated d.c.e. degrees and the nonisolated d.c.e. degrees are dense in the c.e. degrees.

Theorem 4 says that the isolated degrees could be as close to the isolating degrees as wanted. It is natural to ask whether the isolated degrees can also be far away from the isolating degrees, in terms of the high/low hierarchy. Ishmukhametov and Wu provided an affirmative answer in [22].

Theorem 5. (Ishmukhametov and Wu [22], Li, Wu and Yang [29]) There is a high d.c.e. degree **d** isolated by a low c.e. degree **c**. Such a c.e. degree **c** can be found below any nonzero c.e. degree **a**.

There are two new features in the proof of Theorem 5. One feature is that to make **d** high, we apply the standard highness strategy to code TOT into $\mathbf{d}'$, and when an isolation strategy α acts, that is, when we want to recover a computation to stage s_0, we remove all the numbers enumerated into D by lower priority strategies after stage s_0. The other feature is that to satisfy the lowness requirements, we give the lowness strategies the highest priority. That is, at each stage, we first check whether there is a lowness strategy requesting to preserve a computation. If yes, then we do it by initializing all the strategies with lower priorities. Such actions are performed off the true path.

The following theorem of Chong, Li and Yang follows Theorem 5 immediately, if we let **a** in Theorem 5 be a Lachlan nonbounding degree.

Theorem 6. (Chong, Li and Yang;[7] Li, Wu and Yang[29]) There exists a high d.c.e. degree bounding no minimal pair of c.e. degrees.

Note that, by Cooper,[9] the high d.c.e. degree in Theorem 6 cannot be c.e., as every high c.e. degree bounds a minimal pair of c.e. degrees. Another question along this topic is whether a nonisolated d.c.e. degree is really nonisolated.

Definition 7. (Wu[34]) A d.c.e. degree **d** is pseudo-isolated by a d.c.e. degree $\mathbf{b} < \mathbf{d}$ if **b** itself is nonisolated, and **b** bounds all c.e. degrees below **d**.

It is easy to see that any pseudo-isolated d.c.e. degree is nonisolated, as if **a** is a c.e. degree below **d**, then **a** is also below **b**, and as **b** is nonisolated, there is a c.e. degree **e** below **b** (hence below **d**) but not below **a**. The construction of a pseudo-isolated degree **d** involves a construction of nonisolated degree **b** below **d** with the isolating property.

Theorem 8. (Wu[34]) The pseudo-isolated d.c.e. degrees are dense in the c.e. degrees.

As the construction of pseudo-isolated degrees is similar to the one for constructing isolated degrees, many results proved for the isolated degrees are also true for the pseudo-isolated degrees.

We end this section with the question of whether the class of pseudo-isolated degrees is properly contained in the class of nonisolated degrees.

Question 9. Are all the nonisolated d.c.e. degrees *pseudo-isolated*?

3. Intervals of isolating degrees vs. nonisolating degrees

We call a c.e. degree $\mathbf{c}$ an isolating degree, if there is a d.c.e. degree $\mathbf{d} > \mathbf{c}$ isolated by $\mathbf{c}$. If $\mathbf{c}$ is not isolating, then $\mathbf{c}$ is called a nonisolating degree. As the isolated degrees are dense in the c.e. degrees, the isolating degrees are also dense in the c.e. degrees. We are interested in the question whether the nonisolating degrees are dense in the c.e. degrees.

In [5], Arslanov, Lempp and Shore showed the existence of the nonisolating degrees, and proved that these degrees are downwards dense in the c.e. degrees, and can occur in every jump class. Salts proved in [31] that the nonisolating degrees are not dense in the c.e. degrees.

Theorem 10 (Salts [31]) There is an interval of c.e. degrees, each of which isolates a d.c.e. degree.

Sketch of the proof: The construction of such an interval is in essence a modification of the one constructing an isolated degree. We will construct c.e. degrees $\mathbf{e} > \mathbf{c}$ such that any c.e. degree between $\mathbf{c}$ and $\mathbf{e}$ isolates a d.c.e. degree. That is, we will construct c.e. sets C, E, infinitely many d.c.e. sets, D's, and partial computable functionals, Γ's, satisfying the following requirements:

$$\mathcal{P}_e\colon E \neq \Phi_e^C;$$
$$\mathcal{Q}_{e,i,j}\colon W_e = \Phi_i^{C,E} \Rightarrow D_{e,i} \neq \Psi_j^{C,W_e};$$
$$\mathcal{R}_{e,i,k,l}\colon W_e = \Phi_i^{C,E} \;\&\; V_k = \Omega_l^{C,W_e,D_{e,i}} \Rightarrow V_k = \Gamma_{e,i,k,l}^{C,W_e}.$$

where Φ, Ψ, Ω are partial computable functionals and V, W are c.e. sets.

Let $\mathbf{c}, \mathbf{e}$ be the degrees of C and $C \oplus E$ respectively. By the $\mathcal{P}$-strategies, $\mathbf{c} < \mathbf{e}$. Now let $\mathbf{f}$ be any c.e. degree between $\mathbf{c}$ and $\mathbf{e}$, and let $C \oplus W_e$ be a c.e. set in $\mathbf{f}$, where $W_e = \Phi_i^{C,E}$. Then, the $\mathcal{Q}_{e,i,-}$ requirements ensure that $D_{e,i}$ is not reducible to $C \oplus W_e$, and the $\mathcal{R}_{e,i,-,-}$ requirements ensure that every c.e. set reducible to $C \oplus W_e \oplus D_{e,i}$ is also reducible to $C \oplus W_e$. Let $\mathbf{d}_{e,i}$ be the degree of $C \oplus W_e \oplus D_{e,i}$. It is clear now that $\mathbf{f}$ isolates $\mathbf{d}_{e,i}$.

A $\mathcal{P}$-strategy is a standard Friedberg-Muchnik strategy, and a $\mathcal{Q}$-strategy is a variant of the Friedberg-Muchnik strategy. Here, we only give a brief description of a $\mathcal{Q}$-strategy.

We first select a number x as a witness, and wait for

(1) $\Psi_j^{C,W_e}(x)$ to converge to 0 with a use $\psi_j(x)$,
(2) W_e and $\Phi_i^{C,E}$ agree on all arguments below $\psi_j(x) + 1$.

Obviously, if $W_e = \Phi_i^{C,E}$ *and* $\Psi_j^{C,W_e}(x)$ *converges to 0, then we will have a stage such that both (1) and (2) are true.* In this case, we put x into $D_{e,i}$, and prevent other strategies from enumerating small numbers into C and E. This preservation will keep $C \upharpoonright (\psi_j(x)+1)$ and $E \upharpoonright (\psi_j(x)+1)$, and hence $W_e \upharpoonright (\psi_j(x)+1)$ the same (if $W_e = \Phi_i^{C,E}$). Thus, we will have $\Psi_j^{C,W_e}(x) = 0 \neq 1 = D_{e,i}(x)$, and $\mathcal{Q}$ is satisfied.

Note that here we have control of C and E, and as a consequence, we will have either $\Psi_j^{C,W_e}(x) = 0 \neq 1 = D_{e,i}(x)$ or W_e and $\Phi_i^{C,E}$ do not agree on $\psi_j(x)+1$. *Note that the* $\mathcal{P}$ *and* $\mathcal{Q}$ *strategies do not enumerate numbers into* C.

Now we consider how to satisfy a single $\mathcal{R}$ requirement. Let α be a strategy designed to satisfy this requirement. For simplicity, we omit the subscript, and we will indicate the subscripts if needed. We assume that $W = \Phi^{C,E}$ and $V = \Omega^{C,W,D}$ and we will construct a partial computable functional Γ such that $V = \Gamma^{C,W}$. We extend the definition of Γ at the expansionary stages, which are defined as follows. First we define a function measuring the length of agreement between $W = \Phi^{C,E}$ and $V = \Omega^{C,W,D}$ at stage s:

(1) $\ell(\alpha, s) = \max\{x : \text{for all } y < x,\ V(y)[s] = \Omega^{C,W,D}(y)[s] \text{ and } W_s \upharpoonright (\omega(y)[s]+1) = \Phi^{C,E}[s] \upharpoonright (\omega(y)[s]+1)\}$.

(2) $m(\alpha, s) = \max\{\ell(\alpha, t) : t < s \text{ and } t \text{ is an } \alpha\text{-stage}\}$.

In the following, we call $\ell(\alpha, s)$ an agreement function. Say that a stage s is α-expansionary if $s = 0$ or s is a α-stage with $\ell(\alpha, s) > m(\alpha, s)$. We define $\Gamma(n)$ only when $\ell(\alpha, s) > n$, and when we define it at stage s, we define it as $V_s(n)$ with use $\gamma(n)$ bigger than s. $\Gamma(n)$ can be undefined later only when C or W changes below $\omega(n)[s]$. This ensures the totality of Γ if both $W = \Phi^{C,E}$ and $V = \Omega^{C,W,D}$ hold.

We now consider the correctness of Γ. A nontrivial case is when a lower priority $\mathcal{Q}$-strategy, guessing that $W = \Phi^{C,E}$ and $V = \Omega^{C,W,D}$, enumerates a number d into D, and this enumeration can change computation $\Omega^{C,W,D}(n)$. W.l.o.g., we assume that $\Gamma(n)$ has definition at this stage. Then at the next expansionary stage, $V(n)$ can have been changed, but $\Gamma(n)$ is still equal to 0, because no C or W-change can undefine $\Gamma(n)$. This means that Γ becomes incorrect. In this case, as we know that this incorrectness of Γ is caused by the enumeration of d, we can take d out of D, to recover the previous computation of $\Omega^{C,W,D}(n)$, and by preserving C, D, E, we get a disagreement between V and $\Omega^{C,W,D}$ because $V(n)$ is equal to 1, while

$\Omega^{C,W,D}(n)$ is equal to 0. In this case, $\mathcal{R}$ is satisfied vacuously, and we can satisfy the mentioned $\mathcal{Q}$-strategy via a new witness.

However, the removal of d can cause another Γ' (defined by those higher priority $\mathcal{R}$-strategies) to be incorrect. Suppose that d is enumerated into D at stage s_1 and is removed from D at stage s_2, and that $\Gamma'(n')$ is defined at a stage between s_1 and s_2, where the computation $\Omega'(n')$ depends on "$d \in D$". Then the removal of d will definitely change this computation, and as described before, at the next expansionary stage, $V'(n')$ may have been changed, and $\Gamma'(n')$ becomes wrong. We cannot recover the corresponding computation by enumerating d into D, as we are making D d.c.e.

The main idea for getting around this problem is that when we remove d out of D, we also enumerate s_1 into C, to undefine all $\Gamma'(-)$ defined after stage s_1. In particular, the incorrectness of $\Gamma'(n')$ mentioned cannot happen. Also note that the enumeration of s_1 into C will not cause those $\Gamma'(n'')$, defined before s_1, incorrect, as s_1 is bigger than all the corresponding computations involved.

Recently, Cenzer, LaForte and Wu proved in [6] that the isolating intervals are dense in the c.e. degrees.

Theorem 11. (Cenzer, LaForte and Wu [6]) The isolating intervals are dense in the c.e. degrees, and hence the nonisolating degrees are nowhere dense in the c.e. degrees.

In contrast to Theorem 11, Cooper, Salts and Wu proved in [13] that the nonisolating degrees are upwards dense in the c.e. degrees.

Theorem 12. (Cooper, Salts and Wu [13]) For any incomplete c.e. degree $\mathbf{a} < \mathbf{0}'$, there is an incomplete nonisolating degree $\mathbf{c}$ above $\mathbf{a}$.

The main idea of the approach in the proof of this theorem is different from Arslanov, Lempp and Shore's approach given in [5], who observed that for any c.e. degree $\mathbf{b}$ and d.c.e. degree $\mathbf{d} > \mathbf{b}$, there is a degree $\mathbf{c}$ c.e. in $\mathbf{b}$ such that $\mathbf{b} < \mathbf{c} < \mathbf{d}$, and then applying this observation to prove that there is a c.e. degree failing to isolate any 2-CEA degree c.e. in it. Our construction of a nonisolating degree given below tries to avoid working relative to these 2-CEA degrees, which are hard to control if we work above a given c.e. degree. We will construct c.e. sets A, C satisfying the following requirements:

$\mathcal{P}_e$: $A \neq \Phi_e$;
$\mathcal{Q}_e$: $C \neq \Phi_e^A$;
$\mathcal{R}_e$: $\widetilde{D}_e = \Phi_e^A \;\Rightarrow\; \exists B_e \leq_T D_e \oplus A(B_e \not\leq_T A) \;\vee\; D_e \leq_T A$;

where $\{(D_e, \Phi_e) : d \in \omega\}$ is an effective list of pairs (D, Φ), D is a d.c.e. set and Φ is a partial computable functional. Here, $\widetilde{D}$ is the Lachlan set of D, with respect to an effective (d.c.e.) approximation $\{D_s : s \in \omega\}$:

$$\widetilde{D} = \{\langle x, s\rangle : x \in D_s \;\&\; \exists t > s(x \notin D_t)\}.$$

Obviously, $\widetilde{D} \leq_T D$ and it is a c.e. set. From the approximation $\{D_s : s \in \omega\}$, we have the following effective enumeration of $\widetilde{D}$: $\langle x, s\rangle$ is enumerated into $\widetilde{D}$ at stage t if $x \in D_s$ and t is the least stage such $x \notin D_t$.

We only need to describe the idea for satisfying an $\mathcal{R}$-requirement, a nonisolating requirement. That is, for a d.c.e. set D, if D is not reducible to A, then we want to find a c.e. set reducible to $A \oplus D$, but not reducible to A, so A does not isolate $A \oplus D$. This c.e. set can be $\widetilde{D}$, a natural candidate, or someone else, B, the one we will construct. On the other hand, if D itself is reducible to A (of course, $\widetilde{D}$ is reducible to A), then we need to show this fact.

To satisfy $\mathcal{R}_e$, we need to construct a c.e. set B_e, a p.c. functional Γ_e, such that $B_e = \Gamma_e^{A \oplus D}$. At the same time, we also want to ensure that $B_e \not\leq_T A$, provided that D is not reducible to A. That is, the following requirements should be also satisfied:

$\mathcal{S}_{e,i}$: $B_e \neq \Phi_i^A$ or there is a p.c. functional $\Delta_{e,i}$ such that $D = \Delta_{e,i}^A$.

An $\mathcal{R}_e$ strategy defines Γ_e at expansionary stages, and has two outcomes: finitary outcome f and infinitary outcome ∞ with $\infty <_L f$. Below outcome ∞, we will guess that there are infinitely many expansionary stages, and we will have substrategies $\mathcal{S}_{e,i}$, $i \in \omega$, which will work together to construct B_e, and to satisfy the requirement $\mathcal{R}_e$.

An $\mathcal{S}_{e,i}$-module consists of (infinitely) many steps, where each step n tries to find a number x_n such that either $\Phi_i^A(x_n) \neq B_e(n)$ or $\Delta_{e,i}^A(n)$ is defined. Step n works as follows:

(1) Choose x_n as a big number, and wait for $\Phi_i^A(x_n) \downarrow = 0$.

(2) Suppose that $\Phi_i^A(x_n) \downarrow= 0$ at stage s.

We define $\Delta_{e,i}^A(n) = D(n)$ with use $\delta_{e,i}(n) = \varphi_i(x_n)$. Here again, when we see $\Phi_i^A(x_n)[s] \downarrow= 0$, we do not put restraint on A to preserve this computation. Whenever A changes below $\delta_{e,i}(n)$, go back to (2), in which case $\Delta_{e,i}^A(n)$ is undefined by this A-change.

(Here, n can be in D_s or not in in D_s.
If n is currently not in D_s, then n can enter D later, and leave at a further stage. When n enters D, at stage s', we will use the assumption that $\widetilde{D}$ is equal to Φ_e^A, as at stage s', $\langle n, s\rangle$ is not in $\widetilde{D}$, and we want to restrain A from changing to preserve this computation. If later, n leaves D, then $\langle n, s\rangle$ will enter $\widetilde{D}$, making $\widetilde{D}$ and Φ_e^A disagree at $\langle n, s\rangle$.
If n is in D_s, we do nothing here, as if n leaves D later, this change will allow us to undefine $\Gamma^{A\oplus D}(x_n)$, and we can put $x_{j,n}$ into B_e. Note that axioms in Γ_e before n enters D are all invalid by the A-changes, as otherwise we will be in the situation described in the last paragraph, where n leaving D will make a disagreement between $\widetilde{D}$ and Φ_e^A.)

Wait for $D(n)$ to change, and simultaneously start the step $n+1$.

(3) $D(n)$ changes at stage $t > s$. There are two possibilities.

(a) n enters D at stage t. In this case, this $D(n)$ change undefines $\Gamma_e^{A\oplus D}(x_n)$, and instead of putting x_n into B_e immediately at this stage, we wait for $\widetilde{D}$ and Φ_e^A agree on (all numbers $\leq$) $\langle n, t\rangle$.

(We delay the enumeration of x_n into B_e mainly because n may leave D later, and ensure that $\Gamma_e^{A\oplus D}$ is correct, we are forced to put a number into A to change $\Gamma_e^{A\oplus D}(x_n)$, where little progress is done.)

(b) n leaves D at stage t. Then $\langle n, s\rangle$ enters $\widetilde{D}$ at this stage. As the computation $\Phi_e^A(\langle n, s\rangle) = 0$ is preserved at stage s', by keeping this restraint on A, we get a global win as

$$\Phi_e^A(\langle n, s\rangle) = 0 \neq 1 = \widetilde{D}(\langle n, s\rangle).$$

(4) At stage $t' > t$, $\widetilde{D}$ and Φ_e^A agree on (all numbers $\leq$) $\langle n, t\rangle$. We enumerate x_n into B_e, and put restraint on A to preserve both computations $\Phi_i^A(x_n) = 0$, and $\Phi_e^A(\langle n, t\rangle) = 0$.

(Here at stage t, n enters D, which undefines $\Gamma_e^{A\oplus D}(x_n)$. From now on, we wait for the agreement between $\widetilde{D}$ and Φ_e^A to exceed $\langle n,t\rangle$, and in this period, we do not define $\Gamma_e^{A\oplus D}(x_n)$. Thus, at stage t', we can enumerate x_n into B_e directly, as $\Gamma_e^{A\oplus D}(x_n)$ is undefined at this stage. We will see that in the whole construction, only $\mathcal{P}$ strategies put numbers into A. This explains why **0** is nonisolating and why the nonzero nonisolating degrees are downwards dense in the c.e. degrees. We will see in the next section that it will not be like this when we prove the upwards density, where a change of K is required.)

(5) At stage $t'' > t'$, n leaves D. Then as in 4(b), $\langle n,t\rangle$ enters $\widetilde{D}$ at this stage. As the computation $\Phi_e^A(\langle n,t\rangle) = 0$ is preserved at stage t', by keeping this restraint on A, we get a global win as

$$\Phi_e^A(\langle n,t\rangle) = 0 \neq 1 = \widetilde{D}(\langle n,t\rangle).$$

We now consider the outcomes of $\mathcal{S}_{e,i}$, which can run for finitely or infinitely many steps. Note that whenever step n_2 is in progress, and we have an A-change so that we come back to 2 of step $n_1 < n_2$, then $\Delta_{e,i}^A(n_2)$ is undefined automatically by this A-change. After this, we need to start step n_2 from 1. That is, choose a new x_{n_2}.

If some step n passes 3, then we win either by 4(b) or 6, which is a global win as $\Phi_e^A \neq \widetilde{D}$, and this disagreement is preserved forever. This means that there will be no more $\mathcal{R}_e$-expansionary stage, or 5 applies where we get a D-change at n, and this n remains in D, ensuring that the previous axioms enumerating x_n into $\Gamma^{A,D}$ is invalid forever.

If no step passes 3, then note that we put no restraint on A. It can happen that there is a least n such that $\Phi_e^A(x_n)$ does not converge to 0, or for each n, $\Phi_e^A(x_n)$ converges to 0. $\mathcal{S}_{e,i}$ is satisfied in both cases, as in this former case, $B_e(x_n) = 0 \neq \Phi_e^A(x_n)$, and in the latter case, as each step stays at 3, and hence $\Delta_{e,i}^A(n)$ is defined and equal to $D(n)$. So $\Delta_{e,i}^A = D$.

So this $\mathcal{S}_{e,i}$ module has two outcomes: w and s with $w <_L s$. Here w denotes for the case that no step passes 3, and s for the case that some step passes 3. In the latter case, we put restraint on A to preserve computations.

The implementation of the foregoing construction above an incomplete c.e. degree is a complicated $0'''$ argument. We suggest the readers read the paper [13] for details.

4. Isolation and lattice embeddings

In [16], Downey showed that the diamond lattice can be embedded into the d.c.e. degrees preserving 0 and 1. By Lachlan's nondiamond theorem, such an embedding cannot be realized in the c.e. degrees.

Theorem 13. (Downey's Diamond Theorem [16]) There are two incomparable d.c.e. degrees $\mathbf{d}_1, \mathbf{d}_2$ such that $\mathbf{d}_1$ cups $\mathbf{d}_2$ to $\mathbf{0}'$ and caps $\mathbf{d}_2$ to $\mathbf{0}$.

In this section, we show how to construct such an embedding by using the isolation phenomenon. The following proposition is crucial for the construction.

Proposition 14. (Wu [33]) Let $\mathbf{d} > \mathbf{a}$ be a d.c.e. degree isolated by $\mathbf{a}$ and $\mathbf{c}$ be a c.e. degree. If $\mathbf{c}$ and $\mathbf{a}$ form a minimal pair, then $\mathbf{c}$ and $\mathbf{d}$ form a minimal pair.

The proof of this proposition is quite easy. Suppose that $\mathbf{c}$ and $\mathbf{d}$ do not form a minimal pair. That is, there is a nonzero d.c.e. degree $\mathbf{e} \leq \mathbf{c}, \mathbf{d}$. By Lachlan's observation mentioned above, $\mathbf{e}$ can be chosen as c.e. As $\mathbf{a}$ is supposed to be the greatest c.e. degree below $\mathbf{d}$, $\mathbf{e} \leq \mathbf{a}$. Now by $\mathbf{a} \cap \mathbf{c} = \mathbf{0}$, $\mathbf{e} = \mathbf{0}$, contradicting the choice of $\mathbf{e}$.

Theorem 15. (Wu [33]) There are c.e. degrees $\mathbf{a}, \mathbf{c}$ and a d.c.e. degree $\mathbf{d}$ such that $\mathbf{a}, \mathbf{c}$ form a minimal pair, $\mathbf{a}$ isolates $\mathbf{d}$ and $\mathbf{c}$ cups $\mathbf{d}$ to $\mathbf{0}'$. By Proposition 14, $\{\mathbf{0}, \mathbf{c}, \mathbf{d}, \mathbf{0}'\}$ is a diamond embedding.

As noted in [33], together with Harrington-Soare's theorem on the continuity of capping degrees (see [20]), Theorem 15 implies that the nondistributive lattice N_5 can also be embedded into the d.c.e. degrees preserving 0 and 1. That is, according to [20], there is a c.e. degree $\mathbf{f} > \mathbf{c}$ such that $\mathbf{f}$ and $\mathbf{a}$ form a minimal pair. Now by Proposition 14, $\mathbf{f} \cap \mathbf{d} = \mathbf{0}$. Thus, $\{\mathbf{0}, \mathbf{c}, \mathbf{f}, \mathbf{d}, \mathbf{0}'\}$ is an N_5 embedding.

The basic idea of proving Theorem 15 is to make the cupping part ($\mathbf{c} \cup \mathbf{d} = \mathbf{0}'$) and the capping part ($\mathbf{c} \cap \mathbf{a} = \mathbf{0}$) separated. That is, to make $\mathbf{c} \cup \mathbf{d} = \mathbf{0}'$, we construct a p.c. functional Γ such that $K = \Gamma^{C,D}$, and in general, when x enters K, we enumerate $\gamma(x)$ into D. We cannot just enumerate the γ-uses into D, as otherwise, D would be complete. In the construction, we enumerate $\gamma(x)$ into C when x is a threshold of an isolation strategy, and this strategy is ready to make a disagreement, and this enumeration preserves the computation we want to recover to from the coding of K. This action increases the coding markers (except for those $y \leq x$, which can happen at most $x+1$ many times) and does not involve enumerating numbers into A. At the next stage when this node is visited again, we complete the disagreement argument (the standard one) by removing numbers from D to recover the computation, and hence to create a disagreement. Remember that we also need to enumerate a number into A for the consistency among the isolation strategies. As no number is enumerated into C at this stage, completing a disagreement argument is consistent with the Lachlan's well-known minimal pair argument, which ensures our construction works.

The idea of separating the cupping from the capping via isolation was used in [17], where Downey, Li and Wu proved that a nonzero c.e. degree is cappable if and only it has a complement in the d.c.e. degrees.

Theorem 16. (Downey, Li and Wu [17]) If $\mathbf{c} > \mathbf{0}$ is cappable, then there are a c.e. degree $\mathbf{a}$ and a d.c.e. degree $\mathbf{d}$ such that $\mathbf{a}$ isolates $\mathbf{d}$ and $\mathbf{c}$ cups $\mathbf{d}$ to $\mathbf{0}'$ and caps $\mathbf{a}$ to $\mathbf{0}$. As a consequence, a c.e. degree is cappable if and only if it has a complement in the d.c.e. degrees.

The idea presented above can also be applied to prove that the lattices $1-n-1$, $n \geq 2$, can be embedded into the d.c.e. degrees preserving 0 and 1, a statement first announced in Downey's paper [16].

Theorem 17. (Downey [16]) The nondistributive lattice M_3 can be embedded into the d.c.e. degrees preserving 0 and 1. In general, the lattices $1-n-1$, $n \geq 2$, can be embedded into the d.c.e. degrees preserving 0 and 1.

So far no proof of such an embedding has been provided so far in the literature. In the remainder of this section, we show how to embed M_3, and hence in general, $1-n-1$, into the d.c.e. degrees preserving 0 and 1 by using the isolated degrees.

Constructing an M_3 in the c.e. degrees is much more complicated than constructing a diamond. For M_3, the interactions between the coding argument and the minimal pair argument is quite subtle, and needs a powerful technique, that of tracing, to deal with it. For embedding M_3 into the d.c.e. degrees, we show how the proof of Theorem 15 can be extended directly to prove Theorem 17 — the following theorem is the main step.

Theorem 18. There are c.e. degrees $\mathbf{c}_0, \mathbf{c}_1, \mathbf{c}_2$ and d.c.e. degrees $\mathbf{d}_0, \mathbf{d}_1, \mathbf{d}_2$ such that for each i, $\mathbf{c}_i$ isolates $\mathbf{d}_i$, and for each $j \neq i$, $\mathbf{d}_i \cup \mathbf{d}_j = \mathbf{0}'$,

$\mathbf{c}_i \cap \mathbf{c}_j = \mathbf{0}$. As a consequence, $\{\mathbf{0}, \mathbf{d}_0, \mathbf{d}_1, \mathbf{d}_2, \mathbf{0}'\}$ form an M_3 embedding into the d.c.e. degrees preserving 0 and 1.

We will construct c.e. sets C_0, C_1, C_2 and d.c.e. sets D_0, D_1, D_2, and p.c. functionals $\Gamma_0, \Gamma_1, \Gamma_2$ such that for each i,

(1) C_i isolates $C_i \oplus D_i$, and
(2) C_i and C_j form a minimal pair, where $j \neq i$, and
(3) $\Gamma_i^{D_j, D_k} = K$, where i, j, k are pairwise unequal.

Again, in general, to rectify Γ_i, we always enumerate the γ-uses $\gamma_i(x)$ into D_0 when $i = 1, 2$, and into D_1 when $i = 0$, and we enumerate $\gamma_i(x)$ into the other associated set only when an isolation strategy is ready to perform a disagreement, to lift the γ_i-uses to bigger numbers. Exactly as in the proof of Theorem 15, when a disagreement is created, a number is enumerated into the corresponding C. Note that here, when an isolation strategy with C_0, D_0 involved is ready to perform the disagreement, $\gamma_1(k)$ and $\gamma_2(k)$ are enumerated into D_1 and D_2 respectively, as we want some computation to be clear of the γ_1, γ_2-uses. At the next stage when this strategy is visited again we perform the disagreement argument by recovering D_0 to a previous stage, and also enumerating a certain number into C_0 for the consistency with other isolation strategies. Obviously, this disagreement argument is consistent with the minimal pair argument.

We leave it to the readers to realize the idea described above.

5. Isolation and infima of d.c.e. degrees

In this section we consider the relation between isolation and infima of d.c.e. degrees. In [14], Cooper and Yi pointed out that the existence of isolated d.c.e. degrees can be obtained from Kaddah's paper [25] that any low c.e. degree is branching in the d.c.e. degrees.

Theorem 19. (Kaddah [25]) Every low c.e. degree is branching in the d.c.e. degrees.

The crucial idea of Kaddah's proof is to avoid the "dangerous intervals", which does happen in the construction of degrees branching in the c.e. degrees. For this aim, Kaddah introduced the concept of "leading strategies", and whenever the given low c.e. set changes, we will have a check at the leading strategy, where a sequence of expansionary stages are listed, and determine from which stage, which elements should be removed, to make sure that each branching strategy on the tree can avoid the "dangerous interval" — in this sense, the minimal pair argument is a special case of Kaddah's argument. Obviously, the lowness requirement has the highest priority. As the given c.e. set is low, the expansionary stages on the list of the leading strategy will become stable from part to part, and hence the construction works.

Now we compare Kaddah's theorem with Fejer's density theorem for nonbranching degrees. Let $\mathbf{c}$ be any low nonbranching degree — Fejer's density theorem for nonbranching degrees says that $\mathbf{c}$ exists. Theorem 19 shows that there are d.c.e. degree $\mathbf{d}_1, \mathbf{d}_2$ such that $\mathbf{d}_1 \cap \mathbf{d}_2 = \mathbf{c}$. Immediately, we have that one of $(\mathbf{c}, \mathbf{d}_0)$ and $(\mathbf{c}, \mathbf{d}_1)$ contains no c.e. degree, which implies that $\mathbf{c}$ isolates one of $\mathbf{d}_0$ and $\mathbf{d}_1$. The following corollary is obvious now.

Corollary 20. (from Kaddah [25]) Every low nonbranching c.e. degree is isolating, and hence the isolating degrees are dense in the low c.e. degrees.

Corollary 20 cannot be obtained by direct construction of isolated degrees. As Kaddah's branching theorem provides another approach to show the existence of isolating degrees, we ask whether the known results about the isolated degrees can be obtained from Kaddah's branching theorem.

Question 21. Is every low c.e. degree the infimum of two high d.c.e. degrees?

Obviously, an affirmative answer to Question 21 has Theorem 5 as a direct corollary.

Question 22. Is each low c.e. degree the nontrivial infimum of two d.c.e. degrees with join $\mathbf{0}'$?

At first glimpse, one might believe that the answer is affirmative and that the proof should be easy, maybe by a simple combination of Kaddah's construction and a splitting of K. However, it turns out to be much harder: both the coding argument (enumerating numbers into sets) and the lowness strategy (Kaddah's removal actions performed according to the leading strategy) require the highest priority and form a direct conflict.

Kaddah's branching theorem shows that the interval of c.e. degrees branching in the d.c.e. degrees are dense in the low c.e. degrees. We ask:

Question 23. Are the intervals of c.e. degrees branching in the d.c.e. degrees are dense in the c.e. degrees?

As discussed above, an affirmative answer to this question, together with Fejer's density theorem of nonbranching degrees, provides an alternative proof of the density of the isolating/isolated degrees.

A dual case of Kaddah's branching theorem is Cooper-Lempp-Watson's cupping theorem.

Theorem 24. (Cooper, Lempp and Watson [12]) For any c.e. degree $\mathbf{c}$ and high c.e. degree $\mathbf{h}$ with $\mathbf{c} < \mathbf{h}$, there is a low d.c.e. degree $\mathbf{d}$ such that $\mathbf{b} \cup \mathbf{c} = \mathbf{h}$.

The case when $\mathbf{h} = \mathbf{0}'$ was first proved by Arslanov in [1], which gives a first structural difference between c.e. degrees and d.c.e. degrees.

By an anti-cupping theorem of Harrington, every high c.e. degree has a high c.e. degree below it witnessing its anti-cupping property. So fix $\mathbf{h}$ as a high c.e. degree and $\mathbf{c} < \mathbf{h}$ as a witness of its anti-cupping property. By Theorem 24, there is a d.c.e. degree $\mathbf{d}$ cupping $\mathbf{c}$ to $\mathbf{h}$, and hence between $\mathbf{h}$ and $\mathbf{d}$, there is no c.e. degree.

Definition 25. A d.c.e. degree $\mathbf{d}$ is isolated from above if there is a c.e. degree $\mathbf{c}$ above $\mathbf{d}$ such that there is no c.e. degree strictly between $\mathbf{c}$ and $\mathbf{d}$. *It is equivalent to saying that* $\mathbf{c}$ *is the least c.e. degree above* $\mathbf{d}$*, as noticed by Arslanov.*

In contrast to this, the isolation proposed by Cooper and Yi is called isolation from below.

Definition 26. (Wu [35]) A c.e. degree is bi-isolating if it isolates two d.c.e. degrees from above and below respectively. A d.c.e. degree $\mathbf{d}$ is bi-isolated if it is isolated from both above and below.

As each high c.e. degree isolates a d.c.e. degree from above, and degrees isolating from below are dense in the d.c.e. degrees, we can see that the bi-isolating degrees are dense in the high c.e. degrees. After seeing this, Arslanov asked whether the bi-isolating degrees occur in every jump class.

Question 27. Do the bi-isolating degrees occur in every jump class?

In [36], we proved that the bi-isolating degrees can be low, providing a partial solution to Arslanov's question.

For a bi-isolated degree $\mathbf{d}$, there are c.e. degrees $\mathbf{a}$ and $\mathbf{b}$ with $\mathbf{a} > \mathbf{d} > \mathbf{b}$ such that no c.e. degree is between $\mathbf{a}$ and $\mathbf{d}$ and between $\mathbf{d}$ and $\mathbf{b}$. That is, no c.e. degree between $\mathbf{a}$ and $\mathbf{b}$ is comparable with $\mathbf{d}$. As both isolated and nonisolated degrees are dense in the c.e. degrees, we ask:

Question 28. Do the bi-isolated d.c.e. degrees dense in the c.e. degrees?

Arslanov, Kalimullin and Lempp recently proved that there exist d.c.e. degrees $\mathbf{c} < \mathbf{d}$ such that all the d.c.e. degrees below $\mathbf{d}$ are comparable with $\mathbf{c}$. This so-called "double-bubble" phenomenon, was first noticed in [3], where the authors were devoted to providing a refutation to the well-known Downey's conjecture.

Question 29. Do the double-bubble phenomenon appear densely in the c.e. degrees?

6. Isolation and Lachlan sets

In this section, we consider the isolation phenomenon in a more general setting of degree structures, in the Ershov hierarchy and the CEA hierarchy.

Definition 30. A set is n-c.e. if there is a computable function $f(x, s)$ such that for every x,

(1) $f(x, 0) = 0$,
(2) $\lim_s f(x, s) = A(x)$,
(3) $|\{s : f(x, s) \neq f(x, s+1)\}| \leq n$.

So the 1-c.e. sets are precisely the c.e. sets, and the 2-c.e. sets are the d.c.e. sets — as they are also the differences of c.e. sets. Similarly, the n-c.e. sets are those given by starting with the c.e. sets and alternating the Boolean operations of difference and union. This hierarchy can be carried into the transfinite to define α-c.e. sets for computable ordinals α. By the Cooper et al. nondensity theorem, when $n \geq 2$, the n-c.e. degrees, and also the ω-c.e. degrees are not dense.

The concept of isolation in the CEA-hierarchy was proposed in LaForte's paper [28], where the CEA-hierarchy was developed in [24]. In this hierarchy, the 0-CEA degrees are the Turing degrees of computable sets, and the 1-CEA degrees are exactly the c.e. degrees, the 2-CEA degrees are the Turing degrees of sets that are computably enumerable in and above some c.e. set, and so on. Again, this hierarchy can be carried into the transfinite to definite α-CEA sets for computable ordinals α. By Sacks density theorem, Shore pointed out that the n-CEA degrees are also dense.

Theorem 31. (Shore, see LaForte [28]) For any $n \geq 1$, if $B <_T A$ are n-CEA sets, then there is an n-CEA set such that $B <_T C <_T A$.

Sacks density theorem (when $n = 1$) is the essential of the proof of Theorem 31. For $n+1$, $n \geq 1$, we consider two n-CEA sets A_0 and B_0 such that A is A_0-CEA and B is A_1-CEA respectively. Let $\emptyset = A_n \leq_T A_{n-1} \leq_T \cdots \leq_T A_0$ be a sequence of sets such that for each i with $0 < i \leq n$, A_i is CEA in A_{i-1}. Let j be the greatest index such that $A_j \leq_T B$. If $j = 0$, then both A and B are n-CEA in $A_0 \oplus B_0$, and hence by relativized Sacks density theorem, there is an n-CEA set C with $B <_T C <_T A$. If $j \neq 0$, then $B <_T A_{k-1} \oplus B$. As both these two B and $A_{k-1} \oplus B$ are $(A_k \oplus B_0)$-CEA, by relativized Sacks density theorem, there is a set C with $B <_T C < A_{k-1} \oplus B$ and C is $(A_k \oplus B_0)$-CEA, and hence $(n+1)$-CEA. Obviously, $B <_T C <_T A$.

While each n-c.e. degree is an n-CEA degree (according to the generalized version of the Lachlan set mentioned before), Jockusch and Shore pointed out that for any $n \geq 2$, there is an $(n+1)$-c.e. degree but not n-CEA, and that there is a 2-CEA degree but not ω-c.e.. In [4], Arslanov, LaForte and Slaman proved that if a 2-CEA degree is ω-c.e., then this degree is 2-c.e..

Theorem 32. (Arslanov, LaForte and Slaman, [4]) (1) The set of 2-c.e. degrees equals to the intersection of the set of 2-CEA degrees and the set of ω-c.e. degrees.

(2) There is a 3-CEA set A having properly 4-c.e. degree. *So (1) is the best possible result.*

The following is a direct extension of Ishmukhametov ([21]) and Cooper and Yi's result ([14]).

Theorem 33. (Arslanov, LaForte and Slaman, [4]) For any c.e. set B and n-c.e. set A with $n \geq 1$ and $B <_T A$, there is a d.c.e. set C with $B <_T C <_T A$.

Again, when $n = 1$, it is exactly the Sacks density theorem. So we assume that $n > 1$. Let E be an $(n-1)$-c.e. set such that A is CEA in E.

If $B <_T B \oplus E$, then as $B \oplus E$ is $(n-1)$-c.e., by the induction hypothesis, there is a d.c.e. set C with $B <_T C <_T B \oplus E$. Otherwise, $B \equiv_T B \oplus E$, and A has 2-CEA degree. As A is n-c.e., we know from Theorem 29 that A has d.c.e. degree. Now applying the result of Ishmukhametov and Cooper and Yi, the wanted d.c.e. set C is obtained.

The notion of isolation was extended to general cases by LaForte in [28], where for a class of subsets of ω, $\mathcal{C}$, we say that a set B isolates a set A w.r.t. class $\mathcal{C}$ if $B <_T A$ and there is no $C \in \mathcal{C}$ with $B <_T C <_T A$. According to this definition, the notion of nonisolated cannot be extended to higher levels in the Ershov hierarchy, as each properly 3-c.e. degree $\mathbf{e}$ has a predecessor $\mathbf{d}$ such that no c.e. degree is between $\mathbf{d}$ and $\mathbf{e}$.

Theorem 34. (Arslanov, LaForte and Slaman, [4]) Fix $n \geq 2$. Every properly $n+1$-c.e. degree is isolated by an n-c.e. w.r.t. c.e. degrees.

Again, we consider the Lachlan set, $\tilde{A}$ of the given set A with properly n-c.e. degree. Then between $\tilde{A}$ and A ($\tilde{A} <_T A$ as A has properly n-c.e. degree), there is no c.e. set, as otherwise A would be 2-CEA, and hence has d.c.e. degree, which is impossible.

In contrast to this, by a modification of the construction of isolated d.c.e. degrees, we can construct for any $n \geq 0$, an $(n+1)$-c.e. degree $\mathbf{d}$ and an n-c.e. $\mathbf{c}$ such that $\mathbf{c}$ is the greatest n-c.e. degree below $\mathbf{d}$. The basic idea of such a construction is still "wait and catch an error" — to satisfy an isolation requirement by a diagonalization argument. The new feature of this construction is that one isolation strategy may act several times, each of which we need to change D and C to recover a computation to a previous stage — we let s_0 be the last such a stage, and also we need to enumerate to enumerate s into C for the consistency of this strategy and the other isolation strategies with higher priority. We let such an action happen at stage s_1. Then, at the next time, when we want to recover the computation to stage s_1, s_0 will be extracted and s_1 will be enumerated into C. As n is fixed, each such an isolation can act like this at most n many times. Because of this, the following density result is not surprising.

Theorem 35. (Wu, in preparation) Given c.e. degrees $\mathbf{b} < \mathbf{a}$ and $n \in \omega$. There are $(n+1)$-c.e. degree $\mathbf{d}$ and n-c.e. degree $\mathbf{c}$ such $\mathbf{b} < \mathbf{c} < \mathbf{d} < \mathbf{b}$ and $\mathbf{c}$ bounds all n-c.e. degrees below $\mathbf{d}$.

In [28], LaForte considered an extended version of the isolation above — he was able to prove the existence of an $(n+1)$-c.e. degree $\mathbf{a}$ and an

n-c.e. degree $\mathbf{c} < \mathbf{d}$ such that $\mathbf{c}$ bounds all n-CEA degrees below $\mathbf{d}$. The ideal candidate for this isolating set is the Lachlan set of the constructed $(n+1)$-c.e. set. LaForte proved that such isolation pairs are downwards dense in the c.e. degrees.

Theorem 36. (LaForte [28]) Given n and a nonzero c.e. degree $\mathbf{a}$, there are an $(n+1)$-c.e. degree $\mathbf{a}$ and an n-c.e. degree $\mathbf{c} < \mathbf{d} < \mathbf{a}$ and $\mathbf{c}$ bounds all n-CEA degrees below $\mathbf{d}$.

The following question was proposed by LaForte in [28].

Question 37. (LaForte [28]) Are the isolation pairs in Theorem 36 dense in the c.e. degrees?

One difficulty of showing such a density is that we do not have control on the number of changes of $W^{B,C}$ if B is given as a lower set and C is an n-c.e. set constructed by us. Such a situation also happens in Arslanov, Lempp and Shore's approach for the upwards density of nonisolating degrees.

We list two well-known open questions below to end this paper,

Question 38. (1) (Downey [16]) For any $2 \leq m < n$, are the structures of the m-c.e. degrees and of the n-c.e. degrees elementarily equivalent? Arslanov, Kalimullin and Lempp have announced a negative answer to the case when $m = 2$ and $n = 3$.

(2) (following Downey) For any $2 \leq m < n$, are the structures of the m-CEA degrees and of the n-CEA degrees elementarily equivalent?

(3) (Arslanov [2]) Is the $\forall\exists$-theory of $\mathcal{D}_n$ decidable for $n \geq 2$, where $\mathcal{D}_n$ is the structure of n-c.e. degrees? Is the first-order theory of $\mathcal{D}_n$ the same as the $\forall\exists$-theory of arithmetic for each $n \geq 2$?

References

1. M. Arslanov. Structural properties of the degrees below $0'$. *Dokl. Nauk. SSSR* **283** (1985), 270-273.
2. M. Arslanov. Open questions about the n-c.e. degrees. In *Computability Theory and Its Applications: Current Trends and Open Problems* (Eds: Lerman Cholak, Lempp and Shore), volume 257 of *Contemporary Mathematics*, 15–22, Providence RI, 2000. AMS.
3. M. Arslanov, I. Kalimullin and S. Lempp. On Downey's conjecture. Available at URL: *http://www.math.wisc.edu/ lempp/papers/2-3ce.pdf*.
4. M. Arslanov, G. LaForte and T. Slaman. Relative enumerability in the difference hierarchy. *J. Symbolic Logic* **63** (1998), 411–420.

5. Marat M. Arslanov, Steffen Lempp, and Richard A. Shore. On isolating r.e. and isolated d-r.e. degrees. In *Computability, enumerability, unsolvability: Directions in Recursion Theory* (Eds: Cooper, Slaman and Wainer), volume 224 of *London Math. Soc. Lecture Note Ser.*, 61–80. Cambridge Univ. Press, Cambridge, 1996.
6. D. Cenzer, F. LaForte and G. Wu. The nonisolating degrees are nowhere dense in the computably enumerable degrees. Submitted.
7. C. T. Chong, A. Li and Y. Yang. The existence of high nonbounding degrees in the difference hierarchy, *Ann. Pure Appl. Logic* **138** (2006), 31–51.
8. S. B. Cooper. Degrees of Unsolvability. Ph.D. Thesis, Leicester University, 1971.
9. S. B. Cooper. Minimal Pairs and High Recursively Enumerable Degrees, *J. Symbolic Logic* **39** (1974), 655–660.
10. S. B. Cooper. Local degree theory. In *Handbook of Computability Theory* (Ed: Griffor), 121–153. North-Holland, Amsterdam, 1999.
11. S. B. Cooper, L. Harrington, A. H. Lachlan, S. Lempp, and R. I. Soare. The d.r.e. degrees are not dense. *Ann. Pure Appl. Logic* **55** (1991), 125–151.
12. S. B. Cooper, S. Lempp, and P. Watson. Weak density and cupping in the d-r.e. degrees. *Israel J. Math.* **67** (1989), 137–152.
13. S. B. Cooper, M. Salts and G. Wu. The non-isolating degrees are upwards dense in the computably enumerable degrees. In *Theory and Applications of Models of Computation (TAMC08)* (Eds: Agrawal, Du, Duan and Li), Lecture Notes in Computer Science **4978** (2008), 588–596.
14. S. B. Cooper and X. Yi. Isolated d.r.e. degrees. University of Leeds, Dept. of Pure Math., 1995. Preprint series, No.17.
15. D. Ding and L. Qian. Isolated d.r.e. degrees are dense in r.e. degree structure. *Arch. Math. Logic* **36** (1996), 1–10.
16. R. Downey. D.r.e. degrees and the nondiamond theorem. *Bull. London Math. Soc.* **21** (1989), 43–50.
17. R. G. Downey, A. Li, and G. Wu. Complementing cappable degrees in the difference hierarchy. *Ann. Pure Appl. Logic* **125** (2004), 101–118.
18. Y. L. Ershov, A hierarchy of sets, Part I. Algebra i Logika **7** (1968), 47-73 (Russian); Algebra and Logic **7** (1968), 24–43 (English translation).
19. Y. L. Ershov, A hierarchy of sets, Part II. Algebra i Logika **7** (1968) 15-47 (Russian), Algebra and Logic **7** (1968), 212–232 (English Translation).
20. L. A. Harrington and R. I. Soare. Games in recursion theory and continuity properties of capping degrees. In *Proceedings of the workshop on set theory and the continuum* (Ed: Moschovakis). Mathematical Sciences Research Institute, Berkeley, 1989. 39–62.
21. S. Ishmukhametov. *D.r.e. sets, their degrees and index sets.* PhD thesis, Novosibirsk, Russia, 1986.
22. S. Ishmukhametov. and G. Wu. Isolation and the high/low hierarchy. *Arch. Math. Logic* **41** (2002), 259–266.
23. C. Jockusch and R. Shore. Pseudojump operators I: the c.e. case. *Trans. Amer. Math. Soc.* **275** (1983), 599–609.
24. C. Jockusch and R. Shore. Pseudojump operators II: transfinite iterations,

hierarchies, and minimal covers. *J. Symbolic Logic* **49** (1984), 1205–1236.

25. D. Kaddah. Infima in the d.r.e. degrees. *Ann. Pure Appl. Logic* **62** (1993), 207–263.
26. A. H. Lachlan. Lower bounds for pairs of recursively enumerable degrees. *Proc. London Math. Soc.* **16** (1966), 537–569.
27. G. LaForte. The isolated d.r.e. degrees are dense in the r.e. degrees. *Math. Logic Quart.* **42** (1996), 83–103.
28. G. LaForte. Isolation in the *CEA* hierarchy. *Arch. Math. Logic* **44** (2005), 227–244.
29. A. Li, G. Wu and Y. Yang. Bounding computably enumerable degrees in the Ershov hierarchy. *Ann. Pure Appl. Logic* **141** (2006), 79–88.
30. G. E. Sacks. The recursively enumerable degrees are dense. *Ann. of Math.* **80** (1964), 300–312.
31. M. Salts, An interval of computably enumerable isolating degrees. *Math. Logic Quart.* **45** (1999), 59–72.
32. R. I. Soare. Recursively enumerable sets and degrees. Springer-Verlag, Heidelberg, 1987.
33. G. Wu. Isolation and lattice embeddings. *J. Symbolic Logic* **67** (2002), 1055–1064.
34. G. Wu. On the density of the pseudo-isolated degrees. *Proc. London Math. Soc.* **88** (2004), 273–288
35. G. Wu. Bi-isolation in the d.c.e. degrees. *J. Symbolic Logic* **69** (2004), 409–420.
36. G. Wu. Intervals containing exactly one c.e. degree. *Ann. Pure Appl. Logic* **146** (2007), 91–102.
37. G. Wu. Jump operator and Yates degrees. *J. Symbolic Logic* **71** (2006), 252–264.
38. G. Wu. Cupping computably enumerable degrees in the Ershov hierarchy. In *Proceedings of the workshop Computational Prospects of Infinity* (Eds: Chong, Feng, Slaman, Woodin and Yang), part II, 393-420, World Scientific, 2008.

On Π^1_1 conservativity for Π^1_2 theories in second order arithmetic

Keita Yokoyama

Mathematical Institute, Tohoku University,
Aoba, Sendai, 980-8578, JAPAN,
E-mail: y-keita@math.tohoku.ac.jp

In this paper, we give a positive answer to the question whether a union of Π^1_2 theories which are Π^1_1 conservative over a theory T is also Π^1_1 conservative over T or not. Then, we generalize this result and give some applications of it.

Keywords: second order arithmetic; conservative extension.

1. Introduction

In this paper, we study Π^1_1 conservativity for Π^1_2 theories of second order arithmetic. We first recall the definition of ω-submodel. Let (M, S) and (M', S') be models of RCA_0. Then, (M', S') is said to be an *ω-submodel* of (M, S) if $M = M'$ and $S' \subseteq S$. This study is inspired by the following questions.

Question 1.1 (Cholak, Jockusch, Slaman Ref. 1). *Let T_0, T_1 be Π^1_2 theories such that $T_0 \supseteq \mathsf{RCA}_0$. Let T_1 be a Π^1_1 conservative extension of T_0. Then, is every countable model of T_0 an ω-submodel of a model of T_1?*

Question 1.2 (Cholak, Jockusch, Slaman Ref. 1). *Let T_0, T_1, T_2 be Π^1_2 theories such that $T_0 \supseteq \mathsf{RCA}_0$. Let T_1, T_2 be Π^1_1 conservative extensions of T_0. Then, is $T_1 + T_2$ Π^1_1 conservative over T_0?*

We say that a theory T_1 satisfies *ω-extension property* over a theory T_0 if every countable model of T_0 is an ω-submodel of a model of T_1. Harrington showed that WKL_0 satisfies ω-extension property over RCA_0, and thus it is Π^1_1 conservative over RCA_0 (see e.g. Ref. 2). Similarly, Brown and Simpson Ref. 3 showed that WKL_0^+ ($= \mathsf{WKL}_0 +$"a version of the Baire category theorem") is Π^1_1 conservative over RCA_0, and Cholak, Jockusch and Slaman Ref. 1 showed that $\mathsf{WKL}_0 + \mathsf{RT}^2_2 + \mathrm{I}\Sigma^0_2$ is Π^1_1 conservative over $\mathsf{RCA}_0 + \mathrm{I}\Sigma^0_2$ using ω-extension property. We can easily show that the positive answer to Question 1.1 implies the positive answer to Question 1.2, thus it seems to be natural that each of these questions has a positive answer. However, Avigad Ref. 4 showed that the answer to Question 1.1 is no. In this paper, we show that the answer to Question 1.2 is yes. Actually, we can prove this by easy model theoretic arguments, and we can strengthen this result. For a model of second order arithmetic, we usually consider a pair (M, S) where $S \subseteq \mathcal{P}(M)$ and treat the special relation $\in$ as the absolute membership relation, but in this paper, we do not consider $\in$ as the special relation, and this plays a key role. We have some interesting consequences for Π^1_2 theories from this though our method is very simple.

In this paper, we only deal with theories which contain RCA_0, but this restriction is not important. In fact, we only need a base system in which we can define classes of formulas Π^i_n and Σ^i_n correctly.

2. Answers to the main questions

In Ref. 4, Avigad constructed a counter example for Question 1.1, but the following question is still open:

> is Question 1.1 still false if one requires T_0 and T_1 involved to be true?

We first construct true Π^1_2 theories which deny Question 1.1. For this, we treat models of second order arithmetic absolutely, *i.e.*, we only need to consider $\in$ as the absolute membership relation.

Theorem 2.1. *There is a true Π^1_2 theory $T_0 \supseteq \mathsf{ACA}_0$ and a true Π^1_2 theory T_1 such that T_1 is Π^1_1 conservative over T_0 and there is a countable model of T_0 which is not an ω-submodel of any model of T_1.*

Proof. Let $T_0 = \Pi^1_1(\mathsf{ACA}_0^+) + \mathsf{ACA}_0$ and let $T_1 = \mathsf{ACA}_0^+$ where $\Pi^1_1(\mathsf{ACA}_0^+) = \{\varphi \mid \varphi \text{ is a } \Pi^1_1 \text{ sentence and } \mathsf{ACA}_0^+ \vdash \varphi\}$. ($\mathsf{ACA}_0^+$ consists of ACA_0 plus $\forall X \exists Y\, Y = X^{(\omega)}$ where $X^{(\omega)}$ denotes the ω-th jump of X.) Then, T_1 is a Π^1_1 conservative extension of T_0. Since $T_1 \vdash \exists X$ 'X is a full satisfaction class for $\mathbb{N}$ (as an L_{PA} structure)', the first order part of a non-standard countable model of T_1 must be recursively saturated. Note that a recursively saturated model is not a short model. (A model $K \models \mathsf{PA}$ is said to be short if there is an element $a \in K$ such that if $a \in K' \prec_e K$ then $K' = K$.) On the other hand, we can construct a countable model of T_0 whose first order part is a short model. To show this, let T' be the first order part of T_0, *i.e.*, T' be the set of all L_{PA} sentences which are proved in T_0. Then, there exists a non-standard countable model M of T' which is short. (For example, a model which is generated by a single non-standard element is a short model. For recursively saturated models and short models of PA, see Ref. 5.) Thus $(M, \mathrm{ARITH}(M))$ is a required countable model of T_0. □

Now, we give a positive answer to Question 1.2. To show this, we need to treat models of second order arithmetic not absolutely. Thus, from now on, we consider structures for $\mathrm{L}_2 = \{0, 1, +, \cdot, =, <, \in\}$ as two-sorted structures. An L_2-structure is an 9-tuple $(M, S; 0^M, 1^M, +^M, \cdot^M, =^M, <^M, \mathsf{E}^{M\times S})$ where M and S are disjoint sets, 0^M and 1^M are constants of M, $+^M$ and $\cdot^M$ are functions from $M \times M$ to M, $=^M$ and $<^M$ are binary relations on $M \times M$, i.e., $=^M, <^M \subset M \times M$ and $\mathsf{E}^{M\times S}$ is a binary relation on $M \times S$, i.e., $\mathsf{E}^{M\times S} \subset M \times S$. We usually write (M, S) for $(M, S; 0^M, 1^M, +^M, \cdot^M, =^M, <^M, \mathsf{E}^{M\times S})$. Note that we do not consider $\in$ as a special symbol, so $\mathsf{E}^{M\times S}$ is not the absolute $\in$ relation and we do not identify $A \in S$ with $A^M = \{a \in M \mid (a, A) \in \mathsf{E}^{M\times S}\} \subseteq M$.

Lemma 2.1. *Let T_0 and T_1 be L_2 theories such that $T_0 \supseteq \mathsf{RCA}_0$. Then, the following are equivalent.*

(1) T_1 is a Π^1_1 conservative extension of T_0.
(2) If (M, S) is a model of T_0, then there exist $M' \supseteq M$ and $S' \supseteq S$ such that $(M', S') \models T_1$ and (M', S') is an elementary extension of (M, S) with respect to arithmetical formulas.

Proof. To show $2 \to 1$, assume $T_0 \not\vdash \forall X \varphi(X)$ where φ is arithmetical. Then, there exists a countable model $(M, S) \models T_0 + \exists X \neg\varphi(X)$. Then, there exists $(M', S') \models T_1$ such that (M', S') is an elementary extension of (M, S) with respect to arithmetical formulas. Since $(M, S) \models \exists X \neg\varphi(X)$, $(M', S') \models \exists X \neg\varphi(X)$. Hence, $T_1 \not\vdash \forall X \varphi(X)$.

For the converse, let (M, S) be a model of T_0. Let Θ be all arithmetical $\mathrm{L}_2 \cup M \cup S$ sentences which are true in (M, S). Then, $\Theta + T_1$ is consistent. We show this by contradiction. Assume $\Theta + T_1$ is inconsistent, there exist an $\mathrm{L}_2 \cup M$ formula $\psi(\vec{X})$ and $\vec{Z} \in S$ such that $(M, S) \models \psi(\vec{Z})$ and $T_1 \vdash \neg\psi(\vec{Z})$. Then, $T_1 \vdash \forall \vec{X} \neg\psi(\vec{X})$. By Π^1_1 conservativity, $T_0 \vdash \forall \vec{X} \neg\psi(\vec{X})$, but it contradicts $(M, S) \models T_0 + \psi(\vec{Z})$. Thus, $\Theta + T_1$ is consistent and there exists a model $(M', S') \models \Theta + T_1$. Then, (M', S') is an elementary extension of (M, S) with respect to arithmetical formulas. □

Lemma 2.2. *Let Γ be a class of L_2 formulas which is closed under negation, conjunction, disjunction and subformulas. Let $(M_0, S_0) \subseteq (M_1, S_1) \subseteq$*

$(M_2, S_2) \subseteq \cdots$ be an elementary chain with respect to formulas in Γ. Define an L_2-structure $(\bar{M}, \bar{S})$ as $\bar{M} = \bigcup_{i<\omega} M_i$, $\bar{S} = \bigcup_{i<\omega} S_i$, $0^{\bar{M}} = 0^{M_0}$, $1^{\bar{M}} = 1^{M_0}$, $+^{\bar{M}} = \bigcup_{i<\omega} +^{M_i}$, $\cdot^{\bar{M}} = \bigcup_{i<\omega} \cdot^{M_i}$, $<^{\bar{M}} = \bigcup_{i<\omega} <^{M_i}$, $=^{\bar{M}} = \bigcup_{i<\omega} =^{M_i}$ and $\mathsf{E}^{\bar{M}\times\bar{S}} = \bigcup_{i<\omega} \mathsf{E}^{M_i \times S_i}$. Then, for any $i < \omega$, (M_ω, S_ω) is a elementary extension of (M_i, S_i) with respect to formulas in Γ. Moreover, if $\eta(X, Y) \in \Gamma$ and $\exists^\infty i < \omega$ $(M_i, S_i) \models \forall X \exists Y \eta(X, Y)$, then $(\bar{M}, \bar{S}) \models \forall X \exists Y \eta(X, Y)$ (here, "$\exists^\infty i < \omega \ldots$" is an abbreviation of "$\forall j < \omega \exists i < \omega\ i \geq j \wedge \ldots$").

Proof. By the definition of $(\bar{M}, \bar{S})$, (M_i, S_i) is a substructure of $(\bar{M}, \bar{S})$.

We will show that $(M_i, S_i) \models \varphi(\vec{a}, \vec{A}) \Leftrightarrow (\bar{M}, \bar{S}) \models \varphi(\vec{a}, \vec{A})$ for any $\vec{a} \in M_i, \vec{A} \in S_i$ and for any $\varphi(\vec{x}, \vec{X}) \in \Gamma$ by induction on the complexity of formulas. We only need to show that $(M_i, S_i) \models \exists x \varphi(x, \vec{a}, \vec{A}) \Leftrightarrow (\bar{M}, \bar{S}) \models \exists x \varphi(x, \vec{a}, \vec{A})$ and $(M_i, S_i) \models \exists X \psi(X, \vec{a}, \vec{A}) \Leftrightarrow (\bar{M}, \bar{S}) \models \exists X \psi(X, \vec{a}, \vec{A})$ for any $\vec{a} \in M_i, \vec{A} \in S_i$ when $(M_i, S_i) \models \varphi(b, \vec{a}, \vec{A}) \Leftrightarrow (\bar{M}, \bar{S}) \models \varphi(b, \vec{a}, \vec{A})$ and $(M_i, S_i) \models \psi(B, \vec{a}, \vec{A}) \Leftrightarrow (\bar{M}, \bar{S}) \models \psi(B, \vec{a}, \vec{A})$ for any $b \in M_i, B \in S_i$.

Let $\varphi(x, \vec{y}, \vec{X})$ be a formula such that $\exists x \varphi(x, \vec{y}, \vec{X}) \in \Gamma$, and let $\psi(X, \vec{y}, \vec{X})$ be a formula such that $\exists X \psi(x, \vec{y}, \vec{X}) \in \Gamma$. Since (M_i, S_i) is a substructure of $(\bar{M}, \bar{S})$, $(M_i, S_i) \models \exists x \varphi(x, \vec{a}, \vec{A}) \Rightarrow (\bar{M}, \bar{S}) \models \exists x \varphi(x, \vec{a}, \vec{A})$ and $(M_i, S_i) \models \exists X \psi(X, \vec{a}, \vec{A}) \Rightarrow (\bar{M}, \bar{S}) \models \exists X \psi(X, \vec{a}, \vec{A})$ hold by induction hypothesis.

For the converse, assume $(\bar{M}, \bar{S}) \models \exists x \varphi(x, \vec{a}, \vec{A})$. Then, there exist $j < \omega$ with $j \geq i$ and $b \in M_j$ such that $(\bar{M}, \bar{S}) \models \varphi(b, \vec{a}, \vec{A})$. By induction hypothesis, $(M_j, S_j) \models \varphi(b, \vec{a}, \vec{A})$. Thus, $(M_j, S_j) \models \exists x \varphi(x, \vec{a}, \vec{A})$. Hence $(M_i, S_i) \models \exists x \varphi(x, \vec{a}, \vec{A})$ since $(M_i, S_i) \prec (M_j, S_j)$ with respect to formulas in Γ.

Similarly, assume $(\bar{M}, \bar{S}) \models \exists X \psi(X, \vec{a}, \vec{A})$. Then, there exist $j < \omega$ with $j \geq i$ and $B \in S_j$ such that $(\bar{M}, \bar{S}) \models \psi(B, \vec{a}, \vec{A})$. By induction hypothesis, $(M_j, S_j) \models \psi(B, \vec{a}, \vec{A})$. Thus, $(M_j, S_j) \models \exists X \psi(X, \vec{a}, \vec{A})$. Hence $(M_i, S_i) \models \exists X \psi(X, \vec{a}, \vec{A})$ since $(M_i, S_i) \prec (M_j, S_j)$ with respect to formulas in Γ.

We finally show that $\exists^\infty i < \omega$ $(M_i, S_i) \models \forall X \exists Y \eta(X, Y)$ implies $(\bar{M}, \bar{S}) \models \forall X \exists Y \eta(X, Y)$ for any $\eta(X, Y) \in \Gamma$. Let $\eta(X, Y) \in \Gamma$ and $\exists^\infty i < \omega$ $(M_i, S_i) \models \forall X \exists Y \eta(X, Y)$. Let $A \in \bar{S}$. By the definition of $\bar{S}$, there exists $i < \omega$ such that $A \in S_i$ and $(M_i, S_i) \models \forall X \exists Y \eta(X, Y)$. Thus, there exists $B \in S_i$ such that $(M_i, S_i) \models \eta(A, B)$. Since $(M_i, S_i) \prec (\bar{M}, \bar{S})$ with respect to formulas in Γ, $(\bar{M}, \bar{S}) \models \eta(A, B)$. This completes the proof. □

This lemma is just a special case of usual the chain theorem for general first order model theory. The difference between the above proof and the proof for the same theorem for first order single-sorted logic is that we have to check the conditions not only $(M_i, S_i) \models \exists x\varphi(x, \vec{a}, \vec{A}) \Leftrightarrow (\bar{M}, \bar{S}) \models \exists x\varphi(x, \vec{a}, \vec{A})$ but also $(M_i, S_i) \models \exists X\psi(X, \vec{a}, \vec{A}) \Leftrightarrow (\bar{M}, \bar{S}) \models \exists X\psi(X, \vec{a}, \vec{A})$.

Theorem 2.2. *Let T_0, T_1, T_2 be Π^1_2 theories such that $T_0 \supseteq \mathsf{RCA}_0$ and let T_1 and T_2 be Π^1_1 conservative extensions of T_0. Then, $T_1 + T_2$ is a Π^1_1 conservative extension of T_0.*

Proof. Let (M, S) be a model of T_0. By Lemma 2.1, we can construct an arithmetical elementary chain $\{(M_i, S_i)\}_{i\in\omega}$ such that $(M_0, S_0) = (M, S)$, $(M_{2i+1}, S_{2i+1}) \models T_1$ and $(M_{2i+2}, S_{2i+2}) \models T_2$. By Lemma 2.2, $(\bar{M}, \bar{S})$ is an arithmetical elementary extension of each of (M_i, S_i). Since T_1 and T_2 are Π^1_2 theories, $(\bar{M}, \bar{S}) \models T_1 + T_2$. Thus, by Lemma 2.1, $T_1 + T_2$ is a Π^1_1 conservative extension of T_0. □

Using this theorem, we can combine some conservation results on BCT, COH and RT^2_2. Here, BCT is an axiom for a version of Baire category theorem, COH is an axiom for the existence of a cohesive set and RT^2_2 is an axiom for Ramsey's theorem for pairs. We define $\mathsf{RCA}^+_0 = \mathsf{RCA}_0 + \mathsf{BCT}$ and $\mathsf{WKL}^+_0 = \mathsf{WKL}_0 + \mathsf{BCT}$. For the precise definitions of these axioms, see Ref. 1,6.

Corollary 2.1. $\mathsf{WKL}^+_0 + \mathsf{RT}^2_2 + \mathrm{I}\Sigma^0_n$ *is* Π^1_1 *conservative over* $\mathsf{RCA}_0 + \mathrm{I}\Sigma^0_n$ *for* $n \geq 2$, *and* $\mathsf{WKL}^+_0 + \mathsf{COH}$ *is* Π^1_1 *conservative over* RCA_0.

Proof. Straightforward direction from the previous theorem and the following conservation results: $\mathsf{WKL}^+_0 + \mathrm{I}\Sigma^0_n$ is Π^1_1 conservative over $\mathsf{RCA}_0 + \mathrm{I}\Sigma^0_n$ (the case of $n = 1$ is due to Brown and Simpson Ref. 3 Corollary 6.5 and the case of $n \geq 2$ is due to Yamazaki (unpublished)), $\mathsf{RCA}_0 + \mathsf{COH}$ is Π^1_1 conservative over RCA_0 (Cholak, Jockusch and Slaman Ref. 1 Theorem 9.1) and $\mathsf{RT}^2_2 + \mathrm{I}\Sigma^0_n$ is Π^1_1 conservative over $\mathsf{RCA}_0 + \mathrm{I}\Sigma^0_n$ for $n \geq 2$ (the case of $n = 2$ is due to Ref. 1 Theorem 10.2 and the case of $n \geq 3$ is due to Yamazaki (unpublished)). □

Remark 2.1. Tsuboi and Ikeda pointed out that the previous lemma and theorem can be generalized to a similar theorem for general first order theories. They also prove it by proof theoretical arguments (see Ref. 7).

3. Generalization

In this section, we will strengthen the previous theorem. Let $\mathfrak{Uniq}$ be the set of all sentences of the form $\forall X \exists! Y \varphi(X, Y)$ where φ is arithmetical. In Ref. 6, Yamazaki showed that RCA^+_0 is a $\mathfrak{Uniq}$ conservative extension of

RCA_0. In Ref. 8, Simpson, Tanaka and Yamazaki showed that WKL_0 is a $\mathfrak{Uniq}$ conservative extension of RCA_0, and then, they showed that $\mathsf{WKL}_0^+ = \mathsf{WKL}_0 + \mathsf{RCA}_0^+$ is also a $\mathfrak{Uniq}$ conservative extension of RCA_0. Then, is $T_1 + T_2$ a $\mathfrak{Uniq}$ conservative extension of T_0 if T_1 and T_2 are $\mathfrak{Uniq}$ conservative extensions of T_0? The answer is yes. We can generalize Theorem 2.2 as follows.

Theorem 3.1. *Let T be a Π^1_2 theory such that $T \supseteq \mathsf{RCA}_0$. Let Γ be a class of Π^1_3 sentences which satisfies the following:*

(∗)
- *Γ contains all Π^1_1 sentences;*
- *for all Σ^1_2 formula $\psi(X)$ and for all Σ^1_0 formula $\theta(X, \vec{Z})$, if $\forall X \psi(X) \in \Gamma$, then, there exists $\varphi \in \Gamma$ such that*

$$T \vdash \forall X \forall \vec{Z}(\theta(X, \vec{Z}) \to \psi(X)) \leftrightarrow \varphi.$$

If T_1 and T_2 are Π^1_2 theories and each of T_1 and T_2 is a Γ conservative extension of T, then, $T_1 + T_2$ is a Γ conservative extension of T.

Proof. Let T be a Π^1_2 theory such that $T \supseteq \mathsf{RCA}_0$, and let Γ be a class of Π^1_3 sentences which satisfies (∗). Let T_1 and T_2 be Π^1_2 theories such that each of T_1 and T_2 is a Γ conservative extension of T, and let $\forall X \psi(X) \in \Gamma$ such that $T \nvdash \forall X \psi(X)$ where $\psi(X)$ is a Σ^1_2 formula. Then, there exists a model $(M_0, S_0) \models T$ such that $(M_0, S_0) \models \neg\psi(A)$ for some $A \in S_0$. We only need to show that $T_1 + T_2 \nvdash \forall X \psi(X)$. As in the proof of Theorem 2.2, we first show the next claim.

Claim 3.1. *If (M, S) is an arithmetical elementary extension of (M_0, S_0) such that $(M, S) \models T + \neg\psi(A)$, then, for each $j = 1, 2$, there exists an arithmetical elementary extension (M', S') of (M, S) such that $(M', S') \models T_j + \neg\psi(A)$.*

Let Θ be all arithmetical $\mathrm{L}_2 \cup M \cup S$ sentences which are true in (M, S). Then, $\Theta + T_j + \neg\psi(A)$ is consistent. We show this by contradiction. Note that $\Theta + T_j$ is consistent as in the proof of Lemma 2.1 since T_j is a Π^1_1 extension of T and $(M, S) \models T$. Assume $\Theta + T_j + \neg\psi(A)$ is inconsistent. Then, there exist an $\mathrm{L}_2 \cup M$ formula $\theta(X, \vec{Z})$ and $\vec{B} \in S$ such that $(M, S) \models \theta(A, \vec{B})$ and $T_j \vdash \theta(A, \vec{B}) \to \psi(A)$. Then, $T_j \vdash \forall X \forall \vec{Z}(\theta(X, \vec{Z}) \to \psi(X))$. By (∗), there exists $\varphi \in \Gamma$ such that

$$T \vdash \forall X \forall \vec{Z}(\theta(X, \vec{Z}) \to \psi(X)) \leftrightarrow \varphi.$$

Then, by Γ conservativity, $T \vdash \varphi$, thus, $T \vdash \forall X \forall \vec{Z}(\theta(X, \vec{Z}) \to \psi(X))$. This contradicts $(M, S) \models T + \neg\psi(A) + \theta(A, \vec{B})$. Thus, $\Theta + T_j + \neg\psi(A)$

is consistent and there exists a model $(M', S') \models \Theta + T_j + \neg\psi(A)$. Then, (M', S') is an elementary extension of (M, S) with respect to arithmetical formulas such that $(M', S') \models T_j + \neg\psi(A)$. This completes the proof of the claim.

By this claim, we can construct an arithmetical elementary chain $\{(M_i, S_i)\}_{i\in\omega}$ which starts at (M_0, S_0) such that $(M_{2i+1}, S_{2i+1}) \models T_1 + \neg\psi(A)$ and $(M_{2i+2}, S_{2i+2}) \models T_2 + \neg\psi(A)$. Let $\bar{M} := \bigcup_{i\in\omega} M_i$ and let $\bar{S} := \bigcup_{i\in\omega} S_i$. Then, $(\bar{M}, \bar{S}) \models T_1 + T_2 + \neg\psi(A)$ since T_1 and T_2 are Π^1_2 theories and $\neg\psi(A)$ is a Π^1_2 sentence. Hence, $T_1 + T_2 \not\vdash \forall X\psi(X)$ and this completes the proof. □

Corollary 3.1. *Let T_0, T_1 and T_2 be Π^1_2 theories such that $T_0 \supseteq \mathsf{RCA}_0$ and each of T_1 and T_2 is a $\mathfrak{Uniq}$ conservative extension of T_0. Then, $T_1 + T_2$ is a $\mathfrak{Uniq}$ conservative extension of T_0.*

Proof. Let $\theta(X, Z_1, \ldots, Z_k)$ and $\psi(X, Y)$ be Σ^1_0 formulas. Within RCA_0,

$$\forall X \forall Z_1, \ldots, Z_k(\theta(X, Z_1, \ldots, Z_k) \to \exists! Y \psi(X, Y))$$

is equivalent to

$$\forall X' \exists! Y'((\theta((X')_0, (X')_1, \ldots, (X')_k) \to \varphi((X')_0, Y'))$$
$$\wedge(\neg\theta((X')_0, (X')_1, \ldots, (X')_k) \to Y' = \emptyset))$$

where $(X')_j = \{x \mid \langle x, j\rangle \in X'\}$. Thus, the class $\mathfrak{Uniq}$ satisfies the condition $(*)$. □

Corollary 3.2. $\mathsf{WKL}_0^+ + \mathsf{COH} + \mathrm{I}\Sigma^0_n$ *is* $\mathfrak{Uniq}$ *conservative over* $\mathsf{RCA}_0 + \mathrm{I}\Sigma^0_n$.

Proof. Straightforward direction from the previous corollary and the following conservation results: $\mathsf{RCA}_0^+ + \mathrm{I}\Sigma^0_n$ is $\mathfrak{Uniq}$ conservative over $\mathsf{RCA}_0 + \mathrm{I}\Sigma^0_n$ (due to Yamazaki Ref. 6), $\mathsf{WKL}_0 + \mathrm{I}\Sigma^0_n$ is $\mathfrak{Uniq}$ conservative over $\mathsf{RCA}_0 + \mathrm{I}\Sigma^0_n$ (due to Simpson, Tanaka and Yamazaki Ref. 8), and $\mathsf{RCA}_0 + \mathsf{COH} + \mathrm{I}\Sigma^0_n$ is $\mathfrak{Uniq}$ conservative over $\mathsf{RCA}_0 + \mathrm{I}\Sigma^0_n$ (the case of $n = 1$ is due to Kihara Ref. 9 and the case of $n \geq 2$ is due to Yamazaki (unpublished)). □

We can easily generalize Theorem 3.1 for Π^1_n theories.

Theorem 3.2. *Let T be a Π^1_n theory such that $T \supseteq \mathsf{RCA}_0$. Let Γ be a class of Π^1_{n+3} sentences which satisfies the following:*

(∗) – *Γ contains all $\Pi^1_n \cup \Sigma^1_n$ sentences;*
– *for all Σ^1_{n+2} formula $\psi(X)$ and for any Boolean combination of Π^1_n formulas $\theta(X, \vec{Z})$, if $\forall X\psi(X) \in \Gamma$, then, there exists $\varphi \in \Gamma$ such that*

$$T \vdash \forall X \forall \vec{Z}(\theta(X, \vec{Z}) \to \psi(X)) \leftrightarrow \varphi.$$

If T_1 and T_2 are Π^1_{n+2} theories and each of T_1 and T_2 is a Γ conservative extensions of T, then, $T_1 + T_2$ is a Γ conservative extension of T.

Proof. Easy modification of the proof of Theorem 3.1. □

Remark 3.1. In Ref. 4, Avigad showed that there exist Π^1_2 sentence θ such that

- both $\theta + \mathsf{ACA}_0$ and $\neg\theta + \mathsf{ACA}_0$ are Π^1_1 conservative over ACA_0,
- $\neg\theta$ is equivalent to a sentence of the form $\exists n \forall X \exists Y \eta(n, X, Y)$.

Let $T_0 = \mathsf{ACA}_0$, $T_1 = \mathsf{ACA}_0 + \theta$ and $T_2 = \mathsf{ACA}_0 + \forall Z \exists Y \exists n \eta(n, Z_n, Y)$, where $Z_n = \{m \mid (n, m) \in Z\}$. Since $\neg\theta \to \forall Z \exists Y \exists n \eta(n, Z_n, Y)$, T_1 and T_2 are Π^1_1 conservative over T_0. However, $T_1 + T_2 + \Sigma^1_2$-AC is inconsistent. This shows that we cannot weaken the assumption that T_0 is Π^1_2 in Theorem 2.2. Similarly, there exist Π^1_{n+2} theories T_0, T_1 and T_2 which satisfy the assumption of Theorem 3.2 but $T_1 + T_2 + \Sigma^1_{n+2}$-AC is inconsistent.

We derive some more consequences of Theorem 3.1. Define classes of comprehension axioms, axioms of choice and collection axioms[a] as

$$\begin{aligned}
\Gamma_1 &:= \mathfrak{CA} = \{\forall X \exists Y \forall n(n \in Y \leftrightarrow \varphi(n, X)) \mid \varphi(n, X) \text{ is arithmetical}\},\\
\Gamma_2 &:= \mathfrak{AC} = \{\varphi(n, X, Y)\text{-AC} \mid \varphi(n, X, Y) \text{ is arithmetical}\},\\
\Gamma_3 &:= \mathfrak{Coll} = \{\varphi(n, X, Y)\text{-Coll} \mid \varphi(n, X, Y) \text{ is arithmetical}\}
\end{aligned}$$

where

$$\begin{aligned}
\varphi(n, X, Y)\text{-AC} &\equiv \forall X(\forall n \exists Y \varphi(n, X, Y) \to \exists Z \forall n \varphi(n, X, Z_n)),\\
\varphi(n, X, Y)\text{-Coll} &\equiv \forall X(\forall n \exists Y \varphi(n, X, Y) \to \exists Z \forall n \exists m \varphi(n, X, Z_m)).
\end{aligned}$$

Then, the class $\Gamma_i + \Pi^1_1$ satisfies the condition (∗) over RCA_0. (Π^1_1 denotes the class of all Π^1_1 sentences.) For given theories T_1, T_2 and a class of L_2 sentences Γ, we write $T_1 \equiv_\Gamma T_2$ if $T_1 \vdash \psi \Leftrightarrow T_2 \vdash \psi$ for any $\psi \in \Gamma$. Applying

[a]Note that arithmetical collection axioms are Π^1_1 conservative over RCA_0 (due to Montalban and Yamazaki (unpublished)).

Theorem 3.1, for $\psi \in \Gamma_i$, $T_1 + T_2 \nvdash \psi$ if $T_1 \nvdash \psi$, $T_2 \nvdash \psi$ and $T_1 \equiv_{\Pi^1_1} T_2$. This means that sentences in Γ_i cannot be divided into two sentences which are not in Γ_i unless Π^1_1 consequences are changed. For example, if $\varphi \wedge \psi \equiv \mathsf{ACA}_0$, then $\mathrm{I}\Sigma^0_\infty + \varphi$ or $\mathrm{I}\Sigma^0_\infty + \psi$ implies some (non trivial) comprehension axiom. Similarly, axioms of choice, collection axioms, and some other axioms.

Next, we consider the Lindenbaum algebra of second order arithmetic. Let T_0 and T_1 be Π^1_2 theories such that $\mathsf{RCA}_0 \subseteq T_0 \subseteq T_1$, and let Γ be a class of Π^1_3 sentences. We define sets of (equivalence classes of) Π^1_2 sentences $L(T_0, T_1)$ as

$$L(T_0, T_1) = \{[\varphi]_{T_0} \mid \varphi \text{ is a } \Pi^1_2 \text{ sentence such that } T_1 \vdash \varphi \text{ and } \varphi + T_0 \equiv_{\Pi^1_1} T_0\},$$

$$L(T_0, T_1, \Gamma) = \{[\varphi]_{T_0} \mid \varphi \text{ is a } \Pi^1_2 \text{ sentence such that } T_1 \vdash \varphi \text{ and } \varphi + T_0 \equiv_{\Pi^1_1 \cup \Gamma} T_0\},$$

where $[\varphi]_{T_0} = \{\psi \mid T_0 \vdash \varphi \leftrightarrow \psi\}$ (an equivalence class of sentences over T_0) and $\varphi + T_0 \equiv_\Gamma T_0$ means that $\varphi + T_0$ and T_0 prove the same sentences in Γ. Then, by Theorem 3.1, $[\varphi], [\psi] \in L(T_0, T_1, \Gamma)$ implies $[\varphi \wedge \psi] \in L(T_0, T_1, \Gamma)$ if $\Gamma \cup \Pi^1_1$ satisfies $(*)$. Thus,

Corollary 3.3. *Let T_0 and T_1 be Π^1_2 theories such that $\mathsf{RCA}_0 \subseteq T_0 \subseteq T_1$. Then, $L(T_0, T_1)$ is a sublattice of Lindenbaum algebra $\mathcal{A}(T_0, T_1) = \{[\varphi]_{T_0} \mid \varphi$ is an L_2 sentence such that $T_1 \vdash \varphi\}$ with operations $\sup([\varphi], [\psi]) = [\varphi \wedge \psi]$ and $\inf([\varphi], [\psi]) = [\varphi \vee \psi]$. Moreover, $L(T'_0, T'_1, \mathfrak{Uniq})$, $L(T'_0, T'_1, \mathfrak{CA})$, $L(T'_0, T'_1, \mathfrak{AC})$ and $L(T'_0, T'_1, \mathfrak{Coll})$ are convex sublattices of $L(T_0, T_1)$ for any Π^1_2 theories T'_0 and T'_1 such that $T_0 \subseteq T'_0 \subseteq T'_1 \subseteq T_1$. (A sublattice $L \subseteq L(T_0, T_1)$ is said to be a convex sublattice of $L(T_0, T_1)$ if $a, b \in L$, $\sup(a, c) = a$ and $\inf(b, c) = b$ imply $c \in L$.)*

We finally present some open questions. It seems that Π^1_1 conservative Π^1_2 theories over RCA_0 and $\mathfrak{Uniq}$ conservative Π^1_2 theories are very similar by Corollary 3.2.

Question 3.1. Is a Π^1_1 conservative Π^1_2 extension of RCA_0 $\mathfrak{Uniq}$ conservative over RCA_0? In particular, does $L(\mathsf{RCA}_0, \mathsf{ACA}_0, \mathfrak{Uniq}) \subsetneq L(\mathsf{RCA}_0, \mathsf{ACA}_0)$ hold?

Note that $L(\mathsf{RCA}_0 + \mathrm{I}\Sigma^0_2, \mathsf{ACA}_0, \mathfrak{Uniq}) \subsetneq L(\mathsf{RCA}_0 + \mathrm{I}\Sigma^0_2, \mathsf{ACA}_0)$ holds. There exists a $\Sigma^0_0(\Sigma^0_1)$ formula $\varphi(n)$ such that $\mathsf{RCA}_0 + \mathrm{I}\Sigma^0_2 \nvdash \exists Y \forall n (n \in Y \leftrightarrow \varphi(n))$ and $\mathsf{RCA}_0 + \mathrm{I}\Sigma^0_2$ prove that "$A = \{n \mid \varphi(n)\}$ is a low set", *i.e.*, for any Σ^0_1 formula $\psi(x, X)$, there exist a Σ^0_2 formula $\theta_0(x)$ and a Π^0_2 formula $\theta_1(x)$

such that $\psi(x, A) \leftrightarrow \theta_0(x) \leftrightarrow \theta_1(x)$ (see Ref. 10 Theorem 3.8 of Chapter I). Then, $\mathsf{RCA}_0 + \mathrm{I}\Sigma^0_2 + \exists Y \forall n(n \in Y \leftrightarrow \varphi(n))$ is Π^1_1 conservative over $\mathsf{RCA}_0 + \mathrm{I}\Sigma^0_2$, but it is not $\mathfrak{Uniq}$ conservative over $\mathsf{RCA}_0 + \mathrm{I}\Sigma^0_2$.

In Ref. 11, Mummert constructed the correspondence between Σ^1_1 sentences in $L(\mathsf{RCA}_0, \mathsf{WKL}_0)$ and Π^0_1 classes of subsets of ω. We can find the same correspondences for $L(\mathsf{RCA}_0, T, \Gamma)$ for any $T \supseteq \mathsf{RCA}_0$ and a Π^1_3 class Γ which satisfies $(*)$. However, we don't know much about these lattices.

Question 3.2. Let T be a Π^1_2 theory which is not Π^1_1 conservative over RCA_0. Then, does $L(\mathsf{RCA}_0, T)$ have the maximal element? In particular, is $\tilde{T} = \{\varphi \mid [\varphi] \in L(\mathsf{RCA}_0, \mathsf{ACA}_0)\}$ finitely axiomatizable?

There are several more interesting questions. The following questions were raised in discussions with P. Cholak and S. Walsh.

Question 3.3. Suppose that T_0 is a true Π^1_2 axiomatizable Π^1_1 conservative extension of RCA_0. Is there a countable model of RCA_0 which is not an ω-submodel of any model of T_1?

Definition 3.1. Suppose that $T_0 \supseteq \mathsf{RCA}_0$ and that $T_1 \supseteq T_0$ is a Π^1_{n+2} axiomatizable Π^1_{n+1} conservative extension of T_0. Then $T_1 \supseteq T_0$ is the *maximal* Π^1_{n+2} axiomatizable Π^1_{n+1} conservative extension of T_0 if for every Π^1_{n+2} axiomatizable Π^1_{n+1} conservative extension $T_2 \supseteq T_0$ of T_0, it is the case that $T_1 \models T_2$.

Corollary 3.4. *Suppose that $T_0 \supseteq \mathsf{RCA}_0$ and that T_0 is Π^1_{n+2} axiomatizable for some $n \geq 0$. Then there is a maximal Π^1_{n+2} axiomatizable Π^1_{n+1} conservative extension of T_0.*

Proof. This is by compactness and Ikeda and Tsuboi's[7] generalization of Theorem 2.2. □

Question 3.4. What is the maximal Π^1_2 axiomatizable Π^1_1 conservative extension of RCA_0? Is it $\mathsf{WKL}^+_0 + \mathsf{COH}$? (cf. Corollary 2.1).

Question 3.5. The theory Σ^1_1-AC_0 is a Π^1_3 axiomatizable Π^1_2 conservative extension of ACA_0 (cf. Simpson Ref. 2 IX.4.4 p. 385). Is it the maximal Π^1_3 axiomatizable Π^1_2 conservative extension of ACA_0?

Question 3.6. Avigad proved that there is a recursive function f and a polynomial p such that if π is a proof of a Π^1_1 sentence φ from WKL^+_0, then $f(\pi)$ is a proof of φ from RCA_0 and $|f(\pi)| \leq p(|\pi|)$, where $|\cdot|$ denotes the length of the proof (Ref. 12 Theorem 1.1 p. 167). Suppose that T_1 and T_2 are Π^1_2 axiomatizable Π^1_1 conservative extensions of RCA_0 with such functions f_1, f_2 and polynomials d_1, d_2. Is it the case that $T_1 + T_2$ has such a function and polynomial? (Work in Ref. 7 shows that there is a function f but, since cut elimination was used, it is unclear if $|f(\pi)|$ can be polynomially bounded.) Does the maximal Π^1_2 axiomatizable Π^1_1 conservative extension of RCA_0 have such a function f and polynomial d?

Question 3.7. Hájek showed that WKL_0 is interpretable in RCA_0 (cf. Ref. 13 Theorem 3.10 p. 192). Does the same hold for the maximal Π^1_2 axiomatizable Π^1_1 conservative extension of RCA_0?

Acknowledgments

The author would like to thank Prof. Peter Cholak and Mr. Sean Walsh for several interesting questions and useful comments that have improved the exposition in this paper very much. The author would like to thank Prof. Akito Tsuboi and Dr. Kazuma Ikeda for useful comments mentioned in Section 2. The author would like to thank Prof. Kazuyuki Tanaka and Prof. Takeshi Yamazaki for useful discussions and comments. This work is partially supported by the global COE program of MEXT, Japan.

References

1. P. A. Cholak, C. G. Jockusch and T. A. Slaman, *The Journal of Symbolic Logic* **66**, 1 (2001).
2. S. G. Simpson, *Subsystems of Second Order Arithmetic* (Springer-Verlag, 1999).
3. D. K. Brown and S. G. Simpson, *The Journal of Symbolic Logic* **58**, 557(June 1993).
4. J. Avigad, Notes on Π^1_1-conservativity, ω-submodels, and the collection schema, in *Carnegie Mellon Technical Report CMU-PHIL-125*, 2001.
5. R. Kossak and J. Schmerl, *The Strctures of Models of Peano Arithmetic* (Oxford University Press, 2006).
6. T. Yamazaki, *Mathematical Logic Quarterly* **46**, 105 (2000).
7. K. Ikeda and A. Tsuboi, On conservativity for Π^1_2 theories, March 26, 2007, unpublished note.
8. S. G. Simpson, K. Tanaka and T. Yamazaki, *Annals of Pure and Applied Logic* **118**, 87 (2002).
9. T. Kihara, Degree structures of mass problems and formal systems of Ramsey-type theorems Master's Thesis, Tohoku University, February 2009.
10. P. Hájek and P. Pudlak, *Metamathematics of First-Order Arithmetic* (Springer, Berlin, 1993).

11. C. Mummert, *Archive for Mathematical Logic* **47**, 205 (2008).
12. J. Avigad, *Annals of Pure and Applied Logic* **82**, 165 (1996).
13. P. Hájek, Interpretability and Fragments of Arithmetic, in *Arithmetic, Proof Theory, and Computational Complexity*, , Oxford Logic Guides Vol. 23 (Oxford University Press, New York, 1993) pp. 185–196. Edited by Peter Clote and Jan Krajíček.

AROUND ROSY CM-TRIVIAL THEORIES

IKUO YONEDA

Department of Mathematics, Tokuyama Colledge of Technology
Gakuendai, Shunan 745-8585, Japan
E-mail: yoneda@tokuyama.ac.jp

We present results in "Some remarks on CM-triviality"[11] to appear in J.Math.Soc.Japan, which shows that any rosy CM-trivial theory has weak canonical bases and any CM-trivial O-minimal theory having elimination of imaginaries is modular. In the last section we show that one-basedness is equivalent to weak one-basedness which is introduced by B.Kim and R.Moosa in "Stable definability and generic relations" JSL 72 (2007) 1163-1176. Weak one-basedness was defined by using canonical bases in stable theories' context. We characterize it without using weak canonical bases and show the above equivalence in rosy theories.

Keywords: Rosy theories, CM-triviality, O-minimal theories

1. Weak Canonical Bases and GEI in Rosy Theories

We begin with basic properties on rosy theories. Our notation is standard. Let T be a complete L-theory, and let $\mathcal{M}$ be a sufficiently saturated model of T. We work in $\mathcal{M}^{\mathrm{eq}}$, consisting of imaginary elements, which are classes of equivalence relations definable over the empty set. $\bar{a}, \bar{b}, \ldots \subset_\omega \mathcal{M}^{\mathrm{eq}}$ denote finite sequences in $\mathcal{M}^{\mathrm{eq}}$. $A, B, \ldots$ denote small subsets of $\mathcal{M}^{\mathrm{eq}}$ and AB denotes $A \cup B$. For $a \in \mathcal{M}^{\mathrm{eq}}$ and $A \subset \mathcal{M}^{\mathrm{eq}}$, we write $a \in \mathrm{dcl}^{\mathrm{eq}}(A)$ if a is fixed by any automorphism fixing A pointwise. And we write $a \in \mathrm{acl}^{\mathrm{eq}}(A)$ if the orbit of a by automorphisms fixing A pointwise is finite. We write $B \equiv_A C$ for $\mathrm{tp}(B/A) = \mathrm{tp}(C/A)$ in T^{eq}.

Adler[1] introduces a strict independence relation among triplet of small subsets of $\mathcal{M}^{\mathrm{eq}}$:

(1) invariance: If $A \mathop{\smile\!\!\!\!\!\mid}_B C$ and $ABC \equiv A'B'C'$,
(2) monotonicity: If $A \mathop{\smile\!\!\!\!\!\mid}_B C, A' \subseteq A$ and $C' \subseteq C$, then $A' \mathop{\smile\!\!\!\!\!\mid}_B C'$.
(3) (right) base monotonicity: If $A \mathop{\smile\!\!\!\!\!\mid}_B D$ and $B \subseteq C \subseteq D$, then $A \mathop{\smile\!\!\!\!\!\mid}_C D$.
(4) (left) transitivity: If $B \subseteq C \subseteq D, D \mathop{\smile\!\!\!\!\!\mid}_C A$ and $C \mathop{\smile\!\!\!\!\!\mid}_B A$, then $D \mathop{\smile\!\!\!\!\!\mid}_B A$.

(5) (left) normality: $A \mathop{\smile\!\!\!\!\!\!\,|}_B C$ implies $AB \mathop{\smile\!\!\!\!\!\!\,|}_B C$.
(6) extension: If $A \mathop{\smile\!\!\!\!\!\!\,|}_B C$ and $C \subseteq D$, then there exists $A'(\equiv_{BC} A)$ such that $A' \mathop{\smile\!\!\!\!\!\!\,|}_B D$.
(7) (left) finite character: If $\bar{a} \mathop{\smile\!\!\!\!\!\!\,|}_B C$ for each $\bar{a} \subset_\omega A$, then $A \mathop{\smile\!\!\!\!\!\!\,|}_B C$.
(8) local character: For any A there is a cardinal $\kappa(A)$ such that, for any B there exists $B_0 \subseteq B$ with $|B_0| < \kappa(A)$ and $A \mathop{\smile\!\!\!\!\!\!\,|}_{B_0} B$.
(9) anti-reflexivity: $A \mathop{\smile\!\!\!\!\!\!\,|}_B A$ implies $A \subseteq \mathrm{acl}^{\mathrm{eq}}(B)$.

(1)–(8) imply symmetry [1, Theorem 1.14]: $A \mathop{\smile\!\!\!\!\!\!\,|}_B C \Leftrightarrow B \mathop{\smile\!\!\!\!\!\!\,|}_A C$. T is said to be *rosy* if there exists a strict independence relation on $\mathcal{M}^{\mathrm{eq}}$.

Weak canonical bases

An algebraically closed set C is said to be a $\mathop{\smile\!\!\!\!\!\!\,|}$-*weak canonical base* of $\mathrm{tp}(\bar{a}/B)$ if C is the smallest algebraically closed subset of $\mathrm{acl}^{\mathrm{eq}}(B)$ with $\bar{a} \mathop{\smile\!\!\!\!\!\!\,|}_C B$. $\mathrm{wcb}(\bar{a}/B)$ denotes the $\mathop{\smile\!\!\!\!\!\!\,|}$-weak canonical base of $\mathrm{tp}(\bar{a}/B)$ if it exists. We say that a rosy T has $\mathop{\smile\!\!\!\!\!\!\,|}$-weak canonical bases if there exists a $\mathop{\smile\!\!\!\!\!\!\,|}$-weak canonical base for each type. The following are Adler's results.

[1, Theorem 3.20]: A rosy T has $\mathop{\smile\!\!\!\!\!\!\,|}$-weak canonical bases if and only if $\mathop{\smile\!\!\!\!\!\!\,|}$ has the eq-intersection property.(i.e. $\bar{a} \mathop{\smile\!\!\!\!\!\!\,|}_A B$ and $\bar{a} \mathop{\smile\!\!\!\!\!\!\,|}_B A$ imply $\bar{a} \mathop{\smile\!\!\!\!\!\!\,|}_{A \cap B} AB$ for any $\bar{a}, A, B \subset \mathcal{M}^{\mathrm{eq}}$ such that $A = \mathrm{acl}^{\mathrm{eq}}(A)$ and $B = \mathrm{acl}^{\mathrm{eq}}(B)$.)

[1, Theorem 3.3]: If $\mathop{\smile\!\!\!\!\!\!\,|}$ has the eq-intersection property, then $\mathop{\smile\!\!\!\!\!\!\,|}$ coincides with thorn independence relation.

Geometric elimination of imaginaries

T has GEI if for any $e \in \mathcal{M}^{\mathrm{eq}}$, there exists $\bar{b} \subset_\omega \mathcal{M}$ such that $e \in \mathrm{acl}^{\mathrm{eq}}(\bar{b})$ and $\bar{b} \in \mathrm{acl}^{\mathrm{eq}}(e)$. We say that $\mathop{\smile\!\!\!\!\!\!\,|}$ has the intersection property if $\bar{a} \mathop{\smile\!\!\!\!\!\!\,|}_A B$ and $\bar{a} \mathop{\smile\!\!\!\!\!\!\,|}_B A$ imply $\bar{a} \mathop{\smile\!\!\!\!\!\!\,|}_{A \cap B} AB$ for any $\bar{a}, A, B \subset \mathcal{M}$ with $A = \mathrm{acl}(A)$ and $B = \mathrm{acl}(B)$. Now we have the following characterization of the intersection property [11, Propostion 3.4]: T has a strict independence relation having the intersection property $\Leftrightarrow$ T has GEI and a strict independence relation having the eq-intersection property $\Leftrightarrow$ T has a strict independence relation having weak canonical bases in the real sort. (i.e. weak canonical bases are interalgebraic with real elements)

2. CM-triviality in Rosy Theories

CM-triviality is a geometric notion of forking in stable theories, which is introduced by Hrushovski[3] where he disproves Zilber's conjecture on strongly minimal sets. CM-triviality forbids a point-line-plane incident system. Many generic relational structures are known to be CM-trivial.[10,13] Rosy theories do not necessarily have weak canonical bases, we choose the following definition for CM-triviality.

CM-triviality

A rosy theory T is said to be *CM-trivial* with respect to a strict independence relation $\mathop{\smile\!\!\!\!\!\!|}$ if $\bar{a} \mathop{\smile\!\!\!\!\!\!|}_A B$ implies $\bar{a} \mathop{\smile\!\!\!\!\!\!|}_{A \cap \mathrm{acl}^{\mathrm{eq}}(\bar{a},B)} B$ for any $\bar{a}, A, B \subset \mathcal{M}^{\mathrm{eq}}$ such that $A = \mathrm{acl}^{\mathrm{eq}}(A)$ and $B = \mathrm{acl}^{\mathrm{eq}}(B)$. As CM-triviality implies eq-intersection property, we have the following [11, Theorem 2.4]: If T is CM-trivial with respect to $\mathop{\smile\!\!\!\!\!\!|}$, then T has $\mathop{\smile\!\!\!\!\!\!|}$-weak canonical bases, and $\mathop{\smile\!\!\!\!\!\!|}$ coincides with thorn independence relation. As CM-triviality implies the existence of weak canonical bases, we can characterize CM-triviality as follows [11, Remark 2.5]: T is CM-trivial with respect to $\mathop{\smile\!\!\!\!\!\!|} \Leftrightarrow T$ has $\mathop{\smile\!\!\!\!\!\!|}$-weak canonical bases and $\mathrm{wcb}(\bar{a}/A) \subseteq \mathrm{wcb}(\bar{a}/B)$ holds for any $\bar{a}, A, B \subset \mathcal{M}^{\mathrm{eq}}$ such that $\mathrm{acl}^{\mathrm{eq}}(\bar{a}, A) \cap B = A$ with $A = \mathrm{acl}^{\mathrm{eq}}(A)$ and $B = \mathrm{acl}^{\mathrm{eq}}(B)$.

CM-triviality in the real sort

A rosy T is said to be *CM-trivial in the real sort* with respect to $\mathop{\smile\!\!\!\!\!\!|}$ if $\bar{a} \mathop{\smile\!\!\!\!\!\!|}_A B$ implies $\bar{a} \mathop{\smile\!\!\!\!\!\!|}_{A \cap \mathrm{acl}(\bar{a},B)} B$ for any $\bar{a}, A, B \subset \mathcal{M}$ such that $A = \mathrm{acl}(A)$ and $B = \mathrm{acl}(B)$. It is known that generic relational structures are CM-trivial in the real sort. Since CM-triviality in the real sort implies the intersection property, we have the following [11, Theorem 4.2]: T is CM-trivial in the real sort with respect to $\mathop{\smile\!\!\!\!\!\!|}$ if and only if T is CM-trivial with respect to $\mathop{\smile\!\!\!\!\!\!|}$ and has GEI. There exists a CM-trivial SU-rank one structure without GEI, so we see that CM-triviality does not imply *CM-triviality in the real sort*: Evans[2] gave an ω-categorical CM-trivial structure $\mathfrak{C}$ of SU-rank one without weak elimination of imaginaries. Firstly Evans constructed an ω-categorical SU-rank two generic structure M (a countable binary graph with a predimension $\delta(A) = 2|A| - |R^A|$) such that no triangles, no squres in M, and points and adjacent pairs of points are closed in M, and $\mathrm{cl}(*) = \mathrm{acl}(*)$ in M. Fix $a \in M$. Let C be the set of vertices at distance 1 from a. The structure $\mathfrak{C}$ above is the canonical structure on C such that $\mathrm{Aut}(\mathfrak{C})$ is homeomorphic to $\mathrm{Aut}(M/a)$. As $\mathfrak{C}$ and (M, a) are biinterpretable, $\mathfrak{C}$ is of SU-rank one and CM-trivial. It is shown[11] that $\mathfrak{C}$ does not have GEI. Therefore $\mathfrak{C}$ is not CM-trivial in the real sort.

Reducts of CM-trivial theories

Nübling[6] shows that CM-triviality is preserved under reducts in finite U-rank theories. Unlike finite U-rank theories, neither local modularity nor CM-triviality is preserved under reducts in finite U^{p}-rank theories. An example is found in O-minimal theories. (Any O-minimal theory has U^{p}-rank one.[7]) Let $T' = \mathrm{Th}(\mathbb{R}, +, <, \pi(*))$, where $\pi(x) = \pi x \in \mathrm{dcl}(x)$ for each x. Then T' is O-minimal, locally modular and CM-trivial. Consider the following reduct $T = \mathrm{Th}(\mathbb{R}, +, <, \pi(*)|(-1,1))$, where $\pi(x) = \pi x \in \mathrm{dcl}(x)$ for each $x \in (-1,1)$ [5, Example 4.5]. T is linear (CF-property) but non-locally modular[5] and T does not have weak canonical bases.[8] So T is not CM-trivial.

Non-CM-trivialty of fields Of monomial U^{p}-rank

Supersimple fields have monomial SU-rank.[12] Nübling[6] shows that any supersimple field is non-CM-trivial, moreover n-ample for each $n < \omega$. His proof works for fields of monomial U^{p}-rank (Refer to [11, the fifth section]). It is unknown whether any superrosy field has monomial U^{p}-rank.

3. The Geometry of O-minimal Theories Having E.I.

In this section we work in O-minimal theories having elimination of imaginaries. (Any O-minimal theory having a group-operation eliminates imaginaries by definable choice.) Note that dcl $=$ acl^{eq}. Pillay[8] offered variants of weak canonical bases by germs of definable functions as follows: Let $f(\bar{x}, \bar{y})$ be an $\emptyset$-definable function and let $\bar{a}$ be such that $\dim(\bar{a}) = |\bar{a}| = |\bar{x}|$. Let $E_{f,\bar{a}}$ be an $\bar{a}$-definable equivalence relation defined by $E_{f,\bar{a}}(\bar{b}_1, \bar{b}_2) \Leftrightarrow$ **either** there exists an open neighborhood U of $\bar{a}$ such that $f(\bar{x}, \bar{b}_1), f(\bar{x}, \bar{b}_2)$ are defined on U and $f(\bar{x}, \bar{b}_1)|U = f(\bar{x}, \bar{b}_2)|U$ **or** neither of $f(\bar{x}, \bar{b}_1), f(\bar{x}, \bar{b}_2)$ is defined on an open neiborhood of $\bar{a}$.

Fact 3.1 (Pillay[8]). *Suppose that* $C = \mathrm{dcl}(C), \dim(\bar{a}, \bar{b}/C) = \dim(\bar{a}) = |\bar{a}|$. *Let* $\bar{a}$ *and* $\bar{c} \subset_\omega C$ *be tuples and a function* f *be witnesses for* $\bar{b} \subset \mathrm{dcl}(\bar{a}, C)$, *i.e.* $\bar{b} = f(\bar{a}, \bar{c})$.
(1) If there exists $\mathrm{wcb}(\bar{a}, \bar{b}/C) \subseteq C$, *then there exists* $\bar{d} \subset_\omega \mathrm{wcb}(\bar{a}, \bar{b}/C)$ *such that* $\mathrm{dcl}(\bar{d}, \bar{a}) = \mathrm{dcl}(\bar{c}_{E_{f,\bar{a}}}, \bar{a})$ *holds.*
(2) If there exists $\bar{d} \subset_\omega C$ *such that* $\mathrm{dcl}(\bar{d}, \bar{a}) = \mathrm{dcl}(\bar{c}_{E_{f,\bar{a}}}, \bar{a})$, *then* $\mathrm{dcl}(\bar{d}) = \mathrm{wcb}(\bar{a}, \bar{b}/C)$.

Proof. (1): Put $D = \mathrm{wcb}(\bar{a}, \bar{b}/C) \subset C$: Suppose that $\dim(\bar{a}) = \dim(\bar{a}/C) = |\bar{a}|, \bar{b} \subset_\omega \mathrm{dcl}(C\bar{a})$. Let f, g be $\emptyset$-definable functions such that

$f(\bar{a},\bar{c}) = \bar{b}, g(\bar{a},\bar{d}) = \bar{b}$, where $\bar{d} \subset_\omega D$.

Claim 1. $\bar{c}_{E_{f,\bar{a}}} \in \mathrm{dcl}(\bar{a},\bar{d})$.

As $\bar{a} \perp \bar{c}, d$ and $f(\bar{a},\bar{c}) = \bar{b} = g(\bar{a},\bar{d})$, we have $f(\bar{x},\bar{c}) = g(\bar{x},\bar{d})$ on some open neighborhood U of $\bar{a}$. For any $\sigma \in \mathrm{Aut}(\mathcal{M}/\bar{a},\bar{d})$, we have $f(\bar{x},\sigma(\bar{c})) = g(\bar{x},\bar{d})$ on $\sigma(U)$. Thus we see that $f(\bar{x},\bar{c}) = f(\bar{x},\sigma(\bar{c}))$ on $U \cap \sigma(U)$, it follows that $\bar{c}_{E_{f,\bar{a}}} = \sigma(\bar{c})_{E_{f,\bar{a}}}$.

Claim 2. $\bar{d} \in \mathrm{dcl}(\bar{a},\bar{c}_{E_{f,\bar{a}}})$.

Let $\sigma \in \mathrm{Aut}(\mathcal{M}/\bar{a},\bar{c}_{E_{f,\bar{a}}})$. We have $f(\bar{x},\bar{c}) = f(\bar{x},\sigma(\bar{c}))$ for some open neighborhood U of $\bar{a}$. Let U' be an open neighborhood of $\bar{a}$ such that $U' \subseteq U \cap \mathrm{tp}^{\mathcal{M}}(\bar{a}/\bar{c},\bar{d}) \cap \mathrm{tp}^{\mathcal{M}}(\bar{a}/\sigma(\bar{c}),\sigma(\bar{d}))$, and take $\bar{a}' \in U'$ such that $\dim(\bar{a}'/\bar{c},\bar{d},\sigma(\bar{c}),\sigma(\bar{d})) = |\bar{a}'|$. By $\bar{a}' \in U$, put $\bar{e} := f(\bar{a}',\bar{c}) = f(\bar{a}',\sigma(\bar{c}))$. As $f(\bar{a},\bar{c}) = g(\bar{a},\bar{d}) = \bar{b}$ and $\bar{a}' \models \mathrm{tp}(\bar{a}/\bar{c},\bar{d})$, we have $g(\bar{a}',\bar{d}) = \bar{e}$. Then we have $\mathrm{wcb}(\bar{a},\bar{b}/C) = \mathrm{wcb}(\bar{a},\bar{b}/\bar{d}) = \bar{d} = \mathrm{wcb}(\bar{a}',\bar{e}/\bar{d})$ and $\sigma(\bar{d}) = \mathrm{wcb}(\bar{a}',\bar{e}/\sigma(\bar{d}))$. On the other hand, by $\dim(\bar{a}',\bar{e}/\bar{d},\sigma(\bar{d})) = |\bar{a}'|$, we see $d = \mathrm{wcb}(\bar{a}',\bar{e}/\bar{d}) = \mathrm{wcb}(\bar{a}',\bar{e}/\bar{d},\sigma(\bar{d})) = \mathrm{wcb}(\bar{a}',\bar{e}/\sigma(\bar{d})) = \sigma(\bar{d})$.
(2): We need to show that if $E = \mathrm{dcl}(E) \subseteq C, \dim(\bar{a},\bar{b}/E) = \dim(\bar{a},\bar{b}/C)$ then $\bar{d} \subset_\omega \mathrm{dcl}(E)$. Firstly we have $\bar{c}_{E_{f,\bar{a}}} \in \mathrm{dcl}(\bar{a},E)$. (By $\bar{b} \in \mathrm{dcl}(\bar{a},E)$, for any $\sigma \in \mathrm{Aut}(\mathcal{M}/\bar{a},E)$, we have $f(\bar{a},\bar{c}) = b = \sigma(b) = f(\bar{a},\sigma(\bar{c}))$ and $\bar{a} \perp \bar{c}, \bar{a} \perp \sigma(\bar{c})$. So we see $\bar{c}_{E_{f,\bar{a}}} = \sigma(\bar{c})_{E_{f,\bar{a}}}$.) By our assumption, we have $\bar{d} \subset_\omega \mathrm{dcl}(\bar{a},E)$. As $C \perp \bar{a}$, $\bar{d} \perp_E \bar{a}$ follows. So we see $\bar{d} \subset_\omega \mathrm{dcl}(E)$. □

Without using weak canonical bases, Pillay defined linearity in O-minimal theories. An O-minimal theory is *linear* (equivalent to CF-property defined by Peterzil), if for any $\emptyset$-definable function $f(\bar{x},\bar{y})$ and any $\bar{a}$ and $\bar{b}$ with $\dim(\bar{a}/\bar{b}) = |\bar{a}|$, we have $\bar{b}_{E_{f,\bar{a}}} \in \mathrm{dcl}(\bar{a}, f(\bar{a},\bar{b}))$.

If weak canonical bases always exist, as

$$\mathrm{dcl}(\bar{a}, \mathrm{wcb}(\bar{a}, f(\bar{a},\bar{b})/\mathrm{dcl}(\bar{b}))) = \mathrm{dcl}(\bar{a}, \bar{b}_{E_{f,\bar{a}}}),$$

we have that linearity "$\mathrm{wcb}(\bar{a}, f(\bar{a},\bar{b})/\mathrm{dcl}(\bar{b})) \subseteq \mathrm{dcl}(\bar{a}, f(\bar{a},\bar{b}))$" is equivalent to modularity "$\dim(\bar{a}, f(\bar{a},\bar{b})/\bar{b}) = \dim(\bar{a}, f(\bar{a},\bar{b})/\mathrm{dcl}(\bar{a}, f(\bar{a},\bar{b})) \cap \mathrm{dcl}(\bar{b}))$".

In Zariski geometries (strongly minimal strucutures having a good Noetherian topology), we have CM-triviality = linearity = local modularity = one-basedness.

Theorem 3.2. *In O-minimal theories with EI, we have*

CM-triviality = linearity with weak canonical bases = modularity = one-basedness.

Proof. Let T be CM-trivial and O-minimal with E.I. As T does not interpret fields of monomial $U^{\mathfrak{p}}$-rank, by the trichotomy theorem of Starchenko and Peterzil,[9] T must be linear, otherwise T would have a definable field of $U^{\mathfrak{p}}$-rank one on some interval. Since T has weak canonical bases, T must be modular. □

4. Final Remarks

Weak one-basedness was introduced by Kim and Moosa in stable theories' context [4, p. 1170]: T is said to be weakly one-based if for any c, a, b such that $c \mathop{\smile\!\!\!\!\!|}_a b, \mathrm{wcb}(c/b) = b$ and $c \not\subseteq \mathrm{acl}^{\mathrm{eq}}(a) = a$, we have $b \subseteq a$.

Without using weak canonical bases we characterize weak one-basedness in rosy theories, and then we see that weak one-basedness is equivalent to one-basesness.

Proposition 4.1. *Let T be rosy. The following are equivalent.*
(1) If $c \mathop{\smile\!\!\!\!\!|}_a b$, $a = \mathrm{acl}^{\mathrm{eq}}(a)$, $b = \mathrm{acl}^{\mathrm{eq}}(b)$ and $c \not\subseteq a$, then $c \mathop{\smile\!\!\!\!\!|}_{a\cap b} b$.
(2) T has weak canonical bases and if $c \mathop{\smile\!\!\!\!\!|}_a b, \mathrm{wcb}(c/b) = b, a = \mathrm{acl}^{\mathrm{eq}}(a)$ and $c \not\subseteq a$, then $b \subseteq a$.
(3) If $a = \mathrm{acl}^{\mathrm{eq}}(a)$ and $b = \mathrm{acl}^{\mathrm{eq}}(b)$, then $a \mathop{\smile\!\!\!\!\!|}_{a\cap b} b$: T is one-based.

Proof. (1) ⇒ (2): Suppose (1). Firstly we show that T has the eq-intersection property (So T has weak canonical bases.): If $c \subseteq a$ (or $c \subseteq b$) and $c \mathop{\smile\!\!\!\!\!|}_a b$ and $c \mathop{\smile\!\!\!\!\!|}_b a$, as $c \subseteq a \cap b$, $c \mathop{\smile\!\!\!\!\!|}_{a\cap b} ab$ clearly holds. Now suppose that $c \not\subseteq a, c \mathop{\smile\!\!\!\!\!|}_a b$ and $c \mathop{\smile\!\!\!\!\!|}_b a$. By (1) we have $c \mathop{\smile\!\!\!\!\!|}_{a\cap b} b$ and $c \mathop{\smile\!\!\!\!\!|}_b a$, so $c \mathop{\smile\!\!\!\!\!|}_{a\cap b} ab$ follows. Now suppose that $c \mathop{\smile\!\!\!\!\!|}_a b, \mathrm{wcb}(c/b) = b$ and $c \not\subseteq a$. As $c \mathop{\smile\!\!\!\!\!|}_{a\cap b} b$, we have $a \cap b = b$, so $b \subseteq a$ follows.
(2) ⇒ (1): Suppose $c \mathop{\smile\!\!\!\!\!|}_a b$, and let $d = \mathrm{wcb}(c/b)$. As $d = \mathrm{wcb}(c/d)$ and $c \mathop{\smile\!\!\!\!\!|}_a d$, we have $d \subseteq a$. So we have $d \subseteq a \cap b$, $c \mathop{\smile\!\!\!\!\!|}_{a\cap b} b$ follows.
(3) ⇒ (1): From $c \mathop{\smile\!\!\!\!\!|}_a b$ and $a \mathop{\smile\!\!\!\!\!|}_{a\cap b} b$, $ca \mathop{\smile\!\!\!\!\!|}_{a\cap b} b$ follows.
(1) ⇒ (3): Take d such that $a \subseteq d \not\subseteq a$ and $d \mathop{\smile\!\!\!\!\!|}_a b$. By (1) we have $d \mathop{\smile\!\!\!\!\!|}_{a\cap b} b$, so $a \mathop{\smile\!\!\!\!\!|}_{a\cap b} b$ follows. □

References

1. H.Adler, Explanation of independence, Ph.D thesis, Freiburg, 2005, arXiv:math.LO/0511616 v1, 24 Nov 2005.
2. D.Evans, $\aleph_0$-categorical structures with a predimension, Annals of Pure and Applied Logic 116 (2002), 157-186.
3. E.Hrushovski, A new strongly minimal set, Annals of Pure and Applied Logic 62 (1993), 147-166.

4. B.Kim and R.Moosa, Stable definability and generic relations, The Journal of Symbolic logic 72 (2007), 1163-1176.
5. J.Loveys and Y.Peterzil, Linear O-minimal structures, Israel J. Math. 81 (1993), 1-30.
6. H.Nübling, Reducts and expansions of stable and simple theories, PhD thesis, the University of East Anglia
7. A.A.Onshuus, Properties and consequences of thorn-independence, The Journal of Symbolic Logic 71 (2007), 1-21.
8. A.Pillay, Canonical bases in O-minimal and related structures, preprint, 2006.
9. Y.Peterzil and S.Starchenko, A trichotomy theorem for O-minimal structures, Proc.London Math.Soc. 77 (1998), 481-523.
10. I.Yoneda, CM-triviality and generic structures, Archive for Mathematical Logic 42 (2003), 423-433.
11. I.Yoneda, Some remarks on CM-triviality, J.Math.Soc.Japan, 61 (2009), 379-391.
12. F.Wagner, Hyperdefinable groups in simple theories, Journal of Mathematical Logic, 1 (1), 152-172, 2001.
13. V.Verbovskiy and I.Yoneda, CM-triviality and relational structures, Annals of Pure and Applied Logic 122 (2003) 175-194.